France, under its modern name, was born around the year 1000 AD; but its territory, loosely delineated by the natural boundaries of mountains and rivers, was first marked out and unified by the Roman conquest. For centuries it was only the strong tradition of national feeling that united this large amorphous body. France of the modern age was little more than an ideological concept. Its development as a more complex geographical area was not achieved until a relatively late date. The industrial revolution, the development of transportation, and increasing centralization led to the emergence of agricultural and industrial specialization, and the appearance of hierarchically ordered urban networks. The homogenization of the top cultural stratum had been assured by the Counter-Reformation and the main royal routes, but beneath this there was a profound diversity of popular culture. The fragmentations caused by the traumas of revolution brought about a religious and electoral geography which had astonishing stability. Maximum cultural diversification was achieved between 1875 and 1914 despite political centralization. In the twentieth century the picture becomes simpler. The second agricultural revolution has put paid to many regional specializations, and the triumph of the audiovisual is bringing about a profound unification of cultures and behaviours. The resulting unity will overwrite the fragmentations of the past.

Cambridge Studies in Historical Geography 21

AN HISTORICAL GEOGRAPHY OF FRANCE

Cambridge Studies in Historical Geography

Series editors:
ALAN R. H. BAKER   RICHARD DENNIS   DERYCK HOLDSWORTH

Cambridge Studies in Historical Geography encourages exploration of the philosophies, methodologies and techniques of historical geography and publishes the results of new research within all branches of the subject. It endeavours to secure the marriage of traditional scholarship with innovative approaches to problems and to sources, aiming in this way to provide a focus for the discipline and to contribute towards its development. The series is an international forum for publication in historical geography which also promotes contact with workers in cognate disciplines.

For a list of titles in the series please see the end of the book.

This book is published as part of the joint publishing agreement established in 1977 between the Fondation de la Maison des Sciences de l'Homme and the Press Syndicate of the University of Cambridge. Titles published under this arrangement may appear in any European language or, in the case of volumes of collected essays, in several languages.

New books will appear either as individual titles or in one of the series which the Maison des Sciences de l'Homme and the Cambridge University Press have jointly agreed to publish. All books published jointly by the Maison des Sciences de l'Homme and the Cambridge University Press will be distributed by the Press throughout the world.

Cet ouvrage est publié dans le cadre de l'accord de co-édition passé en 1977 entre la Fondation de la Maison des Sciences de l'Homme et le Press Syndicate de l'Université de Cambridge. Toutes les langues européennes sont admises pour les titres couverts par cet accord, et les ouvrages collectifs peuvent paraître en plusieurs langues.

Les ouvrages paraissent soit isolément, soit dans l'une des séries que la Maison des Sciences de l'Homme et Cambridge University Press ont convenu de publier ensemble. La distribution dans le monde entier des titres ainsi publiés conjointement par les deux établissements est assurée par Cambridge University Press.

AN HISTORICAL GEOGRAPHY OF FRANCE

XAVIER DE PLANHOL
University of Paris–Sorbonne

with the collaboration of
PAUL CLAVAL
University of Paris–Sorbonne

translated by
JANET LLOYD

CAMBRIDGE UNIVERSITY PRESS

EDITIONS DE LA MAISON DES SCIENCES DE L'HOMME

CAMBRIDGE UNIVERSITY PRESS
Cambridge, New York, Melbourne, Madrid, Cape Town, Singapore, São Paulo

Cambridge University Press
The Edinburgh Building, Cambridge CB2 2RU, UK

With Editions de la Maison des Sciences de l'Homme
54 Boulevard Raspail, 75270 Paris Cedex 06, France

Published in the United States of America by Cambridge University Press, New York

www.cambridge.org
Information on this title: www.cambridge.org/9780521322089

Originally published in French as Géographie Historique de la France
by Librairie Arthème Fayard, Paris 1988
and © Librairie Arthème Fayard, Paris 1988

First published in English by Editions de la Maison des Sciences de l'Homme and Cambridge University Press 1994 as An historical geography of France

English adaptation © Maison des Sciences de l'Homme and Cambridge University Press 1994

This digitally printed first paperback version 2006

A catalogue record for this publication is available from the British Library

Library of Congress Cataloguing in Publication data
Planhol, Xavier de.
 [Géographie historique de la France. English]
 An historical geography of France / by Xavier de Planhol; with the collaboration of Paul Claval: translated by Janet Lloyd.
  p.   cm. – (Cambridge studies in historical geography: 21)
 Includes bibliographical references and index.
 ISBN 0 521 32208 1
 1. France – Historical geography. 2. Human geography – France. I. Claval, Paul. II. Title. III. Series.
DC20.5.P5813 1994
911'.44 – dc20   92–45912   CIP

ISBN-13 978-0-521-32208-9 hardback
ISBN-10 0-521-32208-1 hardback

ISBN-13 978-0-521-03135-6 paperback
ISBN-10 0-521-03135-4 paperback

To the memory of Roger Dion (1896–1981)
diis manibus

Contents

List of figures	*page* xviii
Preface	xxi

Part I   The genesis of France   1

1 The isthmus of Gaul	3
The Celtic space	5
Before the Celts: prehistoric civilizations	5
Celts, Ligurians, Iberians	6
Gaul	8
The geographical face of Gaul: The peoples; The frontiers; Population density; Rural settlement and agrarian landscapes; The social structure and system of land-ownership; The towns	10
A united Gaul?	22
Marseilles and the tin routes	25
The earliest trading routes between the Mediterranean and north-western Europe	26
The geographical significance of the foundation of Marseilles	26
The expansion of Marseilles and its colonies	27
Marseilles, trade and the tin routes	29
Greek influence in Gaul	32
2 The impact of Rome	34
The spatial organization of Gaul	34
The Province	34
The Rhine frontier	35
The network of roads	37
Internal subdivisions: The Three Gauls; Peoples and cities	39
Adjustments to the urban fabric: New creations; Site switches; Changes in the hierarchy of cities	41
The unification of Gaul	45

	Rome and the geographical face of Gaul	46
	The structure and design of the cities	46
	The mark left by Rome upon the rural landscape: The 'centuriation' system; The spread of large estates (*villae*); The rural Gallo-Roman countryside: the predominance of open fields	47
	New types of cultivation: the slow expansion of vineyards	53
3	**From Gaul to France**	**58**
	The evolution of settlement	58
	The new linguistic frontier to the north and the east	58
	Germanic settlement beyond the linguistic frontier: The archaeological data; The evidence of toponymy; Local situations: conclusions	60
	The impact on land use and the landscape	67
	The decrease in population and its consequences	67
	The stability and the decline of towns: name changes	68
	Changes in the rural landscape	70
	The new organization of the territory	73
	The fragmentation of the Late Empire	73
	The persistence of Aquitaine	75
	The state of the Burgondes and the origins of Burgundy	76
	The Franks: a shift in the centre of political gravity	79
	Peripheral constructions of an ethnic nature: The failures: the Saxons and the Muslims; The successes: the Bretons and the Normans	83
4	**The birth of France**	**90**
	The Verdun division	90
	A vertical division	90
	Was the aim to achieve a geographical balance?	91
	Or was it purely a matter of historical chance?	92
	France: the name and the image	93
	The name	93
	The concept: its territorial evolution	94
	The scholarly and religious aspects of the concept	98
	The image and its symbols	101
	The emergence of a sense of nationhood	104
	The frontiers of France	108
	The frontier of the 'four rivers'	108
	The absence of any idea of a natural frontier in the Middle Ages and at the beginning of the modern period	109
	The natural frontiers: the concept and the realization	111
	An artificial boundary: the northern and north-eastern frontier	115
	Conclusion to Part I	117

Part II	**The traditional organization of the territory of France**	119
5	The major divisions	122
	The great cultural divisions	122
	Linguistic zones: The medieval genesis of the written languages; The linguistic map of France; *Langue d'oc* and *langue d'oïl*; Franco-Provençal and the concept of a mid-Roman area of influence; The structure of the *langue d'oïl* dialects; The *langue d'oc* dialects	122
	North and south	129
	The establishment of the major contrasts in landscape and settlement	134
	The development of the openfield community system: factors and processes	135
	The geographical extension and limits of the communal openfield system	137
	The west and the south: Residual landscapes of dispersed medieval settlement; Late medieval forms of collective organization in settlements in the west and the south; Two particular regional cases: southern Aquitaine and Brittany; The special features of the Mediterranean countryside: the development of hilltop settlements; Plot-patterns and landscapes	141
	Social and economic repercussions	151
	Literacy	152
	Societies and living standards: regions of affluence and regions of poverty	153
6	The secondary divisions	159
	Spatial organization: territorial units	159
	Two levels of organization: 'pays' and 'provinces'	159
	The historic 'pays': from the *pagus* to the 'pays'	163
	The provinces: From the Gallo-Roman cities to the provinces; The structure of the historic regions; the model of northern France. A productive central area with unproductive margins; The structure of the historic regions: the ill-defined area of Aquitaine and its margins; The structure of the historic regions: route intersections in the Rhône valley and the Alps; The structure of the historic regions: between the mountains and the Mediterranean	167
	The range of a jurist and the range of a horseman	186
	Perceived space	197
	The organization of space as it was perceived by its inhabitants	197

xiv   *Contents*

Natural names: Unnamed space; The meaning and resonance of natural names; From the territory as officially organized to the natural territory as humanly perceived; Conclusion: a utilitarian concept	197
The spatial framework of daily life	206
The framework of rural life	206
The framework of life for city-dwellers. Relations between town and countryside; The distribution of property owned by urban landlords; The basins of demographic attraction; Commercial fields of influence and sources of supply; The loose control exercised by the towns over the countryside	208
The beginnings of agricultural specialization	216
Monoculture and polyculture	216
Areas devoted to staple crops: The earliest estates under cereal cultivation: wheat, rye and buckwheat; The maize 'revolution'; The chestnut groves; The potato	217
Sectors of specialized agriculture: The necessary conditions for the introduction of specialized crops; Flanders; Alsace; The Loire valley; Urban agricultural regions	223
The evolution of regions devoted to viticulture: The early Middle Ages: the vineyards survived but trade declined or disappeared; The revival of south–north trade in the late Middle Ages: the first high-quality vineyards; The maritime expansion of northern sources of supply: the Atlantic seaboard and the development of the vineyards of Bordeaux; The absence of commercial viticulture in Mediterranean regions up until the modern period; The decline of the northern vineyards in the seventeenth and eighteenth centuries; The origin of the greatest wines	226
Dispersed industry	233
Spontaneous industry and organized industry	233
The distribution of the principal branches of industry	234
The industrial scene in the regions	237
Traditional industry and the countryside	240
Conclusion to Part II: The beginnings of a territory with a spirit of its own?	240

Part III   The centralization and diversification of the French space   245

7	Paris and the Parisian centralization	247
	The development of Paris	247
	The geographical conditions: The gap routes; The route across the Seine	247
	Stages in the city's political destiny: Paris before the Germanic invasions; After the Germanic invasions: the emergence of a political role; The Capetian capital	251

The heart of France	254
Paris' sphere of influence: the formation of its population	256
The geographical impact of the growth of Paris	259
The Parisian agricultural sphere of influence: A fruit- and vegetable-producing belt; The evolution of the vineyards; The market for dairy produce: butter and cheeses; The large agricultural units of the Ile-de-France; The large wooded massifs of the Parisian region	259
The centralization of communication routes: The roads; The railways	269
Paris, the cultural capital of France	273
Conclusion	275

8	Cultural action and reaction: unity and diversity	276
	The achievement of cultural unity	276
	Songs and stories	276
	Enlightenment and language	279
	Rationalized space: the organization of the departments	281
	The Tour of France	285
	The differences	293
	Variations on a theme: the arts, costumes and dances: Popular art; Costumes; Dances	293
	From regional cultures to regionalism: Local cultures and regional cultures; The birth and development of peripheral cultures; The appearance of cultural regionalism; The late emergence of political regionalism and its significance	307
	The emergence of different regional and political attitudes: The map of religious practices; The map of political attitudes; A tentative interpretation	326
	Rejections: Protestantism in the Cévennes; The revolts in the west: Vendée	338

9	The economic differentiation of space	345
	A new map of land use: the development of agricultural specializations	346
	Transformations in the agriculture of southern France	346
	The development of the vineyards of Languedoc: Before the end of the seventeenth century: hillside vineyards designed to supply local needs; The first extension of the vineyards: the beginnings of large-scale trade; The appearance of vineyards on a massive scale (1830–75); The phylloxera crisis and the definitive establishment of mass vineyards	348
	Repercussions on the vineyards of northern France	350
	Dairy farming and livestock rearing: The selection of pastoral areas; Systems of pastoral production; The establishment of different breeds of animals	353

	Reafforestation: The mountains of southern France; The plains; The Landes of Gascony; The Sologne area; The Orléans forest and the Champagne plain	359
	From dispersed industry to industrial regions	365
	The decline of dispersed industry	366
	The new industrial centres	369
	The large urban industrial complexes	371
	A great market: the geographical development and establishment of commercial brands	374
	The development of the fame of local products	374
	The expansion of manufacturing areas	377
10	The rural exodus and urbanization	380
	The completion of the urban network	380
	The modern period: providing a frame for the national territory	380
	The nineteenth and twentieth centuries: industrial towns and leisure towns	383
	From the countryside to the towns	385
	Temporary migrations	385
	The rural exodus: Rural depopulation; Emigration and natural population change; The rhythms of migration	393
	The geographical aspects of urban growth	400
	The process of urbanization: rhythms and categories	400
	The factors of urbanization	402
	Geographical consequences: new regional dynamism	405
	The formation of the urban networks	406
	The historical development of the process	406
	The principles of hierarchical development	410
	Conclusion: The France of Ardouin-Dumazet	412
11	The France of large organizations (by Paul Claval)	415
	Modernization: the specific features of the French experience	416
	Conditions of unequal favourability	416
	Under the Second Empire, advantages began to be exploited	417
	Enduring social divisions and modernization	418
	Complex and relatively divided élite groups: The land was still an essential source of wealth and prestige; The nineteenth-century élites included some civil servants; For a long time, trade constituted a world apart; The manufacturing world	419
	Strategies of continuity and survival	422
	Popular classes passionately devoted to independence yet with little ambition	423
	Checks upon the homogenization of the territory	426
	The stages of social decongestion	428
	A new geography of political attitudes	430

The geographical conditions of modernization	432
Specialization and social structures	432
The role of the local environment	433
The effects of historical legacies and accidents	435
The stages of progressive modernization and the establishment of big business: The hesitancies of the early nineteenth century; The turning point of the beginning of the Second Empire; 1865–85: twenty difficult years; The major wave of innovations: 1885–1914; The First World War and the inter-war period	438
The 'glorious thirty years' or France at last won over to modernity	442
Eminently favourable circumstances	442
The new face of agriculture	444
An industrial France at last	449
A new universe of movement and communications	453
The art of living and the landscape	457
General conclusion	466
Notes	469
Guide to further reading	504
Bibliography	506
Index	533

Figures

1	A plan of the maps relating to different areas	page 2
2	France in prehistoric times	4
3	Gaul at the time of the Roman conquest	12
4	Types of frontier between the peoples of Gaul	14
5	Traces of large circular crop-marks in the French countryside	19
6	The distribution of toponyms derived from *Mediolanos*	25
7	Marseilles and its colonies	28
8	The principal tin routes in Gaul	30
9	The network of roads in Roman Gaul	36
10	Three stages in the diffusion of Roman pottery in Gaul	38
11	The administrative divisions of Roman Gaul	40
12	Gallo-Roman settlement in the plain of Picardy	50
13	The expansion of the vineyard in Roman Gaul	54
14	The Germano-Roman linguistic frontier	59
15	The distribution of different types of ansated fibulae from Rhineland regions	62
16	The areas of distribution of various Germanic and Germano-Roman series of place-names	63
17	The toponymy of the Beauce	65
18	Types of rural vehicle in France	72
19	The persistence of Aquitaine after the Germanic invasions	74
20	The cult of Saint Félix de Girone in Gaul	77
21	The historical development of Burgundy	79
22	The Merovingian territorial divisions	80
23	Constructions of an ethnic nature in peripheral territories at the time of the invasions	82
24	The birth of France at the time of the division of Verdun	91
25	The major linguistic divisions within the French territory	124
26	North and south	128

27	The major types of land division and rural landscapes within the territory of France	138
28	The areas of the communes of France	144
29	Late medieval grouped settlements in south-western France	145
30	The open fields of Brittany	149
31	Literacy in France	150
32	Sharecropping in France	154
33	Social contrasts in traditional France	156
34	The level of agricultural production (per department and per agricultural worker)	157
35	The network of communication routes at the end of the eighteenth century	158
36	The *gouvernements* in France in 1789	161
37	Corsica: the traditional agrarian landscapes (late eighteenth and early nineteenth centuries) and urban developments	185
38	The range of influence of the schools of Romanesque architecture	193
39	The rural settlements in Lorraine: appearance, genesis, limits	194
40	City landowners in Upper Normandy in 1825–35	210
41	Traditional areas of property in the Alps	212
42	The development of the catchment area of Bordeaux in the eighteenth century	213
43	Travellers putting up in Caen in 1790–91, expressed department by department	214
44	The chestnut groves of France at the time of the old cadastral survey	220
45	The evolution of the rural landscape of the area known as Les Chaussadenches, commune of Vesseaux (Ardèche)	221
46	'Factories' at the time of the old cadastral survey (first half of the nineteenth century)	236
47	The industries of the duchy of Lorraine, according to the declarations made by its communities in 1708	238
48	The 'gap route' and the position of Paris	248
49	The site of Paris	249
50	The origins of the population of Paris (late eighteenth to late nineteenth centuries)	257
51	The development of the belt of market gardens around Paris	260
52	Suburban cultivation around Paris from the late eighteenth to the late nineteenth century	262
53	The traditional areas of cheese production	264
54	The division of land in the Ile-de-France	266
55	The forests around Paris	268
56	The postal routes in 1632	270

List of figures

57	The development of the railways	272
58	From the provinces to the departments	282
59	The origins of the Tour of France	287
60	The Tour of France of the *compagnons* at the beginning of the nineteenth century	289
61	Didactic itineraries of the nineteenth century	290
62	The differentiation of costumes in Brittany: general features	296
63	The differentiation of costumes in Brittany: local aspects	300
64	Aspects of the differentiation of dances in Lower Brittany	306
65	The origin and stability of religious attitudes	327
66	The development of political attitudes	332
67	Aspects of the meridional political contest	337
68	Reformed churches 'set up' in 1562	338
69	'Desert' churches around 1788	339
70	The development of the Languedoc and Roussillon vineyards	351
71	The development of grassland in Normandy	354
72	The development of the Landes forests	362
73	The concentration of the cotton industry in Rouen in the nineteenth century	367
74	The concentration of the Barrois iron and steel industry	368
75	The concentration of the French iron and steel industry in the nineteenth century	370
76	Aspects of the establishment of local trade-names (*aires d'appellation d'origine*) in the nineteenth and twentieth centuries	375
77	The completion of the urban network in the modern period	381
78	Temporary migrations under the First Empire	386
79	The rural exodus	394
80	The rural exodus and mobility in the nineteenth century	398
81	The evolution of the urban picture in the second half of the nineteenth century	403
82	The rate of urban growth in the nineteenth century	404
83	The presidential elections of 1974 and 1981	431
84	The consolidation of fields and the suppression of hedgerows in western France	446
85	Contemporary land clearance in 'Champagne pouilleuse'	448
86	Rail traffic (freight)	450
87	Dencentralization in France	451
88	Contact potentials in France in 1954 and 1975	453
89	Changes in the geography of urban growth between the nineteenth and the twentieth centuries	456
90	Plan of Arcachon in 1926	460
91	The evolution of 'grands ensembles' (huge housing estates)	463
92	Paris – La Défense	465

Preface

This is the first systematic attempt to study the organization of French space in the context of time. Abroad, historical geography, which involves applying to the past the methods of geographical analysis, has long been accepted as a discipline in its own right. In English-speaking countries, in particular, it has produced works of major importance. In France, however, its development has for many years been paralysed by — paradoxically enough — the close association of geographical and historical studies within university courses. The reason for that is the manner in which the French geographical school originated in the late nineteenth century, emerging as it did from the teaching of Paul Vidal de La Blache, who was himself an historian by training. The effect of this secular symbiosis was to permeate the work of French geographers with history, thereby discouraging them from striking out boldly to establish separate studies in historical geography.[1] In France, historical geography was for many years, following Auguste Longnon (1844–1911), conceived simply as the inventory of a sequence of territorial divisions. It was a field in which the erudition of archivists was at a high premium, but the significance of those divisions for the life of the inhabitants was never seriously considered. The perspective of a substantial textbook[2] entitled *Géographie historique de la France*, which appeared in 1947, was equally limited and a more recent work,[3] with an identically worded subtitle, was also unsuccessful in breaking that mould, despite the fact that almost half the book is devoted to a retrospective study of the people involved and their activities. The present work is certainly not intended to supersede those conscientious and meticulous works, which remain useful. The cartography of the present volume is deliberately thematic, being designed simply to illustrate the text[4]: it does not aim to provide the tool of exhaustive and critical chorographic research which is still lacking despite the recent progress made in this area.[5] Its purpose is to fill a lacuna of another kind.

It is a lacuna that goes virtually unnoticed in France itself, but it has deeply shocked foreign observers, who are scandalized that there has as yet appeared no synthesis of the huge mass of works produced by the French geographical school over the past century. Although fragmentary and sporadic, the contribution of this school to the reconstruction of the past has, for the very reasons mentioned above, been particularly fruitful and its findings underpin the thought of French historians in general.[6] It seems somewhat surprising that it should have been in England, fifteen years ago, that the first collective work to tackle general themes of the geographical history of France[7] should have been produced and that, in that collection, the contribution of French authors, albeit important and brilliant, was certainly not predominant. It was, furthermore, a British geographer, Alan R. H. Baker, who initiated the idea of the present volume, when he did us the honour of suggesting that we should in the first instance write it for the English-speaking public and that it should be part of a series which already includes many syntheses of this kind relating to a variety of countries and geographical areas.[8] We have thus aimed to present to the general educated public an overall view which publicizes a range of major discoveries, all too many of which would otherwise remain inaccessible, buried either in university theses or in articles in scholarly journals with a limited readership. We also hope that for the non-French educated public, which may know relatively little about France, and for French university students, who ought to know more than they do, this book will – thanks to its comprehensive critical data – prove to be a useful working tool and a prelude to more advanced studies.[9]

The subject has deliberately been divided into large chronological sections. We are well aware that historians in particular may well regard as excessively sweeping the diachronic shorthand of the expression 'traditional France', which is used to represent a period of time that spans close on one thousand years, from the late Middle Ages down to the end of the first third of the nineteenth century. However, that was precisely one of the concepts which, in the early years of the twentieth century, proved most fertile in the work of the French geographers following in the footsteps of Vidal de La Blache: in their major regional studies, these scholars always sought to set up a contrast between, on the one hand, the new aspects which they recognized to be emerging and, on the other, a past which was still very much present everywhere. Furthermore, the many retrospective studies which have appeared since that time show that in many respects the late Middle Ages, which was a particularly formative period, witnessed the establishment – particularly in the countryside – of landscapes and frameworks of life which were to remain unchanged right down to the industrial revolution.[10] Having, myself, little specialized knowledge of the contemporary period, I should like to express my warmest thanks to my colleague Paul Claval, who has been so kind as to take responsibility for the last chapter of this work, being infinitely better qualified than I in this domain.

The task as a whole was an immense one, as I fully realized: it demanded a vast programme of reading constantly brought up to date yet inevitably still incomplete, the subject matter of which had been densely researched, fervently and meticulously reworked and passionately inspired. In truth, only one man would have been capable of rising fully to the challenge: the scholar to whom the present volume is dedicated. Roger Dion, alone of his generation, chose to dedicate himself totally to the historical geography of France. His original and imaginative talent constantly revealed new perspectives in the subject, while his sometimes unexpected yet invariably productive comparisons shed light upon many of its most obscure and controversial aspects.[11] The pages that follow refer frequently to his hypotheses and discoveries. I venture to present the synthesis which this peerless master, who was more interested in research than in publication, preferred not to undertake, in the hope that I have not been unfaithful to the methods and spirit of the man whose lectures it was my privilege to attend.

PART ONE
The genesis of France

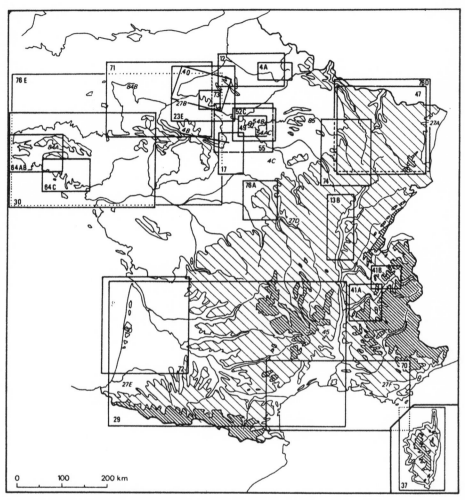

Figure 1. A plan of the maps relating to different areas used in this volume (the numbers correspond to those of the figures). Framed with bold numbers: general maps. In italics: detailed maps.

1
The isthmus of Gaul

A glance at a map of the European peninsula shows that it narrows progressively towards the west. Odessa is 1,200 kilometres from the Baltic Sea, while only 900 kilometres separate Trieste from the North Sea. Yet there are very few routes leading from the Mediterranean to the Atlantic or to the seas bordering upon it. In central Europe, the Alps constitute a barrier that is difficult to cross. To the south-west, the path is blocked by the compact and forbidding mass formed by high plateaux of the Iberian peninsula. The only two easy routes start from the Golfe du Lion and lead to the Atlantic coast, passing either side of the French Massif Central, the one by way of the Lauraguais pass and the low-lying valley of the Garonne, the other along the Rhône corridor and thence, having crossed a few modest hills, through the Seine valley.

Over the centuries, the territory of France was constructed around those routes. France, a country of both the north and the south, was – long before Castille – the first modern state to stretch from the Mediterranean seaboard to the Atlantic, its deepest urge being to develop such relations and cultural contacts. The spirit of France, more than of any other country in Europe, reflects those contacts and that double orientation. France is the European country situated where those two seas are the least far apart. Moving westward from Langres, not a single point in the territory of France is over 400 kilometres from the sea.[1] Even in antiquity, geographers had noticed this peculiarity. Strabo[2] calls the country an 'isthmus'. He notes 'the harmonious arrangement of the country with reference, not only to the rivers, but also to the sea, both the outer sea and the inner'. This is an important factor in 'the excellence of the region' (τῶν τόπων ἀρετή). The rivers, which facilitate communications between the seas, account for the way in which 'the necessities of life are with ease interchanged by everyone with everyone else'. Elsewhere, he remarks:

the river-beds are by nature so well situated with reference to one another that there is transportation from either sea to the other; for the cargoes are transported only a short distance by land, with an easy transit through plains, but most of the way they are carried on the rivers – on some into the interior, on others to the sea.[3]

4   *The genesis of France*

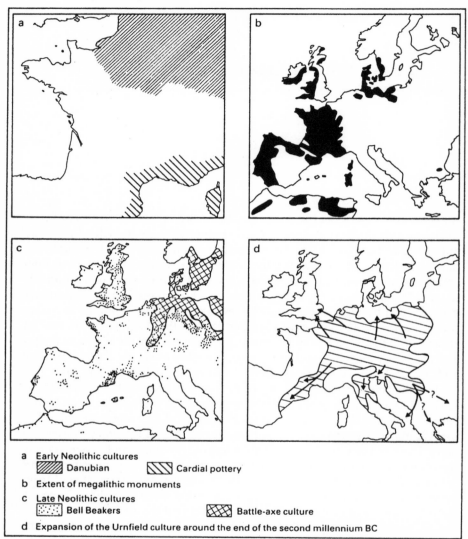

a   Early Neolithic cultures
    ▨ Danubian   ▧ Cardial pottery
b   Extent of megalithic monuments
c   Late Neolithic cultures
    ⋮ Bell Beakers   ▨ Battle-axe culture
d   Expansion of the Urnfield culture around the end of the second millennium BC

Figure 2. France in prehistoric times (after Camps, 1982, p. 301; Niel, 1970, p. 99; Smith, 1978, pp. 21, 25).

That harmony, which is so rare in the Mediterranean region, suggests to Strabo[4] that 'the regions are laid out, not in a fortuitous way, but as though in accordance with some calculated plan' (οὐχ ὅπως ... ἀλλ' ὡς ἂν μετὰ λογισμοῦ τίνος).

It was not until trade routes appeared across the continent, linking the

Atlantic seaboard of Europe with the Mediterranean basin, that this favourable geographical situation realized its full potential. This did not happen until relatively late.

The Celtic space

Before the Celts: prehistoric civilizations

In the Neolithic period, there was no differentiation between the territory which corresponds to France and the rest of the map of Europe. The evolution of the Neolithic civilization seems to have been a complex and diffuse process which began as early as the seventh millennium BC in the western Mediterranean.[5] The first Neolithic products, of 'cardial' pottery, are limited to strictly coastal regions, whilst inland, from the fifth millennium on, the Danubian civilization, with incised pottery, made its way into the Rhineland and thence further westward, eventually reaching much of northern France (fig. 2a). From the end of the fifth millennium on, there developed in southern France the first farming civilization, known as the Chassean civilization, with burnished pottery, which soon spread to the south-west, the Armorican regions, the Paris Basin and even into Franche-Comté, where it flourished during the fourth and third millennia (3800–2300 BC), finding expression in many large megalithic constructions. At this point it covered virtually the entire territory of France, and one eminent prehistorian has even referred to this as 'the very first France'.[6] To a geographer, however, the expression seems somewhat exaggerated. The territory of France at this period was part of a vaster area. Civilizations closely related to the Chassean one occupied northern Italy (the Lagozzian civilization), northern Spain, the western Alps and Switzerland (the Cortaillod civilization). Areas with inter-communications do not seem to have depended at all on the major valleys of the isthmus of Gaul. Flint was present throughout the region and some workshops served quite wide areas (dolerite axes from Seledin in northern Brittany, dating from 4000 to 2000 BC, have been found throughout western France and as far away as the mouths of the Escaut and the Meuse rivers, in the Massif Central and even on the other side of the English Channel). However, it is not possible to make out any coherent pattern of communication routes. The area in which megalithic monuments are scattered covers virtually the whole territory of France but is by no means contained by it (fig. 2b), and it has long been recognized that the main communication routes must have followed the Mediterranean and Atlantic coastlines, skirting the European peninsula by sea. Cultural exchanges took place essentially around the edges of the continental bloc, not across it.

In the Chalcolithic period, the extent of diffusion remained largely pan-European. Towards the end of the third millennium (from 2300 BC on), the

Bell Beaker culture, whose place of origin is unknown but which seems to have spread essentially along the coasts and major river valleys, still covered the whole of western and central Europe. However, in eastern and central Europe, from 2400 BC on it came into confrontation with the Corded Ware or Battle-axe culture, which possibly corresponded to the arrival of the first Indo-Europeans, and in the west stopped short at the Rhine (fig. 2c). The development of an economy largely based upon a mobile pastoral life, which characterized the earliest of the metal ages, now encouraged the expansion of central European networks of cultural communication, as opposed to the peripheral and maritime communication routes of earlier times. In the Bronze Age, from 1400 BC onwards, the Urnfield culture, probably an expression of the individualization of the proto-Celts, began to spread in central Europe, reaching eastern France, Spain and northern Italy (fig. 2d). Built on the foundations of this culture, the Hallstatt and La Tène cultures of the Iron Age, which also originated inland, began to develop during the first millennium BC. These developments testify to the settlement of Celtic peoples over a vast territory covering most of France but also spilling over into large areas of northern Italy, Spain and the British Isles.

Celts, Ligurians, Iberians

Before the Roman conquest, the territory which now corresponds to France possessed no ethnic uniformity. Most of the inhabitants certainly belonged to the Celtic peoples or, to be more precise, to the second wave of Celtic peoples, those speaking languages known as Brittonic ('P languages'), as opposed to the first wave of Goidels (speaking 'Q languages'), who are now confined to Ireland, north-western Scotland and the Isle of Man. The advance of these Celtic groups proceeded in several stages, the most important of which seems to have been the arrival of the *Belgae* in the north from the second half of the fifth century or the beginning of the fourth century until about 150 BC. These were probably proto-Germanic, more or less Celticized people, who overran today's Belgium, the Paris Basin on the north bank of the Seine (the upper and middle reaches of which were left in the possession of peoples from earlier waves of invasion) and spread out along the coasts of Normandy and Brittany. From another direction, they poured through the Belfort gap and the Jura passes, reaching as far as Valais. But these groups speaking Brittonic languages spread far beyond the territory of present-day France. The *Belgae* eventually overran the whole of the south-east of the British Isles. At the time of the Roman conquest, tribes with the same names as those on the continent were living there[7] – the *Atrebates*, for example, after whom Artois, on the continent, is named. Other groups, probably fleeing before the *Belgae*, had also previously gathered on the other side of the Channel. The *Parisii* were to be found not only in the French

Parisian region but also in eastern Yorkshire, where they had settled in about 250 BC[8] and, according to Ptolemy, were still to be found in the second century AD. The Celts were also to penetrate Iberia in large numbers, occupy the whole of the Po Valley (Cisalpine Gaul) and colonize not only the Balkans but even Asia Minor (Galatia). The effect of all these invasions was to disperse various tribes over an extensive area. The *Boii* were thus established in Bohemia, around Arcachon in Aquitaine, in Italy and in Galatia; the *Tectosages* around Toulouse and in Galatia, and so on.[9]

However, the Celts did not totally swamp or assimilate the peoples who had settled earlier in the territory of present-day France. In the south-east in particular, throughout the region of the Alps and Provence, the peoples already established there survived. The authors of antiquity called them Ligurians, a term of doubtful etymology[10] but whose geographical meaning is clear: it was a derogatory name, the equivalent of 'barbarian', which the Greeks used to denote the less civilized peoples with whom they came into contact in the course of their colonizing activities in the western Mediterranean.[11] Meanwhile, in the south-west, to the south of the Gironde and to the west of the Rhône, in the Garonne basin, Iberians survived. In the plain of Languedoc and in Roussillon,[12] they were probably for the most part autochthonous although some had migrated there from their Spanish homeland in the Ebro basin as recently as the early fifth century, fleeing before the Celts' advance into the Iberian peninsula.[13] Having arrived, they had superimposed themselves upon the native 'Ligurian' substrata. At any rate, throughout the south-west, the south and south-east of France the urban network is largely pre-Celtic. Bordeaux (*Burdigala*), Auch (after the *Ausci* tribe; the ancient name had been *Elimberrum, Iliberris* or *Illiberris*, meaning 'new town'), Oloron (*Iluro*), Luchon (*Ilixone*), Tarbes (*Tarbelli, Turba*), Bayonne (*Lapurdum* = Labourd, the name of a Basque province), Foix (*Fuxus*), Lectoure (*Lectora*), Toulouse (*Tolosa*), Elne (*Iliberris*, another 'new town'), Carcassonne (*Carcasso*), Narbonne (*Narbo*), Collioure (*Caucoliberi* in the seventh century AD = (*Cau)co-[il]liberris*), Tarascon, Béziers (*Bessara*), Toulon (*Telo*), Cavaillon (*Cabellio*), among others, all bear pre-Celtic names, some of which are easily explained by the modern Basque language, which is derived from Iberian. The name of *Corbilo* has also been traced back to the Iberian: this was the large centre of the tin trade, situated on the lower Loire in the second century BC.[14] Its exact location is still unknown, but in truth the etymology of the word is more likely to be Celtic, and the Iberian expansion probably never extended so far.[15]

However, the Iberians and the Ligurians did not block the Celtic advance for long. Following the Iberian invasions and as early as the fourth century, it resumed. The Celts now occupied Bordeaux, where the *Bituriges Vivisci* installed themselves on the poorer land which was probably still vacant:[16] the marshy or pebbly areas of the valley and the siderolithic clay or sandy areas of the plateaux. Other Celtic or Celticized peoples settled on the left bank of the river: the Medulli from Médoc, the *Boii* or *Boiates* from the Arcachon basin and the *Basates* (after whom the town of Bazas is named) from the Beuve valley. Upstream, the Celts reached the Garonne, which they crossed, and occupied the important port of Langon, while the *Nitiobroges* settled in the region of Agen, which is a Celtic name. In the second half of the third century,[17] the *Volcae* settled in the Toulouse region and the Languedoc plain from which they dominated the coastal areas of the Golfe du Lion and the Lauraguais pass, occupied Narbonne and Béziers, founded Nîmes or *Nemausus* (named after a Celtic god) and spread as far as the Rhône and, at the time of Hannibal, even beyond. In the Rhône corridor and the adjoining river valleys a number of powerful nations established themselves: the *Allobroges* between Geneva and the confluence of the Isère, the *Tricastini* at the mouth of the Drôme and the *Cavares* in the plains of Comtat Venaissin. However, the Iberian peoples retained most of the land on the left bank of the Garonne. The Ligurians, who were more or less intermingled with the Celts in the great confederation of the *Salluvii* contained by the Rhône, the Durance and the Hyères islands, only survived as pure communities beyond Toulon, in the Maures, the Esterel and the Alpes de Provence. In these marginal sectors, the population was very much aware of its ethnic origins. At the time of Caesar's conquest, all the tribes of the south-west, between the Garonne and the Pyrenees, whether of Celtic, Iberian or Ligurian origin, were known as the *Aquitani*, a name certainly borrowed from the Iberian language.[18] The Ligurians and the Iberians are estimated[19] to have occupied territories of about 15,000 square kilometres and 28,000 square kilometres respectively at the time of the Roman conquest. That accounts for no more than 8 per cent of the area of present-day France, the remaining nine-tenths of which had been Celticized.

Gaul

The proportion of land held by Celts was greater here than in any other area of western Europe, so understandably enough it was from the Celts that what was to become the territory of France derived its most ancient name: Gaul.[20] This was an ethnic name before it came to be applied to a particular area. Up until the domination of the Celts there had been no name for the geographical area which corresponded to present-day France. At the end of the sixth century we find only the name Ligystica, used by Hecataeus to denote the region in which Marseilles is situated, but there is no evidence to suggest that this 'land of the Ligurians' extended any further than the area immediately surrounding the town. In the third century BC, the Greek poet Apollonius of Rhodes[21] is the first to speak of 'the thousand Ligurian and Celtic peoples' of what is present-day France. The name Celts (κελτοί; later κελταί), possibly meaning 'warriors', or 'great ones, superior ones' (in the moral sense), applied only to mainland Europe; it had no equivalent in the British Isles. Initially, however, it applied to the peoples of a much vaster area than present-day France. It may have been the navigator Pytheas, from Marseilles, in the fourth century BC,[22] who first gave a general name, κελτική (*Keltike*) to the Celtic countries as a whole, but at this time it was considered to apply to the area bounded by Marseilles, the Pyrenees, the Atlantic Ocean and the Elbe. A little later, in the second century, the Greek historian Polybius twice[23] refers to κελτία (*Keltia*), drawing a distinction between it and Liguria (Λιγυστίνη – *Ligustinè*).

But at the same time Polybius also uses the name *Galatia* (Γαλάτια) to denote Cisalpine Gaul usually, but on one occasion[24] Gaul proper, that is to say the country bounded by the Pyrenees, Italy and the two seas. *Galatai* (Γαλάται) is also a Celtic name which some scholars consider to be another form of κελτοί, but which is probably a different word meaning 'hero', 'brave'. As used by the Greek authors, the terms *Galatai* and *Keltai* are related in complex ways: sometimes they denote the same entity or peoples, sometimes different ones. 'Keltai' was often used to refer to the nearer peoples, 'Galatai' to those more distant, in the north.[25] Greek authors continued to use both *Keltike* and *Galatia* until well into the Christian era.

Anyway, *Galatus*, the Latinized form of *Galatos*, is thought to be the word from which the Latin form *Gallus* (Gaul), a slang diminutive, is derived. It was assimilated to the Latin word for cockerel (*gallus*) on account of the raucous and aggressive behaviour of both this creature and also the Gauls. Alternatively, the name of the animal, said to come from Gaul and for which there was originally no word in Latin, may be derived from the ethnic name.[26] *Gallia*, the geographical form of the word, is known to have been used from the late third century onwards to denote Cisalpine Gaul (the Po Valley). But it was not until the Roman victory of 120 BC, after which the province of Transalpine Gaul was created, that the term was applied to the Gauls of Gaul, in documents relating to Roman triumphs. The name was still an ethnic one, with no specific geographical application. In the Latin texts of the first half of the first century BC, *Gallia* continued to be mostly used to denote Cisalpine Gaul; to refer to Transalpine Gaul, it was considered necessary to add on epithets such as *transalpina*, *ulterior* or *ultima*. Not until Caesar came upon the scene was the term Gaul applied for the first time (in the first sentence of *Bellum Gallicum*, probably written in 51 BC at the latest) to independent Gaul, excluding the Roman province: '*Gallia . . . omnis*'. Caesar meanwhile referred to the Province and independent Gaul, taken together, as *Gallia transalpina*, *ulterior* or *interior*. At the height of the Roman conquest, in 55–54, Catullus[27] used the expression *Gallia comata*, 'long-haired Gaul', to refer to independent Gaul, where the men wore their hair long, unlike the Italians and as opposed to *Gallia togata*, the Gaul where togas were worn, that is to say the Gaul of the Roman citizens of Cisalpine Gaul. To judge by the Latin texts at least, the geographical identity of Gaul was clearly recognized at the time of the Roman conquest.

The geographical face of Gaul

The peoples

The territory known as Gaul was made up of a conglomeration of different peoples. It would be useless to try to reconstruct an accurate map of these groups in all their variety for the periods which antedate the Roman conquest. However, the limited archaeological data may to a certain extent be supplemented by a study of the names of these peoples, some of which seem to have established themselves before the Celts arrived in Gaul, while others appeared after that invasion and only acquired their separate identities at this point.[28] The unintelligible names are probably the more ancient ones (the *Boiates, Volcae, Helvetii* and *Turoni*; the *Nervii* and the *Suessiones* among the *Belgae*; the *Veneti, Aulerci* and *Pictones* in the west, and so on). To these we should probably add the *Bituriges*, 'the kings of the world', who won an initial supremacy. The more intelligible names seem to be newer ones which were adopted later. Some have numerical implications (the *Remi*: 'the first ones'; the *Petrocorii*: 'the four groups'; the *Voconti*: 'the twenty clans'); some are nicknames (the *Ruteni*: 'the blond ones'; the *Leuci*: 'the shining ones'; the *Aedui*: 'the ardent ones'; the *Caleti*: 'the hard ones'; the *Lemovices*: 'the warriors of the elm' [a fetishistic tree]); some are geographical and relate to the place of settlement (the *Nantuates*: 'the people of the valley'; the *Allobroges*: 'those of the other land'; the *Morini*: 'those of the sea'). Some of these geographical names no longer related to the latest dwelling place of those who bore them, which proves that these peoples had recently moved: the *Sequani*, 'the people of the Seine', were thus to be found in Franche-Comté at the time of the Roman conquest.[29] The deformation of certain names proves them to be relatively ancient (as in the case of the *Osismii*, the people of Finistère, whose name formerly came from Uxisama, 'the last island'). Gaul was thus inhabited by peoples who were the remnants of larger nations which had been dispersed across the Celtic world but had in some cases retained their original names. Then there were also peoples formed at some later date. These groups were constantly splitting apart and reforming and their historical geography almost totally escapes us. In some cases, it is only through later texts that we learn of the common origins of peoples who later separated at dates unknown. Thus Ptolemy, writing in the second century AD tells us of the confederation of the *Aulerci*, which formerly comprised the *Diablintes*, the *Cenomanni* and the *Eburovices*.[30] At the time of Caesar's conquest, the nations of Gaul appear to have consisted of various composite groups between which ties of friendship or vassaldom were still in a state of flux and in some cases had nothing to do with any original consanguinity.

Bearing in mind the above reservations and recognizing that it would be impossible to reconstruct any definitively fixed order in independent Gaul, we can nevertheless be certain that there were two different levels of ethnic and tribal organization, which the Roman conquerors recognized and stabilized. At the lower level there were individual tribes inhabiting territories of on average about 1,500 square kilometres. To refer to these, the Romans used the word *pagus*, the root of which implies the notion of being 'fixed' somewhere and which eventually produced the French *pays*. A document presented to the Roman Senate after the death of Augustus mentions 305 of these peoples. In Caesar's day there would have been 300.[31] These tribes were more or less federated into larger groups which the Romans turned into 'cities' (*civitas*) but also referred to using a number of other terms (*populus, gens, natio*).[32] There were about sixty of these. The largest of them covered a territory of 20,000 square kilometres, with a population of one million inhabitants. These large peoples were situated mainly between the Seine, the Rhône and the western slopes of the Massif Central. The *Arverni*, the *Aedui*, the *Pictones*, the *Santones*, the *Ruteni*, the *Lemovices*, the *Bituriges*, the *Carnutes*, the *Senones* and the *Lingones*, a group of ten, made up half of all Gaul. The smaller tribes seem to have been virtually swallowed up by these larger groups some time before the Roman conquest. The long and complex process of unification and concentration defies analysis but probably took place in the last few centuries before Caesar's conquest. It was far more advanced in the central regions, which had been occupied early on and had remained unaffected by the upheavals provoked by the arrival of the *Belgae*. Amongst the latter, the 'city-nation' system was still far from established at the time of the Roman conquest. In the more southern sectors, amongst the *Treveri*, the *Bellovaci* and the *Remi*, some true 'cities' did exist; but further to the north, in regions still for the most part thickly wooded, a number of smaller tribes with variable connections predominated. Amongst the peoples of the south-east and the south-west, that is to say the Ligurians and the people of Aquitania, who had not been affected by the Celtic conquest, the scattered groups were even smaller.[33]

The frontiers

Although the general location of the peoples of Gaul, or at least of the major groups, is known thanks to a whole series of sources (mostly of a literary or numismatic nature), the precise boundaries of their territories escape us for the period of independent Gaul. Reconstructions are perforce based upon the assumption that the medieval dioceses followed on from the towns of the Late Empire; and (the bishops having taken over from the priests of Augustus) these are themselves considered to be identical to those

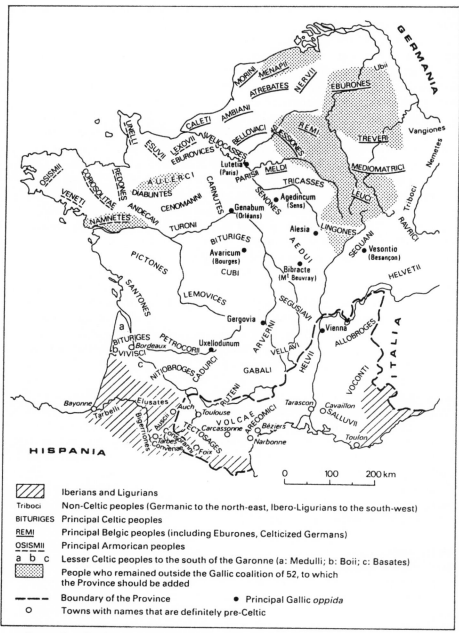

Figure 3. Gaul at the time of the Roman conquest (partly based on Jullian and Duval, 1963).

of the Early Empire, which for their part continue to observe the territorial divisions separating the various peoples of Gaul. 'On to the map of pre-Roman Gaul one thus projects the frontiers of cities whose boundaries are not attested either in general or in detail before the seventh century AD even in the most favourable cases and in most instances not until the eighth or tenth century AD.'[34] Yet those frontiers must have varied not only in the course of the period of independence but also during the Roman period, as the relations of power between neighbouring cities underwent changes.[35] Imprecision was, besides, a feature of the very nature of tribal boundaries in Gaul. With few exceptions, they tended to take the form of frontier zones or 'frontier-deserts', not material boundaries marked out in a concrete fashion.[36] That conception of territory and its limits may have been adopted quite deliberately, with each tribe seeking to maintain the greatest possible distance between itself and its neighbours, for in the Germania of Caesar's period the various peoples took pride in the empty spaces surrounding them, regarding them as a symbol of their power.[37] For the Celts, the emptiness around them was a 'beyond' at once close and distant, an area consecrated to the gods, the world of the Elect, reached by way of sacred hillocks, stretches of water and dense forests.[38]

The various methods of geographical analysis make it possible to recognize where these frontier-deserts were situated. Most consisted of forests, many of which still exist in a more or less residual form, even today. The positions of others are recorded in medieval texts: one of these is the frontier-forest of Arrouaise, which separated the *Viromandi* (Vermand – Saint-Quentin) from the *Atrebates* (Arras) and the *Nervii* (Bavai – Cambrai),[39] and which medieval and modern land-clearance has still not completely eliminated (fig. 4A). The normal territory of a Gallic tribe should be imagined as a cultivated clearing surrounded by more or less unbroken forests. The situation of these forests was usually dictated by the type of soil, which would be unsuitable for cultivation. In the Paris Basin, there would in many cases be a hard siliceous upper layer, testifying to an ancient erosion surface, which remained wooded, whilst a clearing developed below, on either side of the valleys which had worn away the surface layers (fig. 4B). The frontier-forest might thus appear to form a kind of natural wall (*nativus murus*), to borrow the expression which Caesar[40] applied to the forest in Germania which separated the *Suevi* from the *Cherusques*.[41] When the frontier-forests disappeared, more or less sterile heathland, which remained relatively uninhabited, developed along these heights. Another type of border, which seems to have been less common, was constituted by river valleys which were marshy and overgrown by natural vegetation, and were consequently hard to cross. One example is the Mayenne valley, which separated the *Diablintes* from the *Redones*;[42] another is the Moselle valley, which formed the frontier of the *Leuci* territory at the beginning of the first

14   *The genesis of France*

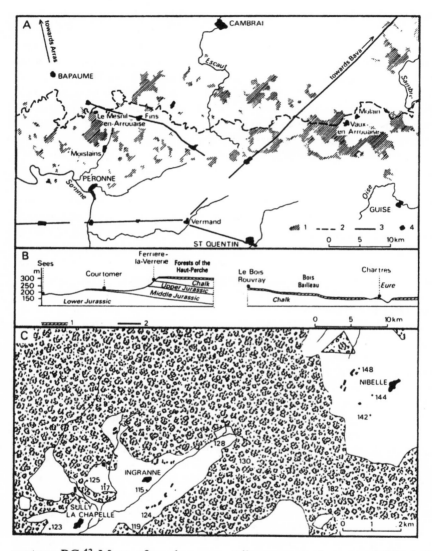

century BC.[43] More often, however, tribes seem to have established their territories along both banks of a river[44] or sought to set up bridgeheads, as the *Andecavi* did on the southern bank of the Loire.[45]

These marginal deserts nevertheless contained landmarks which made it possible for travellers, hunters and woodcutters to see that they were approaching the outer limits of the territory. Sometimes megaliths, very old trees or springs would serve this purpose.[46] Defences would be set up at various points. Finally, certain pre-Latin toponyms, derivations of which have survived to the present day and which are certainly associated with the idea of a frontier, indicate that permanent human edifices were once erected here. One is *mediolanos* (Latinized into *meliolanum*) with its French forms of Meillant, Mélian, Meylan, Montmélian, Molain, Moislains and so on. The positioning of these along the frontiers seems to have taken place at the time when the Celtic peoples first took possession of these areas (see below, pp. 22–5). The etymology and meaning of the term *equoranda* is still a subject of dispute. It used to be thought to mean 'a water-course along a frontier', but the *equo-* element is now believed to relate not to water but to the horse, giving the term the meaning of a 'staging-post'. However, that may be, many names derived from the term (Ingrandes, Ingranne, Aigurande, Yvrande, and so on) certainly do recur along the boundaries of these peoples' territories, probably at the points where they intersected with paths or waterways, and they indicate that constructions designed to facilitate active relations between neighbouring peoples used to exist at these points (fig. 4C). Such meeting places (*conciliabula* in Florus, I, 45, 21) made it possible for representatives of these peoples to confer together. In Germania, the frontier-desert remained a hostile and perilous place, but in independent Gaul it was already on the way to serving a human purpose.[47]

Whatever their form, these frontier-deserts certainly still represented major political and cultural divisions. The sense of ethnic identity was undoubtedly extremely strong among not only the various peoples who stood in vigorous opposition to one another, but also the individual tribes of which each people was composed and which were constantly quarrelling amongst themselves. The territories of these different peoples and tribes must have been differentiated from one another in all kinds of ways. It has been suggested that the boundaries between zones displaying different rural

---

Figure 4. Types of frontier between the peoples of Gaul (after Dion, 1947, pp. 14–15, 18).
A. Remains of the Arrouaise frontier-forest separating the *Viromandi* (Vermand–Saint-Quentin) from the *Atrebates* (Arras) and the *Nervii* (Cambrai).
1. Forest. 2. Present-day departmental border. 3. Visible remains of Roman roads. 4. Settlements. Note the toponyms 'Fins' (from *fines*), 'Moislains' and 'Molain' (from *mediolanum*). B. Frontier-forests in the upper Perche region between the *Sagii* (Sées) and the *Carnutes* (Chartres). An infertile and wooded siliceous crust formed by the disintegration of chalk covers the upper ground. The city capitals are situated in cultivated clearings where the siliceous crust has disappeared either as the result of erosion (the *pays* of Sées) or of being covered over by silt (the *pays* in the Chartres region). 1. Siliceous crust. 2. Silt.
C. Frontier-forest between the *Carnutes* (Chartres), to the south-west, and the *Senones* (Sens), to the north-east. The toponym Ingranne (*equoranda*) indicates the frontier. Settlements are situated in the valleys hollowed out by running water, below the old wooded heights along which the frontier passes. In 1789, the villages in the north-eastern corner belonged to the diocese of Sens, while those in the south-west were attached to the diocese of Orléans.

building techniques may go back to the period of the Gallic cities – for example, the frontier between the *Meldi*, a tribe belonging to the Belgic *Suessiones* people, and the *Senones* people.[48]

Population density

However, we should not be misled by these relatively empty spaces and the constant population movements. Gaul was in fact already densely populated. A close critical comparison between the data available on a number of peoples, relating in particular to the size of their military forces at the time of the Roman conquest, as reported by Caesar, enabled Camille Jullian to estimate the total population contained between the Pyrenees, the Alps, the Atlantic Ocean and the Rhine to be between 20 and 30 million, in other words about 20 million for the territory of present-day France.[49] This is a figure comparable to that reached in the late Middle Ages, before the Black Death, and then again in the seventeenth century. Even if that estimate is too high and, as seems likely (given that it was in Caesar's interest to exaggerate the size of the peoples he had just conquered), Caesar's figures should be corrected to 10 million or even lower, it is nevertheless clear that in the second century BC Gaul was considered one of the most densely populated countries in the world. It is accordingly understandable that the Greek authors, who lamented the low figures of their own compatriots and who were, moreover, somewhat traumatized by the Celtic invasions, reckoned that the Gallic tribes suffered from excessive numbers (ἀνδρῶν πλῆθος).[50] The relative demographic strength of the territory of France within Europe was to be maintained until the early nineteenth century. The explanation for this has yet to be determined. Perhaps it is quite logical to regard it simply as the result of an accumulation at the extreme point of the great continental route which passed over the plain of northern Europe, that is to say to the north of the Alps, at the time of the Indo-European expansion.

Rural settlement and agrarian landscapes[51]

We know very little about the manner in which this relatively large population was settled. However, Caesar's texts testify to the fact that, on the territory of Gaul, there were, apart from more or less urban settlements (*oppida*), two basic forms of rural habitat, the *vici* and the *aedificia* (*privata*). *Vici* clearly denoted something like 'villages' or 'country towns'; *aedificia* meant more scattered dwellings, probably taking the form of hamlets constituted by the homes of at least an extended family. Caesar's description of these isolated dwellings, situated in small clearings close to water (*aedificio circumdato silva ut sunt fere domicilia Gallorum, qui ... plerumque silvarum ac fluminum petunt propinquitates*),[52] may suggest that the *aedificia* constituted the dominant form of habitat, while the plateaux remained relatively empty, as in Neolithic times. However qualification is called for here. We may take it that some kind of balance existed between the centres of grouped inhabitants and those who lived in scattered dwellings. The *Helvetii*, who numbered 263,000,[53] had 12 *oppida*, 400 *vici* and *reliqua privata aedificia*.[54] It is not possible to form any definite conclusions since the average population of the *vici* is not known. Nevertheless, it is reasonable to suppose that the overall figure must at least have matched that of the *aedificia*. Besides, although lack of data makes it impossible to attempt any statistical calculations, it is significant that the terms *vici* and *aedificia* tend to appear together in Caesar's account. They do so eight times, *vici* being also mentioned five times on their own and *aedificia* three times on their own.[55] Unsatisfactory though these data may be, it at least indicates the likelihood of a mixed habitat, composed of a combination of grouped and scattered homes.

The social structure and system of land-ownership

Social structure can be analysed clearly on the basis of this type of settlement.[56] Large estates belonging to the Gallic aristocracy, who are known to have played a dominant social role, must have been common.[57] This aristocracy had already mastered some remarkably efficient techniques, such as that of the horse- or ox-drawn harvesting-machine, which filled the Roman agronomists with wonder.[58] On the other hand, slave labour was scarce in Gaul[59] and work on the great estates must have depended largely upon the populations of the neighbouring *vici*, a proportion of the land being exploited directly by the owner, the rest parcelled out in small lots to tenant farmers.[60] In conjunction with the *aedificia*, the manors of the rich landowners, the *vici*, or rural towns, 'formed an inseparable whole'.[61] Theoretical though it is, that is a perfectly plausible reconstruction. However, the Gallic social stratification is known to have been very complex and it is hard to believe that all of these numerous *aedificia* were truly extensive properties which depended upon outside labour. It seems reasonable to suppose that in the case of medium-sized or smaller properties the exploitation of the land around the *aedificia* could be undertaken by the owner's large extended family, possibly with the assistance of a few slaves. At any rate, comparing the data with the evidence that exists for Ireland,[62] it is tempting to believe that the *aedificia* represented the family and social structures of the invading Celtic aristocrats and took the form of large, isolated estates situated in between the more ancient clusters of villages where Neolithic or Bronze Age traditions were maintained even though the inhabitants were progressively won over to Celtic ways.

How did this pattern of social grouping and these structures of landownership affect the division of land and the landscape in general? The recent progress made in aerial photography used as a means of detection has provided some answers, although they only relate to a few locations. In

Picardy, it has proved possible to make out a small number of systems of irregular enclosed parcels of land similar to the 'Celtic fields' to be found in the British Isles.[63] Elsewhere, the rural countryside surrounding the *aedificia* included numerous enclosures, many of which are detectable on the aerial photographs and have yielded finds from the Iron Age when excavated.[64] But most of these quite vast enclosures appear to be isolated; it is rare to find them arranged in networks. There must have been, above all, enclosing barriers around the farms themselves and separate paddocks for the animals. Some of these still show traces of a funnel-shaped entrance designed for the herds.[65] Of course, aerial photography only reveals stable perimeter lines with deep foundations, usually ditches or steep banks, while the light barriers that surrounded cultivated fields are unlikely to have left any traces. But at any rate, the larger tracts of land, where horse- or ox-drawn harvesting-machines were used, must have covered vast areas in which enclosed parcels of land were few and far between. On the other hand, it is reasonable to suppose that stretches of country made up of open fields kept constantly under cultivation, similar to the clusters of *esch* or *infields* which represent the most ancient layer of permanent agricultural structures in Europe, were commonly to be found in between areas of temporary and, no doubt, enclosed fields carved out of the vegetation surrounding the *vici*. Traces of such arrangements are preserved in the elliptical or circular crop-marks left in areas where progressively extensive land clearance has taken place. These show up to varying degrees in many present-day field patterns.[66] They have been shown to date from a period between the Neolithic and the arrival of the first Celts, that is to say the period between about 3500 and 1500 BC; and their persistence seems to obey certain coherent geographical rules. In the eastern fringe of land that was laid waste, which corresponds to the area in which clashes took place with the people of the Urnfield culture, these crop-marks appear only in an incomplete, semi-circular form and in isolated locations, whereas they have survived longest in areas further to the west, where the land was occupied without interruption (fig. 5).[67] They were an integral feature of the landscape of Celtic Gaul where, in the vicinity of hamlets (*vici* ?), they must have constituted noticeable patches of open fields, probably surrounded by barriers. A careful reading of Caesar's text provides confirmation of the existence of a considerable expanse of open fields or at least of a general scarcity of enclosures. It is significant that, in his careful descriptions of military operations, the Roman conqueror never mentions any barriers or hedges which might have hampered the movements of his troops, particularly the cavalry. He describes as an exceptional isolated case the countryside encountered in the region of the *Nervii*, where a network of thick hedges, irresistibly suggesting a copse of quickset hedges, covered the area where battle took place: *Quum, diversis legionibus, aliae alia in parte ... sepibusque densissimis ... interjectis, pros-*

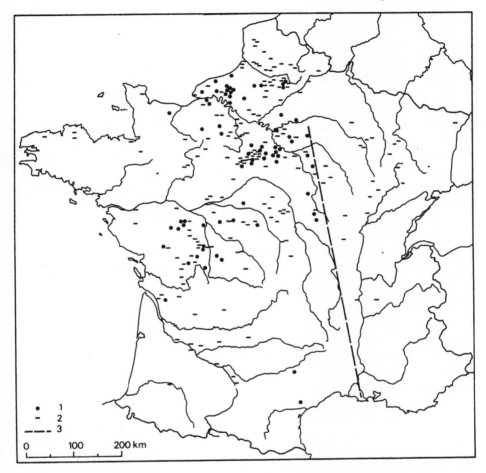

Figure 5. Traces of large circular crop-marks in the French countryside (after Soyer, 1970)
1. Demarcated crop-marks, situated close to settlements.
2. Undemarcated crop-marks, in isolated positions, or semi-circular in shape.
3. The eastern limit of dolmens.

pectus impediretur.[68] The fact that the wooded countryside of the *Nervii* is considered so exceptional suggests that open fields were already a widespread feature of the rural scene in pre-Roman Gaul.

The towns

Gaul was already relatively urbanized. As well as *vici* and *aedificia*, Caesar mentions *oppida*, fortified centres of population with, for the most part, a variety of truly urban functions. They were religious centres, with sanctuaries devoted to the gods; political centres, with institutions set up to serve the local peoples, which also functioned as mustering points and as places of refuge in times of war; and economic centres, with busy markets (toponyms with a *-magus* element) and workshops. We do not know how many of these urban centres there were. The figures provided by Caesar (twelve *oppida* among the *Helvetii*,[69] and twelve among the *Suessiones*[70]) were probably rounded out and the figure of eight hundred towns captured by him in independent Gaul, which Plutarch gives,[71] has the air of being equally artificial, arrived at by multiplying the number of cities in Gaul by twelve. But we do know that these Gallic population centres provided the basis of most of the urban fabric of France and certainly of most of its sizeable towns. The only areas that were not urbanized must have been the Alpine valleys in the Ligurian regions and the territories of the *Belgae* in the north, beyond the Ardennes forest, inhabited by the *Nervii*, the *Morini*, the *Menapii* and the *Treveri*, where only a very few anonymous *oppida* are mentioned and where there were little more than temporary refuges, which Caesar calls *castella*. However, throughout the Armorican region, in the west and the north-west, the population centres must have been very small, hardly rating the name of 'towns'. Between Bordeaux and Boulogne (*Portus Itius*) no names of ports are known to us apart from the mysterious *Corbilo*. These north-western regions, so distant from Greek and Roman influences, were still at a stage of proto-urbanization. In most of Gaul, however, the *oppida* could already be accounted true towns, where the aristocracy accumulated houses and wealth. The *Arverni* nobles maintained winter houses[72] at *Gergovia* and, as for *Avaricum* (Bourges), the pride of the *Bituriges* nation, Caesar went so far as to pronounce it 'beautiful': *pulcherrimam prope totius Galliae urbem, quae et praesidio et ornamente sit civitati*.[73] All the same, they must have looked very primitive, the only stone-built constructions being the outer walls. Some of the dwellings were semi-subterranean. The wooden houses with thatched roofs, which predominated, were mostly scattered in a disorderly fashion over a relatively large space, generally only partially built up, which in times of war would be filled by large crowds of refugees: (40 hectares at *Avaricum*, 75 at *Gergovia*, 97 at *Alesia*, 135 at *Bibracte* and even 162 at Saint-Désir, near Lisieux).[74] These towns have been compared to those of pre-colonial black Africa where, in similar fashion, the precinct would contain a mixture of huts, workshops, wasteland and cemeteries.[75] These more or less disorganized spaces were, however, already acquiring a coherent structure in certain specific districts. In *Alesia*, the best known of the Gallic towns, it is clear that the sanctuary and the market were located at the top of the hill, surrounded by the quarter where the *Aedui* nobles lived. Below and around were the workshops and dwellings of the artisans. *Avaricum* even boasted a forum. In the Mediterranean south, the *oppida* of the Celto-Ligurians, which were exposed to Greek influences, were much smaller (on average between 1 and 3 hectares), but far more highly developed.[76] At Entremont, an *oppidum* of 3.5 hectares belonging to the *Salluvii*, situated on a plateau above Aix-en-Provence and destroyed in 123 BC, the entire area was built up. The houses, roofed by branches and each consisting of a single room, were arranged in a quadrangular pattern which allowed for straight streets designed to give on to a view: at least a hint of a sense of town-planning was certainly present here. At Ensérune, 10 kilometres to the west of Béziers, which enjoyed a brilliant period of development starting in 425 BC and then, after being sacked in the third century, achieved another golden age in the second century, the town plan was based on a chequerboard pattern and the lay-out of the houses, which boasted a number of rooms and a central *impluvium* courtyard, was inspired by a Graeco-Roman model. Overlapping the great trade route between the Iberian peninsula and Italy, this town was not simply 'a refuge-*oppidum* but also, and more importantly, a market-*oppidum*, a centre for the trade that linked the Mediterranean world and the Celtic and Celticized interior'.[77] All the same, such comparisons should not be exaggerated. At Ensérune, there was no trace of the major monumental complexes (temple, forum, theatre, baths and so on) which characterize Graeco-Roman towns. Furthermore, the populations of these towns are very hard to calculate, for the vast areas of the *oppida* which sometimes served as places of refuge may be misleading. When Caesar took *Avaricum*, it contained 40,000 inhabitants, 10,000 of whom were defenders sent in to assist the *Bituriges*.[78] However, those 30,000 inhabitants could easily have included crowds of refugees. Besides, Caesar may well have exaggerated the numbers.

We do not know at what date these population centres began to be developed and to fulfil their urban functions. The names of most of these towns in Gaul are easy enough to understand by reference to the Celtic language. The endings -*briga* and -*dunos* (Latinized form, -*dunum*) suggest the idea of heights or fortifications; -*ritos* (-*ritum*) = ford; -*briva* = bridge; -*nemetos* (-*nemetis* or -*nemetum*) = temple. However, these *oppida* may have been a legacy from pre-Celtic times. Quite apart from the Iberian and Ligurian toponyms in the southern region, mentioned above, some of these population centres – and by no means the smallest of them – (*Lutetia*, *Genabum*, *Gergovia*, *Vesontio*, and so on), in central Gaul, bear names which it is impossible to explain in terms of the Gallic language, or only with difficulty and incompletely.[79] These towns would appear to have taken over from earlier centres which, being situated in favourable positions (Paris, Orléans), may already have fulfilled commercial functions of a proto-urban nature even before the arrival of the Celts. There is, however, no archaeological data to support such a hypothesis. On the other hand, the many centres known as *Noviodunum*, meaning 'new foundation', which date from the Celtic period, certainly do constitute a later group, as do toponyms ending in *-bona* (= foundation). The development of this urban network had certainly already been proceeding for many centuries.

Within this already complex urban fabric, in which many towns were by now situated at the important junctions of rivers or of overland routes (*Lutetia, Genabum*), cities were, however, still above all designed to serve the particular peoples for whom they provided a '*chef-lieu*' and, if necessary, a place of refuge. The most successful towns of independent Gaul were those advantaged by good natural defences and a position that commanded a fertile territory. '*Oppidum munitissimum ... fertilissima regione*': that is Caesar's explanation for *Avaricum* becoming an important capital.[80] The best towns were so easy to defend and favoured by an abundance of resources. The vast majority of sites were fortified, usually perched on a hilltop but sometimes situated on marshy ground or, as in the case of *Avaricum*, surrounded by waterways. Of the twenty-eight names of localities in independent Gaul cited by Caesar, five end with the suffix *-dunum*. A combination of the right conditions would sometimes be found not in the middle of the inhabited, cultivated clearing, but on the edge, close to the woods which would provide a last refuge if the town itself fell and which in times of peace supplied wood, large quantities of which must have been used for building. Such was the case for the *Silvanectes* (Senlis), the *Parisii* (*Lutetia*), the *Carnutes* (Chartres) and the *Remi* (Rheims). The wooded heights around Paris (probably those of Meudon and Sèvres) protected the flight of the Gauls who were defeated at *Lutetia*: '*Silvae montesque texerunt*'.[81] Smaller *oppida* situated close to frontier-deserts must have had a purely defensive function.[82]

A united Gaul?

Gaul was a country still divided between a large number of peoples, but it was densely occupied and was already known by a single name, at least by foreigners and its first conqueror. To what extent did it constitute an inhabited geographical unit at the time of the Roman conquest and to what extent were the people who lived there conscious of that unity?[83]

From a cultural point of view, there can be no doubt that the Celtic peoples of Gaul, including the *Belgae*, were as a whole very conscious of their close consanguinity and common origins: '*propinquis consanguineisque nostris*'.[84] Their traditions represented Gaul as the centre of the Celtic world, the focal point for the migrating Celtic conquerors. When it was a matter of resisting the Romans, they tended to speak of 'the whole of Gaul', as if it constituted a single entity: '*In consilio capiendo omnem Galliam respiciamus*'.[85] But what of the political situation? There can be no denying, despite attempts to do so,[86] that the great revolt of 52 BC was indeed a manifestation of Gallic patriotism. At this point the tribal chiefs 'lamented the common misfortunes of Gaul': '*Miserantur communem Galliae fortunam*' (Caesar, *Bellum Gallicum*, VII, 1, 5). Some scholars[87] have believed that that mass uprising and the ensuing decisive defeat accounts in part for the swiftness of the Roman conquest in Gaul, in contrast to the lengthy resistance in the Iberian peninsula, which was fostered by its dispersed and anarchic nature. However, this largely unificatory movement could not, in itself, unite all the peoples of Gaul. Of the *Belgae*, the group consisting of the *Remi, Suessiones, Leuci, Lingones, Eburovices, Treveri* and *Menapii*, and also possibly the *Diablintes* and the *Namnetes* in the west, dissociated itself. Even when imminent danger threatened, the sense of political unity among the Gauls was never more than an incomplete aspiration, as inconstant and unstable as their general temperament. Caesar, to whom we owe all the texts cited above which convey such sense of unity as existed, himself writes categorically as follows in the very first paragraph of the *Bellum Gallicum*: '*Hi omnes lingue, institutis, legibus inter se differunt* (They differ from one another in language, customs and laws)'.

In earlier periods, the diversity must have been even greater. If we discount mythical references to a general primitive royal line, such as the Hellinized version according to which Hercules was supposed to have founded *Alesia* and ruled as king over all the Celts[88] and those which refer to Ambigat, the king of the *Bituriges*, chief of all the Celts, enthroned at the very centre of the country,[89] the only real political system to establish itself on two occasions over more or less the whole of Gaul was the *Arverni* Empire.[90] This seems to have been an extremely loose hegemony which certainly did not last for long, but did gather the various tribes together in a federation based on no more than links of clientage that were 'diverse and imprecise ... a vast, tumultuous and incoherent fellowship of warriors'.[91] Before the Roman conquest, Gaul never enjoyed any political unity; indeed, the plural form of *Galliae*, 'the Gauls', often used by Latin authors, Caesar in particular, conveys a sense of the forces, predominantly centrifugal, that were constantly at work. Active relations certainly existed between these different peoples. Their aristocracies, at least, intermarried and pursued interests and established links far beyond the confines of their own tribal territory.[92] A well-maintained road network was used by numerous merchants, both native and foreign. News travelled fast along these highways, carried by relays of messengers or possibly passed on by means of optical signals operating between one hill and the next.[93] Clearly, all this presupposes some kind of inter-tribal organization, although nothing appears to have been established on an institutional footing.

The only permanent and attested link between the peoples of Gaul in fact appears to have been religious: to wit, the annual assembly of the Druids.[94] The very spot chosen for it testifies to a measure of geographical unity. It was held at a fixed time (*certo anni tempore*), at a consecrated place (*in loco consecrato*), in the territory of the *Carnutes*, chosen as the reputed central point of the whole of Gaul ('*Regio totius Galliae media habetur*')[95] – a fact which implies a very precise perception of the dimensions and structure of the area involved.[96] In truth though, the choice of this spot was probably dictated more by its commercial importance and the convergence of a number of routes at this bend in the Loire than by the determination of a geometrically central point. Here, religious ceremonies took place and legal and possibly even political decisions were taken, affecting the relations between the various tribes, which is why this assembly has sometimes been compared to the *amphictyonia* of Delphi. But however important it may have been, there is no evidence to indicate the exact extent of the area from which participants were drawn. The tribes which attended were probably all Celtic, with the addition of the peoples from the Armorican peninsula, but no *Belgae*.[97] Intense and continuous relations were, however, maintained between the Druids of Gaul and those of Great Britain. Priests and neophytes had no hesitation in crossing the straits in order to learn from one another.[98] In fact, it is not beyond the bounds of possibility that Druids from Britain may also have taken part in the assembly held on the territory of the *Carnutes*. Inter-Celtic relations of a religious or political nature thus stretched beyond the confines of Gaul, even if not throughout the Gallic territory itself. However, that limitation is no reflection upon the technical abilities of the Celts. We know that the scientific level of the Gallic civilization was high enough for it to possess an accurate perception of the territory and it could certainly have coped with a centralized state exercising control over a precisely bounded and polarized territory. The fact that Gallic terms continued to be used for two of the most common measures in modern France, the league or *lieu* (*leuga*) and the *arpent* (*arpennum*),[99] testifies to a long-established proficiency in the exact and coordinated measurement of distances. C. J. Guyonvarc'h has analysed the distribution of toponyms derived from *mediolanos* (Latinized as *mediolanum*) = 'centre of plenitude'. These places are mostly situated on hills in frontier areas[100] and are distributed in such a way that each one is equidistant from two others (fig. 6); they are far more numerous to the east of a line drawn between Cherbourg and Toulon. The analysis reveals a complex and strictly ordered network of 'geodetic points' established at the time of the Celts' definitive installation in eastern, northern and central Gaul (probably in about the fifth century BC: the foundation of *Mediolanos* – Milan – in the Po valley dates from the same period).[101] Even as early as this period of the Celts' first massive penetration, the regions that they occupied were measured and contained within a pattern of points of reference and regular link-lines organized by the Druids.[102] This 'sacred geography' played a

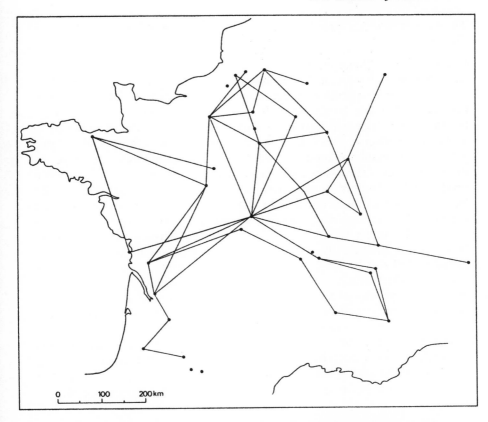

Figure 6. The distribution of toponyms derived from *Mediolanos* (after Vadé, 1976).

useful role at the time of the initial Celtic invasion but, in the absence of any means of political organization thereafter, was not applied systematically to the Gallic territory in its entirety.

Marseilles and the tin routes

The lack of unity in these Celtic lands and the persistent uncertainties and variations in the names used to refer to them by the Graeco-Roman authors testify to the absence of any structured organization to unify the Gallic isthmus and underpin the precarious fragility of its trading relations, which had not yet acquired the kind of permanence which fosters the emergence of stable political institutions.

The earliest trading routes between the Mediterranean and north-western Europe

For a long time, the fraction of Celtic territory which corresponded to what was eventually to become France does not appear to have been crossed by any particularly strongly established routes. Commercial relations between the eastern Mediterranean, which was the principal centre of civilization at the time, and the north-western seaboard of Europe did exist in the early years of the Bronze Age, but at first they did not pass through the Gallic isthmus. The main commodity in demand in the second millennium BC, in the civilization of the eastern Mediterranean, was amber, the fossilized resin produced by conifers. It was extremely fashionable in the Minoan and Mycenaean civilizations and was to be found in abundance in the river and seashore deposits (mainly oligocene sands) of the great plain of northern Europe. The amber traffic passed by way of the German valleys from the Weser to the Elbe and the Vistula, the Brenner pass and thence to the Po valley and the Adriatic Sea. The segmented pottery beads made in the eastern Mediterranean, which were exported in exchange, were certainly transported by way of the amber routes to reach their destinations on the North Sea coast or even as far afield as Britain.[103] At this period there are no signs of any regular trading routes further to the west.

The earliest prefiguration of the land of France, with trans-continental routes developed across the Gallic isthmus, was a consequence of the trading relations established by a Greek town, Marseilles, and of the production of tin ore, one of the principal commodities of trade in the ancient world.

The geographical significance of the foundation of Marseilles

The foundation of Marseilles seems to be the first decisive occurrence in the historical geography of what can truly be regarded as France. Marseilles was first and foremost a port. The town was born from the port and, even today, its urban functions cannot easily be dissociated from it. The very position of the town testifies to its importance as a port. It reflects a compromise between demands which, in the Mediterranean region, often prove contradictory: the need for a safe harbour and also to attract major highways, to facilitate communications with the interior.

The situation of the town was certainly in part dictated by the proximity of the great Rhône corridor, the essential route leading inland. It was inevitable that a large town would develop close to the Rhône delta, but there was no site for a port either there or along the sandy coast to the west, where it would have been easier to develop communications with the interior. Marseilles is positioned at the first possible site for a port that exists on the coast of Provence close to the Rhône valley. The Old Port, the Lacydon of the

Greeks, is situated immediately to the north of the calcareous ridge of Notre-Dame-de-la-Garde. It is an exceptionally fine shelter, comprising about 40 hectares in a depression hollowed out by differential fluvial erosion and subsequently flooded by the sea, and is the most westerly of a string of Provençal *calanques* (coves) strung out towards the east.

It was here, on a rise overlooking the sea, that, in about 600 BC, Greeks from Phocis in Asia Minor established a settlement. In the general movement of Greek colonization, the Phocians, from a thalassocracy founded on the use of a particular type of fast long boat, were the first to reach the western Mediterranean. It was not possible for them to settle along the coasts of North Africa or Spain as these were already under Carthaginian occupation. Carthaginian influence extended as far as the Balearic Islands and southern Sardinia. The Phocians attempted briefly to settle at Cadiz, but were unable to maintain their position there. In the Tyrrhenian Sea, the Etruscans predominated. Only on the dangerous, rocky coastline of the bays of Provence and Liguria, still inhabited by disorganized barbarian peoples, did there remain room for a major trading nation to settle. But these steep cliffs lacked communication routes with the interior and the Phocians accordingly sought out a harbour site as close as possible to the Rhône delta. Legend has it that they had no difficulty in coming to an agreement with the local tribes, although for a long time communications between the town and the interior remained under threat from brigands.[104]

The expansion of Marseilles and its colonies

Reinforced by Phocians who left Phocaea after its capture by the Persians (540), the people of Marseilles soon established a large number of colonies along the neighbouring coasts. The alliance between the Etruscans and the Carthaginians prevented them from maintaining a position on the western coast of Corsica and at Mainake, near Malaga, but they succeeded in imposing their domination over the mainland coasts of the north-western Mediterranean, thanks to a whole series of colonies and trading posts strung out between the Alpes Maritimes and the Balearic Sea.[105] To the east of the Rhône the most important of these were Monaco (Μόνοικος), probably the site of an earlier Phoenician settlement; possibly Nice (Νίκαια, 'which brings victory');[106] Antibes (*Antipolis*, 'the town opposite', on the other side of the Var delta); Olbia ('the fortunate', now Saint-Pierre-d'Almanarre); *Tauroeis* ('the town of the bull', now Sanary (?), a local hellenized toponym); *Citharista* ('the town of the lyre', now La Ciotat, another false etymology based on a Ligurian name). To the west of the Rhône was Agde (Ἀγάθη, 'the good' – good town or good fortune), situated on a volcanic rock, at an anchorage serving the sandy coastline of the Languedoc plain and commanding the junction constituted by the mouths of the Aude, the Orb and the Hérault.

Figure 7. Marseilles and its colonies.

Then, to the south of the Pyrenees, there were Rosas and Ampurias, commanding the Ampurdan plain, which retained its native name of 'Εμπόρια, meaning 'warehouses' or 'markets'. Finally, further south still, there were a few far-flung outposts as distant as the Nao headland, to the south of Valencia, controlling the Balearic Straits.

However, this Phocian expansion remained strictly coastal. The large centres of Narbonne and Arles (*Arelate*) remained Celto-Ligurian. The short-lived settlement of *Theline* ('breast'), founded in the sixth century (?) on the knoll at Arles, had to be abandoned, so all that remained here was a trading post, *Rhodanousia* ('Ροδανουσία), probably positioned on the other side of the river, facing the barbarian town. The people of Marseilles did not establish themselves any further inland. The town nevertheless enjoyed great commercial prosperity, evidence of which is provided by the minting of exceptionally high-quality silver coins which circulated widely. Ancient authors believed that this prosperity was based solely upon trade relations which extended only into the immediately neighbouring regions,[107] but it was in truth essentially a consequence of far-flung trading connections which had been established with the entire Gallic interior. Tin was the major commercial commodity.

Marseilles, trade and the tin routes

Tin, a component of bronze, was one of the metals most sought after by the Mediterranean civilizations of the Near East. Because it was relatively scarce in the Near East, it had to be brought from central Europe (the mountainous areas of Bohemia), Etruria and, above all, north-western and western Europe, where there were many seams (in the north-west of Spain, the north-west of the French Massif Central, Brittany and, above all, Cornwall, in Britain).[108]

The tin trade, which started as early as the end of the second millennium BC, was certainly exclusively maritime at first, when it was in the hands of the Phoenicians, who had established commercial relations with southern Spain, while the Carthaginians took over further west, towards the Atlantic. But as Marseilles developed, trans-continental routes began to be used. Around the beginning of the first millennium, even before the foundation of Marseilles, the Greeks may possibly have been bringing tin from north-western Europe down through the valleys of the Aude and the Garonne. So much, at least, is suggested by the presence here of triangular scraping implements and cross-shaped sheep-tags dating from the late Bronze Age.[109] At any rate, there can be no doubt about the subsequent decisive role played by Marseilles. In the first century BC, immediately prior to the Roman conquest, Marseilles was the destination of virtually all this tin traffic.[110]

How and by what routes was it transported? The Greeks certainly tried to use the sea routes, but came up against the Carthaginian monopoly, which denied them access to the Atlantic Ocean. In 324–323 BC, at the time of Alexander and probably at his instigation, Pytheas of Marseilles led an expedition to the lands of the North,[111] but this was during a period when the power of Carthage was temporarily weakened, after Alexander's capture of Tyre, and the venture remained without sequel. It has been suggested (Roman, 1977) that the use of sea routes leading to the Mediterranean, almost certainly for the transport of tin, may account for the coins of the *Ambiani* tribe of northern Gaul (Picardy). They date from the third century BC and were imitated from the currency of Tarentum, at a time when there were no land routes between the two places. But apart from this there is no evidence for any such links. There can be no doubt that virtually all the tin

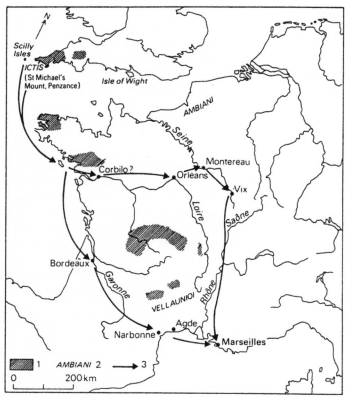

Figure 8. The principal tin routes in Gaul
1. Tin-bearing deposits. 2. Ancient peoples and places.
3. Probable itineraries.

traffic passed overland. The question is: what routes were used? This is an extremely complex problem, still not altogether resolved.[112]

The first problem concerns the starting point of these routes, islands which the ancient texts called Cassiterides but whose situation varies startlingly from one author to another, even if all locate it in north-western Europe: Galicia, the Armorican peninsula, the British Isles... In truth, these contradictions are less puzzling once one realizes that 'Cassiterides' is not a precise geographical name but a generic term meaning 'islands of tin' (*cassiteros* in Greek = tin) which was applied to a whole series of regions or islands in which the metal could be found: particularly the Scilly Isles, but also the area around the mouth of the Loire, the south coast of the Armorican peninsula and even north-western Spain.[113]

A similar uncertainty surrounds the itineraries. They appear to have varied widely, possibly according to fluctuations in Marseilles' political relations with the various tribes of Gaul. The most detailed, but also the latest, relevant text is that of Diodorus Siculus. Written in about 60 BC but based on earlier sources, it makes it possible to reconstruct the situation on the eve of the Roman conquest. It states that the tin produced in Britain was loaded on to chariots and taken at low tide to an island called *Ictis*. Here it was purchased by merchants who proceeded to transport it to Galatia. With their wares carried by pack-horses, it took them thirty days to reach the mouth of the Rhône.[114] In another passage,[115] he describes the tin arriving at Marseilles and Narbonne. There has been disagreement over the location of *Ictis*, which was at first thought to be the Isle of Wight, the Latin name for which was *Vectis*. From here the tin was thought to be transported by way of the valleys of the Seine, the Saône and the Rhône. Philologically, however, it is difficult to identify the Latin *Vectis* with the Greek *Ictis* and more importantly, the 5 kilometre-wide straits which separate the Isle of Wight from the mainland are never uncovered at low tide. The most likely hypothesis, backed most recently by R. Dion, identifies *Ictis* with Saint Michael's Mount, in the bay of Penzance, at the extremity of Cornwall.[116] From here the tin is thought to have been carried by sea to Bordeaux and thence overland to the mouth of the Rhône. The use of pack-horses is explained by the fact that the Garonne is practically impassable for barges. The length of this overland route (550 kilometres) is more compatible with thirty pack-horse stages (18 kilometres per day) than a journey from the mouth of the Seine to that of the Rhône (900 kilometres, 30 kilometres per day, an excessive distance for heavily laden horses). Besides, it is hard to see why the tin should have been carried overland through the valleys of the Seine (which is easily navigated upstream) and the Saône and the Rhône (which are easily navigated downstream). Such, accordingly, appears to have been the general picture immediately before the Roman conquest.

It is quite clear, however, that other routes were used at other times, in response to fluctuations in the internal political situation of Gaul. The most important of these were the routes from the Loire to the Saône and from the Loire to the Seine (via Orléans and Montereau), starting from the major depot of *Corbilo*, in the lower Loire valley, which was prosperous in the fourth century BC and, according to Pytheas and Polybius,[117] still thriving in the second. The existence of an archaic route passing by way of the Seine, the Saône and the Rhône is reliably attested by the distribution of Mediterranean objects introduced by the Marseilles trade.[118] The map of Greek exports to the Celts during the Hallstattian D period testifies to the importance of the Rhône valley route from as early as the fourth century BC.[119] It is true that the overall distribution of objects exported from Greece during the La Tène A period (fifth century BC) seems to indicate even greater activity along the Rhine and through the Alpine passes.[120] However, only control of the passes between the Seine and the Saône basins could explain the extraordinary wealth of the Gallic peoples of the Châtillonais, attested by the splendid discoveries of the 'princess' burial of Vix dating from the first quarter of the fifth century BC. Most impressive of all is the huge bronze Vix *crater*, of Mediterranean origin, the largest that we possess from the ancient world. It stands 1.64 metres high and weighs 208 kilograms and must certainly have been transported in pieces and only assembled upon arrival. Then there were other, even stranger routes. In the second century BC, when the *Arverni* were the dominant people in Gaul, the trade routes certainly passed through their territory. A bronze hand was discovered here, an object which served as an authorized Massaliot pass among the *Vellaunioi*,[121] the Celts of the Velay region. It testifies to the activities of traders from Marseilles in this mountainous region of the Massif Central. The whole of Gaul was criss-crossed by trade routes leading to the great Greek city of Marseilles. We cannot be sure of the precise details of their itinerary or to what extent they varied from one period to another; but there can be no doubt that they existed and that there were many of them.

Greek influence in Gaul

In exchange for tin, the products of the Mediterranean world flooded into Gaul by way of these routes: manufactured objects, salt and, above all, wine,[122] for which the independent Celts were already manifesting an excessive partiality (φιλοινία),[123] to such a degree that in Rome[124] their drunkenness was proverbial, whilst a persistent tradition ascribed their motive for the Gallic invasion of Italy in the early fourth century BC to their frantic desire to gain possession of land which produced wine.[125] Despite the existence in Gaul of wild vines, which were later domesticated and which made a by no means negligible contribution to present-day vintages, there can be no doubt that in independent Gaul these indigenous vines were not cultivated[126] and that all vine cultivation was closely linked with the Greek markets on the Mediterranean coast and their immediate surroundings. Only with Greek colonization was wine cultivation established in Gaul. That tradition has even been preserved in a fourteenth-century fable entitled *Le Martyre de saint Baccus* ('The Martyrdom of Saint Bacchus'):

> .. Saint Baccus selon l'ystoire
> D'oultre mer vint, c'est chose voire[127]
> (According to history, it is quite clear
> That Saint Bacchus came from elsewhere [overseas]).

Even today, in the wine production of Provence, everything, vines and techniques alike, seems to go back to the Greek colonizers and the relations that they maintained with their native country.[128] However, the wine produced by Marseilles and its colonies did not suffice to supply the huge market of Gaul. Marseilles itself served essentially as a distribution centre which despatched to the Gauls of the interior wines imported from Magna Graecia or other countries around the Aegean Sea.[129] It is clear that, from an early date, wines from these countries were exported far and wide, as is shown by the presence of the Vix *crater*, wineflasks and cups dating from the early fifth century and the marks of the Campanian amphorae which have been found at Besançon and Lezoux in the Limagne area.[130] In the mid-first century BC, this trade was proving remarkably lucrative for Italian merchants, who took their wine as far afield as northern Gaul and Germania and would sometimes exchange an amphora for a slave: in return for wine, buyers were even prepared to trade their servants.[131]

Along with the products of the Mediterranean came the Mediterranean spirit. Above all, Greece exerted an important cultural influence upon Gaul. Such inscriptions and legends on coins as have preserved a few fragments of the Gallic language are written in Greek characters. The literacy of Gaul began with Greek letters. The arts took Greece as their model. It is interesting to speculate what Gaul would have been like if it had continued to be influenced by the superior civilization of Greece rather than that of Rome. Camille Jullian has written brilliantly on this subject, and readers are urged to refer to his work.[132]

The political influence of Greece and Marseilles was, however, negligible. These commercial and cultural relations played no role in structuring the political division of the territory. The fact that, for reasons that remain obscure, the centre of gravity of power lay in the *Arverni* Empire in the second century BC seems to have had nothing to do with relations between it and the Mediterranean. It was the Romans who undertook the task of setting up more sophisticated forms of territorial organization.

2
The impact of Rome

Within this territory of Gaul, a land of great wealth and potential, but still troubled and disordered, the work of Rome was of capital importance – a matter not so much of creation but rather of organization and rationalization.

The spatial organization of Gaul

The Province

Rome's seizure of Gaul began in 121 BC when, after her victory over the *Arverni*, she established a Roman Province incorporating present-day Provence and Languedoc. It extended up the Rhône valley to the confluence of the Saône, as far as the southern shore of the Lake of Geneva (in *Allobroges* territory) and beyond the border of the Lauraguais region to the west. It included the region of Toulouse (the territory of the *Volcae*) and later annexed the heights of the Cévennes and the upper plateaux of Vivarais (*Helvii* territory), and the Massif Central balcony overlooking the plain of Languedoc and the Rhône valley. Rome thus controlled the overland route between Spain and Italy and thereby dominated the whole of the western Mediterranean. The capital was established in the Celtic town of Narbonne, the entire population of which was shifted so that a Roman colony could be created. The Roman province, at first called *Gallia transalpina*, or *Gallia citerior*, was later to become simply *Gallia Narbonensis*.[1] Originally known as *Gallia braccata* ('in breeches'), as opposed to Cisalpine Gaul, at the moment of Caesar's conquest it too became *Gallia togata*, as opposed to *Gallia comata* ('long-haired Gaul'), which was still independent.[2]

The creation of the Province involved no more than a limited territorial restructuring of its conquered peoples. The existing organization was in the main respected. However, the *Volcae* were divided into the *Volcae Tectosages* (in the Toulouse and Narbonne regions) and the *Volcae Arecomici* (between the Hérault and the Rhône), apparently as a result of the Roman policy of weakening peoples considered to be too strong.[3] In similar fashion, the territory of the *Ruteni* (around Rouergue) was divided into two, the southern portion, around Albi, becoming a part of the Province, while the northern portion, around Rodez, remained independent.[4]

The Rhine frontier

At the time of the conquest of independent Gaul (58–52 BC), Rome's first preoccupation was to define a frontier to contain its area of intervention within the Celtic territory. Only one border was not a totally natural one, that running from north to east. Caesar was the first to define what, in the French monarchy of the modern period, was to become the Rhine frontier.[5] It was on the basis of this that the land lying to the east of the Rhine became systematically known as *Germania*. This name, which had first appeared in the Latin vocabulary no more than fifteen years earlier,[6] was that of a Celtic tribe (of *Belgae*), part of which was still indubitably settled on the right bank of the river. Before Caesar's invasion, the people of Germania do not seem to have been clearly distinguished from the people of Gaul.[7] Tribes speaking both Celtic and Germanic languages were scattered indiscriminately on both sides of the Rhine.[8] It was Caesar who, at the start of his operations in Gaul, in 58 BC, arbitrarily fixed upon the Rhine as the frontier to contain his activities. Fastening upon the pretext of a westward migration of the *Helvetii*, he forced them to reoccupy their native territory, justifying this as follows:

> His chief reason for so doing was that he did not wish the district which the Helvetii had left to be unoccupied, lest the excellence of the farmlands might tempt the Germans who dwell across the Rhine to cross from their own into Helvetian borders, and so to become neighbours to the Province of Gaul and to the *Allobroges*.[9]

It was in order to safeguard the barrier constituted by the river that, in 58, he crossed it to defeat a chief of the *Suevi* tribe, who bore the non-Germanic name of Arioviste. He was fluent in the Celtic language, which may have been his mother tongue, and his troops certainly included a large number of Celts.[10] As a geographic division delimited by the Rhine, Gaul was a Roman creation within the Celtic space. Soon, that concept was absorbed into Roman literature. Not long after 39 BC, Sallust[11] was referring to '*Gallia cis Rhenum atque inter mare nostrum et oceanum* (Gaul, bounded by the Rhine, the Mediterranean and the Ocean)'. The vast Celtic territory was for the first time precisely defined and the framework of Roman influence clearly specified as relating to the country bounded by the two seas. Later, under the Empire, the *limes* was to be drawn beyond the Rhine and the name

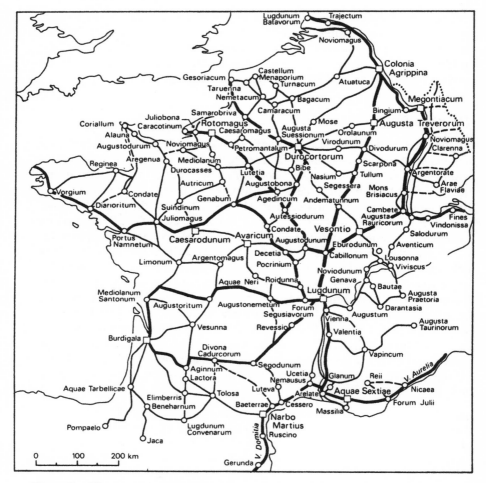

Figure 9. The network of roads in Roman Gaul (after Daremberg, Saglio and Pottier, *Dictionnaire des Antiquités grecques et romaines*, vol. V, article *Via* by M. Besnier).

Germania was to be given, possibly as early as the reign of Augustus,[12] to the provinces which lay on both banks of the Rhine. But in the Late Empire, the Romans pulled back before barbarian pressure, re-establishing the frontier at the river. The two provinces of *Germania prima* (Alsace and the Palatinate) and *Germania secunda* (the northern Meuse basin), known through the *Notitia* of the late fourth century, were at this point limited to the left bank of the Rhine, which now marked the definitive line at which the Roman withdrawal stopped.

The network of roads

Strategic control of this large area was ensured by the network of roads that the Romans developed.[13] A good system of roads already existed in independent Gaul and communications were relatively rapid, which no doubt partly explains the swiftness of the Roman conquest here. But there can be no doubt that the Romans rethought the overall organization of the network, with a view to maintaining their dominion over the entire area. The first major (new) artery was the *via Domitia*, which the Romans began to construct in 120 BC, to link Spain with Italy by way of the Province. Then, after the conquest (from 27 BC on), Agrippa's system was developed, most of it reaching completion during the reign of Claudius. To the north, the great road up the Rhône valley was continued through the Saône valley towards Langres, Trèves and Cologne, where it joined the Rhine highway. To the west, there were several major routes leading to the Atlantic Ocean, Bordeaux, Saintes and Poitiers. One passed through Orléans and along the Loire valley, leading to the Armorican peninsula; other important roads led from Chalon and Langres, by way of Lutetia, to the Seine estuary, and through Rheims to the port of Boulogne (*Gesoriacum, Portus Itius*) from which ships left for the island of Britain. The Roman plan is quite clear. Apart from the links with the Rhine frontier and with Spain, the network was centred upon Lyons, from which a whole series of roads fanned out, diffusing the Roman influence to even the most remote corners of the land and as far afield as the Atlantic coast and the Channel (fig. 9). But Rome's intentions were not so much commercial as political. The Romans were not merely concerned to intensify the exploitation of the resources of the isthmus of Gaul. Rather, they were intent upon safeguarding and promoting their dominion over the country. It was not so much a matter of setting up a general system of service communications, as of getting from A to B as quickly as possible, along routes for the most part set out in straight lines. First and foremost, the Roman roads were planned strategically: they avoided valleys and large dense forests, being instead routed along bare peaks which it was easy to guard. It is both remarkable and significant that the Romans' colossal achievement in respect of the roads of Gaul was not matched by any equivalent development of waterways and ports (with the exception of Boulogne).[14] The preoccupations of Rome were political rather than commercial. Nevertheless, the Roman roads did facilitate the swift introduction of Italian goods into the isthmus: within less than a century, these were to be found even in the most distant parts of Gaul (fig. 10). It is not possible to determine precisely how much traffic passed along the various roads. Different representations of the road network give different evaluations of the importance of each of the routes.[15] Nevertheless, there can be no doubt

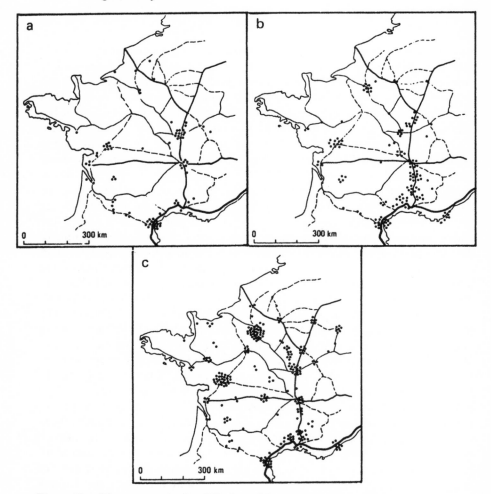

Figure 10. Three stages in the diffusion of Roman pottery in Gaul (after Hofmann, 1976).
a. The distribution of the marks of Arezzo potters datable to the period of Caesar and the beginning of the reign of Augustus. b. The distribution of products with Arezzo marks, from the last quarter of the first century BC. c. The distribution of products from the Italic workshop of Ateius, in operation from 20 BC to AD 20.

about the essential role played by the major transversals and the cluster of highways fanning out from Lyons.

Internal subdivisions

The Three Gauls
In the interests of administrative efficiency, Rome was obliged to divide this vast possession into different sections. She was thus also responsible for the first internal structuring of Gaul. Caesar did not interfere with the traditional unity of independent Gaul. It was Augustus who divided it into the three provinces known as the Three Gauls.[16] But those three provinces essentially reproduced ethnic divisions which had been clearly visible before the conquest, the names of which are already to be found in Caesar's text dating from that time. However, the divisions were now defined more precisely and were slightly modified. To the north – essentially to the north of the Seine – was Belgic Gaul, which corresponded to the northern group of Celtic peoples, who were the latest arrivals. To the south-west was Aquitaine, which took over the originally Iberian name of the peoples settled to the south-west of the Garonne. But because this area was considered too small to constitute a major sector,[17] Aquitaine was artificially extended to include nearly all the regions to the south of the Loire and thus incorporated a large number of different Celtic tribes. Between these two sectors lay Celtic or Lyonnaise Gaul. By and large, this comprised the land between the Seine and the Loire, that is to say all that was neither Aquitaine nor Belgic Gaul, and also the entire area served by these two great natural waterways. This fell directly under the influence of the metropolis newly developed at Lyons (see below, pp. 43–4) with its cluster of roads fanning out to dominate the whole central province (fig. 11A). All in all, Rome simply adapted, codified and fixed major divisions that were distinguishable even before the conquest. The extra provinces created in the high Alpine valleys, once these had been subjugated under Augustus (the Alpes Maritimes, the Alpes Cottiennes, the Alpes Grées and the Alpes Pennines) and around the Rhine in Germania were minor, smaller sectors, additions which did not affect the basic structure of the country.

Peoples and cities
Similarly, by and large, the secondary subdivisions were respected. The territories of the Gallic tribes remained generally untouched, although there were a few alterations.[18] Apart from those connected with the creation of colonial towns (see below, pp. 41–2), the changes were mostly determined by reasons connected with local politics. In some cases it was deemed best to disband tribes which were too powerful, by detaching client groups from

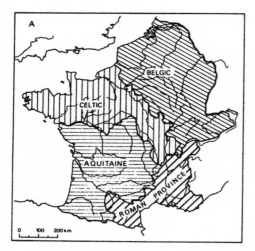

Figure 11. The administrative divisions of Roman Gaul (after Mirot, 1947, pp. 27 and 36). A. In 10 BC. B. At the end of the fourth century AD. 1. The border of the diocese of the Gauls (to the north) and the diocese of Vienne (to the south). 2. and 3. Province borders. The numbers correspond to subdivisions of the names written in large capitals

them, as where the Romans created for the *Vellavi* (Velay) a city quite separate from that of the *Arverni*, and one for the *Segusiavi* separate from that of the *Aedui*. In other instances, they sought to create more homogeneous groups by bringing together tribes which were too small: thus the thirty or so peoples of Aquitaine which Caesar encountered south of the Garonne, were collected into nine groups, probably those known in the second century as the *Novempopulania*.[19] But there was nothing particularly systematic about these policies and many of the considerable inequalities between Gallic peoples dating from before the conquest remained virtually unchanged. They provided the framework for the original sixty 'cities' (*civitates*)[20] existing in the Three Gauls (in 12 or 10 BC, at the time when altars to Rome and to Augustus were set up), which provided the basis of the Roman administration.

Here again Rome endeavoured not to change the former pattern but, on the contrary, to establish it more firmly and clarify the divisions within it. It is true, however, that many frontier-forests and 'deserts' remained. These areas on the city outskirts, which were not included in the census, were known as *subcesiva*, wasteland considered infertile. They continued to be part of the *ager publicus* and tended to be settled by marginal native peoples who were rejected by the great estates (*villae* – see below, pp. 48–52) and forced into them. The Vosges massif was one such marginal area.[21] On the other hand, it was also during this Gallo-Roman period that, by and large, precise frontiers were drawn for the territories hitherto ill-defined, inside independent Gaul. All this colonization and construction in the marginal sectors provided the territory of Gaul with a variety of new reference points. Boundary posts sprang up throughout the region. A whole series of Latin toponyms developed, underlining the permanence of the Roman occupation: some were derived from *fines* ('frontiers', 'borders'), for example Fismes, Fains, Fins, and so on; others from *basilica* ('tribunal'), such as Bazoche, Bazonge, evoking the idea of the legal assemblies held in border areas to adjudicate the litigations that arose from disagreements between one city and another; yet others from *oratorium* ('oratory'), for example Ozoir, Orior, Oradour, indicating, in the Christian period, the position of the chapels erected by the anchorites who settled in these border regions. The task of urbanizing the frontier wastelands was vigorously tackled by the Romans. They set up sanctuaries (*fana*), in many cases alongside a complex of monuments and theatres, which took over from the *conciliabula* of independent Gaul[22] and constituted centres of imperial propaganda in these non-urbanized sectors where the territories of two or more tribes met.[23] These localities also became a focus for festivals and conviviality and this facilitated economic relations between the inhabitants of the large estates scattered throughout the neighbouring region.[24] Essentially, the credit for stabilizing the territories of the tribes of Gaul must go to Rome.

Adjustments to the urban fabric

New creations
The Gallic urban or proto-urban pattern thus survived within these sectors inherited from independent Gaul and is, accordingly, pre-Roman throughout virtually the whole of Gaul. Except in one case (*Forum Julii* – Fréjus, founded by Caesar from whom it still takes its name), Roman colonies were set up in towns which already existed, where they constituted new districts, or alternatively they were set up immediately alongside such towns, keeping the name of the native people and that of the old town. In the *Provincia*[25] there were many of these colonies: every tribe of any standing here received one or more (a total of twenty were founded by Caesar and Augustus, seven of them among the *Volcae* and six among the *Cavares*, in the Vaucluse region). As the older cities disintegrated, they came to constitute so many colonial communes, a network designed to protect the approach to Italy and communications between it and Spain. In 'long-haired' Gaul, in contrast, colonies were few and far between and in the case of some, which remained completely native, the term 'colony' was in effect a purely honorific title.[26]

Even apart from these rare colonial foundations, very few centres were developed in the Roman period. The most important were those in the northern regions, in Belgic Gaul or along the borders of Germania, where few towns had previously existed. Here, new towns gradually formed around the Roman military camps (some of them alongside pre-existing native sites, others not), which in the Augustan period were used as bases from which the Romans launched their conquests of Germania and other territories beyond the Rhine, and around the marches (*canabae*) naturally associated with these border regions.[27] There thus developed along the Rhine frontier a number of urban centres out of which grew large towns which now lie outside the territory of France (Nijmegen, Bonn, Mainz, Cologne and Xanten, the last two being colonies founded, respectively, under Claudius, and under Trajan in 105 AD) and, on the Rhine itself, Strasbourg (whose old Celtic name was *Argentorate*). There is, however, one very unusual type of town which goes back to the Roman period: the spa, designed for invalids and others wishing to take the waters. The Gauls knew about making use of spring waters, but do not appear to have built towns of any size at such sites. Dax, a humble woodcutters' hamlet, where Augustus was convinced that he had found a cure, became *Aquae Augustae*, the *chef-lieu* of the city of the *Tarbelli* and the real capital of the part of the plain of Gascony that lies at the foot of the Pyrenees. Toponyms based on *Aquae* > Aix (Aix-en-Provence = *Aquae Sextiae*; Aix-les-Bains, in Savoy = *Vicus Aquarum*, in the territory of the *Allobroges*, and so on) owe their origins to the Romans. They have no equivalent in the Gallic language. However, the name of the Celtic deity of springs, *Borvone*, is the basis for many other toponyms such as Bourbon, Bourbonne, towns which also prospered essentially during the Gallo-Roman period.[28]

As can be seen, the new towns known for certain to have been developed by the Romans are relatively few in number. However, certain city *chefs-lieux* named for Caesar or Augustus may be of Roman origin: no Gallic names are known for them, nor is there any archaeological evidence suggesting the existence of earlier *oppida* here. Among major towns, the main examples are Troyes (*Augustabona Tricassium*), Tours (*Caesarodunum Turonum*), Angers (*Juliomagus Andecavorum*), Bayeux (*Augustodurum Baiocassium*), Lillebonne (*Juliobona Caletorum*), Limoges (*Augustoritum Lemovicum*), Trèves (*Augusta Treverorum*) and Senlis (*Augustomagus Silvanectum*).[29] It is impossible to say whether these were completely new towns founded by the Romans, or whether the latter simply upgraded already indigenously occupied sites to the rank of major centres, giving them new names, within the framework of a systematic policy of urban organization, the purpose of which was to provide each city with an undisputed *chef-lieu* that would serve both as a regional centre and as a stronghold of Roman domination. In some cases, however, evidence exists to show that the new, renamed capital definitely did take over from an earlier Gallic *oppidum*.

Site switches
These name changes, betokening the official establishment of a network of major towns, also affected the siting of local population centres. Many of the Gallic *oppida*, fortified towns where the people of the particular tribe would gather in times of danger, were built on hilltops. Systematically seeking to do away with these mountain retreats, Rome moved them down to the neighbouring plains and valleys. The same policy had already been adopted in the Province, where Aix (*Aquae Sextiae*) had, in 122 BC, taken over from the *oppidum* of the *Salluvii* tribe, positioned up on the Entremont hill. In about 12 BC,[30] after the conquest of independent Gaul, the capitals of two of its most important peoples were similarly resited. That of the *Aedui* was shifted from Bibracte in the Morvan region (Mont Beuvray, 810 metres) to *Augustodunum* (Autun), the 'town of Augustus', in the Permian basin at the foot of the *massif*; that of the *Arverni* was moved from the Gergovia plateau (720 metres)[31] down to *Augustonemetum* (the 'sanctuary of Augustus'), the future Clermont-Ferrand, on a slight volcanic hillock just emerging from the Limagne plain. In the latter case, the shift was probably made even before the name was changed.[32] Many other similar displacements no doubt took place: at Soissons, Beauvais and Saint-Quentin in Belgic Gaul; Cahors, Limoges and Agen in Aquitaine; Toulouse in the Narbonensis Gaul (shifted down from the hill known as 'old Toulouse'). Many other *oppida* declined spontaneously: one such was Alesia, which had provided a refuge for Vercingetorix, a hilltop town whose importance seems gradually to have dwindled away during the Roman period. But many others, some of them of considerable size, survived: those which owed their strength to their insular or marshy sites, such as *Lutetia* (Paris) and *Avaricum* (Bourges), the respective capitals of the *Parisii* and the *Bituriges*, and also certain of the most prestigious hilltop towns, such as Langres (*Andemantunnum*), the capital of the *Lingones*, Besançon (*Vesontio*), the capital of the *Sequani*, Poitiers (*Limonum*), the capital of the *Pictones*, and Angoulême (*Ecolisma*), the *chef-lieu* of a *pagus* that was to become a 'city' under the Late Empire. Sometimes a new town, at the foot of the hill, would be added to the hilltop one, as at Langres and Besançon.

Changes in the hierarchy of cities.
Above all, during the Roman period changes took place in the relative importance of certain cities. As we have seen, in independent Gaul a town's importance depended upon good natural defences and a fertile surrounding territory (see above, pp. 19–22). In Roman Gaul, the towns that thrived were those situated favourably from the point of view of trade: Chalon-sur-Saône (*Cabillonum*), a trans-shipment point from which it was easy to join the navigable waterways of the Loire and the Seine; Orléans (*Genabum*), a market-town for the *Carnutes*, situated particularly favourably at the bend

in the Loire where the river flows closest to the Seine and whose importance was, in the Gallo-Roman period, to eclipse that of Chartres, the city capital, to such a degree that it no longer depended upon the latter and was itself promoted to the status of a city during the Late Empire; Tours (*Caesarodunum*), where the major overland route to the south coincides with the Loire and which, after a period of slow expansion, emerged at the time of the Germanic invasions as one of the most important centres in the whole country; Trèves (*Augusta Treverorum*), which commanded the convergence of the valleys of the Moselle, the Sarre and the Sûre, set back from the frontier constituted by the Rhine and equidistant from the two strongholds of Cologne and Mainz; Arles (*Arelate*), a great emporium on the Rhône, at the head of its delta, where the overland route crossed the river; Bordeaux (*Burdigala*), the gateway from Aquitaine to the Atlantic and the island of Britain, whose commercial potentialities ensured that it grew spontaneously even without being promoted to the status of a Roman colony, as Arles and Trèves were.[33]

Finally, two other major towns of strategic as well as commercial importance were developed during the Roman period. They were set up as colonies at an early date (43 BC), as a means of controlling the country as a whole.[34] Situated on a bend of the Rhine, at the gateway to Burgundy and Alsace, one was the *Colonia Raurica*, later known as *Augusta Rauricorum* (Augst, where Basle is now situated), which blocked the access to Gaul from Germania. The other was Lyons (*Lugdunum* – 'city of light' in Celtic, or 'town of the god Lug'), situated on the Fourvière hill, above the confluence of the Saône and the Rhône. Before the Roman conquest, this had been the site of a popular sanctuary which, however, had developed into no more than a very modest frontier town between two of the most powerful peoples of Gaul, the *Allobroges* to the south and the *Aedui* to the north, and had since suffered from its vulnerable situation. Originally, it had probably been an *Allobroges* outpost, on the edge of the *Aedui* territory,[35] but at the time of the Roman conquest it had passed into the hands of the *Segusiavi*, who were clients of the *Aedui*. With its exceptional position,[36] at the point where the Three Gauls met, commanding the routes leading to the Atlantic Ocean across Celtic Gaul and those which converged upon the Saône from the Rhine frontier, it was ideally situated to control Gaul as a whole or to impose Roman domination, if need be, in the event of a revolt. As Strabo observes: Ἐν μέσῳ τῆς χώρας ἐστίν ὥσπερ ἀκρόπολις' (it stands in the middle of the country, like an acropolis).[37] This was the town which the Romans chose for the principal junction in the network of roads that they built across the country (see above, p. 37).[38] It was to become the undisputed capital of the Gallic peoples: *caput Galliarum*.[39]

The unification of Gaul

The development of Lyons as the metropolis of Gaul reflects a fact of major geopolitical importance, namely the unity of the Three Gauls. The work of Rome would not have been complete had she not introduced higher structures of unification over and above the individual cities and provinces that she had set up and stabilized. As early as the reign of Augustus, institutions serving Gaul as a single community were introduced[40]: the Council of the Three Gauls brought together delegates from all the tribes and exercised control over the actions of the provincial governors; a high priest, the authority for all sacrifices, presided over the Altar of Rome and Augustus, upon which were inscribed the names of the sixty cities; and major annual festivals (*panegyria*) were held in the month of August. These activities took place in what was a veritable holy city, a sort of autonomous canton, positioned near the confluence of the Saône and the Rhône, but placed under the aegis of Rome itself, below the Roman colony of Fourvière. The model was provided by the provincial council of Narbonnaise Gaul, which also performed religious functions, and it was one that was also copied in the non-Celticized or partially Celticized provinces of Germania, the Alps and the Novempopulania area, all of which were provided with similar institutions.

The people of Gaul were thus indebted to Rome for the political and moral unity which, despite the assembly of the Druids, they had never really achieved at the time of their independence. It was a unity that several times manifested itself in efforts to obtain autonomy or liberation. In 69 AD, a national conspiracy led to Classicus, the chief of the *Treveri*, proclaiming an 'Empire of Gauls', although the assembly of cities, meeting at Rheims, swiftly dissociated itself from this move. More serious and longer lasting was the dynasty of Gallo-Roman emperors from Postumus to Tetricus (258–273), set up in dramatic circumstances in reaction to the threat posed by barbarians. The individuality of this territory bounded by the Rhine and the Pyrenees was clearly recognized by the rest of the Empire. In his description of the Roman Empire, written under the reign of Tiberius, Strabo devoted a special chapter to Gaul because, he claimed, it was the homeland of 'a single family': τὸ συμπαν φῦλον.[41] That tradition was observed right down to the last geographers of the Late Empire.[42] There were also to be anthropomorphic representations of Gaul. The very name, which progressively fell into disuse for the Gallic provinces of Italy, survived all the better on the other side of the Alps, because it gradually came to encapsulate all the memories evoked by the word 'Gallic'.[43] The Gauls did not possess any truly national feeling but they did share with the purely Roman inhabitants of their territory a provincial sense of belonging to Gaul.

It was a sentiment essentially bounded by a consciousness of the framework of the Empire, within which Gaul was quite simply felt to deserve a special place. Over the centuries, since the beginning of Roman rule, this feeling grew stronger.[44] During the Early Empire, patriotism had been limited to the framework of individual cities, and the different peoples of Gaul had still been divided by terrible quarrels. But by the Late Empire, a 'Gallic motherland' had come into being, unified by its attachment to Rome and the defence that it put up against the barbarians. The concept informed the thoughts and hearts of the Gallic soldiers fighting on the eastern frontiers of the Empire. Without it, the actions and success of Julian would have been inconceivable. Gaul was Rome's favourite daughter, 'tua Gallia', as Sidonius Apollinaris called it in the poems that he addressed to Rome.[45] In contrast to the Celtic language, which had no word to express territorial loyalty to anything vaster than a '*pays*' of extremely limited dimensions (the *bro* of the Britons), Rome gave Gaul the word for – and the very concept of – a motherland: *patria*. But even as they passed through the Roman mill, the Gauls retained their own originality and temperament. The Celtic substratum was suddenly to resurface in the period of invasions from Germania and was to prove a determining factor in the creation of the French language (see below, p. 125). In the sixth century, Gaul was to disappear as a political entity (see chapter 3), but the idea and memory survived with a persistence that was later (see below, pp. 98–9) to play a decisive part in the emergence of France.

Rome and the geographical face of Gaul

Rome set about fashioning the territory thus carved out according to her own particular design. Her marked concern to rationalize its general organization was matched by her attention to more detailed means of organizing the occupation of this land.

The structure and design of the cities

The cities constituted the primary key to the Romanization of Gaul. Their appearance underwent a profound change, chiefly as a result of the reorganization of the Gallic centres of population, involving either the addition of new quarters or the destruction of the existing buildings to make way for new ones. There is still some disagreement as to whether these transformations were systematically planned to conform to a uniform model or whether town-planning followed a more flexible pattern characterized, rather, by spontaneity and pragmatism.[46] Although archaeology has as yet revealed no more than a dozen or so towns organized according to a geometrical plan based upon the intersection of a *decumanus* and a *cardo*, the various techniques of aerial photography suggest that forty to fifty such cities existed. It is thus highly probable that most Gallo-Roman towns were organized, or reorganized, according to a definite plan. However, it is quite clear that some (Vienne and Vaison, for example) were set out in an irregular pattern although it may have incorporated certain quarters designed according to a grid-system. Quite a variety of models seems to have existed. Some *insulae* are known to have been rectangular, others – probably of later date – more square. In any event, the Roman towns were marked by their grandiose monumental appearance,[47] with their *fora* surrounded by buildings of a religious and public nature: porticoes and shops, temples, colossal altars, theatres (a hundred or so are detectable) and amphitheatres (above fifty), public baths and imposing aqueducts. The most spectacular complex was naturally that of Lyons, which has been described as 'a mirror of Rome in Gaul'.[48]

The permanent populations of these towns were certainly larger than those of the Gallic *oppida*, but it is hard to determine by how much. Calculations can be made on the basis of the capacity of the arenas and theatres, which vary between 15,000 and 35,000 spectators for the larger ones (Nîmes, Arles, Autun, Bordeaux and Lutetia) or on the basis of the load carried by the aqueducts which brought the water supply. Historians differ in their calculations: Jullian attributes at least 80,000 to 100,000 inhabitants to the larger Gallo-Roman towns and at least twice as many to Lyons, while the estimates of F. Lot[49] are much lower, perhaps excessively so: less than 10,000 inhabitants for most cities. It has recently been suggested, reasonably enough, that even the larger towns (Narbonne, Lyons, Autun, Rheims and Trèves) are unlikely to have exceeded 30,000 to 50,000 inhabitants.[50] What Gaul owed to the influence of Rome was not so much a quantitative increase in city population levels, but rather the establishment of a truly urban civilization.

The mark left by Rome upon the rural landscape

The Roman conquerors' work of systematization and rationalization affected the countryside no less deeply, for this was radically remodelled.[51]

The 'centuriation' system
In Gaul as elsewhere, the most striking manifestation of the Roman occupation was the creation of a landscape divided into centuries.[52] The aim was to reorganize the territory in accordance with a general plan of colonization and cultivation, usually by dividing the land into large plots of about 710 square metres, some of which were subdivided into smaller parcels (an *actus* measured 35 square metres, a *jugerum* 35 by 70 metres). These would mostly be measured out starting from the edge of the urban network of roads or

from the border of the major highways. We possess only one epigraphic source of evidence for this system, but it is certainly a spectacular one: the cadastral plan of Orange,[53] a fragment consisting of three immense marble tablets covering more or less the entire region stretching from Carpentras to Orange and Montélimar in the Rhône valley. But many century borders are retained in the modern network of rural roads and are respected by other boundaries and numerous attempts have been made to reconstitute the arrangement, particularly with the help of aerial photography, for which excessive claims have sometimes been made. These have recently been severely criticized.[54] In the case of the Province, it nevertheless seems certain that the century pattern covered virtually all the plains, both the Languedoc plain and the small valleys of Provence.[55] Here, the reorganization of the land was systematic and the entire region was affected.

As for the Three Gauls, the situation is more problematical. Even if one accepted the authenticity of all the centuries claimed to have been discovered, they are essentially discontinuous, scattered in separate isolated areas in the vicinity of towns or along important highways. It is not possible to tell whether the century system truly never was widespread here or whether that only appears to be the case because most traces of it disappeared at the time of the Germanic invasions or in the Middle Ages, when the system here suffered far graver disruptions than it did in southern France. The problem remains unresolved but no doubt both hypotheses contain a measure of truth. At any rate, even if it is true that the century system was less well preserved in northern France, there can be no doubt that it was also far less general than it was in the Province.

The spread of large estates (villae)
The Three Gauls thus do not appear to have been affected by a general reorganization of land in a pattern based on centuries. On the other hand, the spread of large properties does appear to have been a general feature of the Roman period here. Continuing a trend that had already started when Gaul was still independent, the great estate, or *villa*, became the normal and certainly the most widespread form adopted to cultivate the land.

Textual and archaeological testimony has long been available to show that the *villae* were spreading all over Gaul at this period.[56] All the data point in the same direction. The literary and epigraphical sources contain innumerable mentions of *villae*, but far fewer references to *vici*. 'We know of thousands of estates by name, but only of sixty or so villages', Fustel de Coulanges observed at the end of the nineteenth century.[57] In many cases, the Gallo-Roman *villa* was the direct topographical successor to the *aedificium* of the previous period,[58] but there was a major difference, of a social order, between the two, for both slavery and direct cultivation of the land were now general. On the other hand, in all probability many of these *villae* were set up in the course of a new phase of land colonization under the Late Empire, when far more large cereal-producing estates were developed on the plains, where a powerful commercialized economy could be established. The evidence of toponomy and the systematic distribution of some of these estates along the Roman roads suggest that within the framework of these large properties a powerful increase in land cultivation took place.[59]

The part played by the *vici*, as opposed to the *villae*, is hard to determine. Even their exact nature is hard to gauge. Were they agricultural villages, or mainly small centres of craftsmen and merchants, positioned at the intersection of important roads? The problem, which has been the subject of many debates of a somewhat theoretical nature, is further complicated by the fact that the terms *villa* and *vicus* seem quite often to have been used indiscriminately. The fact that a *vicus* had no precise legal status and frequently took the form simply of a random 'grouping'[60] means that the term itself was certainly applied to a number of different kinds of settlement. About twenty years ago, one author of a general survey reached the conclusion that the whole question was due for reappraisal.[61] Since then, a systematic investigation has taken place, the first conclusions of which have already been published.[62] It is clear that the vast majority of the clusters of habitation detectable in the Gallo-Roman period were not agricultural but centres devoted to crafts, trade or religious functions.[63] Where it is possible to determine the area of the land under cultivation around them, this seems to have been very restricted.[64] They are often situated on highways and the fact that some are designed on a grid-system suggests that they were deliberately planned.[65] Nevertheless, the populations of the *vici* of Roman Gaul seem to have remained essentially indigenous. Such villages increased in number, replacing more scattered dwellings, in the late second and early third centuries, when the first incursions from Germanic peoples had already affected the conditions of life in the countryside, and they thus contributed towards the 'Celtic renaissance' that is detectable at this period[66] and so stand in contrast to the cradles of Romanization constituted by the *villae*. At the time of the Roman takeover, their role seems to have been more modest. Whether the spread of the *villae* took the form of further land clearance on the cereal-bearing plateaux or a reorganization or seizure of land already more or less cultivated by isolated individuals, there can be no doubt that, compared to those large estates which dominated the countryside of the Early Empire, the relative number of *vici* had waned considerably since the days of independent Gaul.

But exactly how much? Neither exhaustive and concerted regional archaeological surveys and records, nor detailed excavations of a more individual nature have been able to produce any accurate answers to that question. However, the recent progress in detection made by aerial photography has for the first time made it possible to produce a regional map of

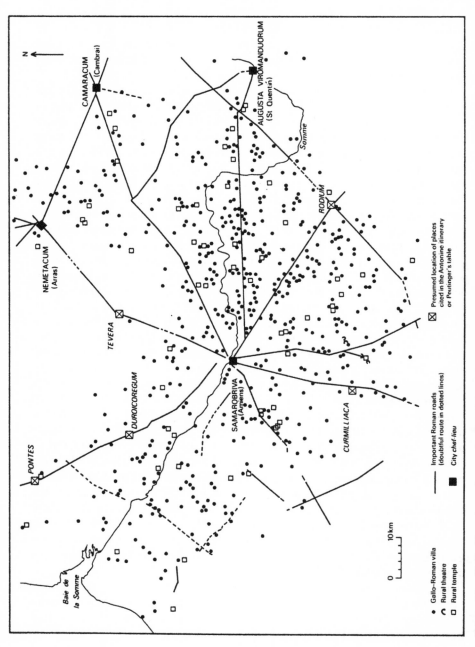

Figure 12. Gallo-Roman settlement in the plain of Picardy (after Agache, 1975, 1978).

the network of *villae* and, thanks to this, we now possess striking confirmation of the views of Fustel de Coulanges, which have in the past sometimes been dismissed as out of date.

Roger Agache's remarkable field-work in a region formerly considered not to have contained many *villae*[67] has shown that they were, on the contrary, densely distributed, occupying a much greater area than did the few urban complexes and *vici* (fig. 12).[68] Admittedly, the apparent scarcity of *vici* is no doubt partly due to the fact that the flimsy constructions of wood or mud of which these native hamlets were composed may have escaped aerial detection. Other *vici* are undoubtedly hidden beneath the existing villages which succeeded them and mask them.[69] But even if that first impression needs modification, the fact remains that *vici* 'appear to have been rare, wretched and, all in all, of little importance compared with the great abundance of large *villae* which are scattered throughout the countryside and are truly the dominant feature of the ancient habitat'.[70] In the light of these discoveries it is important to undertake a thorough review of the conclusions that have hitherto been drawn from archaeological explorations in other regions.[71] The problem that remains is to determine to what extent these still partial and fragmentary discoveries should be taken to apply generally, in both space and time. The increase in the number of *villae* in the Early Empire, in the fertile wheatlands, particularly in Picardy on the silty plateaux of Santerre, was occasioned by the widespread cultivation of cereals on what was clearly a commercial scale. No doubt that number was, in both absolute and relative terms, considerably lower in less favourable areas, as is indeed the case in Picardy itself along the coastline, where the soil is less silty.[72] As a working hypothesis at least, it seems reasonable to suppose that the *villae* predominated in the great calcareous or silty plains, whilst a higher proportion of agricultural hamlets populated by independent peasants were to be found in the poorer mountainous regions which were still thickly wooded. Essentially, the main geographical location of the *villae* must have been the '*pars Galliarum planior*',[73] where the Roman agronomists marvelled at the horse- or ox-drawn harvesting machines of the *latifundia*. But only with more systematic aerial research can we hope to resolve this crucial matter with certainty. However, it is perfectly possible that some *villae*, those not situated on the great cereal-bearing plains, may have been based upon quite different economic structures. *Villae* are in fact known to have existed in all kinds of geographical environments. In Languedoc, in the Béziers area, where their numbers increased, displacing an early centuriation system, many were situated in the hilly regions where they were devoted to the cultivation of a wide range of Mediterranean crops, in which tree crops probably predominated over cereals. Overall, however, they were less densely distributed here than on the plains.[74]

The rural Gallo-Roman countryside: the predominance of open fields

The rural landscape of the *villa* was certainly characterized by a preponderance of open fields.[75] Close to the buildings there were, to be sure, generally one or more enclosed areas, taking the form of courtyards or walled paddocks, to which Sidonius Apollinaris sometimes applies the term '*vestibulum*'.[76] For both economic and aesthetic reasons, this area surrounding the *villa* buildings was probably quite extensive. Writing of the *villa* of Avitacus (Aydat, 20 kilometres to the south-west of Clermont-Ferrand), Sidonius calls it a *campus*, so it must have been quite large: '*donec domicilio competens vestibuli campus aperitur*'.[77] Aerial photography reveals many examples of these enclosures close to the *villa* buildings. Surrounded by open fields, they were to remain an integral part (known as the *curtis* or *curticula*[78]) of farms in the Carolingian period and indeed down to the present day.

But what of the arable fields themselves, stretching away beyond the immediate vicinity of the *villa*? Sidonius also uses the word *campus* to describe the land stretching from the house and its *vestibulum* out to the encircling hills: '*vestibulo, campo, colle amoenissimo*'.[79] The *villa* was clearly surrounded by a flat, open space ('*in plana patentiaque*'[80]). Within its own territory, there was no need for enclosing fences or walls. The *villa*'s staff would include shepherds and cowherds (*pastores, bubulci*[81]) to guard the herds and restrict them to the borders of the cultivated fields or to the wastelands, keeping them away from the crops. The dimensions of the fields must have been vast, in keeping with the ample workforce and the possibilities afforded by the horse- or ox-drawn harvesting machine. The land surrounding these *villae*, the larger ones at least, must essentially have been made up of extensive cultivated fields, no doubt generally of a more or less regular rectangular shape. The countryside must have resembled that surrounding the present-day large farms of the central Paris Basin, for which the classic description nowadays is a 'mosaic' open field.[82]

Similar conclusions may be reached with respect to the borders separating one *villa* from another. The entire Roman legal tradition, as expressed in particular in the works of the *gromatici*, produced and codified essentially during the Early Empire, the hey-day of the *villa*, suggests a landscape in which enclosures played little part and property boundaries were in the main indicated by markers and reference points of various kinds.[83] It is thus reasonable to suppose that, at least in regions where a virtually uninterrupted series of *villae* abutted one upon the other, they more or less filled the entire rural landscape, composing a veritable regional open field. That picture of a uniformly open landscape is certainly conveyed by the splendid image of a sea of wheat that Sidonius uses in his description of the Limagne area: '*illud aequor agrorum, in quo sine periculo questuosae fluctuant in segetibus undae*'.[84] In the early fourth century, a speech of thanks composed by the pseudo-Eumenius and addressed to Constantine describes the landscape of the Saône valley, at the foot of the Côte d'Or, in the days before it had been devastated by barbarian pillaging. The picture painted is one of open valleys (*vallis patentibus*).[85] This seems to have been the preponderant pattern in many regions, in Picardy, for instance[86] and also in Auvergne. Here, in the Merovingian period, 'a few hedges are indicated ... to enclose the parcels of land cleared by hermits deep in the forests covering the mountainous areas. But the cultivated areas, apart from those in pioneer zones, seem to have been left unfenced.'[87] Similarly, in Brittany it has been shown that we must discard the idea that the network of thickets (*bocages*) goes back to prehistoric boundaries. At the most, they can only have been used at a few points in later times and the rare scattered traces of hedges or earth-banks which may be ancient do nothing to contradict the predominant impression of an openfield system which survived right down to the early Middle Ages.[88]

In these openspaces, the numerous and imposing *villa* buildings would draw the eye. They would be grouped around a couple of courtyards of varying designs[89] – usually of a symmetrical rectangular or trapezoidal shape – in an arrangement in which the *pars urbana*, where the master lived, would be carefully distinguished from the *pars rustica*, which incorporated the work-sheds and the living quarters for the overseer and the slaves. These stereotyped, rational designs bear the stamp of the Roman style. Under its influence, the Celtic *aedificia*, from which many of the *villae* had developed, underwent a profound transformation, in a series of distinguishable stages. Rome shaped the rural settlement pattern no less than the urban.

It is, however, much harder to determine the exact spatial sphere of influence of this clearly defined model and the extent of the area covered by the *villae* which embodied it. There is no evidence to suggest that the countryside surrounding the *vici* underwent substantial transformations after the period of independent Gaul. Above all, we know that discontinuous, more fragmented landscapes continued to predominate in many regions of hills and valleys and those where the soil was poor. The uniformly open pattern of the great *villae* must have been applied in the first regions to be uniformly cultivated, where the limestone or silty soil was well-drained and fertile,[90] but it must have been much less common, if present at all, in ill-drained clayey or siliceous terrains which were still partly covered by forests. One suspects that Ausonius had just such a distinction in mind when, choosing his words carefully, he drew a contrast between on the one hand the '*campos ... santonicos*', the open fields of the chalky limestone belt of the Saintonge region[91] and, on the other, the '*tarbellica arva*' of the sands of Chalosse.[92]

New types of cultivation: the slow expansion of vineyards

Though the mark left by Rome on the division of land and on the settlement pattern was of capital importance, it affected systems of cultivation less

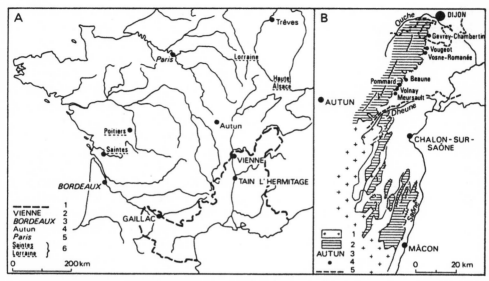

Figure 13. The expansion of the vineyard in Roman Gaul
A. Stages in the expansion (after Dion, 1959, pp. 156–7).
1. The boundaries of the Roman Province.
2. Vineyards attested or probably in existence before the conquest of independent Gaul.
3. The same in the first century AD.
4. The same in the third century.
5. The same in the fourth century.
6. Localities or regions without vineyards attested in Antiquity.
B. The positions of the great vintages, along the Côte d'Or, within the boundaries of the ancient diocese of Autun (after Dion, 1959, p. 47).
1. Crystalline rocks and primary shale.
2. Jurassic limestone.
3. *Civitates* and episcopal seats in the early Middle Ages.
4. Great vintages (*grands crus*).
5. The boundaries of the medieval diocese.

deeply. The agricultural pattern of independent Gaul, even then based on the large-scale production of cereals such as the soft-grained wheat (which could be used for fine-quality bread), barley and millet,[93] seems to have undergone no substantial changes.

The essential innovation was the introduction of Mediterranean crops. The traditional fruits of Gaul, apples and nuts, were supplemented by cherries and peaches from the East. Fig trees grew as far north as Paris, where they were bearing fruit in the fourth century, thanks, in truth, to the straw matting in which they were wrapped for protection in the winter.[94] Above all, even before Caesar's conquest, olive trees and vines spread rapidly in the Province, where the Roman Senate had ruled that their cultivation and likewise the profits made by exporting their products to independent Gaul should be restricted to Roman citizens with full rights or colonials with Latin rights.[95] Vineyards were developed on the northern and western borders of the Province, where they were particularly well placed for sales to the independent Gauls: the Gaillac vineyard in the lower Tarn valley, attested as early as the reign of Augustus, but which must have existed well before that period; and the vineyards of Côte-Rôtie, near Vienne and Tain l'Hermitage, in the Rhône valley, *Allobroges* territory, where the aristocracy may have taken a hand in cultivating them even before Caesar's conquest. Certainly, as early as the mid-first century, one vineyard in this region already enjoyed a great reputation: '*vitifera Vienna*' is how Martial[96] describes the capital of the *Allobroges* people.

After Caesar's conquests, the expansion of the vineyards continued to be very gradual. The principal reason was no doubt the persistence of restrictive measures imposed by the Roman administration. At the end of the first century AD, an edict issued by Domitian,[97] to deter over-production, forbade the development of new vineyards in Italy and ordered the uprooting of at least half the vines in the provinces, that is to say all those that had displaced cereal crops. In 212, Caracalla's edict granted Roman citizenship to all inhabitants of the Empire and this in principle did away with any surviving legal discrimination limiting the right to cultivate vines to Roman citizens. Nevertheless, not until 276[98] did an edict promulgated by Probus extend authorization to plant vines to all the Gauls: '*Gallis omnibus*', thereby rendering general a privilege hitherto granted only to a restricted number of cities.

We know of at least the major stages in this progressive expansion. At the time when Strabo was writing, under Augustus, vine cultivation was still contained by the barrier constituted by the Cévennes[99] and he considered the Atlantic seaboard of Gaul to be characterized by an absence of vines: 'τὴν Κελτικὴν τὴν παρωκεανῖτιν˙ οπου μηδὲ φύεται˝αμπελος'.[100] By the first century AD, the Bordeaux region was planted with *biturica* slips, possibly imported (*arcessita*)[101] from northern Spain, and the vineyards of Bordeaux were certainly exporting to the British Isles, that is to say Britain and Ireland.[102] Meanwhile *allobriga* slips,[103] adapted to hard winters, were being introduced in the Vienne territory, thereby helping vine cultivation to spread northwards. However, the origin of the vineyards of Burgundy poses a problem. Vine cultivation must have been introduced here well before 312, when the *rhetor* (the pseudo-Eumenius) wrote praising the reputation of the regional wines but reporting the Burgundian vineyards to be choked with tangled roots and their vines to be 'old and exhausted'. It is not hard to deduce that this lack of care and failure to replace old vines with new was a consequence of the great Germanic invasions which assailed Gaul between 250 and 280. But exactly how far back did vine cultivation go? E. Thévenot has dated the origin of the Burgundian vineyards to between 50 and 150 AD, on the grounds that the last traces of wine amphorae from the Mediterranean countries disappear from Burgundy around the end of the first century AD, indicating that massive wine importation was discontinued at this point. R. Dion favours the later date of the third century, believing the disappearance of Mediterranean wine amphorae in the first century to be neither significant nor peculiar to Burgundy: it simply reflects the fact that at this date wooden casks were introduced as wine receptacles in all non-Mediterranean regions of France. Furthermore, numerous inscriptions testify to the great power wielded in the second century by the corporation of *negociatores vinarii* of Lyons, which appears to have dominated a trade clearly oriented towards more northern districts. Such a situation would hardly have been compatible with the existence of important Burgundian vineyards at this date.[104] The vines of Trèves and the Moselle valley, whose importance is attested in 291, probably also go back to the third century.

The extension of vine cultivation northwards thus seems to have taken place at a relatively late date and even then it was a privilege restricted to a handful of cities particularly favoured by Rome: Bordeaux, Trèves and Autun. Only the sector of Burgundy dependent upon Autun had already been planted with vines, and this was a sign of the wealth of the *Aedui* aristocracy.[105] On the Atlantic seaboard, there is no sign of any vineyards to the north of Bordeaux in the Antonine period. In Lorraine, the toponymic evidence and the distribution of Gallo-Roman archaeological remains[106] indicates that at this period the foothills had, as yet, not attracted many settlers, from which it is reasonable to deduce that no vineyards existed there.

It was Probus' decree that signalled a general expansion of vine cultivation in the northern provinces. This emperor filled Gaul with vineyards: '*Galliam ... vinetis replevit*', as Aurelius Victor put it.[107] Thus, it was not until the late third and the fourth century that vines spread as far as their cultivation was feasible. In the second half of the fourth century, they were flourishing around Paris[108] and had certainly already become established all over the Seine and Loire basins. Uncertainty remains in the case of northern Alsace, where there is no textual evidence for the presence of vines until the ninth century. Probably the difficulties presented by the navigation of the Rhine, which precluded northward exportation by way of the river, paralysed the development of vine cultivation here.[109] In other areas too, the spread of vines was irregular. The relative rarity of vines in Saintonge and Poitou and the absence of medieval traditions of vine cultivation around Saintes and Poitiers may perhaps be explained by a sudden expansion of Bordeaux vines in the Loire valley, leaving the intervening area neglected.[110] However that may be, by the very end of the Empire, Rome had stamped its mark firmly upon the agricultural landscape even in the northernmost parts of Gaul. The influence of the Mediterranean civilization was thus reaching its limits precisely at the moment when the new Germanic influences intervened – influences which were to affect Gaul deeply.

3
From Gaul to France

The system of land organization that had been established during the centuries of Roman peace was profoundly disrupted by a series of great invasions.[1] First came the wave of Germanic invasions and the consequences that these provoked. After a first massive invasion occasioning considerable devastation as early as the mid-third century, the frontier was more or less re-established along the Rhine up until 406, when it was swept aside by the arrival of further massive hordes. It was not to be restored until the Merovingian state was established in the early sixth century. The ninth and tenth centuries witnessed further invasions, this time of Scandinavians along the Atlantic seaboard and Hungarian peoples in inland Europe, whilst in the south, the Muslim incursion reached its most advanced position in the eighth century and did not diminish until the eleventh.

Paradoxically, these dark and troubled centuries were decisive ones in the shaping of France. It was now that the Germanic name which it still bears first made its appearance. It was also now that the last important additions were made to its peoples and that the last shifts took place in its ethnolinguistic groupings. These changes were to fix its centre of gravity and once and for all determine the future evolution of a number of its territories, several of which – including some major ones – first emerged at this period. Let us now turn our attention to these new elements.

The evolution of settlement

The new linguistic frontier to the north and the east

How can we evaluate the new influx of peoples and analyse its consequences? The linguistic evidence, which is the best known and the most significant, at least proves that the influx of Germanic peoples, although certainly of no mean proportions, was by and large contained within relatively limited bounds. The most spectacular manifestation of it was the establishment of a

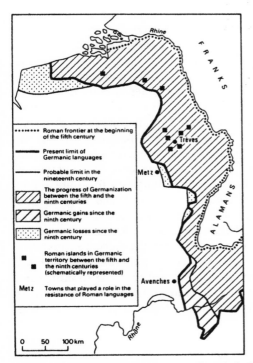

Figure 14. The Germano-Roman linguistic frontier (after Musset, 1965, p. 175).

new linguistic frontier to the west of the Rhine. This constituted a major feature in the human face of western Europe, where no more than changes of detail were to take place after the ninth century. That is when this frontier can first be distinguished on the map and from the thirteenth century on it remained almost totally stable. In fact, the precise location of this line (fig. 14) poses problems which remain unresolved.[2] It cuts more or less horizontally across the plain of northern Gaul, running from west to east as far as the point where, to the west of Aix-la-Chapelle, it bends sharply southward despite the absence of any natural obstacles at this point. The first such obstacles only arise further south, in the shape of the Vosges, which were probably still more or less uninhabited at this period and served as a boundary to the settlements of Alaman peoples in the plain of Alsace.

What is the explanation for the line traced by this frontier? It has been suggested that in the great northern plain it coincided with a line of Roman fortifications set up in the Late Empire to the west of the Rhine, a line which crosses Belgium not far from the present-day linguistic frontier, the *limes belgicus*. Scorn was poured upon that idea when it was produced in the late nineteenth century, but over the last decades, following a number of archaeological discoveries, it has enjoyed a renewed popularity. All the same, the connection between those discoveries and the facts of linguistic geography is by no means assured.[3] We should also reject the notion that the frontier reflected on the one hand a linguistic Roman reconquest of territory with limits that remain unexplained and, on the other, a collapse of Germanism supposed to have taken place towards the end of the Carolingian period, after an interval of more or less total bilingualism throughout an area which extended as far as the Loire. The few linguistic arguments (essentially of a toponymic nature) that have been offered in support of this theory establish nothing more than the sporadic existence of a number of Germanic nuclei which preserved the use of their language for quite a long time; but that does not amount to true bilingualism.[4] The simplest and probably the surest point to note is that the new linguistic frontier ran very roughly 100 kilometres to the west of the course of the Rhine.[5] Its position could be explained by the fact that large numbers of the Germanic invaders settled densely in the first strip of land they came to after crossing the Rhine, and here the countryside was thoroughly laid waste to make room for the new arrivals. Further into Gaul, the only Germanic settlements were more or less compact, isolated pockets, which were integrated into the existing agricultural and territorial frameworks – islands which may have survived for a while but were eventually assimilated.

Germanic settlement beyond the linguistic frontier

The archaeological data
Is it possible to form a precise idea of the size of the Germanic settlements established beyond the linguistic frontier? In this domain, neither archaeology nor anthropology is of much help.[6] Between the early fifth and the mid-sixth centuries, new types of cemetery known as 'linear', where the graves are set out in rows, in open country, heads pointing west and feet pointing east, became common throughout northern Gaul, but were also to be found, with variants, as far east as the Elbe and in southern England. However, these are not necessarily a sign of the presence of Germanic peoples. They represent a new cultural form which developed in the context of invasions west of the Rhine, but which was initially unknown to the east of that river and which owes as much to Roman customs (coffined burial) as to those of the newcomers (burial of the dead together with their arms: the wearing of armour was a sign of a combination of insecurity and wealth, rather than the exclusive mark of a Germanic culture). An analysis of the distribution of certain types of jewelry, the origin of which may be traced to the invaders, such as the ansated fibulae from the Rhineland (fig. 15) may give some idea of the location of some of the pockets of Germanic peoples in

Gaul. But such evidence is no more than fragmentary. The presence of all kinds of material artefacts (furnishings, jewelry, pottery) certainly testifies to an increasingly widespread tendency to return to the Celtic past. It was accompanied by a new style of ornamentation which had certainly originated in the Eurasiatic steppes, but this simply testifies to the emergence of a new fashion, a new aesthetic ideal that resulted from a meeting of different cultures; it is not a precise indication of the presence of new peoples. The necropolis of Saint-Martin-de-Fontenay, near Caen, contains a wealth of arms and jewelry of a Frankish style; however this is not accompanied by any increase in the range of anthropological data, which remains remarkably constant from the sixth century BC right down to the end of the seventh century AD.[7] Archaeology may be a precious source of evidence for the history of civilizations but it is of little help when it comes to evaluating changes in the settlement of different peoples. As for the physical anthropology of the area, the limited series of statistical data available provide insufficient evidence of any morphological changes of importance.

The evidence of toponymy
Nor can we expect much help from a study of personal names. The fact that during the fifth century names formed from a combination of two Germanic elements were adopted generally is simply indicative of a new fashion that spread through the barbarian states.[8] Place-names, on the other hand, constitute a fundamental source of information, although interpreting them tends to be a tricky and controversial business.[9] The renaming of many rural estates and the appearance of new forms of names composed according to the rules of Germanic syntax are only compatible with the presence and determining influence of new peoples. The general model is that in which the first part of the name is that of a person, in most cases a Germanic chieftain who took over from a Gallo-Roman landowner or created a new estate. There are two forms of this kind of place-name: one truly Germanic, in which both words (the personal name and its suffix) come from the new language; the other a mixed form, in which only one element is Germanic, or in some cases no more than the order of the words. It is above all these Gallo-Roman forms that have been the subject of much scholarly debate.

Apart from a number of appellatives (*fara* > *La Fère, La Fare*, etc. = family; *tal* = valley; *bah* > *-bais, -becque* = stream, and so on), the most interesting section of the purely Germanic layer of names is that which includes names derived from the collective suffix *-ing, -ingen* or variants denoting the group of people in the entourage of the important figure after whom the estate is named. The suffix has a variety of dialect endings: *-ange, -enge* in Lorraine; *-ingue* in Flanders; *-ans, -ens* in Franche-Comté (in the Burgundian area). These forms are mostly to be found in definitely northern or eastern areas, but traces of them appear as far afield as Aquitaine

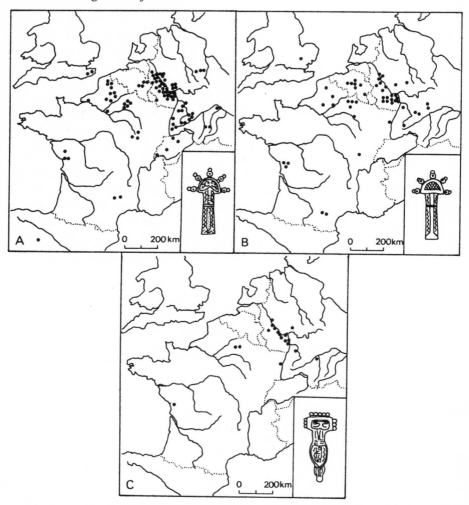

Figure 15. The distribution of different types of ansated fibulae from Rhineland regions (after Salin, 1950–59, I, pp. 218–9). A. Ansated fibulae with a virtually rectangular foot decorated with foliated scrolls on the head. B. The same decorated with cross-hatching on the head. C. Ansated fibulae decorated with wavy lines, with an elliptical foot and a rectangular head.

(brought there by the Visigoths), Savoy and the Rhône valley, although here they are, in fact, extremely rare. Moreover, this kind of evidence from southern locations should be treated with caution. Names ending in *-anges* in these regions are certainly derived from the Gallo-Roman *-anicum* and it has recently been shown that in the Pyrenees of the Ariège district, the names

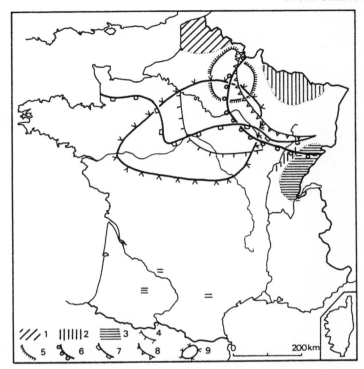

Figure 16. The areas of distribution of various Germanic and Germano-Roman series of place-names (partly after Rostaing, 1945, p. 67 and Bonnaud, 1981, maps 4 and 5). 1–3. The evolution of the collective *-ing*: 1. > *-ingue.* 2. > *-ange.* 3. > *-ans, -ens* (appearing in Aquitaine). 4–5. Germanic name + *iacus*: 4. Main south-western limit. 5. Maximum density. 6–9. Germano-Roman formations. Principal limits: 6. *-court.* 7. *-ville.* 8. *-villers.* 9. *-villiers.*

ending in *-ein*, which used to be considered to belong to this family, in fact probably derive from the Basque language.[10]

The first of the mixed forms is that in which the conquerors succeeded in imposing only the name of the important figure involved, while the suffix remained Gallo-Roman. In the names of estates, the first element, which was Germanic, was thus complemented by the suffix *-iacus* (from the gentilitial in *-ius*), the final result being a toponym ending in *-y*. The variant *ad -iacas* (*villas*), an accusative plural, produces names ending in *-ies*. Such names are distributed most densely in the north and the east (French Hainaut, Picardy, Ardennes) and forms in *-ies* are particularly frequent in the Valenciennes region. There are also a few isolated pockets of them between the Seine and the Loire. On the whole, this is a group which was probably concomitant with that mentioned above but it is more widely diffused beyond the linguistic frontier. Together, these two groups constitute the most ancient layer of names.

The truly Germano-Roman layer, where a Roman appellative is associated with the name of a person, usually Germanic but occasionally Roman, essentially consists of names in *-court* (from the Latin *cohors* > *curtis*, meaning farmyard, farm or village) and in *-ville* (from *villa*) or the series of names in *-villier(s)* and *-viller(s)* derived from *villare* (part of a *villa*), which usually suggest private *villae* which were divided up by the newcomers. These are also to be found in Germanic regions in the form of *-weiler*, the term having passed into the conquerors' own language. Names ending in *-court* and *-ville* are most densely distributed in Lorraine and in Artois–Picardy, although place-names ending in *-ville* also extend further south-west, to the Paris region, the Beauce, where names of this type predominate, and Normandy, where the first part of the name is frequently Scandinavian, which proves that this was a formative model right down to the time of the Norman invasions, being particularly influential in the seventh and eighth centuries. Forms in *-villier(s)*, which predominate in the north, and in *-viller(s)*, which predominate in the east, are also to be found much further afield, appearing in all southern regions in forms such as *Villar, Villard, Villars* and so on. These are purely Roman, some of them very late developments. Apart from these major types, German names are also, but less frequently, added on to other Latin terms: *mansionile* > *mesnil, magny* = peasant's house, with a small plot of land attached; *mansus* > *meix* (same meaning); or terms indicating the altitude of the land, such as *mons* > *mont, vallis* > *val* and so on.

What may we conclude from all this material? Nothing very precise. A series of hypotheses has been advanced concerning the Germano-Roman layer of place-names and relating to the language spoken by the peoples responsible for coining them.[11] The idea, still supported until quite recently, that these were Germanic peoples who then became Romanized, now seems unacceptable. Given the 'victors' psychology' of the conquerors and the precautions that they took in an attempt to safeguard their own ethnic identity, it seems inconceivable that truly Germanic peoples used these types of names. The theory according to which these were Romanized peoples still influenced linguistically by their Germanic masters (F. Lot) seems more valid. However, it has recently been suggested that the so-called 'Germanic' order of composition may in reality reflect a Celtic resurgence or even exist in Latin. In most cases the names seem to have been used by Gallo-Roman peoples at a time when a fashion for Germanic names was developing among the dominant strata. If that is so (Johnson), this entire toponymic layer is without interest where the history of new settlements is concerned and

From Gaul to France 65

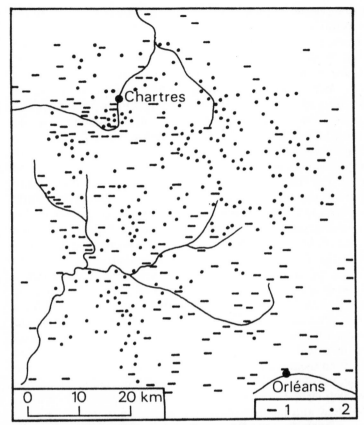

Figure 17. The toponymy of the Beauce (after Dauzat, 1939, pp. 43, 47, 49 and Kroemer, 1976, p. 31).
1. Prehistoric settlements, Celtic toponyms and toponyms ending in *-acum*. 2. Germano-Roman toponyms ending in *-ville* and *-villiers*.

should be disregarded in all discussions on this subject. One very significant example which supports such a view is that of the Beauce region (fig. 17).[12] It was originally thought,[13] on the grounds of the predominance of Germano-Roman place-names ending in *-ville* in the whole central part of the plain, that during the Merovingian period the area had been developed concentrically, spreading inward from the outskirts, at the direction of Frankish overlords. But the very name of the plain, which is of Celtic origin and means 'open plain',[14] shows that land-clearance here dated from the Gallo-Roman period at the very latest. And ancient remains, seldom discernible at ground level, have been revealed throughout the region by aerial photography. The next idea[15] was that the earlier peoples settled here had been wiped out at the time of the invasions and had been almost totally replaced by the newcomers. But in truth, such an extreme hypothesis is not necessary; in fact it is even contradicted by the absence of substantial evidence of any linguistic Germanization. We should no doubt recognize the likelihood of considerable devastation having taken place along such a natural route for the invaders to take (the 'gap route', see below, p. 248) and there probably was a certain hiatus in the history of the land occupancy, but essentially this would simply have involved the replacement of the old traditional landowners by a new layer. As for the Germanic layer itself, here too we must proceed with caution. The large numbers of names ending in *-ans*, which extend from the Porte d'Alsace as far as the neighbourhood of Gray and Besançon and into the northern spurs of the Jura, used to be thought[16] to reflect military and agricultural colonization by the *Burgondes*, anxious to retain control of this extremely important passageway that was under threat from the *Alaman* peoples established in the plain of Alsace. Today, this notion is questioned. It is unlikely that the *Burgondes*, who are known not to have been very numerous, concentrated so many men in one district. Besides, the Burgundian kingdom was eventually destroyed at the hands of Franks from the north. The region is more likely to have been settled by peoples hailing from the vicinity of the Germano-Roman linguistic frontier, a few of whom spread westwards and south-westwards. Hence the idea that the Burgundian chiefs may have taken in a wave of refugees fleeing before the advance of the *Alamans*. They settled them only in the areas that were cultivable, not in the mountainous regions of the Jura, which remained uncultivated at this period.[17] A similar interpretation may be valid for the series of toponyms ending in *-weiler*. Few and far between in the Alsace plain itself, these are concentrated in the foothills of the Vosges, to the west of the plain. They are probably connected with Roman peoples who, having been expelled from the plain by the *Alamans*, took refuge in the bordering valleys and subsequently became Germanized.[18]

Toponymy, like archaeology, thus proves misleading as soon as one seeks precise conclusions relating to any particular regional neighbourhood. The most that can be inferred are general impressions and geographical conclusions of a comparative nature.[19] Common nouns are found hardly anywhere but in the Artois–Picardy region, Lorraine and Normandy, all of which must have been relatively thoroughly Germanized. In general, the Germanic influence is reflected essentially in the personal names that were given to estates. Even if those names do testify to more than just an anthroponymic fashion among the native aristocracy, they cannot be taken as evidence of the settlement of new peoples in appreciable numbers. It should also be emphasized that they are almost totally absent from southern Gaul – not only in the old Roman Province but in Aquitaine too. An analysis of place-names leads to the unexceptional conclusion that the Germanic presence was far greater in northern than in southern Gaul, but it cannot really tell us much about the numbers involved.

Local situations: conclusions
In truth, it is only at a local level, by dint of in-depth research that takes into account the entire complex of factors of different kinds and relates them to one another, and by undertaking not only an intensive geographical investigation but also extensive cemetery excavations, that it might be possible to arrive at reasonably firm conclusions on the composition of populations. There have to date been too few research programmes of this kind for any clear view to emerge.[20] Such results as we have are in the main of a negative nature. In any particular region, the general absence or extreme rarity of any archaeological or toponymical evidence or of any evidence of the arrival of new peoples there justifies the conclusion that Germanic settlement must have been negligible. Such is the case of the Vivarais region, which does not lie in the direct path of invaders following the Rhône valley route.[21] But the same seems to be true of the Loire area.[22] Elsewhere, local analyses sometimes produce a confused picture of flux and reflux, with isolated pockets of Germanic peoples, who were more or less rapidly assimilated, alongside Gallo-Roman populations, some of which remained almost pure, even in areas that were exposed to or close to the linguistic frontier, as Lorraine was. One well-known example is that of the Gallo-Roman village of Sion, which, despite absorbing a few cultural influences, remained completely unadulterated, co-existing alongside the smaller more or less purely Germanic population of Chaouilley.[23] There must have been many other similar cases.

Overall numerical calculations thus remain unreliable, although some have been attempted (for example, 'closer to 10–15% than to 5% for Gaul as a whole'[24]) and seem plausible. The surest thing that can be said is that the settlement of Germanic peoples was weak to the south of the Seine, considerable to the north of it, but only really massive to the east of the new linguistic frontier. Notwithstanding its limited proportions, the influx of Germanic peoples, through the new cultural elements that it brought in its train, the devastation that accompanied it and the imbalance that it introduced between the north and the south, nevertheless played a decisive part in shaping the face and the spatial organization of the country.

The impact on land use and the landscape

The decrease in population and its consequences

The first tangible manifestation of the invasions was a serious contraction of the area of land under cultivation. This reflected a considerable drop in

68   The genesis of France

population, for the arrival of the Germanic peoples had by no means compensated for the general decrease in numbers that was occasioned not only by wars and massacres but also, from the mid-sixth century on, by a series of great plagues. It has been estimated[25] that the population of Gaul in the sixth century AD was one quarter of its size in the second century and that in the seventh to early eighth century, which appears to have constituted the lowest point, numbers dropped to no more than one eighth of the maximum. Innumerable indications point to this massive drop in population numbers. Even today, one may frequently come across Gallo-Roman ruins buried deep in the forests which in many areas have also submerged Roman roads originally built across open land.[26] It is a phenomenon to be found in more or less every part of the territory, from Picardy right across to Burgundy and the Languedoc plain.

However, this picture of land repossessed by the forests should not be exaggerated. Although there can be no doubt of the general phenomenon, some regions were spared. In Limagne, land occupancy is believed to have expanded continuously from the Late Empire through to the Carolingian period.[27] Elsewhere, as in Picardy, the human population soon reclaimed the land that had been lost. Here, the Merovingian period, from the mid-sixth to the mid-eighth centuries, is believed to have been a time of intense rural colonization at the hands of Gallo-Roman communities.[28] In other parts, it was the Germanic invaders who settled in the as yet uncleared sectors, such as the forest of Auberive, in the Langres region, or the region of Gendrey in between the Serre and the Chaux forests, on the plain of the upper Saône,[29] or in the forests bordering the Gallo-Roman settlements in the Albi district.[30] During this period of general confusion, the pattern of development in different places was infinitely varied.

The stability and the decline of towns: name changes

In this chaotic situation, the urban network constituted one point of stability. The network of the larger ancient towns survived virtually unimpaired. Apart from a few exceptions, it was only the smaller cities that disappeared. Almost all the towns that were repeatedly ruined or destroyed by fire in the great invasion waves of the mid-third century or in 406 and the years that followed rose again within their town walls. The majority of barbarians who established themselves definitively in Gaul settled in the countryside. More or less integrated into the administrative framework inherited from the Empire, they admired the Roman civilization and respected its towns. Later on, the new masters moved into the towns and took over the official quarters and public buildings there.[31] In this respect, continuity was unbroken, the Germanic peoples making virtually no contribution at all to the urban structure of Gaul. They were responsible for changing the name of only one town, Argentorate, which became Strasbourg, 'the town of the road', a name that evokes the stone-built highway built by the Romans across the Ill marshes, which must have amazed the tribes that crossed the Rhine. But the name was also known, in its Latin form of *Stratae Burgus*, probably as early as the fourth century, when it may have applied to a particular quarter of the Gallo-Roman town.[32] As for changes of site, these were extremely rare, the result, where they did occur, of peculiar local circumstances rather than of a general trend.

The main change was the considerable decrease in the populations of the towns. These withdrew within their precincts, which soon became too big for them, and fell partially into ruins between which the land was more or less cultivated or used for grazing.[33] The organization of parcels of land and streets was seriously impaired, although vestiges of an orthogonal arrangement are still to be found today, here and there (Rouen, Orléans, Rheims, Metz, Limoges). This decline was accompanied by one of the most remarkable processes in the evolution of the urban topography of Gaul: namely the general discarding of the Late-Empire town names, in which a whole mass of cities lost their names and took on those of the tribes for which they served as *chefs-lieux*.[34] This was the reverse of what had usually happened in the Graeco-Roman world, in which cities tended to fare better than tribes, and in the Roman Province, where the creation of colonies had soon resulted in the disappearance of the names of the native peoples, while in long-haired Gaul, the custom of double names was preserved.[35] Only in Gaul, of all the lands conquered by Rome, did tribe names come to be reapplied to the main towns. Forty-six examples may be cited. Thus *Lutetia* took the name of the *Parisii* > Paris; *Avaricum* took that of the *Bituriges* > Bourges; *Condate* that of the *Redones* > Rennes; *Augustoritum* that of the *Lemovices* > Limoges; *Durocortorum* that of the *Remi* > Rheims; *Caesarodunum* that of the *Turones* > Tours; *Limonum* that of the *pictavi* > Poitiers and so on. There are far fewer examples of the reverse process: twenty-one outside the Narbonnaise (Rouen, from *Rotomagus*, in the city of the *Veliocasses*; Autun, from *Augustodunum*, in that of the *Aedui*; Bordeaux, from *Burdigala* in that of the *Bituriges* and the *Vivisci*; Besançon, from *Vesontio*, in that of the *Sequani*; Toul, from *Tullum*, the capital of the *Leuci*, and so on). The reason for this is clear. The names of *chefs-lieux* remained unchanged wherever Romanization had been strong: that is to say in the Narbonnaise, but also in the Saône and Rhône valleys, around the military centres in the north and the Rhine valley and in a few prestigious cities, such as Autun, where the aristocracy was deeply Romanized. The same is true of a certain number of ports and crossroads with names that were so widely known in trading circles that they could not be changed. However, the dominant tendency in the evolution of place-names implies a reemergence of the Gallic substrata, encouraged by a considerable dwindling of the prestige and influence of the towns, and indicative of the enduring sense of ethnic identity possessed by the peoples of Gaul. That evolution may have begun at the time of the edict of Caracalla which, in 212, granted Roman citizenship to all subjects of the Empire, thereby laying the foundations for the identification of the names of the *chefs-lieux* with those of their surrounding territories. The first known instance of the phenomenon (Rennes) is attested in 237, and others (for example Limoges) occurred in 243, but the great increase in such cases following the upheavals of the mid-third century indicate that the process was essentially a consequence of the general decline suffered by urban life in the context of the invasions.

Thus reduced and diminished, the towns eventually broke up and when timid attempts were made to reanimate them, in early Christian times, their structure changed. Alongside the ancient city, where the bishop would reside within his often tiny precinct, where possible new population centres were built, mainly during the 'Carolingian renaissance' and particularly in the vicinity of votive sanctuaries, many of which evolved into large abbeys (Saint-Germain-des-Prés in Paris, Saint-Rémy in Rheims, Saint-Médard in Soissons, Saint-Valérien in Tournus, Saint-Martial in Limoges and so on). Double towns thus developed, or triple (Auxerre, Sens) or even quadruple ones (Metz, Rheims). From being single entities, towns developed a plurality of districts.[36] Such urban clusters, a reflection of the fragmentation of power in the ninth to tenth centuries, were only to be reunified in the late Middle Ages, when urban life improved and the urban communes established their own identities with greater confidence.

Changes in the rural landscape

Repercussions were greater in the countryside, although here they were gradual and progressive. The general framework of the *villa* remained on the whole unchanged, although many passed into new hands or were now shared with new owners. However, the *villae* were now organized differently. Most of the Gallo-Roman *villae* had been cultivated directly by their owners, with the help of slaves, most of whom were housed centrally, on the estate. There had certainly been a few sectors where *coloni* or *servi casati* would be established on small plots which they themselves cultivated. But these were in the minority. In the Late Empire, the proportion of such units increased. In the immediate aftermath of the first major upheavals, in the mid-third century, attempts were apparently made to recultivate land that had been abandoned: Germanic groups of 'letes' were installed there, *coloni* whom the emperors tied to the land to ensure that they would defend it. This arrangement is believed to account for the presence, in Picardy, of centuries that are detectable from aerial photography, which are arranged in a pattern that has nothing to do with the distribution of the *villae*. These centuries were created considerably later, when recolonization was taking place in the vicinity of the Roman thoroughfares.[37] The process progressively gathered speed, taking a number of different regional forms. In areas shared with *federati et hospites* (Visigoths and *Burgondes*) it was a partial one. In regions conquered by the invaders, where barbarian communities had probably totally expropriated the ancient estates, the system was imposed more suddenly and more completely. Isolated settlements of free workers and small plots of cultivated land multiplied, in accordance with the Germanic tradition of dispersed populations noted by Tacitus.[38] The Carolingian *villae* of northern and north-eastern Gaul may still have been largely under direct cultivation, but this seems to have been exceptional, a phenomenon limited to this particular region and connected with the close proximity of various rulers and the needs of their courts.[39] Overall, the picture of the Carolingian period presented by the relationship between the central unit of the *villa* and the messuages (scattered holdings cultivated by dependent peasants) suggests that these small farms were considerably more common than during the Gallo-Roman period.[40]

While the structure of landownership was changing, so too was the landscape. The proliferation of tenant holdings certainly led to more enclosures. The scene was now quite different from that of the large-scale landowners of Roman Gaul, for in the traditionally egalitarian society of the Germanic peasants, with its initially flexible, then fixed pattern of fields (*kampen*), carved out of the natural vegetation, enclosures were designed to keep out the grazing herds. The general corpus of barbarian laws conjures up a clear image of a rural landscape very different from that suggested by the collections made by the *gromatici*. Alongside terms borrowed directly from the Roman tradition,[41] it is striking to find increasingly frequent mentions of hedges (*sepes*) and measures designed for their protection.[42] Philological proof that the development of hedges was part of the Germanic culture's contribution to Gaul is provided by the fact that the French word for 'hedge' (*haie*) is of Germanic origin (from the Frankish *haga*; cf. the English 'hedge' and the German *Hag*).[43] The term originally denoted the wooded strips, sometimes reinforced with fencing, which separated off and protected clearings in the forests,[44] but it soon came to apply to all forms of enclosure consisting of living vegetation. In the course of the early Middle Ages, enclosures of this kind thus gradually became a general feature of the landscape. It was at this point, during the second half of the ninth century, that the term 'close' (*clausum*) was frequently used in Auvergne, to denote vineyards.[45] However, it was a slow process. In the ninth century, Redon's cartulary reveals a largely open landscape in lower Brittany, where enclosed fields cannot have occupied more than half the total area.[46] It was probably not until later that, as the messuages were progressively broken up, the density of enclosures increased significantly. Precious legal testimony is

72 *The genesis of France*

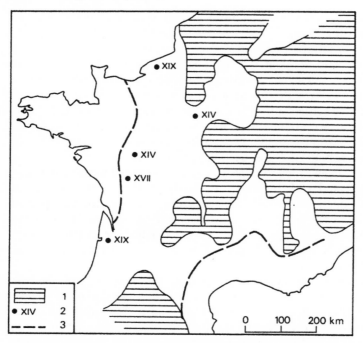

Figure 18. Types of rural vehicle in France (after M. Gautier, 1971, p. 256).
1. The present-day distribution of 4-wheeled carts (used either exclusively or in conjunction with 2-wheeled carts).
2. Places where the presence of 4-wheeled carts is attested in earlier periods (fourteenth to nineteenth centuries).
3. The limits of the Atlantic and the Mediterranean distribution of 2-wheeled carts.

provided by the rules applying to enclosures, stipulating that every tenant farmer was responsible for the yearly maintenance of a particular length of fencing and hedging. These rules began to make their appearance as early as the polyptychs of the Carolingian period and continued to develop until the twelfth century, when there were more of them than at any other time.[47]

All in all, the invasions brought about disintegration and fragmentation in the countryside. The vast open landscapes dominated by the massive structures of landownership of the Gallo-Roman *villa* were replaced by a countryside that was divided up into heterogeneous holdings of a more individual nature. There appear to have been virtually no positive contributions to agricultural, or even to pastoral technology. The invaders introduced important innovations in the techniques of warfare and hunting (falconry, the stirrup), in metallurgical production (weaponry), in mental images, popular beliefs and artistic representations (the names of the rainbow and the Milky Way, for example) but in the domain of agriculture they had no new suggestions for the native peasant population, which was infinitely more advanced. The only noteworthy novelty was the four-wheeled cart, a vehicle which originated in the steppes, and whose distribution appears to have been limited to the more central areas of the European continent, giving place, in the modern period, to the two-wheeled cart of Mediterranean or Atlantic origin. One curiosity is perhaps also worth mentioning: the introduction of the pack-camel which, up until the seventh century, continued to be used as far south as Uzès and along the banks of the Garonne.[48]

The new organization of the territory

Changes in the pattern of settlements and in the urban structure of the territory may have been slight or at any rate slow and gradual during the period of the invasions, but at a political level important transformations took place.

The fragmentation of the Late Empire

The harmonious organization established by Rome had in the main been adapted to pre-existing frameworks and had respected the individuality of the major ethnic divisions of Gaul, at the same time safeguarding and stabilizing them. But it had not been able to withstand the upheavals caused by the Germanic invasions. In the last centuries of the Roman Empire, the need for immediate defences and new strategic perspectives led to the creation of new cities organized into new groups. From the time of Aurelian onwards, towns were surrounded by fortified walls, and *civitates* that were sprawling and consequently ill-suited to a coordinated system of defence were systematically divided into two. The latter process was clearly connected with the former to the extent that all fortified population centres tended to be set up as autonomous administrative centres. In truth, this was not an altogether new phenomenon (see above, pp. 39–41), but from being exceptional it now became the rule.

At the same time, provincial divisions multiplied. Successive lists produced under the Late Empire testify to the progressive and continuing process of fragmentation (fig. 11B). The Verona list (of 316 AD, according to A. Piganiol) recognizes two dioceses and fifteen provinces in Gaul. The diocese of the Gauls comprised all the northern provinces (the first and second Belgic provinces, the first and second Germanic, the first and second provinces of Lyons, the Sequanaise (after the Saône), the Alpes Grées and the Alpes Pennines), which were the most exposed. The diocese of the Vienne region included the seven provinces of the south and the south-west: the first and second Narbonensis,

74 *The genesis of France*

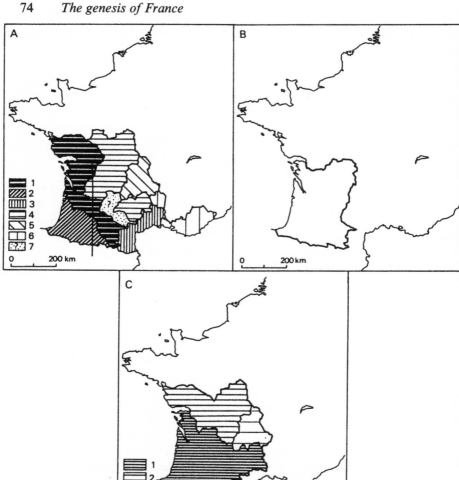

Figure 19. The persistence of Aquitaine after the Germanic invasions (after Rouche, 1977, pp. 7, 57, 96).
A. Formation of the Visigoth kingdom, 419–76. 1. Extent in 419. 2. In 440. 3. In 461. 4. In 472. 5. In 475. 6. In 476. 7. The city which became Visigoth at an unknown date. B. The kingdom of Gondovald, 584–85. C. Independent Aquitaine at the end of the Merovingian period. 1. After 658. 2. 676. 3. 711.

the first and second Aquitaine provinces, *Novempopulania* and the Alpes Maritimes. At the end of the fourth century or the beginning of the fifth, the *Notitia provinciarum et civitatum Galliae* (after 375) listed seventeen provinces, ten in the diocese of the Gauls and seven in that of Vienne.

The third-level units (*civitates* or *castra*) totalled 119. The number of cities, like that of the provinces, had thus roughly doubled since the period of Augustus. The reverse process, of several small cities amalgamating to form a single one, was rare but not unknown. It no doubt occurred in situations where only the new *chef-lieu* had been provided with ramparts.[49]

In truth, these secondary divisions were by no means totally new. Many had certainly existed among the peoples of independent Gaul, in the form of *pagi*. Far more important was the reorganization of the provinces, now split into two dioceses. This constituted a new division in Gaul, setting up an opposition between the north and the south which was the first mark of the influence of the Germanic invasions and one which has never disappeared from the territory of France.

The persistence of Aquitaine

The only territorial unit that seems to follow on from the divisions of the Roman period at this point is Aquitaine,[50] which was to serve as the pivot of the Visigoth state. In the fifth century, a kingdom was set up here, in the ancient province of the second Aquitaine, expanded to incorporate the territory of Toulouse. It was gradually to extend over the entire area between the Loire and the Pyrenees as well as the province of Narbonne and Provence (fig. 19A) and, in the mid-fifth century, it also conquered the greater part of the Iberian peninsula. But in the early sixth century, the power of the Visigoths in Gaul collapsed beneath the onslaughts of the Franks, led by Clovis and his sons. The Visigoths retained a hold over only the Languedoc plain, which stretched from Nîmes to Narbonne and was known as *Septimania* or *Gothia*. Meanwhile the *Vascones* (Basques) came down from the Pyrenees and, from the end of the late sixth century onwards, progressively established themselves throughout the former *Novempopulania* to the south of the Garonne, in such numbers as to deeply mark the anthropological stock of this area (where over 70 per cent of the population even today belongs to the O blood group, constituting the only example in France of a clear correlation between the distribution of blood groups and the historical arrivals of new peoples). But Aquitaine was to be resuscitated several more times in the course of the Merovingian period: during the short-lived success of Gondovald's kingdom, 584–85 (fig. 19B); more lastingly in the form of the kingdom of Caribert II, 629–32; and during another long period of independence which, despite the Muslim invasion of the early eighth century, endured from 658 until the definitive Carolingian conquest in 768 (fig. 19C). This persistent series of rebirths was the consequence of a particularly strong sense of identity based upon a solid Roman heritage (the strength of the Visigoth presence probably amounted to no more than 100,000 newcomers), combined with the wealth of a traditional aristocracy which had remained powerful. It was reflected in the diffusion of certain cultural phenomena, such as the cults of particular saints (fig. 20).

The state of the Burgondes and the origins of Burgundy

Another federated people, the *Burgondes*,[51] created its own state, which has left its name to a province of France that has retained a strong sense of its own identity right down to the present day. The *Burgondes* were an eastern Germanic people situated, *circa* 200 BC, in the Baltic island of Bornholm, which continued to be named after them (*Burgundarholm* in the thirteenth century = 'the high island'). In the early first century AD, they moved into Pomerania, to the west of the Vistula. Their first kingdom was set up to the west of the Rhine, in what is now the Palatinate, to the west of Worms and Mainz (413–36). They were then conquered by the Roman patrician Aetius who, reckoning that they might usefully check the power of the far more redoubtable *Alaman* people, in 443 installed them in *Sapaudia*[52] (from which the name Savoy is derived), an area that probably corresponded to the city of Geneva, the mountainous sectors of which must still have been mostly uncultivated at this date. However, their state soon expanded to take in the entire Rhône valley as well as the upper reaches of the Loire and the Seine, resulting in a territory traversed by a number of important routes converging upon Lyons, its second capital, and thereby controlling communications between the Atlantic seaboard and the Mediterranean. This network of highways was long to remain a major feature of the future Burgundy and a major asset in the development of the isthmus as a whole. The *Burgondes* were probably quite a small tribe, soon assimilated into the Gallo-Roman civilization, if not its language (they continued to speak their own until the seventh century) and, together with the Visigoths, they appear to have been the major heirs to that civilization. They seem to have spread in a relatively peaceful fashion. The division of the area which resulted from the establishment of their state was closely modelled upon earlier frameworks and it would have been particularly favourable to commercial relations if the times had been more opportune. As it happened, however, trade was in deep decline and, in the early sixth century, the kingdom of the *Burgondes*, with its fragile ethnic basis, fell victim to the onslaught of the Franks, just as the kingdom of the Visigoths did.

The problem that remains is the persistence of the name *Burgundia* throughout a succession of different political structures, first within the framework of the Merovingian states (in which the region's boundaries were redrawn on several occasions); then that of the Carolingian states; and finally, following the treaty of Verdun (see below, pp. 90–1), when the area was divided up between France and the Empire and thus extended both sides of the new frontier (fig. 21). Nevertheless, the last attempt to establish

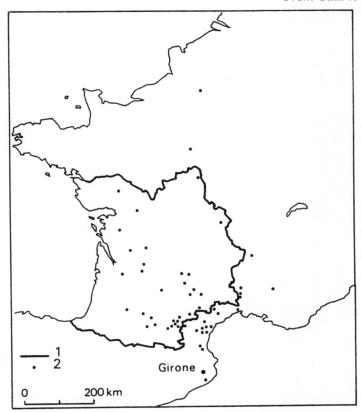

Figure 20. The cult of Saint Felix of Gerona in Gaul (after Rouche, 1977, p. 322).
1. The frontiers of Aquitaine. 2. Cults of San Feliu, Saint Elix and Saint Félix.

an independent Burgundy, undertaken by the patrician Aletheus,[53] took place in 613. The independent spirit of Burgundy, albeit not nearly as strong as that of Aquitaine, certainly cannot be denied within the Merovingian state of the seventh and eighth centuries and it sometimes played a decisive role in the struggles between Neustria and Austrasia.[54] However, it is impossible to regard it as much more than a purely defensive reaction on the part of the local aristocracy.[55] It would certainly be exaggerating to speak, as some do, of a 'sense of Burgundian nationality'[56] at this period. In truth, the Carolingian *Burgundia* was hardly more than a confused image, a vague entity by now very much reduced since the days of the Burgundian kingdom. It lacked any political reality up until the feudal establishment of the territories which went by the name of Burgundy from 879 to 888 and which took the form of a

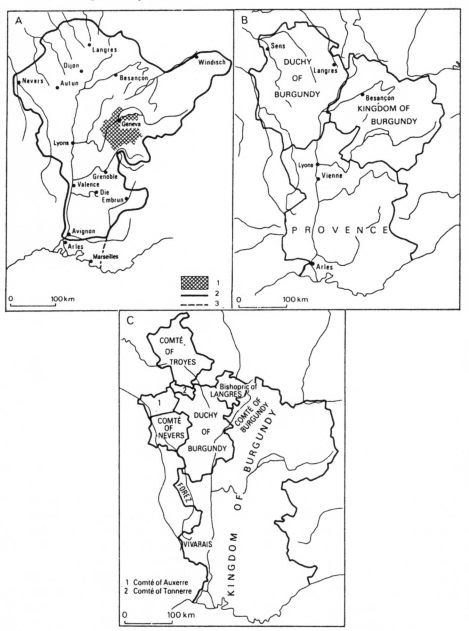

'kingdom' (in the Empire) and a 'duchy' (in France).[57] The prestige of the Burgundian kings and the relatively harmonious co-habitation that they promoted between Romans and barbarians seems nevertheless to have sufficed to perpetuate the name.

The Franks: a shift in the centre of political gravity

Notwithstanding the creation of the territorial units of Aquitaine and the kingdom of the *Burgondes*, both more or less characterized by the integration of barbarians and Romans and continuity in political and social traditions, it was a people established in northern Gaul, the Franks, that was to initiate the reunification of the territory. The Franks were probably of Scandinavian origin,[58] but by the time that they took their place on the historical stage (mid-third century), they had already for some time been established around the mouth of the Rhine, where they appear to have assimilated several purely Germanic tribes (the *Sicambres*, *Cattes* and *Chamaves*). They were mostly organized into small bands of pirates who sailed along the coastline, then made their way inland up the rivers and proceeded thence on foot (in contrast to the *Goths* and the *Alamans*, nomads from the steppes, who invaded on horseback). The Franks finally settled in relatively large numbers in northern Gaul, from the fifth century onwards. Although a few quite large groups must have established themselves in Normandy and in the Maine region, while a few small pockets became established as far south as Aquitaine,[59] by far the majority installed themselves between the Seine and the Rhine. This was certainly a most important new factor. With the arrival of the Franks, the political centre of gravity in Gaul shifted to the plains of the Paris Basin. The Merovingian kingdom, established under Clovis during the last years of the fifth century and the first decade of the sixth, and the states created when it was divided all took northern towns as their capitals: Rheims, Metz, Paris, Orléans, Soissons, Cambrai and Chalon-sur-Saône. After the division of Clovis' kingdom in 511, the capitals of Clovis' four sons were to be found right at the heart of the Paris Basin, at Rheims, Orléans, Paris and Soissons, for the brothers, whose territories extended on every side out to the frontiers of the kingdom, grouped their capitals together, within a limited area, so as to remain close to one another

---

Figure 21. The historical development of Burgundy
A. The kingdom of Burgundia before its Merovingian annexation (after Guichard, 1965, p. 283). 1. The city of Geneva. 2. The kingdom of Gondebaud in 516.
3. The maximum extension of the frontiers of the kingdom of Gondebaud between 500 and 506. B. The situation at the end of the ninth century (boundaries according to Mirot, 1947, p. 93). C. The situation at the beginning of the eleventh century (boundaries according to Mirot, 1947, p. 113).

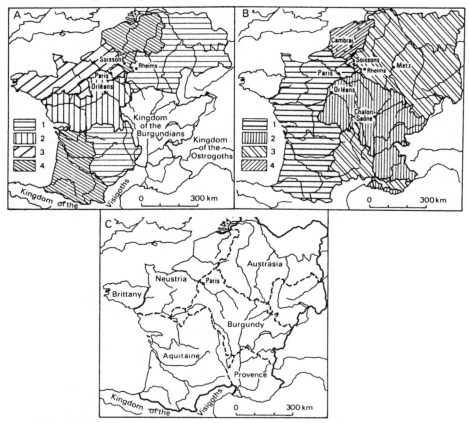

Figure 22. The Merovingian territorial divisions (boundaries as established by the *Atlas classique*, Vidal de La Blache, Paris, A. Colin, 1931, p. 21).
A. The division at the death of Clovis (511). 1. The kingdom of Theodoric I. 2. The kingdom of Clodomir. 3. The kingdom of Childebert I. 4. The kingdom of Clotaire I. B. The division at the death of Clotaire (561). 1. The kingdom of Caribert. 2. The kingdom of Bontran. 3. The kingdom of Sigebert I. 4. The kingdom of Chilperic I. C. the kingdom of Dagobert in 632 (after the death of Caribert). The heavy line indicates the borders of the kingdom as a whole; the dotted lines show the provincial divisions.

at the centre of a state which, despite the division, theoretically remained a single unit. Once the Franks had conquered the Visigoths and the *Burgondes* (in 511 and 534 respectively), their kingdom more or less encompassed a reunited Gaul, and when the Franks of Cologne also recognized Clovis as their king, it extended far beyond the Rhine as well, taking in the whole of Germania and establishing a protectorate over the *Alamans*, the Thuringians and the Bavarians.

Within this *regnum Francorum* (there was as yet no reference to *Francia*, conceived as a territorial unit: see below, pp. 93–108), *Burgundia* may have retained its political identity for a while, being at several junctures a major co-heir. Aquitaine, in contrast, throughout the period during which it was a possession of the Merovingian kings, was never considered as more than a distant dependence definitely not fully integrated within the principal domain, and whenever the kingdom was reorganized into new regional divisions it tended to be carved up in an altogether arbitrary fashion. In the seventh century, the internal lines of division finally crystallized into an opposition between two poles, both northern: Neustria to the west, constituting the north-western side of the territory, centred upon the lower Seine (broadly speaking the kingdoms of Paris and Soissons), and Austrasia to the east (incorporating Rheims, the Meuse, the Rhine and the land beyond the Rhine). The reasons for such a division are none too clear, despite the fact that Neustria was purely Roman, Austrasia bilingual. The two do not appear to have emerged quite simultaneously as territorial concepts. Austrasia (*Austria, Austrasii*, 'the country in the East', 'the people of the East') was already a term with which Gregory of Tours, writing in about 591, was familiar. It first appears as a name for the kingdom of Sigebert, as constituted by the territorial division of 561 after the death of Clotaire I, which took over from the *Francia Rinensis* of Cologne. The first concrete expression of its existence is the transfer of Sigebert's capital from Rheims to Metz, abandoning the principle of a capital located in the central sector of the Paris Basin.

At this same date, Gontran, whose share of the kingdom incorporated Burgundy, similarly shifted his capital from Orléans to the more distant Chalon-sur-Saône. The concept of Neustria (*Neuster*, from *Neuestreich*, 'new empire'), that is to say the regions more recently conquered by the Franks, was of later date. The term is first used only at the beginning of the seventh century, in Jonas' *Life of Saint Colomban* (chapter XLVII). Throughout the seventh century, these two territories were to dominate the internal Merovingian struggles which started after the reign of Clotaire II who, between 613 and 629, succeeded in reuniting the Frankish monarchy and appointed separate palace mayors (chiefs with authority over all officials and other chieftains) for Neustria, Austrasia and Burgundy. But it seems clear that these major divisions were initially of little substance. They simply reflected the *de facto* division of this kingdom with a northern centre of gravity into separate western and eastern regions. It was only as a consequence of years of conflict that, during the seventh century, rivalry developed between them, possibly together with a sense of belonging and patriotism, particularly in the Neustro-Burgundian kingdom during the half-century that followed the reign of Dagobert (629–39). According to *The Life of Saint Didier*, after the death of this king, the *regnum Austrasiorum*

82 The genesis of France

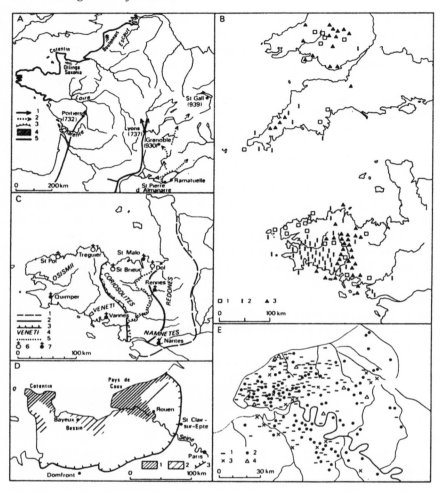

Figure 23. Constructions of an ethnic nature in peripheral territories at the time of the invasions.
A. Saxon and Muslim. 1. Muslim offensives of the eighth century (after Sénac, 1980, p. 22). 2. Muslim incursions of the tenth century (after Sénac, 1980, p. 54). 3. The limits of the *Djabal al-Qilâl*. The names marked are of Arabic origin. 4. Large Saxon settlements attested by texts or toponymy. 5. The *litus saxonicum* in about 428–30. B. Cults devoted to three Britannic saints (after Bowen, 1969 and Fleuriot, 1980, fig. 13). 1. Saint Paul. 2. Saint Winwaloe (Gwenolé). 3. Saint Gildas (Gweltas). C. The formation of Brittany (limits according to Fleuriot, 1980, figs. 6, 9 and 11). 1. Approximate eastern limit of Brittany between 850 and 915. 2. Approximate eastern limit of the Romano-Breton zone, where the Breton language was spoken in the ninth century. 3. The limit of the Breton language in the fourteenth century and down to the present day. 4. The names of

passed to Sigebert III, the *regnum Francorum* to Clovis II. This was the first manifestation of the particular region that was eventually to become France (see below, pp. 94-8). But there was really no geographical basis to the opposition between the two regions: they were purely a product of dynastic history.[60] The opposition became less important in the Carolingian period, when Neustria and Austrasia appear to have been considered simply as peripheral regions flanking the *Francia* contained between the Seine and the Rhine. Such as it was, the opposition expressed the political pre-eminence of these northern regions where the competition for power was concentrated. The northern regions continued to predominate when, at the end of the eighth century, the Carolingian Empire briefly ressuscitated the unity of a Christian West. Its principal capital was at Aix-la-Chapelle in the Rhineland, facing the eastern frontier, where the Empire's main attempts to expand into Saxon territory were concentrated.

This major shift in power to regions north of the Loire was, in the last analysis the most important legacy of the period of the Germanic invasions. It was during this period that the centre of political gravity was definitively fixed, sealing the subsequent fate of a Gaul which had originally been shaped from the south, with Marseilles and Lyons as its main centres of influence, the former a Greek town which was a focus for trade, the latter a stronghold of Roman power. Henceforward Gaul slipped out of the sphere of those southern influences and eventually achieved its territorial unity through forces that originated in the plains of the north. The Germanic peoples transformed the geography of power in Gaul rather than that of its population.

Peripheral constructions of an ethnic nature

Compared to the major impact of the Germanic invasions, the other territorial developments of the period seem of secondary importance. Yet they were by no means negligible. During those troubled times, many peripheral strikes from the sea also threatened the unity of Gaul (fig. 23). In two instances the threat was long-lasting but eventually collapsed, leaving no

---

Gallo-Roman peoples. 5. The eastern limit of the *Coriosolites* territory, which was the approximate limit of Brittany from the fifth century to the eighth (753).
6. Bishoprics reconstituted in the ninth century upon new territorial bases.
7. Bishoprics which survived on their Gallo-Roman sites. D. The formation of Normandy (after Musset, 1965a, p. 131). 1. Very dense Scandinavian population.
2. More dispersed Scandinavian population. 3. The frontier of Normandy (established by the treaty of Saint-Clair-sur-Epte [911], to the east only). E. The distribution of a number of toponymic suffixes of Scandinavian origin in eastern Normandy (after Sion, 1911, and Brunhes, 1920, p. 303). 1. *-tot.* 2. *-mare.* 3. *-bec.* 4. *-dale.*

tangible traces apart from place-names. On the other hand, two ethnic regions, retaining their own particular individualities to differing degrees, have endured – a fact that is reflected in the territorial divisions of contemporary France.

The failures: the Saxons and the Muslims

Mentions of Saxon pirates[61] arriving by sea at various points of the western seaboard of Gaul occur from 286 on. In the *Notitia dignitatum* (about 428–30), the name *litus saxonicum* is given to the coastline between the Escaut and the Loire, and also to the east and south coasts of Britain. The Saxons established settlements in various locations in Gaul (fig. 23A), close to the coastline: in the Boulogne region; around Bayeux where, in the Carolingian period, there still existed an *Otlinga Saxonia* (from '*Edlingen*', 'the nobles, the warriors'; cf. the German '*adlig*'); in the Cotentin region; along the banks of the Charente and around the mouth of the Loire, where the Franks must have fought them. No trace remains of them except in the toponymy: names in *-tûn* > *-thun, -tun* ('hedge, enclosure' and subsequently 'population centre'; cf. the English *-ton, -town*); in *hêm* > *-ham* ('inhabited place'); in *-gate* (= 'door'); in *hou, ou*, from the Saxon *ho* ('spur of land'); in *-naes* > *-nez* ('cape'), to mention only the least controversial, for they are difficult to distinguish from the rest of the Germanic and Scandinavian layer of place-names.[62] It is only through toponymy that we know that there was a large Saxon enclave in the Boulogne region, for there is virtually no archaeological testimony to their presence. The Anglo-Saxon objects of the period that have been discovered along the Normandy coastline may well have found their way there simply in the course of trading.[63]

A series of Muslim occupations took place in various localities in southern Gaul from the early eighth century (before 720) down to the end of the tenth century (fig. 23A): the Visigoths of Septimania and, more importantly, also those of the Narbonne region were conquered. Both areas remained under Muslim rule for several decades, until 759, and provided bases for military raids which affected the whole of the south-west as far north as Poitiers (732) and the Rhône valley as far north as Lyons (735–37). A base was also established in the Camargue in the second half of the ninth century and between 880 and 890 a settlement was founded in the Massif des Maures (referred to in Arabic sources as *Djabal al-Qilâl*), where it survived until 972.[64] The hostility between the civilizations was such that after the reconquest, all traces of the Arab settlements disappeared. Archaeological and numismatic testimony to their presence is also very limited and of uncertain interpretation. A number of place-names indubitably do testify to the Arab occupation: Ramatuelle, near Saint-Tropez (from *Rahmatollâh* = 'the pity of God'); Saint-Pierre-d'Almanarre near Hyères (from *al-manâr* = 'the lighthouse') and so on.[65] A number of Arab terms passed into the Provençal dialects, such as *'goudron'* the word for 'tar' (*Qatrân* > *Qitran*),[66] testifying to heavy exploitation of the local forests. This may be explained by the importance of the Massif des Maures to the Arabs, who were sometimes obliged to search far and wide for timber to build their ships. The Saracens are said to have been responsible for introducing into France the buckwheat which bears their name (*blé sarrasin, sarrasin*).[67] However, new plants of unknown origin are often attributed to some foreign people or other, and it has also been suggested that this type of wheat was brought to France at the time of the Mongol invasions – a hypothesis that seems more probable.[68] Certain breeds of goats and horses from Provence and the Camargue are also believed to be of Saracen origin, but there is no convincing proof to support this. After the reconquest of *Djabal al-Qilâl*, a number of Saracen slaves merged with the local population. But all in all, these traces of the Arab presence amount to no more than a handful of curiosities. The threat posed by Arab raiders established in their strongholds in the Maures and striking out into the Alps may in part account for the northernmost locations of the Mediterranean *habitats perchés* (hilltop villages) strung out in a ribbon along the Alpine ridges (they are particularly numerous in the southern Alps)[69] but recent hypotheses incline to other explanations.[70]

The successes: the Bretons and the Normans
In the end, it was only Brittany and Normandy that developed a distinct regional individualism. The origins of Brittany[71] go back to Britons who emigrated from Britain to the continent (fig. 23B), under pressure from Saxon invaders in the east and the south and also from the Irish, who were ravaging the western coasts of Britain in a series of waves between the second half of the fourth century and the seventh, but which continued sporadically until the tenth century. The emigration from Britain, which took the form of military manoeuvres, was initially sanctioned by the Romans: from 383 onwards and in 407, massive army units from Britain were landed on the continent. These were followed by the arrival of large numbers of clerics, peasants and sailors, so that this first wave of emigration spilled over beyond the confines of the Armorican peninsula, establishing numerous settlements as far-flung as the furthest frontiers of Gaul, concentrated particularly in the north-west (present-day Normandy, but also in the Seine and Loire valleys).

From the end of the fifth century on (after the treaty of 497 between the Franks and the Gallo-Roman peoples of the Armorican peninsula), settlements were tolerated only within the peninsula (in the territories of the ancient cities of the *Osismii*, the *Veneti* and the *Coriosolitae*). Here organized groups, mostly from western Britain (present-day Wales and Cornwall) installed themselves. This wave of emigration appears to have been essentially peaceful, particularly when led by pious figures such as the Breton 'saints', who acted as guides to their communities. The first settlements that they founded bear names in *Lan-* ('consecrated place'), then in *Loc-* (from the Latin *locus* = 'place'). Later, as parishes were created, many place-names in *Plou-* appeared (from the Latin *plebem*, which passed into Welsh with the meaning of 'community of believers'), usually in conjunction with a personal name. The distribution of these is particularly dense throughout the western coastal regions of the peninsula. A peculiar feature of Brittany is that here the central points of parishes coincide not with estates, as they do in the rest of Gaul, but rather with the sites of Christian cults. Many of these parishes originally covered vast areas, so that their subdivisions (toponyms in *Tré*, from *tref* = 'subdivision' derived from the Celtic root *treba*, generally in connection with a second descriptive term) were themselves sometimes converted into parishes (in these cases the second term is a personal name).[72]

These Breton settlements (fig. 23C) were certainly mostly concentrated along the northern coast, where the earlier system of ecclesiastical organization was more or less wiped out.[73] The territorial bases of the bishoprics reconstituted there in the ninth century were entirely new (those of Saint-Pol, Tréguier, Saint-Brieuc and Dol) and very extensive (dependencies of the bishopric of Dol were to be found as far away as the mouth of the Seine), putting one in mind of the bishoprics of Wales and Ireland. On this northern coast, the only see to retain its previous site was that of Alet–Saint-Malo. On the southern coast, in contrast, the bishoprics of Quimper (the city of the *Osismii*), Vannes (the *Veneti*), Nantes (the *Namnetes*) and also that of inland Rennes (the *Redones*), retained their Gallo-Roman structure. However, the Breton language was adopted throughout the peninsula west of a line running from west of Saint-Brieuc to the mouth of the Loire on the northern coastline, which approximately coincided with the western boundary of the city of the *Coriosolitae*. Further east was a mixed Romano-Breton zone which extended as far as the territory of the *Redones* and the *Namnetes*. The furthest limits of the Breton thrust in the ninth century can be roughly determined in particular by reference to the Gallo-Roman place-names ending in *-acum* that evolved into an *-ac* form, in contrast to those in the area which remained Roman, where they evolved into an *-é* form (compare *Sévignac* in the Côtes-du-Nord and *Sévigné*, Ille-et-Vilaine). It penetrated up to a line stretching from the bottom of the Normano-Breton gulf as far as Donges, to the east of Dol, on the estuary of the Loire, between Nantes and Saint-Nazaire, looking eastwards towards the towns of Rennes and Nantes, which certainly constituted major Roman strongholds. Between the tenth and the fourteenth centuries, this mixed linguistic zone was reconquered by the Roman language and the linguistic frontier then stabilized, until the late eighteenth century, along the limit of pure Breton, which even today remains the theoretical linguistic frontier (for the contemporary retreat of Breton has been occasioned by an internal collapse). However, the political frontiers of Brittany extended much further than its ethnic boundaries. These Breton settlements were the basis of a state which expanded in a strong thrust eastwards during the Carolingian period: Noménoé, who became Duke of the Bretons in 826, declared independence in 844 and was crowned king in 848. The kingdom continued to expand to the point where, at the end of the ninth century and the beginning of the tenth, it incorporated both Cotentin and extensive areas of Maine and Anjou. Thereafter, right up until the union with France in the early sixteenth century, the Francophone dynasties of feudal Brittany continued to use either Nantes or Rennes as their capital.

Normandy,[74] which took its name from the *Northmen*, constituted a similar territorial structure (fig. 23D), which developed from coastal settlements of Scandinavian peoples (essentially Danes, some of whom had moved on from England but also – particularly in Cotentin – Norwegians, some of whom arrived by way of Ireland). These peoples soon spread inland. The limits of Norman penetration can be determined on the basis of toponymy[75]: names in *-bec* (from *bekrr* = 'stream'); in *-beuf* (from *budh* = 'shelter, temporary residence'); in *-tot* (from *topt* = 'piece of inhabited land'); in *-mare, -mer* (from *mare* = 'pool, pond'; given that this word then passed into French, only those compounds in which the first part is also Scandinavian are significant); in *-dale, -dalle* (from *dal* = 'valley'); and probably also names in *-fleur* (from the old Norse *flet* = 'space', 'level', rather than from the Saxon[76]) and so on. Such toponyms abound in the Pays de Caux and the lower Seine valley, as they do in northern Cotentin, and are more sparsely scattered round the Seine bay and along the western coast of Cotentin. They are totally absent from the Bocage Normand and the southern part of the Eure department. Inland, they disappear a long way short of the frontiers of the territory ceded to the Norman chief Rolland by the treaty of Saint-Clair-sur-Epte (911), following about a century of raids which had begun around 810 in the Seine valley. This first Norman territory, centred around Rouen and initially bounded by the semi-circle formed by the Bresle, Epte and Avre rivers, was won in an organized campaign of conquest in a sector that was too close to the heart of the French monarchy for the latter ever to have ceded it to the invaders except under absolute duress. The settlements in the west, in the Cotentin and the Avranches regions, much further away from the Carolingian centres of interest and in a sector which in the late ninth century was a dependency of the Bretons, seem to have been established in different circumstances by less organized, individual groups. It was only with difficulty that the dukes of Normandy managed to establish their authority in these localities. Here, new territories were spontaneously carved out by the Scandinavians, whereas elsewhere the boundaries of the old *pagi* survived. When the definitive frontier was drawn, in 924 and 933, it followed an artificial line dictated by the somewhat capricious locations of the Carolingian *pagi*. Only one further modification was made when, following a military clash, the Duchy of Passais (the Domfront region), hitherto a possession of Maine, was also incorporated into Normandy. The frontier of Normandy was thus established in a series of phases, but from an early date the Norman territory possessed a clearer psychological and legal identity than virtually any other region in the feudal period.[77]

It was not only around the Seine bay that the Normans established settlements. Elsewhere, here and there, 'abortive Normandies' were set up, for example in Lotharingia, where a considerable proportion of the feudal aristocracy was of Scandinavian origin (the counts of Louvain, Ardenne and Bouillon, and Luxemburg).[78] There was even one other regular concession at Nantes, where a Scandinavian state existed from 919 to 937. But Rollo's was the only state to survive. It is not hard to see why, for it alone lasted long enough to profit from the second wave of Viking expansion, which began in about 980 or 990, after a lull in Scandinavian migrations that had lasted half a century. The reinforcements that continued to arrive until about 1015 afforded Normandy a second wind.[79] Above all though, its survival was due to the greater density of its settlements and also to the attraction that the great waterway of the Seine held for these seafaring invaders, leading as it did to the rich city of Paris, which they looted in 845 and 856 and again besieged in 885–86.

Brittany and Normandy were thus originally similar political structures, both historical provinces which expanded far beyond the settlements of peoples from overseas who founded them. Their political and cultural destinies were very different however. As described by Robert Wace (*Roman de Rou*, 1, 10448–10452), a writer who belonged to the circle formed around the Anglo-Norman monarchy, Normandy was at first quite distinct from France. However, it was not long before it was made subject to the French monarchy, following the victories of Philip Augustus in the early thirteenth century. Brittany, on the other hand, retained its independence until the sixteenth century when, in 1532, it was united with France. In Normandy, the Scandinavian language soon died out. As early as 920 the future Duke Richard I had to go to Bayeux especially to learn Danish, as it was no longer spoken in Rouen. Scandinavian probably never became a written language in Normandy.[80] In Brittany, in contrast, the Breton language has survived down to the twentieth century, producing a wealth of written literature. The early extinction of the language of the invaders in Normandy is not hard to understand, for there appear to have been very few women amongst them (only three women's names of Nordic origin are known as opposed to eighty men's names).[81] Through their indigenous mothers, the children of the mixed marriages that took place learned the Roman language as their mother tongue. In Brittany, the settlement of the immigrants had taken quite a different form. Instead of being composed exclusively of male warriors, the groups that had arrived here had been demographically balanced and had installed themselves peacefully. But deeper geographical reasons for these differences may be adduced simply by looking at the map of France. Normandy is situated in close proximity to the centre of development of the French state, namely the Parisian region. It represented such a threat, especially after the Norman conquest of England (1066) and the constitution of the Anglo-Norman monarchy, that its annexation became vitally necessary to France. Normandy was, furthermore, cleft in two by the Seine valley, which had always been a great artery of trade and communication along which French cultural influences passed. It did not take long for Normandy to absorb those influences. Brittany, in contrast, was a distant peninsula, seemingly at the world's end. And it was certainly then every bit as difficult to diffuse French cultural influences there as it has been ever since. One essential factor that facilitated the deep implantation of the language of the Britons from overseas was certainly the continuing existence of earlier local Celtic substrata at the time of these migrations from Britain. In the Armorican peninsula, the Breton language was superimposed upon the Gallic, which was still very close to the language spoken in the British Isles. We should reject the notion generally accepted in the late nineteenth century and the early twentieth,[82] namely that the Gallic language was totally wiped out in the Armorican peninsula during the Roman period.[83] But linguists have debated at length as to exactly where the language did survive. At first,[84] it was thought that traces of such Gallic survivals were indicated by certain peculiarities in the regional dialect of the Vannes region. But in truth, a better explanation for those peculiarities appears to be a strong Latin influence. The Vannes dialect is 'Britonnic, strongly influenced by the Roman language'.[85] It is easy to see that it would indeed be within the sphere of influence of the important city of Vannes that Latin would make decisive progress during the Gallo-Roman period. Meanwhile, it is equally logical that the Gallic language should survive better in the western parts of the peninsula, just as the Breton language does today. By virtue of its resistance to Romanization, Brittany must have seemed a marginal province even in Roman Gaul. Throughout its later history it has continued to seem just that.

4
The birth of France

The territorial structure from which present-day France emerged goes back to the Treaty of Verdun which, in 843, divided up the Carolingian Empire between the three sons of Louis the Debonaire. After the division of the complex mass represented by the realm of Charlemagne, who had temporarily reunified the greater part of Christian western Europe, the western territory apportioned to Charles the Bald gradually acquired the name of France. This was the territory that, over the years, gave birth to the France of today. Let us now address ourselves to the following questions: firstly, what, if any, were the initial reasons and motives for the division? Secondly, how was it that one of the sectors thus created, to wit the most western one, gradually became the sole bearer of the name of the Franks and, with that name, forged a particular identity for itself? Finally, how did the France created by the treaty of Verdun, a territory which lay far more to the west than it does today, acquire its present frontiers?

The Verdun division

A vertical division

The Treaty of Verdun[1] created three states, those of Charles the Bald to the west, Lothaire in the centre and Louis of Germania to the east. They were separated along more or less north–south axes. The state of Charles the Bald stretched from the North Sea (the mouths of the Escaut) down to the Mediterranean, at a point west of the Rhône, taking in the marches of Spain and Navarre. The state of Lothaire stretched from Italy to Friesland, bounded in the east by the Rhine and in the west by the Escaut, the Meuse, the Saône and the Rhône. The state of Louis of Germania was concentrated to the east of the Rhône but, remarkably enough, also incorporated an enclave to the west of the river, in the Palatinate, which extended as far as the Moselle (fig. 24). What was the rationale behind this arrangement?

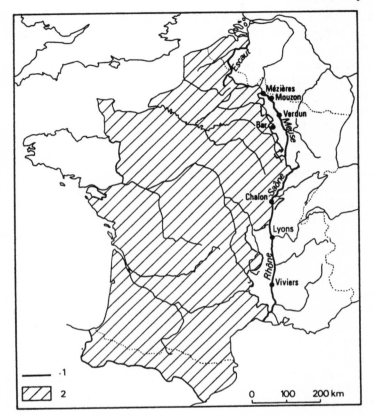

Figure 24. The birth of France at the time of the division of Verdun
1. The 'four rivers'. 2. The territory attributed to Charles the Bald.

Was the aim to achieve a geographical balance?

The interpretation suggested by Roger Dion is most attractive.[2] According to him, the division imposed by the Treaty of Verdun was deliberately designed to allocate to each heir a share that took a fair account of the climatic conditions of the various zones in the Empire and their respective natural resources. There were, at this time, three principal zones that were considered vital to the harmonious economic organization of a territory. The first consisted of the band of coastal alluvial plains along the shores of the North Sea which, between Flanders and the Danish peninsula, comprised rich natural meadowland. Upon these pastures were raised not only the large flocks of sheep which produced the raw material for the incomparable

'Frisian sheets', *pallia fresonica*, such as Charlemagne offered, among other fine gifts, to the Caliph of Baghdad,³ but also – and above all – a powerful breed of horses. By virtue of their strength, these animals, which it was impossible to raise in the south, were unrivalled on the battlefield and also as draught horses, both for agriculture and for transport. It was horses 'such as these, produced on the land of the Franks' (*qualia Francorum gignere terra solet*) that, in 816, Louis the Debonaire presented to Pope Stephen.⁴ The second zone consisted of the band of loessic agricultural plains of hercynian Europe, which were especially well-suited to the production of cereals and food-crops. The third zone was that of the southern regions characterized by the cultivation of the trees and vines which produced the wine and oil indispensable for the celebration of religious cults, and also salt. In 826, the same Louis the Debonaire presented Herold, king of the Danes, with a horse and weapons, fertile land, vineyards and various precious objects necessary for glorifying the deity.⁵ R. Dion emphasizes that, at the time of the division of 843, each of the emperor's sons would naturally have wanted his portion of the territory to include some of the land to the north, where those strong horses were raised, a share of the vineyards and also a share of the southern coastal areas where, around the beginning of the tenth century, the Church of Paris owned olive groves which it valued highly for the oil that they produced for lighting purposes (*arbores oliviferas ad luminaria facienda*).⁶ In the case of Louis of Germania, there is specific evidence to show that in the absence of Mediterranean regions, his portion of the Empire was augmented by an area of land on the left bank of the Rhine, 'because of its great vineyards (*propter vini copiam*)', as we are told by the chronicle of Reginon, Abbot of Prüm.⁷ We know besides that, in contrast to the series of very random divisions relating to no systematic policy (see above, pp. 79–81) that were made during the Merovingian and Carolingian periods, now, for the first time, the task of dividing up the territory was based upon a thorough appraisal of the situation. This clearly represented an important step forward. Some 120 high-ranking 'experts (*omni nobilitate praestantes*⁸)' were gathered in Metz on 1 October 842 to elaborate 'as fair a division of the Empire as possible (*ut...regnum prout aequius possent dividerent*⁹)' and only following their deliberations did the apportioning take place. It was determined by an inventory drawn up in order to establish 'a clear idea (*noticiam ad liquidum*¹⁰)' of the Empire, whereas when earlier divisions had been made less account was taken of 'the fertility and comparability of the shares (*fertilitas aut equa portio regni*)' than of 'the affinities and conveniences of each'.¹¹

Or was it purely a matter of historical chance?

Did considerations of economic and geographic equality really play as large a part as R. Dion would have us believe? It must be admitted that his arguments, though valid, are somewhat slender. Some authors have been more sceptical – Lucien Febvre for instance, who speaks of 'anachronism' in connection with the idea of a deliberate plan of division devised by 'experts'[12] – while others even oppose the idea resolutely. F. L. Ganshof,[13] for example, believes that the essential motivation was still a desire to achieve a fiscal equilibrium between the kingdoms, so as to put all three co-heirs in a position to offer material satisfactions to those whom they wished to recompense for their support. As for the north/south pattern of division, this is regarded as essentially fortuitous. Basically, all that was divided was the northern *regnum Francorum*, for each of the three brothers already held an undisputed personal base further south: Lothaire in Italy, Charles in Aquitaine and Louis in Bavaria. The three shares of the Empire were made up by adding to the territories already *de facto* in possession the northern regions closest to them.

In truth, it is not easy to discern the deeper motives of the 'experts' of Verdun. Such economic considerations as were involved may well not have been as important as R. Dion suggests. On the other hand, they certainly did exist. On this point at least the evidence of the text on Louis of Germania's possessions on the left bank of the Rhine is irrefutable. In any event, the Treaty of Verdun certainly recreated a state with both an Atlantic and a Mediterranean seaboard and organized along general lines henceforward definitely fixed, which was a direct legacy of what had once been Gaul.

France: the name and the image

The name

This state was soon to become 'France'. The name comes from a Germanic people, the Franks, who from the mid-third century on appear in late Latin texts,[14] where they are said to hail from beyond the Rhine. The term probably comes from a root meaning 'brave, bold, courageous' with an overtone of 'wild, ferocious' (see the old Norse *frekkr*). That, at any rate, was Isidore of Seville's interpretation as early as the seventh century: '*a feritate morum nuncupatos* (named thus on account of their savage customs)'; and in the ninth century his words are echoed by Ermold the Black: '*Francus habet nomen de feritate sua* (the Frank is named for his ferocity)'.

Another explanation connects the name with *wrang* (wandering), suggesting men who leave their own land in search of adventure.[15] In any case, the link with *frakka* ('stave, lance'), the term for an old Nordic weapon, seems the reverse of that suggested by certain etymologists, for the weapon probably derives its name from that of the people, rather than *vice versa*.[16] The same goes for the double-edged axe (*francisca*), which was only called this in Spain[17] and whose name only passed into the historiography of Gaul in the

94   *The genesis of France*

eighth century.[18] As for *francus = liber*, 'free', which survives in modern French, this is an adjective that was drawn from the name of the people at a later date (in the Merovingian or Carolingian period). Anyway, the two meanings, 'brave' and 'free', are by no means exclusive, but rather complement one another.

This name was applied as a general ethnonym to a number of peoples or tribes (the *Salluvii*, the *Ripuaires*, the *Chamaves*, the *Sicambres* and so on) from the lower Rhineland, some of which make their appearance at a relatively late date, may or may not have existed, or are somewhat obscure.[19] When and in what circumstances did it become the name of what had formerly been Gaul?

The concept: its territorial evolution[20]

The name *Francia* was first applied to the region of Germania occupied by the Franks in the last centuries of the Roman Empire. As late as the fifth century (in about 475–80), the Ravenna Cosmographer was referring to a *Francia Rinensis* centred on the Rhine, the Meuse and the Moselle.[21] From the mid-sixth century, the name *Francia* was to denote the northern part of Gaul, which had indeed been conquered by the Franks. But from the seventh century on, it also applied to all the countries situated under the political domination or influence of the Franks to the north of the Loire: Belgium, the Rhineland, Franconia and Bavaria. The Merovingian kings never took the title 'King of France' (*Franciae*), but always styled themselves 'King of the Franks' (*Francorum*). At this time, the term *Franci* essentially stood in opposition to the *Romani* of the territory south of the Loire, as did the term *barbari* (which initially meant 'barbarians' but soon acquired the meaning of 'strangers').[22] The assimilation of these two elements began through the name of the people. From the reign of Dagobert (around 630) onwards, *barbari* lost its derogative sense of 'pagan', while concurrently the term *Franci* came to be applied to all the Merovingian kings' subjects, whether or not they were of the Frankish race. Down to the eighth century, however, *Romani* continued to be used, with legal or occasionally religious connotations (for example, to denote Catholics as opposed to Aryans).[23] In any case, at this period *Francia* continued to be set in opposition to Aquitaine and Burgundy, as it also was in the Carolingian period. Now, though, Austrasia and Neustria were superimposed upon the concept of *Francia* (see above, pp. 40–1), which consequently sometimes came to apply only to the oldest region occupied by the Franks, namely the area bounded by the Rhine and the Seine, in between Austrasia to the east, beyond the Rhine, and Neustria, which lay further west. In western texts at least, in the seventh century the tendency was to reserve the name *Francia* for the regions to the west of the Rhine.[24] However, in the eighth century, the expansion of the

Carolingian Empire to include the greater part of the western Christian world once more extended the meaning of *Francia* to take in all the territories within the Franks' domain. Under Charlemagne, *Francia* denoted the entire vast Empire established by the son of Pippin, except for Italy, which remained a separate kingdom organized as a distinct entity.[25]

The ninth-century divisions of the Empire determined the future restrictive evolution of the concept of *Francia*. After the *Regni divisio* of 831 between the sons of Louis the Debonaire, distinctive epithets came into use for the various sectors, first for those carved out in the northern part of the Empire, then, following the treaty of Verdun, for the overall shares allotted respectively to the three brothers: *Francia occidentalis* for Charles the Bald's portion, *Francia media* for Lothaire's, *Francia orientalis* for Louis of Germania's.[26] The unity of Lothaire's kingdom was weak. It soon disintegrated and the name of *Francia media* disappeared. The other two were more enduring. But the peculiar individuality of *Francia occidentalis*, particularly in its cultural aspects, soon became apparent, as is expressed in the alternative epithets that came to be applied to it: the terms *Francia latina, gallica* or *romana* alluded to the predominant language spoken, setting this territory in opposition to Germania. And although writers east of the Rhine and the chroniclers of Lombardy long continued to apply the unqualified name *Francia* to the eastern territory, in the chronicle of the Astronomer of Limoges, written in the late ninth century, a clear opposition is already drawn between France and Germania: *populi Franciae et Germaniae*.[27] At the end of the tenth century, after the death in 911 of Louis the Child, the last Carolingian to reign in the east, his successors from time to time, when they were also the kings of Italy, continued to style themselves *rex Francorum et Longobardorum*, but most were simply *reges* who were customarily distinguished by the name of their respective native duchies: king of the Saxons, king of the Alamans, king of the Franconians. The region between the Rhine and the Elbe, the future *Deutschland*, abandoned the name of *Francia*.[28] As for Lothaire's erstwhile subjects, they had become 'Lorrains'. In the space of less than two centuries, the concept of *Francia* had become meaningful only in the west, in the country which had originally been allotted to Charles the Bald. By 987, when the Capetian dynasty definitively took over from the Carolingians, 'France' was already a meaningful concept in the minds and language of its inhabitants; the 'kingdom of France', as opposed to the Empire and the Christian principalities of Spain, already existed. To be sure, it took some time for the expression to be adopted in written texts generally. Some tenth-century texts already use the expression *Regnum Franciae* alongside *Regnum Francorum*, but the latter was long to remain the formula used in official charters. It was not until the second half of the eleventh century that Philip I, in 1083, dated one of his proclamations to the 'reign of Philip in France (*Philippo regnante in Francia*)'.[29] The expression *Regnum*

Franciae was to be used by the chancellery from Philip Augustus' accession to the throne in 1205 onwards, but not until 1254 was *rex Francorum* officially changed to *rex Franciae*, although the expression had been known ever since 1196. Only very gradually did the kingdom become generally identified as *Francia* by scholarly historians. Hugues de Fleury certainly regarded Charles the Bald as the first *rex Franciae*, but in his *Chronique universelle*, written in about 1100, the term *Francia* is still used in the restricted sense of the territory to the north of the Loire or the Seine, and the political framework envisaged is still Carolingian rather than Capetian. In the twelfth century, *Francia* was still rarely used to denote the kingdom as a whole, for which the expression *Francia tota* was preferred.[30] However, in the vulgar tongue, already in the late tenth century, people must have spoken of 'the king of France' rather than 'the king of the Franks'.[31] And there was already a tendency to refer to the *Franci* of the west using the diminutive *Francisci*, from which the word 'français' is derived (it appears in the *Chanson de Roland*, at the end of the eleventh century, in the form of *'franceis'*).

However, the old accepted meanings of the word, the one more restrictive, the other more expansive, continued to live on. At the time of Hugh Capet, the term *Francia* seems to have applied essentially to the area bounded by the Seine, the Meuse and the Escaut, the ancient heart of the Merovingian dynasty, and the literary texts reflect that restricted interpretation, to which the political vicissitudes of the tenth century had restored significant meaning. Lines 1423–1429 of the Oxford text of the *Chanson de Roland* preserve a passage from an earlier version, in which the name of France applies to the area extending from Besançon to Wissant, a port in the Pas-de-Calais, and from the sanctuary of Saint-Michel-du-Péril (Mont Saint-Michel) *'josqu'as Senz'* (or *Seinz*), identified by Hermann Suchier, followed by Ferdinand Lot, as Xanten (*ad Sanctos*) on the Rhine, but by René Louis as the town of Sens, on the edge of Burgundy. In another passage, at line 2910, which is at odds with the rest of the text, it is Laon, not Aix-la-Chapelle, that is the capital of the Empire. The identification of *Seinz* with Xanten (*Colonia Traiana*), assimilated through false etymology with *Troia* and considered, within the framework of the legend of Troy, as the Franks' place of origin), is now accepted more generally (Ewig, 1982) than it used to be. Nevertheless, the passage clearly reflects traces of a tenth-century tradition according to which Charles the Simple and the last Carolingians reigned over a France that consisted of Neustria and Lorraine, in which Laon was their favourite or enforced residence from 936 to 987.[32] Throughout the Middle Ages, that restricted acceptance of the term continued to be used in popular parlance. In a *chanson de geste* of the early thirteenth century, travellers making their way from Languedoc to Paris, through the Auvergne and Berry regions, are said not to reach France until they have crossed the bridge spanning the Loire at Orléans:

> *A Orliens vindrent, si paserent le pont,*
> *Lors entrerent en France.*[33]
> (They came to Orléans and crossed the bridge,
> At which point they entered France.)

And in the vocabulary of the wine trade of the same period, *'vins de France'* exclusively meant the wines of the central regions of the Seine basin, the Paris region, stretching from Sens and Epernay as far as Pontoise and Mantes, as opposed to the wines of Burgundy and the Berry district.[34] Not until the sixteenth century did the expression come to apply to wines from all over the kingdom. Over the years, the restricted meaning was to become even more restricted. Eventually it came to denote no more than a small area to the north of Paris, the 'Ile de France' (first mentioned as such in 1429), bounded by the Seine, the Oise and its tributary the Thève, and the Marne and its tributary the Beuvronne. The expression was to survive in the toponymy of the rich wheatlands of the *'plaine de France'* on the rough limestone of the lower Eocene, to the north-east of Paris (see figs. 49 and 55).[35] Meanwhile however, the wider meaning of the name also continued to be used extensively, particularly in the *chansons de geste*, fed by the literary tradition of the Carolingian period. It is this meaning that predominates in the Oxford text of the *Chanson de Roland*, produced in the eleventh century, after the appearance of a number of earlier drafts. In this text, the term France is at times also used with its more restricted meaning, as we have seen,[36] but no fewer than 170 times it refers to Charlemagne's Empire in its entirety.[37] At the end of the twelfth century, Jean Bodel (who died in about 1205) began his *Chanson des Saisnes* (= Saxons) with a famous couplet in which he refers to France, Brittany and great Rome as the only three places worth considering:

> *Ne sont que trois matères à nul homme entendant*
> *De France, de Bretagne et de Rome la grant.*

Here 'France', the realm of the *chansons de geste*, is still the Carolingian Empire in its entirety. This wide meaning of the term is equally the only one recognized in Arab geography, where the land of the Franks, *Ifrandja*, corresponds initially to the entire Empire of Charlemagne, then to western Europe as a whole, sometimes including Spain.[38]

Several territorial images of France, some of a more, others of a less extensive nature, thus co-existed throughout the Middle Ages. France means sometimes Charlemagne's Empire in its entirety, sometimes the *Francia occidentalis* created by the Verdun Treaty, sometimes the more restricted realm of the early Capetians, centred – as the power of Clovis' Franks had been – on the Paris Basin, sometimes even the tiny France of the 'Ile de France'. But one of these representations was soon to eclipse the rest. This was *Francia occidentalis*, the France which belonged, if only in theory, to the king: the 'kingdom of France'.

The scholarly and religious aspects of the concept

What is the explanation for the success of *Francia occidentalis*? How was it that it ended up monopolizing a term which had formerly denoted a far vaster entity, the common legacy of all the sons of Charlemagne? One essential factor which developed during this period was the identification of France with *Gallia*.[39] The monks of the time invariably use the terms *Franci* and *Galli* indiscriminately to refer to the inhabitants of the Capetian kingdom. *Galli* is the word used by the Venerable Peter, Abbot of Cluny, writing in 1112 to *basileus* Jean Comnène, to commend to his care all the French in the east: '*pro universis Gallis nostris*'. Between the ninth and the twelfth centuries the terms are used indiscriminately by Richer, a monk of Saint-Gall, and other historians. The identification of France with Gaul extends to the name of the country too. Gauzlin, the Abbot of Fleury-sur-Loire, when setting about building his church, expresses the hope that it will be worthy to serve as a model for the whole of 'Gaul': '*opus tale quod Galliae sit exemplum*'. As early as the eighth and ninth centuries, *Francia* is even identified with *Gallia* by writers in Germany (the Reichenau glossary), England and Italy: perhaps the confusion is more understandable on the part of writers in such distant lands. But what is remarkable is that none of these texts ever identifies *Francia* with *Germania*. The western *Francia*, with its more ancient civilization and its far greater prosperity, is the one normally equated with Gaul, where all the capitals and lines of communication vital to the Frankish confederation were to be found, situated in the territory bounded by the Atlantic Ocean, the Loire and the Rhine.[40] It was the monks, the scholars, who were responsible for the designation and that scholarly interpretation proved decisive in the culture of the Middle Ages, which was essentially shaped and transmitted by those monks. At the time of the first Capetian kings, students from all over France flocked to the Abbey of Fleury-sur-Loire, not far from the former sacred centre of Gaul, where they learned from their teachers that France was the daughter of Gaul, and a most valiant and holy nation.[41]

This idea of France's descent from Gaul bears upon another determining factor, namely religion. Initially, the Franks had no reason to consider themselves heirs to the Gauls. Even when, having been assimilated linguistically, they no longer felt themselves to be Germanic, the origins to which they sought to attach themselves were the Trojan ones celebrated in the legend which appeared in the chronicle of the pseudo-Frégédaire (mid-seventh century) and thereafter enjoyed great popularity. The *Histoire des Francs*, written by the monk Aimoin at Fleury-sur-Loire in about 1000, is strongly influenced by it and it remained largely accepted until the erudition and 'gallomania' of the sixteenth-century humanists rediscovered and rehabilitated their Gallic ancestors.[42] When *Francia* was first identified with Gaul, it was as the defender of Christianity, another idea for which men of the Church were initially responsible. Pope Gregory the Great, writing to Childebert II at the end of the sixth century, was the first to apply the name *Francia* to the whole Merovingian kingdom and it was also he who set the seal upon the primacy of the crown of the Franks, the defenders of Catholicism against Aryanism: 'The crown of the Franks is as much superior to the other crowns of the world as royal dignity surpasses ordinary functions. To reign is of little account, since others than you are also kings, but what bestows upon you a title that no other kings deserve is the fact that you are Catholic.'[43] *Francia* was to be the rampart of Christianity. From this period on, the Church assigned it a special role. In this lay the seed of all subsequent developments. In 878, Hincmar, Archbishop of Rheims, in his *Life of Saint Rémi*, pronounced the baptism of Clovis to be the first act in the founding of the kingdom.[44] The history of the Franks is indissociable from their conversion. That idea was to emerge in all its force in the early years of Capetian France, at the time of the Crusades. With the Germanic Emperor constantly opposing the power of the pope, France was favoured as a base for the Crusades. When Pope Urban II came to preach at Clermont in 1095, to muster support for the Crusades, he addressed himself to 'the beloved French, the elect of God'. It is a theme that surfaces frequently in the literature of the period of the Crusades. Guibert de Nogent, for instance, writes: 'Every time the Popes are faced with serious problems, they come to ask the French for help.'[45] By the late twelfth century, the notion that the faith of the *Franci* surpassed that of all other peoples had become a commonplace. The *gens Francorum* was a *beata gens*.[46] The most prestigious formulation of that belief appeared in Gregory IX's papal Bull of 21 October 1239: 'The son of God, to whose orders the whole world is subjected and who is obeyed by the battalions of the heavenly army, has established here below, as a sign of his divine power, a number of kingdoms diversified by their peoples and their languages. And just as in byegone times the tribe of Judah, among all the sons of the patriarch Jacob, received from on high a very special blessing, so the kingdom of France is above all other peoples, crowned by the hand of God himself with extraordinary prerogatives and favours.'[47] In the thirteenth century, the king of France became 'the most Christian king', and Philip IV, 'the Fair', who clashed violently with the papacy from 1296 onwards, was the first to use that title regularly. It was precisely from that idea of a direct link between God and the French king and French kingdom that he derived the strength to fight the pope both for fiscal reasons and in the name of the independence of the Church of France.[48] In the second half of the fourteenth century, under Pope Paul II, the title figures in the protocol of the pontifical chancellery, where it is reserved exclusively for the king of France.[49] France soon came to be regarded as the Church's eldest daughter and the fourteenth and fifteenth centuries produced a stream of declarations of the superiority of the faith of this most Christian nation:[50] not only were the French a people elect, but this was a sacred line of kings of pure, holy and miraculous blood. A whole theological mythology was elaborated around 'the blood of France'. In about 1376, the author of the *Songe du Verger* (Orchard Dream) wrote: 'All the kings of France possess the power to perform miracles'.[51]

This Christianization of the concept of France, developed around a dynasty of northern origins, gradually expanded to include the entire kingdom. Its progress can be gauged from the historiography and the diffusion of royal cults. In 1317, the geography of the christianization of Gaul, as conceived by Yves de Saint-Denis, was already out of date compared to the real expansion of the territory. Very few of the early foundations attributed to this apostle were situated south of the Loire or in the west; nor were there any in Brittany or to the east of the 'four rivers' which at this time formed the border with the Empire. The picture presented is a very 'Parisian' one.[52] The geography of the cult of Saint Louis (formally canonized in 1317 but widely revered very much earlier) is equally instructive. At first, the area involved was also very Parisian. At the beginning of the fourteenth century, the cult did not extend as far as the Flanders frontier but was limited to the area bounded to the north by Arras and Corbie, in Artois. To the south and the south-east it extended no further than Orléans and Sens, except in south-eastern Languedoc and Provence, where it was implanted by Alphonse de Poitiers, Saint Louis' brother, and by the Angevin dynasty. Shrines existed to the west, in Normandy, where the French influence had been strong ever since the conquest of Philip Augustus, and also in eastern Brittany, where the cult had been introduced by the Franciscans. But vast regions remained unaffected: Burgundy and the whole of south-western France. In the fifteenth century, the cult spread northwards to the frontier with the Empire and, above all, southwards throughout Charles the Dauphin's 'kingdom of Bourges'. The unaffected zones of Guyenne and the Pyrenees, Burgundy, Celtic Brittany and the Dauphiné were shrinking fast.[53] The sanctification of Clovis, the model and archetype of the most Christian king, founder of the holy royal line, conqueror of Gaul and the first king of the Franks and Aquitainians, followed a similar pattern. This cult, in which the religious idealization of France found its culmination, first appeared in the fourteenth century, in the north; but in the fifteenth century it spread throughout the south (reaching, for example, the sanctuaries of Moissac in Languedoc, Saint-Pierre-du-Dorat in lower Marche and Sainte-Marthe-de-Tarascon).[54] By the dawn of the modern period it was believed throughout the territory that France and her kings were specially beloved by God.

The image and its symbols

Not surprisingly, this concept of largely scholarly and religious inspiration soon produced a literary image of a highly laudatory nature. The praises of the land of France were already being sung on every page of all the most ancient major texts in the French language – the *chansons de geste*, for example – even though, as we have seen in the *Chanson de Roland*, the geographical concept of France in its limited form was still generally imprecise. These works celebrate *dulce France*[55] (sweet France: an expression that was to enjoy a lasting success and survive in imaginary representations of the nation right down to modern times), *France la bele*,[56] *France la solue* (= free).[57] Sometimes the praises take the form of an invocation:

> *Tere de France mult estes dulz païs*[58]
> (Land of France, you are a most sweet country.)

This was to be a constant theme throughout the twelfth and thirteenth centuries: '*Quand Dieu fonda cent royaumes, le meilleur fut douce France* (When God established a hundred kingdoms, the best was sweet France)'.[59] This incomparable land is endowed with every quality. It is '*la plus vaillant do mont* (the most splendid in the world)'[60] and is rich and prosperous, *France la garnie*.[61] It abounds in woods, rivers and meadows, maidens and beautiful ladies, fine wines and fearsome knights: '*abonde en bois, en rivières et en prés, en pucelles et en belles dames, en bons vins et en chevaliers redoutés*';[62] and its inhabitants are endowed with every virtue:

> *Quar France est un païs el' quel on doit trouver*
> *Honor et loyauté et tout bien savourer*[63]
> (For France is a country in which one will find
> Honour and loyalty, and all to be enjoyed.)

Elsewhere, '*En nul pays n'a gent plus douce ne plus vraie*[64] (No country has a people more gentle and true)' or warriors more impassioned: '*Cil poples est plus fiers que lupart ne lion*[65] (This people is prouder than leopard or lion)'. Such declarations abound, although sometimes they do seem somewhat stereotyped. One school of critics[66] considers '*la dulce France*' to be an imitation of a Latin *topos* which is frequently found in Virgil[67] but in fact goes right back to Greek tragedy. When Roland is dying and

> *De plusieurs choses a remembrere li prist:*
> ... *De dulce France* ...[68]
> (Falls to remembering many things:
> ... Sweet France ...)

one is tempted to detect a direct echo of the famous line in the *Aeneid*,[69] where the dying Antor's thoughts turn to his faraway native land: '...*et dulcis moriens reminiscitur Argos*'. However, perhaps the poet's Virgilian erudition should not be exaggerated. Equally, however striking it may be to find the Saracens using the familiar expression '*France dulce*'[70] even when attacking their French enemies, it would be mistaken to assume this to be an imitation of the passage in which Aeschylus has the Persians speak of 'the divine land of Greece', for it is, in truth, simply a matter of habit and negligence on the part of the writer. Ramon Menéndez Pidal has skilfully shown that these laudatory formulae are in truth not scholarly *topoi* but simply part of the elegant speech of the day. They constitute manifestations of an 'epic polygenesis'[71] rather than classical imitations. They cater for public taste: the reason for the widespread use of these clichés was simply that the uneducated public which listened to the *chansons de geste* had a liking for them. *La dulce France* was part of their attachment to their country, although it would no doubt be anachronistic to call this 'patriotism', in the modern sense of the word (see below).

Gradually, the image was further enriched. From the second half of the twelfth century on, the *roman courtois* took over the simple, rather obvious idea of the *chansons de geste*, lending it a more subtle refinement. In particular, the notion of culture, or *clergie*, that it introduced was new, as was that of a *translatio studii* from Greece and Rome to France. Chrétien de Troyes thus writes:

> *Ce nos ont nostre livre apris*
> *Que Grèce ot de chevalerie*
> *Le premier los et de clergie*
> *Puis vint chevalerie à Rome*
> *Et de la clergie la some*
> *Qui ore est en France venue*[72]
> (This our books have taught us:
> That Greece had the first renown
> In chivalry and in learning.
> Then came chivalry to Rome,
> And the heyday of learning,
> Which now is come into France.)

The *chansons de geste* had presented a France with somewhat stereotyped attributes; the *Domina Francia* had been a noble and Christian reflection of the clerical and chivalric models of the period. Alongside this there was now juxtaposed the image of a scholarly France, whose prestige was founded upon culture. By the end of the thirteenth century, a new picture had already been elaborated in the prologue to Primat's *Grandes Chroniques*,[73] providing a poetic theme that was to continue to be popular down the ages. Ronsard thus writes:

> *Celle qui t'a nourry et qui t'a fait aprendre*
> *La science et les ars dès ta jeunesse tendre*[74]
> (She who has nurtured you and taught you
> Science and arts from your tenderest youth.)

In the fourteenth and fifteenth centuries, the royal jurists were to add the notion of exceptional institutional excellence. France was the best governed of all countries. Salic Law, now rediscovered and endowed with mythical status,[75] was the best of all systems to determine the succession. The picture was now complete, and in 1558 Du Bellay summed it up in his memorable line: '*France, mère des arts, des armes et des loix*[76] (France, the mother of arts, arms and laws)', setting the ultimate seal on the idealization of France. From the mid-fourteenth century on, all those concrete attributes gradually crystallized around the allegory of France personified. Throughout the miseries of the Hundred Years War, France spoke with a voice of a grieving mother; when victory came, she became a radiant princess. The religious theatre seized upon the theme; the miniaturists determined its typology, based on that of the Virtues.[77] France personified was a woman clad in a robe of white, the colour of royalty; a perfect lady, chaste and inviolate, frequently identified with the Virgin. The Joan of Arc story drew heavily upon this image for its inspiration since, as Michelet was to remark in 1841, in a splendid passage of his work, 'The saviour of France had to be a woman. France was herself a woman.'[78]

As this mental construction was being elaborated, a whole group of concrete symbols soon gathered around it: the royal standard evolved from the flag of the Abbey of Saint-Denis, in the twelfth century;[79] the lily of France, derived from a religious symbol of virginity, entered into the heraldic panoply at about the same time, manifesting the devotion of the royal house of France to Our Lady;[80] alongside these, there evolved all the paraphernalia connected with the ceremony of coronation (the crown of Saint Louis; the holy oil and so on) and a belief in special royal powers, such as an ability to cure scrofula.

As can be seen, religious elements were essential to all this symbolism. The French language, for its part, played a more ambiguous role in its elaboration.[81] As early as the thirteenth century it was, to be sure, credited with the same harmony and beauty as France herself; and it enjoyed considerable prestige throughout Europe. It was 'the most beautiful language in the world',[82] 'the most noble language in the world'.[83] Its importance as a means of international communication was well established and foreigners had to learn it: 'Whoever cannot speak French is considered a barbarian.'[84] The mid-fifteenth century saw the emergence of the idea that the French language must belong to the French kingdom and the claim that Burgundy and Savoy, French-speaking regions with the Empire, must be returned to France.[85] But the full nationalization of the language was a lengthy process, given the multilingual nature of the kingdom in reality. Of all the foreign languages involved, the only one which provoked positive hostility was Flemish, for the region where it was spoken lay the closest to the centre of French political power.[86] In the fourteenth and fifteenth and even the early sixteenth centuries, Latin was essentially the language in which the praises of France were sung.[87] Not until 1549 did the French language really become a fundamental component of the concept of the motherland, for this was the date which saw the appearance of Du Bellay's *Deffence et Illustration de la Langue Françoyse*, the work which associated the mother tongue with the cult of the ancient world and demonstrated its potential in the highest forms of literary expression.

The territorial aspect associated with this image in the fourteenth and fifteenth centuries remained relatively abstract. France was seen as a garden, 'le jardin de France':

> *Le beau jardin de grâce plein*
> *Où Dieu par especiauté*
> *Planta le lys de royauté* ...
> *Tel jardin fut à bon jour né*
> *C'est le jardin de douce France*[88]
> (The fine garden full of grace
> Where God especially
> Planted the lily of royalty...
> One happy day such a garden was born
> It is the garden of sweet France.)

Initially, this owed much to the Garden of Eden and the Christian tradition of Paradise. But in the fifteenth century it acquired the specific, secular character of an agricultural land. With all her aesthetic attributes and moral allegories, France was also a land of fertility and abundance, cultivated by the king himself. This 'garden' was imagined as a definite territory, shut off from the outside world, landlocked, despite the fact that in reality the kingdom already possessed extensive maritime frontiers. It was an ideal, unique of its kind, for at this period the city-states of Flanders and Italy were represented as urban landscapes, bristling with ramparts and towers. France, in contrast, was the perfect land, ordered, enclosed and defined,[89] the concept of which lives on even today in the notion that France is a '*pré carré*' (square meadow) or the myth that it is a hexagon.[90]

The emergence of a sense of nationhood

Abstract and vague though it still was, France was nevertheless a land worth defending. Over these formative years, a sense of nationhood undeniably developed. Not that there was a word for it initially. In the Middle Ages 'nation' was applied to many kinds of groups. The expression 'the French nation' was seldom used before the late fifteenth century and 'national sentiment' was not widely used until the eighteenth century, while the word 'nationalism' made its appearance only in 1812. As for '*patrie*', the word was unknown in French until about 1530–40 and was for a long time used only by the humanists. 'Patriots' first appeared in 1568 but was not entered in the *Dictionnaire de l'Académie* until 1762 and the word 'patriotism' did not pass into common parlance until the end of the eighteenth century. In Latin, however, the word *patria*, which had fallen out of use in the early Middle Ages and had, in any case, only ever had the sense of 'small native locality', came back into use at the end of the twelfth century, when it acquired the meaning that it has today.[91]

This clearly marked a turning point. In the ninth and tenth centuries the only politico-geographic units that evoked feelings of attachment in their inhabitants were the *regna*, separate fragments of the Carolingian Empire or of its subdivisions. When Charles the Bold was crowned king of the Lorrainese, Hincmar declared that the sovereign should be crowned in each of the kingdoms of which he took control. The Franks often came to bloody blows with the Aquitainians and the Gascons, whom they despised and who reciprocated with similar sentiments towards them. In the tenth and eleventh centuries, nobody regarded the Aquitainians as French (Werner, 1970, pp. 292–4). Nor do the *chansons de geste* contain any evidence of the existence of truly national sentiments, despite Léon Gautier's eloquent claims to the contrary. The 'France' which inspired affection was simply the land of one's birth, the familiar countryside; it was not yet a nation. It was a mythical land of the days of Charlemagne, but not yet a kingdom.[92] In the twelfth century, new ideas appeared, in keeping with the political organization of this period. As early as the beginning of the twelfth century, Guibert de Nogent, who was very advanced for his day, speaking of the Norman prince Bohémond, who was powerful in southern Italy, considered him to be a '*Francus*', since he came from Normandy '*quam Franciae partem esse constat* (which is clearly a part of France)'. He also referred to the Crusade as *Gesta Dei per Francos*, an expression destined to win great popularity. In the middle of the century, the Abbot of Cluny reminded the king '*est et Burgundia de regno vestro* (Burgundy too is a part of your kingdom)' (cited from Werner, 1970, pp. 294–5). From this point on, the *regna* progressively lost their importance. The first concrete manifestation of a sense of nationality was possibly the unanimous and vigorous response elicited by Louis the Fat's 1124 appeal for solidarity in the face of the coalition formed by Emperor Henry V and the king of England. At any rate, the concept of loyalty to the kingdom of France, seen as a political entity, is certainly present in the mid-twelfth century, in the work of Suger, the great statesman and historian of this period. He calls France '*notre terre* (our land)' and speaks of the need to 'defend the kingdom (*regnum defendere*)'. That expression was also frequently to be used in the reign of Philip Augustus, at the end of the century.[93] It is also in the work of Suger, as well as in official documents of the reign of Louis VII, that we find the new notion of 'the crown'. Jean Bodel was to develop this eloquently in his *Chanson des Saxons* (I, 2) at the end of the century:

The genesis of France

> La corone de France doit estre mise avant
> Quar tuit autre roi doivent estre à lui apandant
> De la loi crestiene qui au Deu sont creant[94]
> (The crown of France must be put first
> For all other kings must be dependent upon it
> By the Christian law of those who believe in God.)

At this point, a 'transfer of affections'[95] to the king of France seems to have taken place and during the thirteenth century the practice of offering up prayers for him seems to have become general.[96] Bouvines Day (23 July 1214), the fame of which became quite widespread, was considered – with reason – to be a crucial date in the development of French feeling for France.[97] It is true that the battle hardly rated a mention south of the Loire[98] and that national sentiment was still particularly a feature of the Paris Basin; but it was spreading rapidly and by the thirteenth century was occasionally manifesting itself against the sovereign himself, when certain territorial vicissitudes arose as a consequence of feudal rights. In 1259, when Saint Louis, prompted by scruples of conscience, restored Aquitaine to the Plantagenets, his action provoked in that region an astonishment that was also expressed by Joinville.[99] In 1274 Primat, translating the *Grandes Chroniques de France* into French, introduced the expression 'foreign nations', which was not in the original Latin, in order to set the French nation apart. Meanwhile, jurists were resurrecting the ancient idea of the Roman *patria*, which included within it all smaller local versions of the concept. In 1270, Jacques de Révigny, of Orléans, wrote as follows: '*Quia Roma est communis patria, sic corona regni est communis patria, quia caput* (Just as Rome is the common country of all Romans, so the crown of the kingdom is the country as a whole, for it is its head)'. In 1318, the expression *natione Gallicus* was used, introducing the earliest concept of a 'French citizen'. At this time a distinction was already made between 'natural' members of the kingdom and foreigners, the former being referred to as '*bourgeois de royaume*' (1312) or '*regnicole*' (1324). The idea of a French nation had first emerged among the monks of Saint-Denis, in the twelfth century, and had been elaborated within certain restricted circles familiar with historical, political and religious concepts. But in the thirteenth century it reached the mass of the population and was now related to the geographical framework of the kingdom. To borrow the felicitous expression of Bernard Guenée (1967), to whom we are indebted for collecting these texts, 'the State created the nation'. From this time on, more and more references to the French nation crop up, peaking in the fourteenth and fifteenth centuries during the Hundred Years War which, with its conquests of Normandy and southern France, definitively completed the process of creating a nation. After the Peace of Brétigny (1360), the towns and provincial lords in the south-west, handed over to the English despite their being francophone at the time and entirely committed to the culture of France, found it hard to accept their lot: '*Nous avouerons les Anglais des lèvres, mais le coeur ne s'en mouvera ja* (We will pay lip service to the English, but our hearts will not be in it)', the community leaders of La Rochelle declared.[100] In his *Quadriloge invectif* (1422), Alain Chartier has a personified France, clad in a soiled, tattered gown, entreat the three orders not to allow her to perish: 'Nature above all commits you to the common salvation of the land of your birth and the defence of this realm under which God has had you born and live.'[101] And the declaration of the humanist, nurtured by the virtues of antiquity, is matched by the words of a common man: 'Must the king be ejected from his kingdom and must we be English?', exclaims an ordinary soldier in 1428, at a time when many a great lord became a turncoat.[102] The whole story of Joan of Arc should be understood in the light of this intense national feeling. Charles of Orléans, a prisoner in England from 1415 on, expresses a similar simple loyalty along with the nostalgia of an exile (Ballad LXXXIX):

> *En regardant vers le païs de France,*
> *Ung jour m'avint, à Dovre sur la mer,*
> .....
> *De veoir France, que mon cueur amer doit.*
> (While gazing towards the land of France
> One day, at Dover on the sea, I happened
> .....
> To catch a glimpse of France, which my heart is bound to love.)

And after the wretched years of war, François Villon's heartrending refrain condemns anyone who would wish ill to France, '*qui mal vouldroit au royaulme de France*'.[103]

From now on, men would lay down their lives for their country. In the thirteenth century, to die in battle was considered a misfortune rather than a glorious fate. Only a crusader could be sure of going to heaven if he died in battle. At this time, the *amor patriae* of Parisian University circles was still simply a form of love for God, a Christian love of the common weal. But from the fifteenth century on, starting in the kingdom of Bourges, it became generally accepted that to die for one's country was one's duty, a sacrifice still deeply Christian but now, increasingly, also of a political nature. 'No death suffered for the State is shameful or wretched', declares the *Assertio Normaniae*, noting the death of the Bastard of Vaurus, an old trooper who was certainly no model knight, hanged by the English in 1422. And the epitaph of Jean de Bueil, who died in 1477, runs as follows:

> *... occis à la grant guerre,*
> *En bataillant pour la France et pour vous*
> (... killed in the great war,
> Whilst fighting for France and for you.)

108   *The genesis of France*

This is not so far from the '*mort pour la France* (died for France)', which is to be seen in today's cemeteries all over the country.[104] In the closing centuries of the Middle Ages, the territorial state of France became emotionally associated with the national culture. This was a totally new development, later to be reproduced in England and long to remain a phenomenon limited to western Europe.[105]

The frontiers of France

The country which thus at an early date took shape in the minds and hearts of its inhabitants had yet to acquire its frontiers. It took a thousand years to do so.

The frontier of the 'four rivers'

At a time when there were no such things as accurate maps, the eastern edge of Charles the Bald's new state had essentially been determined, for convenience, by the courses of four rivers which drew a clear and ineffaceable line across the land. It was known as the frontier of the 'four rivers': the Escaut, the Meuse, the Saône and the Rhône. In the south, it followed the courses of the Rhône and the Saône from the Mediterranean as far as a point situated approximately at the latitude of Dijon; it then ran along the Meuse from its source to the gorges of the Ardennes, to the north of Mézières, where it joined the Escaut, which it proceeded to follow from the upper reaches right down to the estuary. In the east and the north-east, this was by and large the frontier that continued to separate France from the Empire until the sixteenth century.[106]

In truth, the line thus traced provided no more than a rough basis, the very concept of the 'four rivers' being an approximation, a simplification produced by the chroniclers, in particular by Prudence de Troyes where the Meuse was concerned. Even at the time of the Verdun division of the territory, important adjustments were made to the boundary constituted by the rivers. The Cambrésis region in the north, the left bank of the Meuse between Mézières and Neufchâteau, and the entire right bank of the Rhône to the south of Mâcon had been allotted to Lothaire (fig. 24). On the other hand, the division of Lotharingia at Meersen between Louis of Germania and Charles the Bald (in 870) considerably extended the possessions of the latter to the east of the four rivers. However, soon the frontier shifted, as the balance of forces did. The first changes of the feudal period were made at the expense of the king of France. When the Capetian dynasty came to power, in 987, Cambrai on the Escaut was a fiefdom of the Empire, as were Mézières and Sedan on the Meuse. Further south, the border was fixed to the west of the river, at the Argonne forest and the marshes of Champagne Humide, leaving Verdun and Bar-le-Duc within the Empire. Although the line of the Saône was in the main respected, to the south France lost more territory. Lyons and its dependencies and the Forez and Vivarais regions became part of the kingdom of Arles, that is to say possessions of the Empire. Not until the reign of Philip the Fair was the kingdom's border re-extended, at many points as far as the frontier formed by the four rivers and in some places even beyond it. It was only in this period that the Rhône frontier was established, when the Bishop of Viviers and the Archbishop of Lyons, in 1305 and 1307 respectively, recognized the sovereignty of the king of France over the Vivarais and Lyons regions. It was at this point that it became generally customary to refer to the right bank of the river as *Riau* (kingdom) and the left bank as *Empi* (empire).[107]

The absence of any idea of a natural frontier in the Middle Ages and at the beginning of the modern period

For a long time the French monarchy, for all its solid and instinctive sense of landownership,[108] did not evolve the concept of a precise territorial frontier based upon the accidents of natural geography. The sovereign's sense of ownership was at this period essentially personal and familial. His territory varied according to marriages, inheritances and partitions, all of which brought in their train a series of diverse and fluctuating divisions.[109] The very term 'frontier' emerged relatively late. Until the end of the thirteenth century, the much vaguer expression used was 'marches', which denoted a military area, a peripheral war zone under the authority of a particular leader, rather than a precise border. That was a concept which was to survive even when a string of fortified castles was built along the border, creating a concrete frontier. The term 'frontier' only appeared in 1315,[110] when it was used for the border with Flanders. Then, up until the end of the seventeenth century, it was extended to apply to just one other, the one most under threat, namely the maritime frontier with England; but not to any others. Many years were to pass before it was also used to denote the inland frontier with the Empire. Furthermore, the limits established at the time of France's expansion in the Middle Ages by no means corresponded with any natural frontiers. The idea was to recreate the Empire of Charlemagne. That was precisely what preoccupied Philip Augustus, for example, who is recorded as saying: 'I wonder whether God will grant me or my heirs the favour of raising France to the level that it occupied at the time of Charlemagne.'[111] At the height of nineteenth-century Romanticism, Alexandre Dumas produced a magnificent literary rendering of that idea, which even in the fifteenth century was still not totally anachronistic, ascribing the words to Charles VII at the time of the kingdom of Bourges:

> ... *Eh bien, voici mes lois:*
> *La France de Philippe Auguste et de Valois*
> *N'est point mienne: il me faut celle dont Charlemagne*
> *A tracé la limite au sein de l'Allemagne,*
> *Quand ce géant touchait, en maître souverain,*
> *D'une main l'Océan et de l'autre le Rhin.*
> *Or que ma volonté, messeigneurs, soit la vôtre,*
> *Car c'est ma France à moi; je n'en connais point d'autre*[112]
> (... Well then, here are my laws:
> The France of Philip Augustus and the Valois
> Is not mine: I need the France of Charlemagne,
> Who traced out its border through the heart of Germany
> When, giant that he was, as sovereign lord
> He touched the Ocean with one hand, the Rhine with the other.
> Now let my desire, my lords, be also yours,
> For that is my France; I recognize no other.)

Conversely, the Alps and the Pyrenees were by no means regarded as inevitable frontiers. The history of the 'Spanish marches' is significant in this respect. The year 817 saw the creation of a Carolingian military province which encompassed the whole of the coast from the Rhône to the Ebro: this included Septimania (the Languedoc plain) and the Spanish marches (which were separate from the Toulouse marches). In 865, this territory was cut into two, with the line of division fixed at the pass of Salses, allowing Cerdagne and Roussillon to fall into the orbit of Barcelona. After the decisive battle of Muret (1213) had ruined Aragon's designs to the north of the Pyrenees, Louis IX, inspired by feudalistic sentiments which were natural enough at this period, had the border represented by the pass of Salses established by the Treaty of Corbeil (1258) as the frontier of the French state. When Louis XI, between 1463 and 1473, and then again from 1475 onwards, reoccupied the counties of Cerdagne and Roussillon, this was not so much in order to perfect the '*pré carré*' but rather a first step towards extending the influence of France throughout Catalonia.[113] At this point the territory occupied by the French extended much further than the southern slopes of the Pyrenees to incorporate Puigcerda, well beyond the border that was to be established in 1659 by the definitive treaty of annexation. Many similar examples could be mentioned. The two sides of the Pyrenees between Navarre and Béarn were also inextricably linked for many years. Yet, even in the feudal period, these mountain ridges, by virtue of their importance as defensive barriers, were already playing a crucial role in the establishment of the limits of political zones of influence. Efforts at expansion were mainly concentrated upon establishing 'bridgeheads' across the outer edges of these mountain ranges.[114] But distant forays beyond such barriers were constantly being planned: in the fifteenth and sixteenth centuries, at the time of the Italian Wars, the ambitions of the French sovereigns were in the main directed beyond the Alps. At this period the idea of the Rhine constituting a frontier to France was still totally alien to the policies of the kings of France. Contrary to what was once believed, it seems to have played no part in determining Henri II's expedition of 1552, which resulted in the conquest of the three bishoprics of Metz, Toul and Verdun. Commenting upon that expedition, Gaston Zeller writes as follows: 'Seen from a distance, it seems rather to have given birth to that idea, not so much at the time itself, but through the consequences that ensued.'[115] He goes on to explain that so long as the French monarchy continued to nurse the hope of gaining a foothold in the Po Valley, it could not be expected to envisage the Rhine as the limit of French expansion towards the north-east. To have abided by the frontiers of Gaul would have meant rejecting the notion of crossing the Alps.[116]

In the first half of the seventeenth century nothing had changed in this respect. There has been much discussion as to whether Richelieu's policies were inspired by the idea of establishing natural boundaries,[117] but it now seems that they were not. Richelieu, followed by Mazarin, was above all concerned to establish France within the territory of the Empire, thereby acquiring gates of access to neighbouring states. Even at the Treaty of Westphalia, French diplomats were still demanding that France become a member of the Holy Empire and that its king simply take over the rights of the Habsburgs in certain parts of Alsace, without however stipulating that these areas become part of the kingdom of France. It was the Empire's ambassador who, to thwart the manoeuvre, reminded the French that the Rhine had been destined by nature to separate France from Germania,[118] a point that is not without a certain irony in view of later events. In 1667, Antoine Aubery was still defending the medieval concept of a French Empire, in his dissertation significantly entitled *Des justes prétentions du Roi sur l'Empire*. Gaston Zeller, the historian responsible for most of the present demonstration, has also claimed, albeit perhaps with exaggeration, that France continued to nurse aspirations in that direction until a much later date. He suggests that, even in 1792, which was such a decisive turning point in the French Revolution, it was still mainly public opinion that favoured the idea of a Rhine frontier, while those in power remained reticent on that score.[119] Not until the last years of the eighteenth century do the leaders of France appear to have deliberately adopted the policy of acquiring natural frontiers.

The natural frontiers: the concept and the realization

In fact though, alongside this absence of any concrete interest on the part of the sovereigns of France, there had long existed a whole current of ideas which did favour a France bounded by natural frontiers. They made their appearance within circles of urban bourgeois, jurists and men of letters, long before they inspired any royal policies, for these were both more, and less, ambitious. Such ideas were expressed by the loyal inhabitants of Lyons, anxious as they were to justify the annexation of Burgundian territory after the death of Charles the Bold in 1477: 'The King wishes and has always wished to uphold and maintain that the kingdom should extend on the one hand as far as the Alps, where the land of Savoy is enclosed, on the other as far as the Rhine, where the kingdom of Burgundy is enclosed'.[120] The theme of natural frontiers became a literary tradition. Just as the erudite identification of France with Gaul exerted a determining influence on the shaping of national feeling (see above, pp. 98–9), it also presented a geographical model which implied a return to the ancient frontiers. Even at the time when the monarchy's idea of its territory was still that of a heterogeneous patrimony, the collective consciousness already harboured the concept of an ideal France bounded by the seas, the mountains and the Rhine. In his *Louange de la France* (1550), Joachim du Bellay makes no mention of the Rhine, referring as yet only to the mountains and the seas:

> *Comme une grand' coquille creuse,*
> *Qui s'eleve devers ses bords,*
> *D'une double mer fait ses ports*
> *Une province plantureuse.*
> *Ses flancs superbement bornez*
> *Sont doublement environnez*
> *Des Alpes, et des Pyrennees*
> *D'Europe ...*
> (As for a great hollowed shell,
> Curved upward at the edges,
> Two seas provide the ports
> Of this fertile province.
> Its proudly bounded sides
> Are doubly surrounded
> By the Alps, and the Pyrenees
> Of Europe ...)

However, the Rhine does appear in the *Discours au Roy* (= Epistre I), composed by Mathurin Régnier between 1601 and 1608:

> *La mer aux deux côtés cet ouvrage bordoit:*
> *L'Alpes de la main gauche en biais s'espandoit,*
> *Du Rhein jusqu'en Provence; et le mont qui partage*
> *D'avecques l'espagnol le françois héritage,*
> *De Leucate à Bayonne en cornes se haussant,*
> *Monstroit son front pointu de neiges blanchissant.*[121]
> (The sea bounded this creation on two sides:
> The Alps stretched out sideways from the left,
> From the Rhine to Provence; and the mountain which divides

The French heritage from the Spanish,
Rising in peaks from Leucate to Bayonne,
Displayed its jagged brow, white with snow.)

At the beginning of the seventeenth century, the geography text books in use in the colleges run by the Society of Jesus established a close and permanent link between ancient Gaul and France and their respective boundaries. In his *Géographie royale* (1633), Father Philippe Labbe made no attempt to conceal that his aim was to restore Gaul-France 'to its illustrious and initial beauty', with all its ancient frontiers: 'Gaul borders on Spain, being separated only by a ridge of very high mountains. Its ancient borders on the side where the sun rises were the River Rhine, part of the Alps and the little River Var; to the South, the Mediterranean Sea and the Pyrenees Mountains; to the side of the setting sun, the great Ocean, which was called the Gulf of the Aquitaines; and to the North, the Channel or arm of Sea which flows between France and England as far as the mouths of the Rhine.'[122] A France bounded by natural boundaries may not have figured in the imaginary representations of its kings, but it certainly existed in the work of its poets and the teaching of its Jesuits.

Only gradually did these ideas come to affect practical politics. At the beginning of the seventeenth century, when arguments more forceful than the mere rights of inheritance were sought to justify the annexation of new territories, the argument of a common language emerged. That was the argument used by Henri IV in 1602, to justify the reunion of the Bresse and the Bugey regions: 'He told the inhabitants of these two provinces: since your natural tongue is French, it was reasonable that you should be subjects of the King of France. I grant that the Spanish language should remain in Spain, the German one in Germany, but the French language must be mine.'[123] That was also the idea to which Louis XIV appealed in connection with the conquest of Franche-Comté, although its annexation in 1678 could equally be seen as the remarkably successful outcome of a policy designed to acquire natural boundaries.[124]

It was the wars of Louis XIV that introduced the idea of natural boundaries into French politics, for they now came to be regarded as strategic necessities. The basic idea was that of frontiers which would facilitate the defence of the country since they constituted natural barriers and afforded advantageous military positions. It was under the reign of Louis XIV that the notion that the Pyrenees, the Alps and the Rhine represented the boundaries necessary to ensure the country's security acquired an axiomatic force. The first step in the realization of that ideal came with the Treaty of the Pyrenees in 1659. The second tangible manifestation was the annexation of Franche-Comté in 1678, at which point the Treaty of Nijmegen established the frontier along the high Jura range of mountains. Finally, with the Treaty of Ryswick (1697), France gave up the outposts that Richelieu had established on the right bank of the Rhine (Fort-Louis, Fribourg, Brisach and the bridgeheads of Huningue and Kehl) and the Alsace frontier was established along the river. This was given concrete expression in 1698, when a 'New Brisach' (Neuf-Brisach) was built four kilometres to the west of the Rhine.[125] Throughout the eighteenth century, the idea of natural frontiers, now translated into reality, gained in force, particularly under the influence of men such as d'Anville (1697–1782), first geographer to the king, who drew upon the scholarly tradition of the ancient frontiers of France to support the military policies of the day. Now promoted to the status of a theoretical doctrine, it could be used to justify a policy of expansion beyond the limits suggested by the sharing of a common language. That was precisely the line adopted by Danton in the famous speech which he addressed to the Convention on 31 January 1793, proposing the reunification of Belgium and France. 'It is in vain that people seek to make us fear to expand the Republic too much. Its limits are marked out by nature. We shall reach them at all four corners of the horizon, at the Rhine, at the Ocean, at the Pyrenees and at the Alps. Those are the boundaries of France; no human power can persuade us to overstep them.'[126]

Those ambitions have been virtually totally realized along the mountainous edges of France ever since the definitive reunification of Savoy and the comté of Nice in 1860, although in detailed ways the line followed by these frontiers has been affected by complex historical circumstances and a variety of human situations which account for local irregularities.[127]

The frontier of the Alps, the most recent to be established, between France and Switzerland and, more importantly, between France and Italy, finely balanced atop the ridges and peaks, is a triumph of nature. Mont Blanc, situated entirely in French territory, is one of its few anomalies. Another is the plateau of Mont-Cenis, which the agreements of 1860 placed entirely in Italy, along with the alpine pastures of Lanslebourg, on both sides of the peak. A similar situation obtains at the Little Saint-Bernard Pass, because of the hospice there, which belongs to the Piedmontese order of Saint Maurice and Saint Lazarus. In the comté of Nice, the communes of Tende and Briga and parts of the communes of Saint-Martin-Vésubie, Saint-Sauveur and Isola, all situated on the French side, remained Italian in 1860, on the pretext of royal hunting rights which King Victor-Emmanuel possessed there and should retain. This situation was only rectified after the Second World War. But these are small anomalies.

The frontier constituted by the Pyrenees was established in 1659 and was described in 1795 as 'the crest of the mountains which form the watersheds for the rivers of France and Spain'.[128] A few anomalies remain here too, however, the principal one being Spain's possession of the Val d'Aran, an upper valley of the Garonne, which flows towards France, but also had been a dependency of the Aragonese monarchy ever since the Middle Ages.[129] Other remarkable archaisms also survive. Treaties pertaining to *lies et passeries* (places and rights of way), which regulate the amicable common use of the mountain pastures by the communities from both sides and the traditional dues to be paid by them, have been confirmed by diplomatic agreements. By recognizing these pasturing rights and economic relations, they reflect the ancient unity of these mountain regions before the establishment of the state boundaries.[130]

The frontier constituted by the upper Jura range, in a hilly area of gentler slopes and greater domestication, is by far the most complicated of these mountain borders. The line drawn by the mountain ridge is only roughly followed by the frontier, which reflects the balance of sovereignty more or less achieved over the centuries between the communities and inhabitants of both sides. Some sectors of the border certainly do depend totally upon accidents of nature, one such being the fifty-kilometre-long stretch constituted by the gorge of the upper Doubs; but others were drawn in an arbitrary fashion across cultivated fields and even through villages and, up until the mid-nineteenth century, modifications dictated by personal interests on one side or the other were constantly being made.[131]

An artificial boundary: the northern and north-eastern frontier

The Rhine boundary between northern Alsace and the sea has not been achieved. Here, the established frontier reflects a strategic balance and is the product of the victories and annexations which Louis XIV won at the expense of the Spanish Netherlands at the end of the seventeenth century. The present line drawn by the frontier was not determined immediately.[132] It evolved over a century and a half and was finally established by the readjustments made by the Treaty of Courtrai (1820). Initially, under Louis XIV, this was a wartime frontier where capital importance was attached to the military value of the towns won or defended. The limit of the French advance was marked by a series of fortifications, separated by many areas of varying dimensions which were not yet annexed. Such was the 'muddled' frontier of the Treaty of Aix-la-Chapelle (1668). It was not until the Treaty of Nijmegen (1678) that the idea of a regulated boundary emerged and a preliminary exchange of enclaves took place. The realization of an uninterrupted boundary line thus progressed as treaty followed treaty, but such a frontier was not completely established even with the Treaty of Utrecht (1713), after which no more major changes were made for a time and France continued to own three large enclaves situated within the imperial territory, namely Barbançon, Mariembourg and Philippeville, anomalies which did not disappear until the treaties of 1815. With that exception, the series of conferences on national boundaries held in the eighteenth century successfully resolved the outstanding minor irregularities: a major achievement of detailed negotiation. The war frontier was replaced by a peace-time frontier, the lives of the communities involved were set upon a more rational basis and the tax-evasion and smuggling fostered by the indefinite nature of the boundaries were curtailed. The purpose of the countless frontier rectifications which took place during this period was 'to enclose the State, so far as the nature of the country allowed, so as to defend it as well as possible in times of war and in times of peace and to guard against desertion and the defrauding of the king's rights and also to leave no hope of escape to miscreants who sometimes only indulge in the worst excesses when they perceive an assured place of asylum before them, as happens when territories are confused'.[133] In 1788, on the eve of the Revolution, the task was almost completed. Without substantially altering a purely conventional boundary, the Age of Enlightenment had, by dint of slow, pragmatic efforts at compromise and definition, given that frontier an effective reality both on the ground and in people's lives.

Once drawn, that arbitrary line produced geographic effects of considerable importance.[134] It established the pattern of military policies now out of date but still operative up until 1914; it gave rise to a customs system of exceptional density (almost half of France's customs posts are located in the Nord department), thanks to which the protectionist France of the nineteenth century was able to build up a powerful sector of industry, encouraged by the general national market that these secure frontiers guaranteed; finally, it prompted the development of a new system of communications running parallel to the frontier and designed to serve the needs of industry. The courses of the rivers which flowed northwards naturally encouraged the development of commercial outlets at the mouths of the Escaut and the Meuse. However, to maintain circulation within the national territory, it became necessary to construct a number of internal communication routes, in the first place canals. The Escaut canal to Dunkirk and Calais initially took the form of lengths of canal linking the stretches of river which flowed in the right direction (one linking the Scarpe with the Deule, 1683; one between the upper and the lower Deule, passing through Lille, 1750; and the Neuf-Fossé canal between the Lys and the Aa, completed shortly before 1775); but in the nineteenth century this was converted into a continuous waterway along a more direct line. Next came railways, chief amongst them the important track linking the coalfields of the north with the minerals of the Lorraine basin. Elsewhere, the Saint-Quentin canal, linked to the upper Escaut by way of the Riqueval tunnel, the technical marvel of the age, made it possible, as early as 1810, to carry coal from the north to Paris. All in all, this accidental frontier, which evolved in haphazard fashion in the course of a series of wars, powerfully reshaped the countryside and the life of the great northern plain.

Conclusion to Part I

France emerged from the Treaty of Verdun. This kept most of the territories of ancient Gaul welded together. At this time, western Francia did not extend as far eastwards as it does today. The monarchy battled for centuries to establish more satisfactory frontiers in the east.

The apparent stability of France's territorial limits ever since antiquity is somewhat misleading, however, for Rome had arrested a situation which was in the full process of evolution. When the Empire collapsed, other political ways of organizing the territory were both possible and tried out. Some took a limited, provincial form; and those provinces might well have developed into separate states, as Flanders indeed did. A similar pattern might well have evolved in Brittany, Lorraine and Franche-Comté, as Lucien Febvre has skilfully demonstrated (1912). Political units founded upon mastery of the seas were equally viable: straddling the English Channel, one such state more or less survived for four centuries, from William the Conqueror to the end of the Hundred Years War – five centuries even, considering that Calais was not lost until the middle of the sixteenth century. Similarly, in the south, the counts of Barcelona attempted to extend their rule over Languedoc and Provence. The process was interrupted by the Albigensian Crusades, but traces of it still exist.

It is certainly no mere chance that the state which eventually prevailed was positioned squarely on the continent: states whose internal relations depend upon maritime communications are often vulnerable in times of crisis. In this land of France, the languages spoken were mostly closely related, a factor which favoured unification. But it had certainly not always been inevitable that France should be constituted within the boundaries recognized today, nor that Paris should be its national capital: Lyons was the first to assume that role and it retained its economic pre-eminence until the end of the sixteenth century.

The French territory made it possible to construct a solid society and a solid economy. It was a solution that presented certain advantages, but it was not the only possible one. Its success testifies to the intelligence with which its rulers exploited similar cultural foundations and the natural articulations of the area, as they constructed a strongly structured nation and state.

PART TWO

The traditional organization of the territory of France

While France was gradually emerging as a living territorial construction, after the Treaty of Verdun, systems were being set in place for the spatial organization and occupation of the land, systems which were by and large to remain stable until the Industrial Revolution of the nineteenth century and which left many traces that are still perceptible today.

This 'traditional' organization of the territory comprised two fundamental levels of articulation. First, there were major, homogeneous divisions: major contrasts in crops, landscapes and social development, great dissecting forces which produced cracks across the entire edifice of France, dividing it into a number of vast, coherent sectors. At a lower level, there were smaller, regional units, areas of close communications, mostly constituted on the basis of complementary natural elements interacting to a greater or lesser degree within the framework of feudal political institutions and dominated by the urban centres to which they were linked.

5
The major divisions

The great cultural divisions

In the Roman period, the major cultural influences upon Gaul originated in the south. The Germanic invasions repulsed those influences, partially at least, and this was the origin of a whole series of fundamental divisions, the essential effect of which was to set up an opposition between the north and the south.

Linguistic zones

The medieval genesis of the written languages
In the period when the concept of France was emerging, between the ninth and the eleventh centuries, around the year 1000, no coherently distinctive language existed in western Europe apart from Latin, which remained virtually the only written language and also the spoken one for men of the Church. However, it was no longer spoken by the rest of the population, nor had been for many years. A whole crop of vernacular dialects had grown out of the ruins of spoken Latin, from which they were all derived, but as yet none had developed any coherent, distinctive personality of its own.

Three centuries later, at the end of the thirteenth century, the situation was quite different. Little by little, from the eleventh century on, a number of those dialects crystallized and, as more and more texts came to be written in these vulgar tongues, they gradually acquired the status of literary languages, each with its own distinctive characteristics.[1] At this point, it is possible to make out a rough linguistic map, the main lines of which remain unchanged today, or at least did until the recent eclipse of local languages. Of course, we should not credit the boundaries traced on such maps with more validity than they deserve. No firm boundaries separated one dialect from another. A host of distinctive linguistic features developed gradually and unevenly in different areas. One language did not take over neatly from

The linguistic map of France

Where were those areas? Within the French territory, the major contrast is between a northern language and a southern one, the *langue d'oïl* and the *langue d'oc* (from the word meaning 'yes', which is derived in the north from the Latin *hoc ille*, in the south from the Latin *hoc*). Even as the territory progressed rapidly towards political unity during the Middle Ages, paradoxically two clearly distinct languages were establishing themselves. The differences between them were considerable and today the *langue d'oc* may seem closer to the other Roman languages than to the French which developed out of the *langue d'oïl*. Of nineteen linguistic criteria that are considered to be discriminative, sixteen set the two languages of France in opposition, whereas the *langue d'oc* is distinguished from Catalan by only four, from Spanish (Castillian) by seven, and from Italian (Tuscan) by eight. In truth though, the present situation is misleading, for it results to a large extent from the evolution which the French *langue d'oïl* has undergone since the thirteenth century. In the Middle Ages, the difference between the two languages was much less marked.[2] Linguists consider them to be two branches of a single larger entity, Gallo-Roman, spoken over an area roughly bounded by the Alps and the Pyrenees (but which overflows on to the Piedmontese side of the Alps, just as Catalan and Basque encroach on to the northern side of the Pyrenees). The area includes the Roman parts of Switzerland and Belgium and all in all corresponds approximately to the territory of Gaul, although we should remember that there was an overlay of Germanic and Breton languages in some regions. The situation is complicated, however, by the existence of an intermediary area sandwiched between the domains of the *langue d'oïl* and the *langue d'oc* and centred upon the region of Lyons, Savoy and Suisse Romande. Here the language is Franco-Provençal. In the west, this linguistic band peters out and the *langue d'oc* and the *langue d'oïl* come into direct contact meeting at a border which loops a little to the north of the Massif Central, joining the coast close to Bordeaux (fig. 25).

In both the major linguistic areas, a number of more or less strongly marked dialects are also spoken. Within the domain of the *langue d'oïl* the distinctions are less marked, but all the same, alongside the Frankish dialect of the Ile-de-France, which was adopted as literary French, a number do stand out: the Picardian dialect, in the north, which was also a literary language of importance until the late fourteenth century (used, in particular, by the historian Froissart), the dialects of Normandy and Lorraine, and

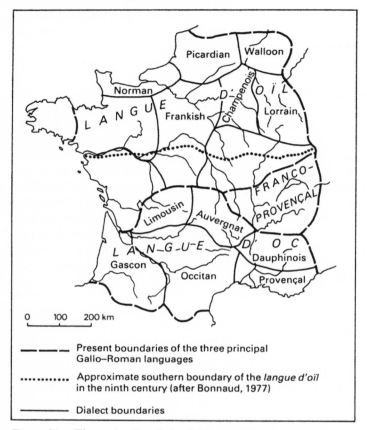

Figure 25. The major linguistic divisions within the French territory.

Walloon in present-day Belgium. The dialects of the west, the central west, the Loire district and Champenois are, on the other hand, much less distinctive. Within the *langue d'oc*, the distinctions are much more marked. The dialects fall into three main groups: the central and eastern group, destined for the most brilliant literary future, includes the Occitan language, the main vehicle of the medieval troubadours, and also Provençal, resuscitated as a literary language by Frédéric Mistral and the Félibre movement in the nineteenth century. This group is sharply distinguished from on the one hand the dialect of Gascony, spoken in the western region of the Aquitaine basin, essentially to the south of the Garonne, and, on the other, the group of northern dialects spoken in Limousin, Auvergne and Dauphiné. These three groups are so distinctive as to be more or less mutually unintelligible.

Langue d'oc and langue d'oïl

What is the explanation for these differences? Broadly speaking, the basic distinctions between *langue d'oc* and *langue d'oïl* are now clear. Scholars agree[3] that they represent two different developments resulting from the situation that existed prior to the Germanic invasions and the subsequent effects of these. In the north, Romanization had never been as thorough a process and here the Latin base of the Gallo-Roman language proved too weak to resist the many forces working to destabilize it. In particular, the Celtic substrata must still have been exerting a powerful influence at the time of the Germanic invasions. By decreasing contacts between regions north of the Loire and the more Romanized areas of the south, those invasions encouraged a resurgence of Celtic influences. Celtic itself could not be reborn, for it had already been too widely wiped out or adulterated and, besides, the social prestige of Latin and the fact that it was the language used by the Church ensured its predominance. Nevertheless, the Gallic language left deep imprints upon *langue d'oïl*, the language which must, *par excellence*, be considered as 'Gallo-Roman'. Germanic influences were also strong, introducing many new words (520 survive in modern French out of the 700 or so originally absorbed, the remainder having disappeared). The new words were, however, relatively specialized, as might be expected of the linguistic contribution of conquerors. (They include terms referring to warfare and seafaring, hunting, orientation, and political, social and legal institutions; also many terms used to convey abstract ideas and many proper nouns expressing the presence and role of the conquerors.)[4] Above all, however, the new contribution worked indirectly to destabilize the language in a region where Romanization had never made a very deep impact. Such were the origins of French which, of all the Romance languages, is the least closely related to Latin. Compared to the domain of the *oïl* tongue, which was characterized by an on-going rapid evolution, the *oc* domain has a more conservative air and remained much closer to its Latin roots.

Franco-Provençal and the concept of a mid-Roman area of influence

A far more thorny problem is raised by the geographical explanation for the Franco-Provençal language.[5] Many hypotheses have been suggested, one being the presence of the Burgundian substrata. But as we have seen, these were of little importance. Another theory is that Lyons exerted a determining influence, but this does not provide a satisfactory explanation for the limits within which this language flourished, for these do not seem to relate to the natural sphere of the town's influence, particularly in that they extend much further eastward. Some scholars regard the language as a proto-*langue d'oïl* of an extremely conservative and archaic nature, which was arrested in the evolution that affected the French language. But that hypothesis is not satisfactory either. Franco-Provençal was at an early date subject to intense pressure from the French language, both from the direction of Lyons and from that of the Savoyard monarchy, whose official language was French. From the early years of the modern period at least, both these regions played an extremely active role in encouraging the development of French. If Franco-Provençal had originally been affiliated to *langue d'oïl*, that affiliation would have been strengthened rather than weakened over the years.

A much more satisfactory solution has been suggested by P. Bonnaud, who eventually came up with the idea of a 'mid-Romania', located in a median band stretching across the French territory and including not only the Franco-Provençal language but also all the other *oïl* dialects to be found in the south: the Poitevin, Berrichon, Nivernais and Comtois dialects, all of which also manifest close affinities with the northern group of *langue d'oc* dialects. This general area corresponds to the median band which is, to be sure, separated from the south by the Massif Central, but which was also probably affected later than the north by diminishing relations with the south and which was, furthermore, already in the late Middle Ages, swamped by the influence of French as a result of its proximity to the expanding Parisian centre. Bonnaud's thesis has been conclusively demonstrated in respect of the Comtois dialect,[6] which manifests unmistakable traces of a Franco-Provençal substratum. Similar vestiges of a linguistic stage prior to the expansion of the French *langue d'oïl*, which can be traced to Franco-Provençal or even to *langue d'oc* itself, are also detectable in the south-western Paris Basin.[7]

Franco-Provençal thus constitutes the sole vestige of this mid-Romania. It survived better than other dialects thanks to its geographical location, at the south-eastern limit of the mid-Roman domain, where it was further away from the French language's centre of diffusion. Its position also afforded it protection from the effects of the Hundred Years War, which were negligible here in comparison to the severe consequences of that conflict in the southern Paris Basin and in the west. There, the war often led to the resettlement of entire communities of *langue d'oïl*-speaking people, originally from more northern regions, and their resettlement played a decisive role in the elimination of the mid-Roman band in these regions. Those movements of recolonization, which also resulted in the implantation of pockets of *langue d'oïl* well within the *langue d'oc* domain (such as the '*gavacherie*' of Entre-deux-Mers, between the Garonne and the Dordogne, to the east of Bordeaux[8]), radically altered the populations of the Saintonge and western Angoumois regions. But they were, in truth, simply one episode – albeit one of the most spectacular – in the permanent flow of people and influences from the north. It was between the seventeenth and the late eighteenth centuries that *langue d'oc* receded in the west in a number of marginal regions of the Massif Central (Montmorillonais, Angoumois, Limousin).

Already in the Middle Ages, the foundation of La Rochelle had created a magnet that attracted many people from the north.[9] At any rate, in the Middle Ages, *langue d'oïl* seems to have made considerable progress towards the south throughout western and central France. Even before the year 1000, the boundary between the evolving *langue d'oc* and *langue d'oïl* appears to have been situated roughly along the Loire.[10]

The structure of the langue d'oïl dialects
The progress made by the French language also explains how it was that the *oïl* dialects were relatively undifferentiated. Had it been possible to map out their spheres of influence around 1000 AD, before the development of a written vulgar language, they would no doubt have presented differences as marked as those of the *oc* dialects of today.[11] The relative uniformity of the *oïl* dialects can be explained by the early impact made upon them by French, the vehicle of communication adopted in the prosperous and densely populated region centred upon the Paris Basin and northern France and characterized by complexes of large village clusters (see below, pp. 135–41). At a very early date, many of these *oïl* dialects disappeared, so that by the seventeenth and eighteenth centuries, the people of these northern regions simply spoke a *patois* or deformed version of French, much as they do to this day.

The langue d'oc dialects
In the *langue d'oc* domain, much more distinctive differences have survived, for the area involved is topographically and historically divided into more independent units. The easiest case to account for is that of the distinctiveness of the Gascon dialect, which was so noticeable in the Middle Ages that the *legs d'Amor* (1356) (which attempted a linguistic codification of the language of the troubadours) considered it to be a '*lengatge estranh* (foreign language)'.[12] The part played by linguistic substrata is clearly crucial here.[13] The Gascon area corresponds exactly to that where Paleobasque (Iberian) substrata survived until a relatively late date, probably until the sixth century as far north as the Garonne, and until the tenth century in the central Pyrenees (see above, pp. 134–44). The terms 'Gascons' and 'Basques' are themselves both derivations of the Latin *Vascones*.

More delicate is the problem of the difference between Occitan (the Languedoc language) and Provençal. An entire linguistic and ideological school of thought, emotionally attached to the idea of a 'great Occitania', holds that the two dialects only diverged in modern times (from the sixteenth century onwards) as a result of the penetration of French by way of the Rhône valley.[14] In support of that theory, it has been alleged that the conservation of the peripheral dialects of Nice and Haute Provence stems from the existence of an initial Occitano-Provençal language which resisted

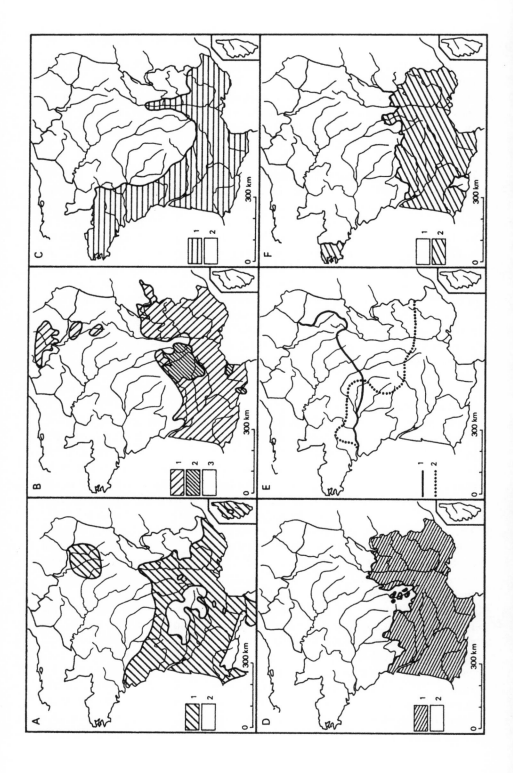

the influence of French. That theory does not appear to be admissable, however. The differences between Occitan and Provençal seem to predate the arrival of the French influence by many years. As early as the Middle Ages, the Languedoc language adopted the bi-labial Ibero-Gascon *b* rather than *v*, so that the Rhône region seems already to have been distinct from the Ibero-Aquitaine region. Within the *oc* domain, the Languedoc language proved to be a dynamic dialect whose dominance could be traced to the powerful centre of attraction and literary evolution constituted, at least up until the Albigensian Crusade, by Toulouse and the political institutions located there. Along the margins of its sphere of influence, particularly throughout the southern regions of the Massif Central, it swamped and supplanted the other native *oc* dialects. The undeniable evolution of the Provençal language, in contrast, as a reaction to the penetration of French, was a quite separate and much later phenomenon.

North and south

The linguistic separation of *langue d'oïl* from *langue d'oc* is but the most obvious of a series of deep parallel divisions which cut across the territory of traditional France. Research into these oppositions is far from complete[15] and has sometimes been impaired by a confusion with socio-economic and topographical contrasts which are of more recent origin (see below, pp. 134–51). Many problems have yet to be resolved. However, in some areas, it is at least possible to determine where they lie and to indicate possible angles of investigation.

Regarding a number of material aspects of the French culture, the facts speak for themselves. The first area to have been systematically investigated is that of differing types of roofs.[16] In general, the roofs of southern France slope gently and are covered by curved titles which are kept in place simply by contact with one another. Northern roofs, in contrast, slope steeply and are covered in thatch, slates or flat tiles which are carefully secured. The main dividing line runs horizontally from Poitou to Bresse, passing through the Marche and Bourbonnais regions (fig. 26A). A number of pockets of

---

Figure 26. North and south.
A. Types of roofs. 1. Shallow sloping roofs and hollow tiles. 2. Steeply sloping roofs, flat tiles, slates and (in the past) thatch. B. The names used for the female horse (after Gilliéron, 1902–20). 1. *Cavale*. 2. Derivations from *equa*. 3. *Jument*.
C. Techniques used for threshing cereals (after Jeanton, 1939, p. 11). 1. Threshing in the open air. 2. Threshing indoors. D. Juridical areas (after Specklin, 1982).
1. Area of written law. 2. Area of customary law. E. The limits of southern influences in Romanesque art (1) and Gothic art (2) (after Brutails, 1923, plate VI).
F. Types of water-mills (after Rivals, 1984). 1. Preponderance of the type with a vertical wheel. 2. Preponderance of the type with a horizontal wheel.

steeply sloping roofs also exist in the south (in the southern Massif Central and as far south as the Monts de Lacaune and the Montagne Noire, in the central Pyrenees and in general throughout the Alps), while a relatively large pocket of gently sloping roofs with curved tiles is to be found in the north, in Lorraine and the Barrois region. The curved title (*imbrex*) is of Mediterranean origin and was propagated by the Romans. Initially designed as a joint-cover to alternate with large flat tiles (*tegula(e)*), but easier to produce than the latter and suited to more rudimentary roofing frames, little by little the curved tile prevailed on all southern roofs. Its enduring popularity clearly testifies to the intensity of Romanization, for no climatic determinism can account satisfactorily for its area of distribution. In the countryside of northern France and in southern mountainous regions relatively unaffected by Romanization, the barbarian invasions and the decline of the organized economy favoured a retention of or a return to roofs mostly covered with thatch, for which steep inclines are necessary: such were the old houses of independent Gaul, 'Gallic' (rather than 'Germanic') houses, as opposed to the 'Latin' houses, which had gently sloping roofs.[17] The operative factor at work here, as in the major linguistic contrasts, was a resurgence of Celtic substrata. In southern mountainous regions, the presence of schistous pockets of slate may have thwarted the readoption of the hollow tile. As for the Lorraine enclave, here most houses were covered by wooden slats (*scindula(e)*) or thatch, in the Middle Ages, although in certain rural areas the use of tiles may have persisted throughout.[18] Essentially, the explanation for the general presence of tiled roofs here today is that in the modern period a return to the use of tiles was prompted by the example set by urban centres which had remained partially faithful to the Roman tradition and had retained constant contact with the south, thanks to the Lyons–Trèves communication artery. In truth, the slope of Lorraine roofs took some time to become less abrupt, adapting only gradually to the new covering material, as can be seen from the existence of hollow tiles equipped with hooks, to cope with steep inclines (still quite commonly used in the nineteenth century).

This type of analysis has also been applied to a number of other aspects of the material culture of France, where again the lines of division turn out to be drawn roughly across the middle of the country. The northern limit of the area in which grain was trodden out in the open air (fig. 26C) can certainly be explained essentially by the role played by the climate, but its extension to the Saône valley can only be the result of the presence of southern influences here. Similarly, it is assuredly remarkable that the bees of southern France must content themselves with relatively rudimentary wooden hives, whereas in the north they are provided with special baskets, as in Germany.[19] The contrast is equally striking in respect of water-mill techniques, which it is possible to analyse thanks to statistics dating from 1809.[20] In northern France, with the exception of Finistère, the vertical mill-wheel was used almost exclusively. In southern France, mills with horizontal wheels predominated, although pockets where the vertical wheel was more common were also to be found in the Béarn region and in Corsica. The distribution limits closely follow the linguistic boundaries, although there is an outcrop of mills with vertical wheels in the northern and central parts of the Massif Central (fig. 26F). The mill with a horizontal wheel is a simpler, more rustic type and the pattern of its distribution in the south and in Finistère is clearly residual in relation to the general expansion of the domain where vertical wheels were used. The vertical wheel was in fact introduced in the south quite early on (mills on the Garonne at Toulouse were equipped with such wheels before the twelfth century). But it was not adopted generally. Right down to the present day the south has culturally resisted this innovation brought from the north. The anomaly of Corsica is explained by the deliberate action taken by French administrators in the eighteenth century; that of the Béarn region coincides with evidence pointing to the early progress of literacy in this area (see below, pp. 137–53). The great diffusion of water-mills took place after the year 1000, so it cannot be regarded as a direct consequence of the Germanic invasions. However, the situation does prove that, even in the Middle Ages, the domain of the *langue d'oc* still constituted a distinct technological area.

These contrasts are all the more significant in that they coincide with many more abstract cultural factors. The most important from the point of view of daily life and also the one that was the first to be studied[21] is the opposition between the areas of unwritten (or common or customary) law, of Germanic origin, and those in the south, of written law, a legacy from the Romans (fig. 26D). Then there are vocabulary differences.[22] Quite a number of concepts are expressed by different words on either side of a frontier running across the middle of France. A case in point is the word meaning 'mare': the Latin tradition is evident in the south, where the words used are *cavale* (from *caballus*) and other terms derived from the Latin *equa* (the feminine form of *equus*), whereas in northern France the word *jument* is commonly used (fig. 26B). The northern pockets where *cavale* prevails coincide remarkably closely with the Lorraine pocket of curved tiles. However, the general picture is far from clear. *Brouillard* (fog) in the Paris Basin stands in opposition to *brume* in Aquitaine, but also to derivatives from *nebula*, which is also Germanic (the German *Nebel* = fog), in the south-east. The chief alternatives to *cercueil* (coffin) are *caisse* to the south of the linguistic boundary and *bière*, which is of Frankish origin, to the north of it. But pockets where *bière* is used are also to be found as far south as the Landes of Gascony and the Pyrenees.[23] Names for people are also revealing and the evolution of attitudes towards these proper nouns indicates that significant differences persisted throughout the Middle Ages. The general tendency to adopt saints' names as Christian or fore-names began in the south. The supremacy of

Germanic names, which were adopted more or less universally throughout the territory of Gaul following the invasions, had earlier been challenged there by the names of, in particular, the Evangelists. Family names emerged from the surnames which were adopted generally between the eleventh and the fifteenth centuries, but again the impulse to make these hereditary began in the south.[24] We are here entering the domain of mentalities or imaginary representations. In the Carolingian period and the early years of the Capetians, *langue d'oc* France undeniably constituted a separate domain. It was the synods of southern France that first pressed for the *Pax Dei* and the *Treuga Dei*, thereby expressing their trust in the lost dream of Christian unity and the greatness of Rome. It was the chivalry of southern France that first embraced the idea of the Crusades with enthusiasm. In the eleventh and twelfth centuries, the lyrical poetry of the troubadours, steeped in Arab influences, provided this region with a prestigious literary mode of expression which was gradually to influence the provinces of the north and to play a crucial role in forming the spirit of chivalry. The south was to provide the French language with a whole collection of words such as *amour* (love), *époux* (spouse), *jaloux* (jealous), *rossignol* (nightingale) and many others too, all connected with feelings and passions.[25] On the other hand, Provençal produced no epic (whatever the claims of Raynouard, Fauriel and Gaston Paris). Epic literature was an exclusively northern phenomenon. All the *chansons de geste* came from the *langue d'oïl* area, including those belonging to the cycle of William of Orange, which unfold in the south, giving an account of the struggle against the Muslims. *Girart de Roussillon*, the most southern of the *chansons de geste*, was produced on the borders of the two linguistic areas, in southern Burgundy and northern Dauphiné, a region where the lyrical poetry of the troubadours had made little impact and did not stand in the way of the spread of epic poetry. Until the end of the twelfth century, the territory of France certainly was divided into two separate cultural areas which expressed themselves through different literary genres.[26] Architecture, the foremost art of the Middle Ages, provides further testimony of that division. The Romanesque style, born in the South, in emulation of Roman building styles, spread northwards, just as the Gothic style, born in the twelfth century in the Ile-de-France, spread southwards. Both established themselves throughout the Christian West. But historians of art have in each case distinguished southern and northern domains which are recognizable from many distinctive features.[27] Between the Romanesque period (which peaked around 1050–1150) and the Gothic period (which started around 1150–80), the architectural boundary dividing the north from the south shifted noticeably southwards over most of central France, advancing northwards only in one particular part of the Loire region (fig. 26E). The shift was symptomatic of the retreat of the Occitan domain of influence, as was the shift of the linguistic frontier which we have already noted.

Can one go further and seek to establish distinctions in the permanent psychological features of the two areas? The volatile, boastful, southern temperament has become a stereotype throughout France. Unforgettable literary examples abound: d'Artagnan (Alexandre Dumas), Tartarin de Tarascon (Alphonse Daudet), Cyrano de Bergerac (Edmond Rostand) ... Wandruska's excellent study (1939) distinguished and classified a number of types: Gascon, Provençal, and Auvergnat. But do they correspond to any reality? Do they really represent particular substrata? It would seem for the most part anyway, that we should regard this typology as an image forged at a relatively late date, a representation of adventurers and emigrants forced into the position of a minority and regarded with varying degrees of severity by the official culture of northern France. R. Lafont (1971, p. 173) even detects a positive policy of exclusion here, designed to thwart the rise of the southern nobility and its accession to positions of national responsibility (during the seventeenth century) and to limit the role played by Occitan politicians (in the nineteenth). But whatever some scholars claim to have detected from this 'ethnic dialogue' in the medieval texts (Lafont, 1971a, pp. 120–2), in truth no signs of such hostility become evident until the sixteenth century, when the Edict of Villers-Cotterêts (1539) imposed the use of French, rather than Latin, in all official documents, thereby relegating the *langue d'oc* culture to a status subordinate to that of the *langue d'oïl*. Besides, those stereotypes have varied considerably over the years and have been identified with very diverse geographical locations. It has thus been argued (Giordan, 1980) that the pejorative image of the loud-mouthed Gascon braggart owes much to its sixteenth-century identification with the figure of the Spaniard, then regarded by the French as their major enemy. Meanwhile, in contrast, at the same period, the positive image of the Provençal, together with that of his passionate identification with the French motherland, was fostered by the proximity and prestige of Italy. As the lyrical poetry of the troubadours was reclaimed as a national asset, Provence came to be regarded as the land of the Muses, poetry's elected home (Emmanuelli, 1977, p. 121); and meanwhile the Gascon stereotype was associated with an area that extended as far as the Rhône and encompassed the whole of what was known as Occitania. The 'southerner' in France is not so much an objective reality as a literary image or a product of northern views and prejudices. And, as Alphonse Daudet himself agreed, does not every Frenchman seem to some extent a native of Tarascon? It is certainly a somewhat paradoxical fact that, in French literature, the brave but boastful and unreliable southerner possesses many of the qualities that Latin antiquity was in the habit of attributing to the Gauls in general.

To dig a bit deeper – albeit somewhat rashly – it is perhaps worth pondering one scholar's considered opinion that all the great poets of the French language have come from the provinces situated in northern

France.[28] Could one really find fundamental explanations here, contrasting the intellectualism of the strongly Latinized provinces with the more dreamy nature of the Germanic and Celtic personalities? Madame de Staël certainly thought so: in her *De l'Allemagne* (1810), she opposed the Romanticism of the north to the Classicism of the south, declaring: 'The North and the South ... shared the empire of literature between them.' It was an idea that Lanson was to echo in his *Manuel d'histoire de la littérature française* (1894). Or is it simply that, in the southern provinces, the literary language had to be a learnt language once the *langue d'oc* lost its prestige and, as such, was considered suitable for analysis, reasoning and prose, but not for giving lyrical expression to feeling?[29]

All in all, it is impossible to make every boundary and contrast mentioned above coincide satisfactorily on the map. The indisputably distinct but complex and fluctuating entities constituted by northern and southern France are separated by not one but a number of boundaries. To the ethnic variations which have been considered to explain certain divisions, a series of historical episodes have been added. We can guess at some of them – for instance, the military frontier between the Visigoths and the Franks, running along the course of the Gartempe, which was recently (in 1979) invoked by Specklin and which explains the origin of the name of the province of La Marche. But there were no doubt many others of which we know nothing. Such as it is, the imprecise division between the north and the south is, even today, the deepest to slice across the land of France.

The establishment of the major contrasts in landscape and settlement

These cultural oppositions between the north and the south, inherited from the Germanic invasions, were amplified and to some extent overlaid by contrasts of a different kind, the economic and social impact of which began to modify the landscape essentially in the late Middle Ages. Between the ninth and the eleventh centuries, rural life was affected by new factors which were to bring about a chain reaction of fundamental changes. Within a few centuries the aspect of the countryside altered radically, acquiring many features which were to endure right down to the twentieth century. Major contrasts developed between two different types of landscape and habitat, each homogeneous over vast tracts of the territory: in the north and the east, open fields dotted with large clusters of villages, associated with strips strung out in parallel lines on land exploited on a collective basis, for instance for the communal grazing of flocks and herds; in most of the south and west, a looser pattern of hamlets and scattered farms (except along the Mediterranean coastline), with larger fields in a landscape featuring many more enclosures, especially in the *bocages* (thickets, copses) of the west.

The development of the openfield community system: factors and processes

Fifty years ago, these contrasts had already been recognized in scholarly works which have recently been republished on the strength of their descriptive merits.[30] But if their descriptive flair is indeed undeniable, their genetic conclusions are totally out of date, as is their interpretation of the historical development of these landscapes. Based solely upon contemporary observation or relatively late texts (from the eighteenth century in particular), they pronounced the landscapes of the north and the south to be 'agricultural civilizations', regarding them as mysterious entities whose origins went back to a distant past: either prehistoric times[31] or the period of the Germanic invasions.[32]

Thanks in particular to the work of the German school of geography,[33] it is today established that these systems of land organization were introduced at a much later date. It was, broadly speaking, in the ninth and tenth centuries, at the time of the upheavals occasioned by the second wave of invasions (Scandinavian, Hungarian and Muslim) that the process was sparked off by phenomena of a social order.[34] From the ruins of the ancient system characterized by direct state intervention in the context of a society based on slave-ownership, the feudal social system now evolved. This was characterized by a general network of personal links of dependence. Within this structure of strictly interlocking hierarchies, a basic system of protection for people and property was gradually restored in the countryside. The first noticeable manifestation of the new social system was a concentration of settlements.[35] Villages made up of substantial groups of houses took over from the pattern of more or less scattered settlements occasionally gathered into small, vulnerable hamlets, by which the early Middle Ages were still characterized. The new system crystallized in the ninth century, when castles began to make their appearance and thereafter multiplied. At first they consisted of no more than large walled precincts; then, from the second half of the tenth century on, keeps were built on natural or artificially constructed mounds.[36] Meanwhile, also mainly during the tenth century, a definitive network of parishes was established throughout the western world. This resulted in the creation of the walled village, which probably first appeared in the Mediterranean region (such villages were widespread in Italy by the tenth century), but became a common feature of the whole territory of France, north and south alike, in the course of the tenth and eleventh centuries.

At the same time, powerful repercussions affected the organization of agricultural work. By roughly the middle of the eleventh century, a certain territorial balance had been achieved between the various feudal estates, and external aggression had ceased or considerably abated. Now the feudal lords began to seek to replace their erstwhile war-profits by the income to be derived from their estates. They were keen to cultivate more of the land and set about organizing this at a time when demographic growth was increasing. It was to gather momentum and peak in the eleventh and twelfth centuries and was reflected in a strong tendency to colonize new tracts of land. On top of this expansion into new areas, cultivation was becoming considerably more intensive, thanks to a veritable revolution in techniques: the 'revolution of the year 1000'. Seeking to produce as much as possible from their land, the feudal lords encouraged the general use of heavy metal ploughs with a complicated traction system. Such machinery had been known for several centuries,[37] but now it became more common, while at the same time new metallurgical techniques were evolving in the vicinity of the feudal castles, in response to the demand for weaponry. Decisive improvements were made in the equipment for draught animals – the head-yoke for oxen[38] and the shoulder harness for horses[39] – and these made it possible to set about cultivating heavy fertile soils which had until then been too laborious to exploit.

As a result, agricultural production increased considerably, particularly the largest crops, namely cereals. The clearest evidence of this is the appearance of many new mills during the eleventh and twelfth centuries. In Picardy, 40 new mills were built between the mid-ninth century and 1080, 40 more between 1080 and 1225, and a further 165 over the following fifty years.[40] The ancient economy had been largely founded upon the pasturing of herds in unproductive forest clearings and hunting in those same marginal areas which surrounded the quite modest tracts of land that were under cultivation. Now a different economy gradually took over. It was founded upon the stable and regular cultivation of fields of cereals which were extended to take over the areas hitherto left uncultivated, thereby increasing overall productivity.

This cerealization[41] of the economy in its turn led to a radical reorganization of the cultivated land. The extension of cereal cultivation made it necessary to find a way for herd-grazing and cultivation, which had hitherto each been restricted to their own domains, to co-exist. The solution found was a system of crop-rotation: the land was divided into large units of rotation (big fields) alternately cultivated and left fallow, to be used as pasture. The system was not fully established until the first half of the thirteenth century which, to our present knowledge, is the earliest point at which large areas of fallow land are attested on the territory of France: between 1220 and 1229 in Picardy, near Cambrai,[42] and in 1248 on the plain of France, near Paris.[43] This overall reorganization of the land, which could hardly have been undertaken except with the authority of the feudal lords, proceeded over several centuries and in many regions was not completed until the sixteenth or seventeenth century. In Alsace, fifteenth-century texts still frequently refer to hedges and at Woerth (in southern Alsace) the big field system was introduced only in 1599.[44] In some places in northern Lorraine, it did not come into practice until the end of the seventeenth century and even then only as a result of reorganization that followed upon the destruction or abandonment of villages caused by the Thirty Years War.[45] The establishment of regular big fields soon left its mark upon cultivated areas and the countryside in general. The division of cultivated land into numerous strips reflected the need for every peasant to have fields to cultivate in each of the units used alternately for cultivation and for pasturing. The adoption of a communal shepherd, to care for the village flock grazing on fallow land, led to the disappearance of enclosures. This became general in the Metz area of Lorraine by the end of the thirteenth century.[46] The overall picture presented is that of a coherent, rigorous system founded upon a set of rules guaranteed and codified by the authority of the feudal lord – rules which were the end-product of a socio-economic process set in motion by the establishment of the feudal system within a context of persistent demographic pressure.

The geographical extension and limits of the communal openfield system

This pattern of open fields and large grouped villages extended throughout northern and north-eastern France over vast, continuously homogeneous expanses. Despite its dynamism and economic efficiency, however, the system soon discovered its limits and never prevailed in the greater part of the French territory.

It was clearly best suited to the areas devoted to cereal crops in inland Europe, where the grain ripened well in the hot summers, as opposed to the misty regions along the Atlantic seaboard, which were better left as grassland. It was particularly well adapted to the sedimentary plains and plateaux of the hercynian zone, with their light permeable soils made up of loessic silts from the borders of quaternary glaciers or derived from the decomposition of limestone. Here, a number of factors combined to produce good results: the labour required of the peasants was relatively easy and harvests were relatively regular on these well drained soils which suffered from neither excessive aridity nor excessive humidity and where agricultural catastrophes were consequently unknown.[47] Without being necessarily the most fertile of all soils, these rewarded the peasants' efforts with the best grain harvests on the most regular basis. To judge by the evidence of prehistoric cemeteries, it was here, on these rich limestone or loessic northern plains, that the *villae* had probably been concentrated and that the population density had certainly been the greatest already in the Roman period and probably also in the days of independent Gaul and even during the Bronze Age. And it was in these regions, which were densely settled at a very early date, that demographic growth soon took off generally, leading to the cerealization of the

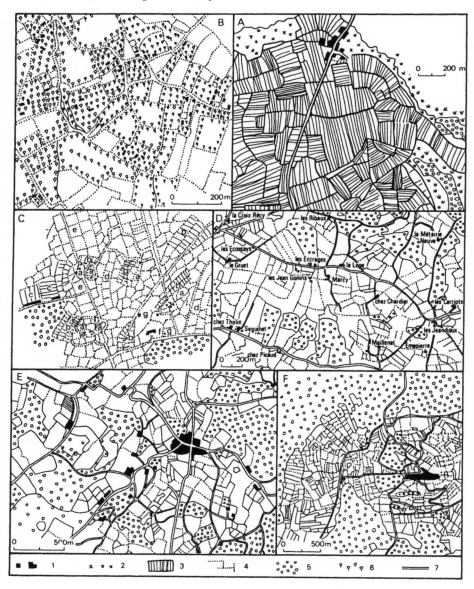

Figure 27. The major types of land division and rural landscapes within the territory of France.
Legend for all six sketch-maps: 1. Dwellings. 2. Pastureland. 3. Open and strip fields. The heavy lines indicate the limits of blocks of parallel strips. 4. Hedges and other enclosures. 5. Woods and thickets. 6. Planted trees. 7. Roads. A. Kertzfeld (Kochersberg, northern Alsace) (after Juillard, p. 35 in Juillard *et al.*, 1957).

economy and all the consequences that this brought in its train for the development of a countryside featuring large village clusters amid open fields submitted to a crop-rotation system.

However, this kind of cereal production was not possible on the poorly drained, mediocre soils of the Hercynian massifs. Nor was such cultivation viable in sedimentary basins, on detrital land subject to weathering that is characteristic of formerly eroded surfaces unaffected by later vertical erosion. Actually, these surface detrital deposits are seldom continuous over large areas. In most cases, they have been broken up by the creation of Pliocene and Quaternary valleys where the scree and alluvial terraces produce soils that are suitable for cereals. Despite that fact, all in all they constituted an obstacle to population expansion and large-scale agriculture. As for surface layers of clay and heavy clayey marls, these proved impossible

---

Grouped dwellings and open fields. In the centre, a block of long parallel strips (*Langstreifenflur*) probably constitutes the original nucleus of the settlement. Around the periphery, short, interlocking blocks (*Kurzkreuzgewannflur*) or ribbons cut into transverse strips (*Breitstreifenflur*) represent later additions.
B. Villers-sur-Mer (Calvados) in the eighteenth century (from National Archives, N IV, Calvados I, Land Atlas of Villers-sur-Mer, section D4, reproduced in Dion, 1934, p. 74). Irregular fields enclosed by hedges, mostly surrounded by or dotted with trees. C. Theoretical sketch of a section of land in Brittany (after Meynier, 1966, p. 606). a. Ellipses, probably original, sometimes circumscribed by a road, with a hamlet at the centre or at the edge. b. Groups of long, narrow fields, many of which lie alongside an ellipse, representing an extension of the cultivated area at a time of demographic growth. Other similar bands, with more or less curved fields, seem to have developed out of an original ellipse. Today most of these fields are enclosed but some plots have remained open. c. Clusters of narrow strips (*méjous, gaigneries*) in the vicinity of a line-hamlet. d. Wide bands representing a later stage of land-clearance. e. Extremely regular plots created in the course of the most recent clearance of the communal territory, in the nineteenth century. f. Castle with its own land. g. Farm tied to the castle. D. Countryside featuring estates created by family communities in the region of Cronat–Vitry-sur-Loire (Sologne bourbonnaise) (after Chiffre, 1985, p. 212). Note the place-names formed with *chez* ... and *les* .... E. Briscous (Pyrénées-Atlantiques) (based on an aerial photograph reproduced in Dion, 1934, reprinted Neuilly-sur-Seine, 1981, extra plate III, fig. 23). Irregular fields, many of which are edged or dotted with planted trees. hamlet and rare isolated dwellings. F. Seillons-source d'Argens (Var) (from an aerial photograph printed in Livet, 1962, plate XI). Cultivated basin surrounded by wooded hills. To the north-east, the wooded plateau of Rians (Portland limestone). To the south and south-east, a spur of the Triassic Var plateau with pockets of cultivated land. At the centre, a dip in which Miocene outliers provide the site for the hill-top village and determine the field-pattern according to the contour lines. The plots become smaller the closer they lie to the village. A few low-lying farms nestle at the foot of the hills.

to cultivate until the introduction, at a late date, of teams of draught animals and heavy ploughs. These too remained mostly outside the sphere of medieval agriculture.

There were thus two kinds of limits to the extension of the communal system of big open fields with regular crop-rotation. The first kind occurred actually within the area where that system was generally adopted. Essentially, it took the form of outcrops of rocks and infertile areas of impermeable soil: Hercynian massifs such as the Vosges and the Ardennes; regions with surface areas of clay in the Pays d'Auge, Champagne Humide and Thiérache; the clayey sand of the Pays de Bray, Puisaye, Sologne and Brenne; and the detrital decalcified soils of the ancient surfaces of Hurepoix (the northern Beauce) and the upper Brie region in the east. In medieval texts, these tended to be referred to as *silva* or *saltus*, as opposed to open neighbouring regions which are called *campania*: the Caen 'Campagne', the 'Champagne' area of Rheims-Châlons. They consisted for the most part of unorganized plot arrangements or of field-patterns that, although organized, were nevertheless not submitted to the big field system, land that was not organized into strips and in many cases belonged to line villages where it was cleared relatively late and individual plots took the form of large rectangles set out on either side of the principal street.

The second type of limit was a general frontier in the west and south-west. It was occasioned by the fundamentally intractable nature of the prevailing natural factors, namely the climate and the distribution of rocks which prevented the extension of large-scale cereal cultivation. Broadly speaking, the frontier ran along the edge of the Paris Basin from the north-west to the south-east, following a line determined in the north-east in particular by the area covered by loessic silts, the edge of which in the main corresponded to the limit of the openfield landscape of the late eighteenth century.[48] To the south of the Paris Basin, the far greater expanse of impermeable detrital soils and tertiary silicose alluvial deposits blocked the development of cereal cultivation with obstacles which for the most part proved insurmountable. Here, already, the heavy mould-board plough gave way to the scratch-plough, which did no more than turn the surface of the soil. However, the boundary was rendered extremely irregular by small variations. The communal system extended furthest south on the limestone plateaux of lower Burgundy, Nivernais and the minor Beauce, and was soonest arrested on the infertile '*gâtines*' (wastelands) of Morvan, Puisaye and Gâtinais. The open landscape extended further southward and westward than the organized big field system with regular crop-rotation did. It took in the outcrops of Jurassic limestone in the Campagne de Caen and the Champagne of the Berry region, where only a very few communes introduced a big field system and even those only in furlongs, never over larger areas. It is true that in these regions communal institutions such as common grazing were fragile and ephemeral, for the villages themselves remained rudimentary, hardly more than small hamlets.[49]

The west and the south

Residual landscapes of dispersed medieval settlement

To the south and west of the line plotted above, open countryside with large village clusters became rare and something of an anomaly. The concentrated population centres constituted by the hill-top villages along the Mediterranean coast were an exception, which we shall have to consider separately. Apart from these, the predominant forms of habitat were of a dispersed or semi-dispersed nature – isolated farms or quite loosely bound hamlets more or less part of a framework of relatively sparsely scattered towns with functions of a mainly tertiary nature. In the main, the countryside was enclosed, although the types of enclosure varied enormously from one region to another: the wooded character of the west explains why it was traditionally described as a '*bocage*' (thicket, copse); in the south and the Massif Central in general, hedges and wooded areas were much more sparse.

Historically, these were residual landscape types. They were peculiar to areas where there had been no village growth and no reorganization of land-divisions such as that which, in inland regions suited to cereal production, accompanied the introduction of the feudal system and the powerful demographic pressure of the eleventh and twelfth centuries.[50] Here, that pressure resulted in different types of land occupancy. Starting from a situation in which the population was much less densely settled and which certainly did not initially feature centres of population such as the large rural population clusters of the fertile northern and eastern regions, here the population expansion and land-clearance which took place between the eleventh and the thirteenth centuries essentially continued to take the form of a dispersed or semi-dispersed colonization of new land. This is reflected in the presence of strata of late medieval place-names whose linguistic form is French. Generally speaking, these are formed with the articles *Le-*, *La-* or *Les-* and they refer to a person or family group. These names fall into several distinct categories. Forms ending in *-ière/-erie* and *-ais/-aie*, in many cases preceded by the article and a personal name (La Gaudinière, La Bernerie and so on), are common throughout the western inland region from Cotentin to the banks of the Loire, where the earliest group has been shown to date from the eleventh century, and the second from the twelfth, although most examples of both date from the thirteenth century. These names were given to individual properties which were not a part of the feudal system, or at any rate to isolated holdings cultivated by family groups rather than organized communities. This type of name continued to be formative until the eighteenth century. Further south, the *Chez-* series is to be found. *Chez-* is followed by a personal or family name and the series appears essentially within a band running from Suisse Romande to the Saintonge region and extending no further southward than a line stretching from the Gironde to the Isère. These forms of toponym are relatively late, most of them having appeared between the fourteenth and the sixteenth centuries. They were particularly fashionable when land was being recolonized following the Hundred Years War.[51] Other forms (for example *L'Huis-*, followed by a personal name, which is common in Morvan) are more localized, but may be similarly explained. These series of toponyms are to be found over an extensive area. It has been calculated that, within a belt crossing central France from Bresse to Poitou, they may be found scattered over 37.7 per cent of the total surface area. In most cases they relate to family communities (so-called 'tacit' communities, 'sharing the same cooking-pot and the same fire'), consisting of several related families who joined forces to cultivate their land together. This was a particularly widespread model for the colonization or recolonization of land with poor soil: that is to say in the Bresse, Morvan, Bourbonnais, Puisaye, Gâtinais and Sancerrois regions during the second half of the fifteenth century and the first half of the sixteenth.[52]

Late medieval forms of collective organization in settlements in the west and the south

Tentative forms of collective organization at a somewhat higher level also existed. They were much less common but did exist among certain populations in these western and southern areas. These too are recognizable from their toponyms. The first series uses *-ville* or *-court* – a French syntactic form which developed in imitation of the Germano-Roman forms in *-court* and *-ville* studied above. These French forms exist in the north and the east, above all in marginal, infertile areas, where they indicate late attempts at colonization which are, furthermore, limited to the outer edges of regions settled earlier. They are to be found throughout a belt stretching from the borders of the Armorican peninsula, where they are particularly dense, across to Burgundy and Franche-Comté. They are clearly manifestations of attempts to provide a framework for a dispersed population of a pioneering nature, an extension of the communal system of the north-east; but they constitute no more than a small minority of the total mass of toponyms.[53] A particularly high proportion of these forms which indicate collective organization consist of toponyms based on saints' names, reflecting the religious framework which helped to integrate populations which were still marginal and unsettled.[54] The fashion for such names began to develop in the Carolingian period, when they used the form *Dominus* > *dom* or *Dam* followed by the name of a saint (Dommartin, Dampierre, Domrémy and so on). These are common along the western, southern and eastern edges of the Paris Basin, where they apply to the later village concentrations in this area. The next phase was that of place-names containing the word *Saint*. These are very rare in northern and eastern France, where most of these village concentrations were already established by the feudal period, which was when this type of name became popular. On the other hand, they are very common in most of western and southern France, where they testify to the decisive role that the network of parishes played in providing a stabilized territorial communal framework. They are particularly numerous in a belt stretching from the Gironde to Dauphiné and Savoy and passing by way of north-western Aquitaine (to the south of Angoumois and Périgord): Limousin, Marche, the southern central part of the Massif Central, Forez, Lyonnais and southern Burgundy. Of more recent date than the ancient toponyms, most of which are strung out along the most favourable hillside sites, these place-names based on saints' names make up the densest layer of names given to the peasant settlements that were then proliferating amid a very sparse scattering of toponyms developed from local dialects.

It should, however, be noted that, like the other series studied above, names using *Saint-* peter out again in the southernmost part of France (to the south of the 'au Bois' areas in Aquitaine and in the Pyrenees, southern Languedoc and the southern edge of the Massif Central, Provence and southern Dauphiné) – a fact that is particularly remarkable, given that names from *Sanctus* > *Saint* were initially disseminated from the south. The explanation for their comparative rarity is that in these southernmost regions, an urban framework for the countryside still existed since the towns, which had suffered much less than those in the north from the Germanic invasions, still controlled vast communal territories and continued to exercise a decisive influence upon the surrounding countryside. In these regions, it had consequently not been so important to develop a network of parishes to provide a framework for the habitat. This is also the reason why the communes in southern France are so much larger than elsewhere (fig. 28). As the map shows,[55] they are in general considerably above average size in the west and the south, but it is only in the extreme south that the giant communes of over 15,000 hectares are to be found, and these are all urban communes (Arles: 1030 km^2; Aix-en-Provence: 289 km^2; Narbonne: 278 km^2; Marseilles: 228 km^2; Montauban: 135 km^2; Toulouse: 118 km^2; and amongst the smaller towns, Millau: 187 km^2; Tarascon: 103 km^2; Lectoure: 84 km^2; Condom: 83 km^2).

Two particular regional cases: southern Aquitaine and Brittany
Amid all these examples of late medieval urban and village development, that of the Aquitaine Basin rates a special mention.[56] From the second half of the eleventh century on, there was a succession of phases in the development of the habitat, all directly connected with the secular or ecclesiastical feudal system. They were responsible for the establishment of virtually all

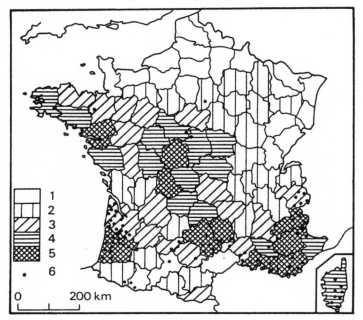

Figure 28. The areas of the communes of France (after Meynier, 1945). Average area by departments: 1. 7–10 km². 2. 10–15 km². 3. 15–20 km². 4. 20–25 km². 5. 25–40 km². 6. Giant communes (over 10,000 hectares).

the small towns and grouped villages which make up the habitat at an infra-urban level, below that of the towns created in the ancient period (fig. 29). The first to appear, in the last decades of the eleventh century, were the *Castelnaux* ('new castles'), which consisted of villages grouped around a central castle. Thirty toponyms belong to this series and many other villages also originated as *Castelnaux*, even though their names do not reflect the fact. Names of this type continued to appear until the thirteenth century, but most date from between 1100 and 1175. The same period saw the creation of *Sauvetés*, agricultural centres, mostly founded by abbeys and placed under the protection of the Church. About one hundred of these were developed, mainly between the mid-eleventh century and about 1130–40. A temporary lull in colonization occurred around the mid-twelfth century, probably as a result of a spate of migrations to areas won from the Arabs of Spain and to the towns, which were growing rapidly at this time. In the thirteenth century, when the Albigensian War was over, colonization picked up again, taking the form of foundations which were granted various franchises and advantages and were known as *bastides* (meaning 'habitation'). Between about 1220 and 1325, these multiplied rapidly and a number continued to be built

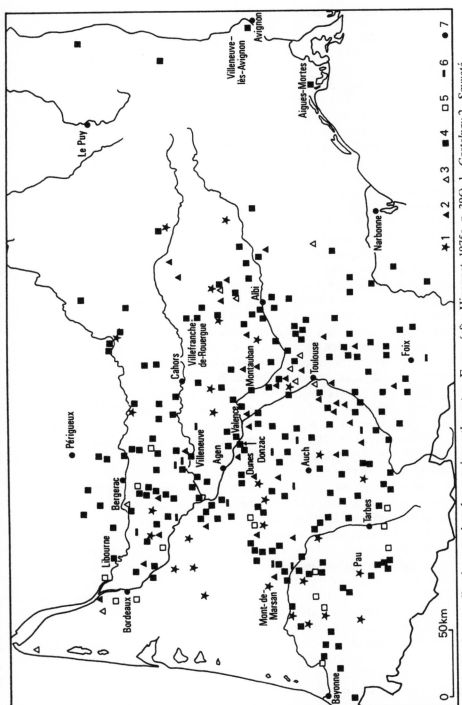

Figure 29. Late medieval grouped settlements in south-western France (after Higounet, 1975a, p. 396). 1. *Castelnau* 2. *Sauveté* 3. *Sauveté* now disappeared. 4. *Bastide*. 5. *Bastide* now disappeared. 6. Enlarged ancient site. 7. Major town.

up until the middle of the fourteenth century, bringing the total to about 350. The *bastides* represent a particularly complex phenomenon by virtue, in the first place, of both their origin and their destiny. In this politically disputed zone, rivalry ran high between the kings of France and England and also between their respective officers, great lords and small ones alike. All entered into the business of building *bastides*, sometimes not so much with a view to agricultural development, but rather in order to seize a strategic advantage or in the interests of military defence (one third of the *bastides* in Gascony were walled precincts). Many clearly served as border fortresses. (In some places we find twinned *bastides*, as where the English Valence d'Agen (1283) stands face to face with Donzac (1265) and Dunes (1269), both built by Alphonse de Poitiers, who was the brother of Saint Louis.) The demographic role played by the *bastides* is also interesting. They certainly attracted people from the outlying countryside, who moved in and settled; but they also absorbed the demographic overspill, as is proved by the fact that the halt in the building progress coincided with the demographic decline caused by the Black Death. Equally, the design of the *bastides* was distinctive. Most were deliberately laid out according to a regular plan, generally around a central square which served as a meeting area and also a place of refuge; but many others were arranged in a haphazard fashion, according to uncoordinated spontaneous plans which were more or less adapted to the local topography. Finally, the subsequent history of the *bastides* is interesting: many were later abandoned or disappeared, some languished on as modest villages, while quite a number of others acquired urban functions and grew into large towns (Libourne, Villeneuve-sur-Lot, Villefranche-de-Rouergue and Montauban; another departmental *chef-lieu*, Mont-de-Marsan, was originally a *castelnau*). Altogether, about 600 villages and new towns were established in this way in southern Aquitaine in the course of the Middle Ages. This development was certainly part of a general European pattern, but in this particular area it was exceptionally dense – a fact which, over and above the local causes mentioned above, prompts one to ponder the significance of the geographical limits of the phenomenon.

In particular, it would be interesting to know why in Brittany as a whole, where the state of the settlement pattern was initially just as dispersed as in the Aquitaine Basin, if not more so, similar developments did not take place, whereas new towns did appear along the inland borders of the Armorican peninsula. In Brittany, the *bourgs* (country towns), with tertiary rather than agricultural functions, which developed during the Middle Ages and in the modern period, in conditions that are somewhat obscure, seem almost all to have evolved spontaneously. There are two possible explanations for this: first, the great movement of colonization in Aquitaine, which was, anyway, concentrated in the centre and south of the Basin rather than along the northern edges, fits in with the persistent southern tradition of providing a framework for the countryside by developing urban settlements of the type described above; settlements whose importance is reflected in the exceptional size of the southern communes of today. In the second place, in Brittany, the lack of connection between the organization of the parishes on the one hand and the feudal structures on the other (see above, p. 83) set up a virtually insurmountble barrier that blocked the development out of parish centres and the countryside could not accommodate any other focal points.

The special features of the Mediterranean countryside: the development of hilltop settlements

It is not hard to see why the development of *bastides* did not extend further eastward.[57] The reason for the absence of *bastides* in these Mediterranean regions is the much earlier concentration of the rural settlement pattern. In the twelfth and thirteenth centuries, when the phenomenon was at its height in the Aquitaine Basin, here settlements had already been concentrated into villages insofar as was possible. The process had taken place as early as the tenth and eleventh centuries, producing villages which were both fortified and high up, no doubt in response to the particular and persistent insecurity (occasioned by the Arab invasions), within the framework of an olive-producing economy. With such an economy, a village needed to be perched on the heights in the midst of its encircling olive groves. The importance of the correlation between perched settlements and the cultivation of trees in the Mediterranean region has quite rightly been emphasized.[58] In regions devoted solely to cereal production, the ravages of war may destroy the harvest of an entire year. But in districts based upon the cultivation of trees, it takes a whole thirty years to repair the destruction. Villages needed to be situated as far as possible from communication routes. It is true that, in the Mediterranean, the '*villages perchés*' appeared much earlier than tree cultivation. (The Celto-Ligurian *oppida* antedate the cultivation of olive trees, which was mainly the result of Greek influences.)[59] However, the fact that the economy was largely based on the cultivation of trees and vines certainly by and large accounts for the phenomenon of hilltop settlements, most of which were established during the tenth and eleventh centuries within the framework of a concentrated settlement pattern linked with feudal structures and in the atmosphere of peculiar insecurity generated by the constant threat of Saracen raids. Hilltop settlements are not unknown in northern France, but there they are few and far between and attributable to particular local circumstances.[60] The furthest limits of the area within which hilltop settlements are to be found overlap with the Alpine ridges and coincide neatly with the boundary of Mediterranean-style cultivation and the innermost limits of medieval Saracen incursions.[61] Mediterranean hilltop settlements continued to be built and inhabited up until the early sixteenth century, when a new tendency developed. The trend which now set in was towards a dispersion of settlements and a removal to lower ground and this has persisted down to the present day, frequently producing double villages or leading to more scattered settlement in the neighbouring plains and valleys.[62]

Plot-patterns and landscapes
All these processes with different origins (the establishment of a network of parishes, the feudal influence, the insecurity peculiar to the Mediterranean) thus created in western and southern France a network of concentrated settlements in the form of small towns and villages. It was certainly less dense and less omni-present than in northern and eastern France, but it nevertheless provided a framework for social life and productive relations for populations for the most part scattered over the countryside. However, the effect of these centres upon the organization of the countryside was minimal. The land continued to be divided into large blocks of fields arranged around isolated farms or hamlets composed of interrelated family groups. Here and there, attempts at openfield cultivation and the creation of communal institutions were made in relatively small areas which were cultivated in a continuous fashion, but were surrounded by vast wastelands dotted with enclosed cultivated fields. These areas, which tended to be associated with small hamlets, were situated in, for example, the south-eastern part of the Massif Central, where the dryer climate was more suitable for cereal cultivation.[63] The collective system of open fields and communal pasturelands was also adopted in the isolated territories of some villages located in the valleys of the Pyrenees, where it was necessary for the winter grazing of the large herds which lived off the mountain pastures in the summer months.[64] But such cases were exceptional. At a regional level, progress towards openfield cultivation and community institutions was extremely limited, extending only to a few spots with particularly favourable natural features, where settlement was reorganized around denser population centres at an early date: the Béarnaise valleys in the Pyrenean foothills, where the soils were rich and well drained; and also – although in no more than a partial and imperfect fashion – in the Grande Limagne, where the plain had been created by subsidence in the Massif Central.[65]

The development of enclosures was very general, but uneven and relatively late. It took its most spectacular form in the west, in the Armorican Massif and around its edges, where it was encouraged by the evolution towards enclosed pasturelands, in a climate that was favourable to grasslands. Here, almost all heaths and peripheral regions were divided up into individual properties marked off by enclosures. These bushy, compact *bocages*, which were extremely common by the second half of the nineteenth century and the early twentieth, took some time to establish, however. In the Gâtine area around Poitiers, the process only really speeded up in the

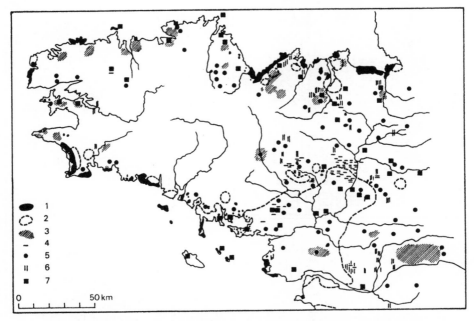

Figure 30. The open fields of Brittany (after Meynier, pp. 86–7, in Juillard *et al.*, 1957).
1. *Méjous* (continuous and isolated). 2. Frequent form of *méjou* in the present plot-pattern. 3. Ancient *méjou* indicated by historical documents. 4. Hamlets set out in lines. 5. Place-names indicating the presence of ancient open fields (*méjou, gaignerie, champagne, plaine*, etc.). 6. Strips of curved plots indicating the location of ancient open fields. 7. Ancient or present-day communal practices, common grazing and the use of the *sillon* as a unit of measurement.

fifteenth and sixteenth centuries when, in the aftermath of the Hundred Years War, land was being recolonized by the establishment of scattered sharecroppers' farms.[66] In Brittany, the division of communal heathland, at a time of demographic pressure in the late eighteenth and the early nineteenth centuries, marked the decisive last phase of the process. In many parts of the Bocage Normand, the process was not completed until the mid-nineteenth century, when the switch to commercialized stock-rearing took place.[67] In all these inland western *bocages*, here and there one comes across the fossilized remains of land submitted to continuous openfield cultivation, masked by the network of enclosures but revealed by the toponymy (*méchou, méjou* in Breton-speaking Brittany; *gaigneries* in the French-speaking regions). Such land is disposed around the hamlets which already existed and is surrounded by an enclosure in the middle of the heathland (fig. 30). Such open fields survived in coastal regions, where fertilization by means of

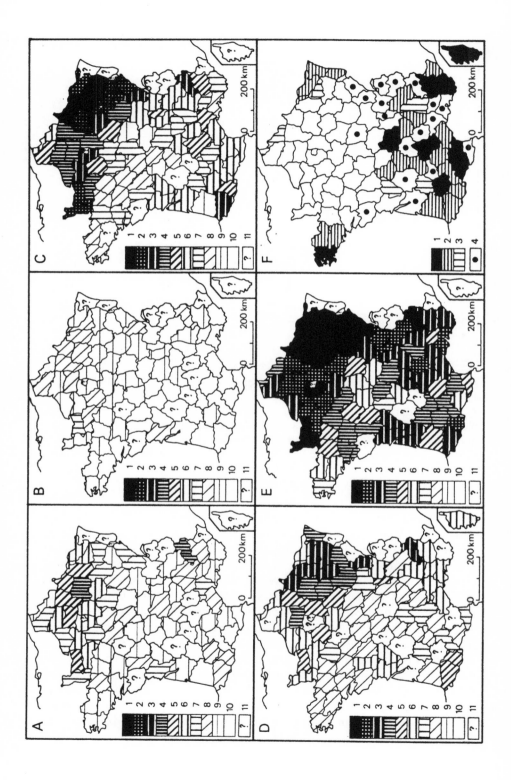

seaweed enabled pockets of intense cultivation around the hamlets to remain under permanent cultivation even after the division of the heathland into individual plots had done away with the organic fertilizers which had formerly been obtained by burning the natural vegetation.[68]

Elsewhere, in the Massif Central and the south-west, the landscape is much more varied than in the west – a combination of sparse woodlands dotted with small open plots and wasteland, which up until the nineteenth century still covered vast areas. Enclosures were less pervasive here in the west, on account of the lower level of demographic pressure and the much more archaic type of economy. But this brings us to contrasts of a different order.

Social and economic repercussions

The contrast between on the one hand the landscape of open fields and organized communities in the north and the north-east, on the other the unproductive and enclosed landscapes with dispersed settlements in the west and the south by and large coincided with major contrasts in the degree of development and the level of living standards in the countryside of the past. Traditional France was composed of two territories: the more developed one, where intellectual progress came relatively early, which was better fed and where life in general was easier, was, broadly speaking, the France of large villages and open fields in the north and north-east; the other, again broadly speaking, more backward, illiterate, poor and under-nourished, was the France of hamlets and isolated farms in the west and the south.

Literacy

A major and most revealing test of those inequalities is provided by the progress of literacy (fig. 31), which has been analysed elsewhere in an

---

Figure 31. Literacy in France. The percentage of spouses who signed their marriage certificates: A. Men, between 1686 and 1690 (after Furet, 1977, I, p. 59). B. Women, between 1686 and 1690 (Furet, p. 59). C. Men, between 1816 and 1820 (Furet, p. 61). E. Men, in 1866 (Furet, p. 62). Legend for maps, A, B, C, E: 1. 100–90. 2. 90–80. 3. 80–70. 4. 70–60. 5. 60–50. 6. 50–40. 7. 40–30. 8. 30–20. 9. 20–10. 10. 10–0. 11. No data. D. Young men eligible for military recruitment between 1827 and 1830, knowing how to read and write (from Le Roy Ladurie and Dumont, 1971, p. 439): 3. Over 70.8%. 4. 59.56–70.79%. 5. 50.11–59.55%. 6. 42.16–50.10%. 7. 35.48–42.15%. 8. 20.53–35.47%. 11. No data. F. Percentage of population of the commune not speaking French in 1863 (after Furet, 1977, I, p. 327). 1. Over 80%. 2. From 40 to 80%. 3. Under 40%. 4. Departments where *patois* are indicated (no numerical statistics). Left blank, the departments where the entire population is recorded as speaking French.

altogether exemplary fashion.[69] The massive spread of literacy began in the sixteenth century, with the printing press and the religious wars. The influence of the Catholic Church of the Counter-Reformation, for which education and the provision of a Christian framework for the population were high priorities, was in fact directed to similar ends as those of the Protestant sects, whose main objective was the reading of the Holy Scriptures. At the end of the seventeenth century, the effects of these influences may be gauged from the percentages of men and women who signed their marriage certificates between 1686 and 1690. At this time there was certainly one large area where literacy was the most advanced: this was northern and north-eastern France, by and large the France of open fields and village communities, generally well-to-do and practising a market economy involving active relations, where one had to be able to read and write if one was to be integrated into social life. With varying degrees of intensity, the figures spell the same story for women as for men. However, the spatial correlation with areas of openfield cultivation is not absolute: as far west as Cotentin, lower Normandy was a region of prosperous *bocages* already deeply integrated into the market economy of northern France and it belonged to the compact block of regions where literacy was well advanced. In contrast, the France of the south and the west was, broadly speaking, less advanced, with a scattered settlement pattern and rural areas incompletely integrated into social life. Yet even here, some more highly evolved sectors constituted exceptions: the valleys of the Béarn region, with their large villages and open fields; and alpine valleys where the standard of literacy was affected by the high level of temporary emigration connected, in part, with the pedlar's trade, for which at least a minimum of schooling was necessary.

Only gradually did these contrasts become less sharp, in a process which was to reveal another set of remarkable regional characteristics. The maps showing the percentage of men who signed their own marriage certificates between 1786 and 1790, and 1816 and 1820,[70] demonstrate that the progress of literacy in the west and the south (apart from the exceptional sectors noted above) began in the Rhône valley and the south-east, the major route whereby the French language and Parisian cultural influences penetrated the region. At this point, what was to become a major feature of the cultural geography of nineteenth-century France becomes apparent, namely the existence of a vast triangle of illiteracy, with its base along the Atlantic coastline and its apex in the eastern Massif Central. It included Brittany, Aquitaine and the western Massif Central. It is still clearly visible on the map for 1866, although it is beginning to shrink at this point, except in Brittany, all of which was still extremely backward. That backwardness was certainly connected with the fact that the French language was not even spoken in lower Brittany; and the same phenomenon may be observed in the Basque country, which stands in sharp contrast to the Béarn valleys (a fact that explains the strong opposition between the departments of the Basses-Pyrénées and the Hautes-Pyrénées). However, the linguistic correlation is by no means a general rule. On the eve of the 1870 war and annexation by Germany, German-speaking Alsace could boast one of the highest levels of literacy in French among its conscripts. (The Bas-Rhin and the Haut-Rhin departments came, respectively, sixth and ninth in the ratings for literacy in 1865; similarly, in 1866, the Bas-Rhin was the department with the lowest level of illiteracy: only 5.44 per cent of the population over the age of five.[71]) The explanation is that this was a well-to-do region characterized by grouped villages. Furthermore, under the Second Empire, the map showing the non-French-speaking percentage of the population differs significantly in the south, from the map showing levels of literacy. Part of the population speaking the *langue d'oc* was also literate in French (in the south-east generally, particularly Provence, and in Béarn), whilst part was not (in Aquitaine generally and in Limousin). Along the Mediterranean coastline, the concentration of settlement favoured the progress of literacy. The most backward regions of all were the sparsely inhabited rural areas of the south-west.

Societies and living standards: regions of affluence and regions of poverty

Illiteracy correlates far more closely with social factors than with the persistence of regional languages. In the mid-nineteenth century, for example, the correlation between ignorance and sharecropping is startling, for sharecropping, as opposed to tenant farming or direct exploitation of the land, is a major indication of agricultural archaism. In 1866, the 'Atlantic triangle' of illiteracy was clearly reproduced on the map showing the percentage of sharecropping as opposed to other types of farming (fig. 32A). In Brittany however, there is already a discernible difference, for although sharecropping was common here under the Ancien Régime, tenant farming took over at a relatively early date. The phenomenon may be regarded as one of several which confirm the Breton ability to adapt to new economic situations, particularly agricultural ones, despite the region's evident cultural backwardness – a point which is again clearly illustrated by the Second Agricultural Revolution of the twentieth century.[72] The other maps (fig. 32B) show the progress and achievement of social emancipation in Brittany. In contrast, the predominance of sharecropping in the Aquitaine Basin and to the south and west of the Loire generally, with lingering vestiges on the borders of the Armorican peninsula (Mayenne), was to persist right down to the eve of the Second World War, constituting clear evidence of the area's general social and economic backwardness. In Provence, a relatively small area was devoted to sharecropping (fig. 32C) yet the percentage of sharecroppers in relation to other kinds of farmers was high, indicating a social

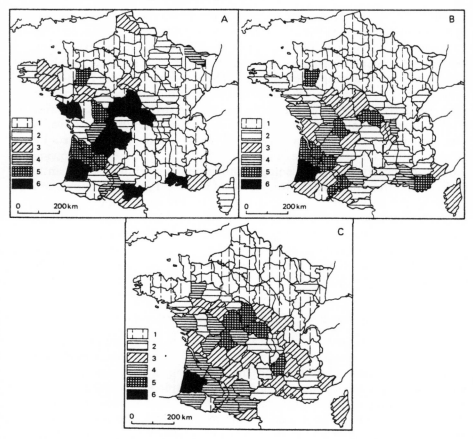

Figure 32. Sharecropping in France
A. The percentage of sharecroppers in relation to other kinds of farmers in 1866 (after Furet, I, p. 178). 1. Less than 6.65%. 2. 6.65–13.28%. 3. 13.29–19.93%. 4. 19.94–23.25%. 5. 23.26–29.90%. 6. Over 29.90%. B. Percentage of sharecroppers in relation to other kinds of farmers in 1931 (after Demangeon, 1946, I, p. 150). 1. Less than 1%. 2. 1–5%. 3. 6–10%. 4. 11–15%. 5. 16–30%. 6. Over 50%. C. Percentage of areas farmed by sharecropper (Demangeon, p. 151). 1. Less than 2%. 2. 2–4%. 3. 5–14%. 4. 15–33%. 5. 34–50%. 6. Over 70%.

system in which sharecropping was limited to a scattering of very small holdings. This points to the economic dualism of the region, reflecting an archaic social past that lingered on despite the relatively early cultural progress that was made, thanks to the establishment of a concentrated settlement pattern.

The social contrasts correspond to disparities in living standards. In northern and eastern France, life was already easy and economically progressive; western and southern France, in contrast, were poor and resistant to progress. By the mid-eighteenth century, the descriptions of rural prosperity featured in Contrôleur Général Orry's 1745 enquiry were already testifying to the higher standards characteristic of the northern and north-eastern open countryside (fig. 33A). In the first third of the nineteenth century, a shattering contrast was revealed by the military records of detailed information concerning the physical and professional characteristics of conscripts. These archives make it possible, for the first time, to produce maps based on a wealth of statistical material relating to the 1819–26 period.[73] The first contrast was of a physical nature. The height of men from the north-east was on average 5 to 6 centimetres greater. There are no genetic causes for this. The disparity has by now virtually disappeared and extremely precise analyses of the evolution of human physical growth in the twentieth century have been produced.[74] The elimination of the disparity may be attributed to the improving standards of contemporary life generally and this also proves, conversely, that the contrasts of the early nineteenth century were due to socio-economic factors, first and foremost levels of nutrition in terms of proteins and calories in early childhood. North-eastern France, which enjoyed a higher level of cultural development at this period, was also better nourished. That opposition was reflected in spectacular fashion in the mid-nineteenth century in the respective levels of output per agricultural worker in the two regions (fig. 34). There is yet more to be said on this subject. Maps showing the professional composition of this batch of conscripts indicate that the prosperous France of the north and north-east had reached a higher level of economic development, in which specialized activities of a non-agricultural nature played a greater part, whilst in the west and south the percentage of conscripts employed as agricultural workers was much higher. In northern and north-eastern France there were many more artisans and industrial workers. And in this part of France, the circulation of traffic was also much greater, as can be seen from the distribution of professions linked with transport: cartwrights and wheelwrights, harness makers and sadlers (fig. 33 C–E). These maps correlate with the late nineteenth-century map of communication routes, which were far denser in the north and the east (fig. 35). There are, to be sure, strategic reasons to account for this situation in these more exposed frontier regions. But it also clearly stemmed from the general level of economic activity. The general prosperity and high level of development were reflected, over and above the elementary level of literacy, in the greater proportion of members of the 'élite' (determined according to the criteria listed above: teachers, students from the Ecole Polytechnique and so on). In the west and the south, the only exceptions that stand out, for the reasons already given, were the valleys of the Béarn region (fig. 33F).

The proto-sociologists of the early nineteenth century were well aware of

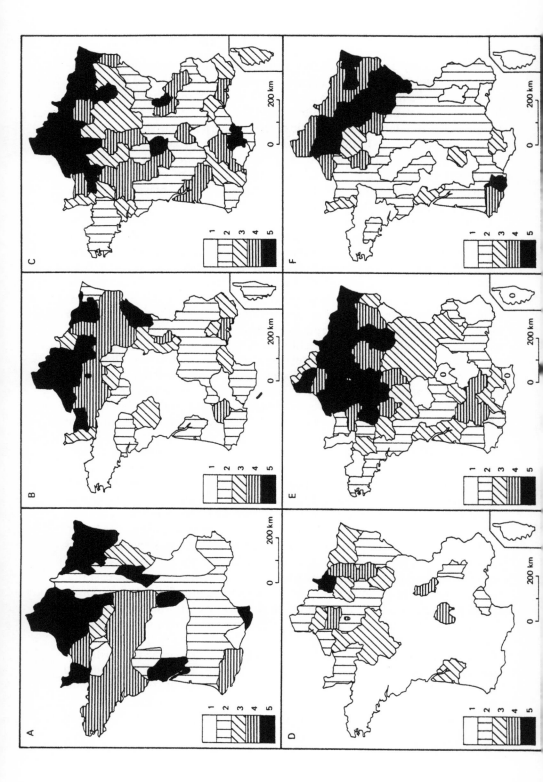

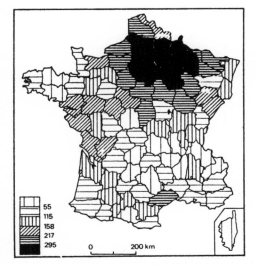

Figure 34. The level of agricultural production (per department and per agricultural worker) in 1862, in francs (after Clout 1977b, p. 434).

these major contrasts.[75] They had already plotted the fundamental line between Saint-Malo and Geneva which divided France into two different areas. The contrasts between the two have only disappeared in the course of the three decades of major development since the Second World War. It was a line which, in the last analysis, expressed contrasts of landscape and settlement which had been established in the Middle Ages.

Figure 33. Social contrasts in traditional France
A. Living standards in 1745, according to the Orry enquiry (after Dainville, 1952 and Fel, 1977, p. 226). 'The living standards of populations'. 1. 'Abject poverty'. 2. 'Poverty'. 3. 'Some do better than others'. 4. 'All make a living'. 5. 'Easy circumstances'. B–F. The proportions of different social categories among conscripts, 1819–26 (after Le Roy Ladurie and Dumont, 1971, maps 1, 4, 5, 8, 22). B. 'Tall' (over 1.65 m.). 1. Under 23.05%. 2. 23.05–29%. 3. 29.01–34.46%. 4. 34.47–40.96%. 5. Over 40.96%. C. 'Proportion of men employed in agriculture'. 1. Over 72.85%. 2. 61.30–72.85%. 3. 51.58–61.29%. 4. 43.40–51.57%. 5. Under 43.40%. D. 'Proportion of workers in leather, wood, iron and other metals'. 1. Under 11.88%. 2. 11.88–15.39%. 3. 15.40–18.29%. 4. 18.30–21.74%. 5. Over 21.74%. E. .'Cartwrights and wheelwrights'. 1. Under 0.20%. 2. 0.20–0.47%. 3. 0.48–0.80%. 4. 0.81–1.36%. 5. Over 1.36%. F. 'Elite' (= teachers, students from the Ecole Polytechnique and military and naval colleges, students from colleges preparing for careers in the civil service, young men with special qualifications). 1. Under 0.28%. 2. 0.29–0.47%. 3. 0.48–0.80%. 4. 0.81–1.36%. 5. Over 1.36%.

158 *Traditional organization of France*

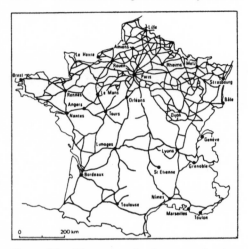

Figure 35. The network of communication routes at the end of the eighteenth century (after the map by Desauches produced in Year V and reprinted by Vidal de La Blache in Lavisse, 1911, I, 1, p. 379 and by Cavaillès, 1946, p. 162).

6

The secondary divisions

Spatial organization: territorial units

Two levels of organization: 'pays' and 'provinces'

The major contrasts studied above, which structured the French territory as a whole, made little or no impact upon the popular consciousness. Even linguistic differences passed more or less unnoticed, so gradual was the transition from one dialect to another. The perception of such differences was a matter of scholarly knowledge. But at a lower level, there were more detailed divisions which were expressed and defined by particular names and were, on that account, more or less perceived as realities. In traditional France, those names fell into two categories, corresponding to a higher and a lower level of organization, both deeply rooted in the past and stemming from the territorial organization of Gaul.

At the lower level, there were the names of *pays*, below which the only toponyms were those designating places connected with particular people, the most immediately perceptible of all geographical units. Many of the *pays* of medieval and modern France were certainly directly derived from the Gallo-Roman *pagi*, themselves more or less a legacy from the divisions of independent Gaul. Actually, only about thirty names of *pagi* dating from the Roman period are known to us and only five or six of those go back as far as the period of independent Gaul.[1] The vast majority are known from later texts, most of them Carolingian, and many others must have appeared even later. But their very number, calculated from the exhaustive lists that have been made (281[2] or 313[3] for Gaul as a whole, 242[4] for the territory of present-day France), is close enough to the total (300 or 305) given for basic divisions in Gaul by ancient sources dating from the time of the Roman conquest (see above, p. 10); and this suggests that the general principle according to which the territory was divided up remained the same and that these were indeed the same kind of elementary territorial units. During the Carolingian period, the term *pagus* was replaced by *comitatus*: *comté* (county). These are estimated to have numbered 258 in the reign of Charlemagne, 214 in the mid-tenth century.[5] Except for that slight fall in numbers occasioned by the state of disorder throughout the territory at this period, the general size of these zones underwent little change. When, at the expense of the feudal lords, the royal administration little by little regained control of the territory, the divisions that it introduced in the various areas of its sphere of activity reconstituted similar units. The bailiffs and seneschals, who had originally been the fully empowered representatives of royal authority, numbered 421 in 1789, on the eve of the Revolution.[6] At this period, the financial administration was based upon 453 constituencies.[7] These administrative divisions of the Ancien Régime were the result of a complicated history involving a number of stages of patchwork fragmentation that testified to the increasingly effective control imposed upon the kingdom; and they varied considerably in size. In 1720, in the Paris Basin, where the average area of the 114 'elections' (financial sub-divisions) was 1,369 square kilometres, the election of Le Mans covered 4,186 square kilometres, that of Bourges 4,184 square kilometres.[8] In those circumstances, the notion of an average area may seem meaningless in concrete terms. But in fact that average size was close enough to the size of the *pagi* of Gaul. These were areas of between 1,200 and 2,000 square kilometres, circles with a 20 to 25 kilometre radius, and they correspond to the elementary unit of organization which, in a variety of forms and with a variety of functions, has continued to be used down to the present day.

Under the Ancien Régime, there existed, above these basic units, a higher level of territorial organization, involving units of much larger dimensions. These took various similar, but not identical, forms, elaborated in an empirical fashion as the great feudal fiefdoms were gradually incorporated into the royal domain and the authority of the monarchy was progressively extended. From the fourteenth century on, the *gouvernements* ('governments') gradually developed.[9] These were initially units of an essentially military nature set up in frontier regions, but by the end of the fifteenth century they had also been introduced in the interior of the kingdom and had taken over many administrative and legal functions from the groups of bailiwicks that had formerly been responsible for these various areas. In the sixteenth century, there were eight of these *gouvernements*, but gradually they multiplied so that by the eve of the Revolution there were forty of them, seven of which were extremely small. By this time, the functions of governor, though still lucrative, were hardly more than honorific (fig. 36). Meanwhile, from the sixteenth century onwards a system of administrators known as *intendants* had also been set up.[10] These were civil servants (*intendants de justice, de police et de finances*) who little by little took over most of the

The secondary divisions 161

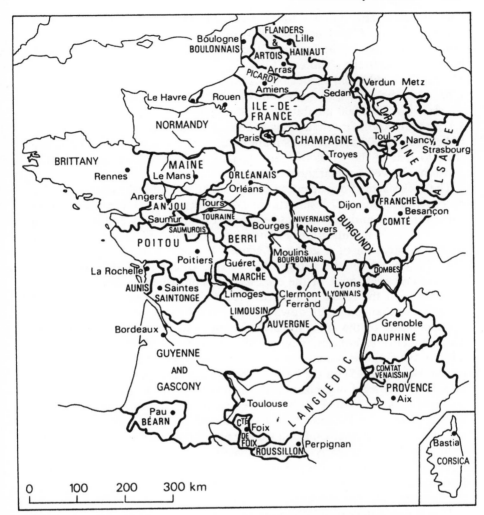

Figure 36. The *gouvernements* in France in 1789 (after Mirot, 1947, II, p. 361).

functions of the governors. By 1789, there were thirty-three *intendances*, a number equal to that of the larger *gouvernements*. By the eighteenth century, the term *intendance* was virtually synonymous with *généralité*, the term used for a financial unit whose more ancient origins went back to the fourteenth century. The *intendances* and the *gouvernements* coincided to a large extent, twelve of them almost exactly. In many cases, a *gouvernement* corresponded almost exactly to two or three *généralités*, or sometimes the converse was true.[11] At any rate, as can be seen, the total number of these upper-level administrative and military units remained close to that of the *civitates* ('cities') of Gaul. (There had been sixty of these within a considerably larger territory than that of the France of the Ancien Régime.) They varied in size, but no more than the *civitates* of Gaul. In 1789, the largest *généralité*, that of Montpellier, embraced an area of 34,252 square kilometres and incorporated 1,700,000 inhabitants; the *généralité* of Rennes covered 28,392 square kilometres with 2,270,000 inhabitants. The smallest, that of Valenciennes, was 4,116 square kilometres in area, with 265,000 inhabitants.[12] With an average of between 10,000 and 20,000 square kilometres, the size of these units was close to that of the tribes of Gaul, so that at this level too the pattern of divisions showed a remarkable consistency.

The ecclesiastical organization of the territory[13] looked rather more disorderly (there were 177 bishoprics in 1789), but was not intrinsically different. Ever since antiquity it had remained extraordinarily stable, except in respect of a few modifications of later date, the only one of any significance being the creation of a group of sixteen bishoprics in the south-west in 1317–18, to combat the Albigensian heresy. Under the Late Empire, the borders and *chefs-lieux* of the ecclesiastical constituencies had taken over from those of the Roman administrative divisions and they had remained virtually unchanged right down to the Revolution, with the bishoprics corresponding to the *cités*, the archbishoprics to the seventeen provinces. The number of bishoprics, in excess of one hundred, reflected the fragmentation of the Gallo-Roman *civitates* which took place under the Late Empire (see above, pp. 73–5) on the basis of the same general principles of division as before. The legal organization of the territory, for its part,[14] had a very different look to it. By 1789 it included no more than thirteen *parlements*, one of which, the Parlement de Paris, served a gigantic area of roughly 160,000 square kilometres. But a similar pattern of smaller units had in fact been established as early as the sixteenth century, with the *présidiaux*, primary courts of appeal operating at a lower level than the *parlements*. Initially, in 1552, thirty-two of these *présidiaux* had been set up, but by 1764 the number had risen to a hundred or so.[15]

This upper category of divisions was based upon a territorial concept which evolved over the centuries of absolute monarchical rule: namely the 'province'.[16] The term never acquired a precise administrative meaning, but from the end of the fifteenth century it was used to denote well-defined parts of the kingdom, tending at this period to supplant the old feudal terms *duché* and *comté*. Rather than speak of the 'duchy of Brittany' or the 'county of Champagne', the term 'province' would be used. The switch in vocabulary reflected a revived interest in antiquity. 'Province' (*provincia*) came back into fashion, as did the military term 'legion' and also 'senate', now used to refer to the *parlements*. The kingdom of France, where the authority of the monarch was expanding and becoming consolidated, was now composed of a collection of subject 'provinces'.[17] Despite its lack of legal definition, the term did acquire an official use, above all in the expression *Etats provinciaux*, applied to the assemblies of the *pays d'Etats* (the most important of which were eventually Brittany, Burgundy, Languedoc and Provence) which have retained the privilege of fixing and voting on their own taxes. Between the seventeenth and the eighteenth centuries 'province' became part of the current vocabulary of administrators, historians and jurists. It was in general use by the end of the eighteenth century, having become such an integral part of French consciousness that the ruling of June 1787 created 'provincial assemblies' in the regions known as *élections* and it was the provinces – despite the fact that they had no existence in real terms – that were ordered to be divided into departments by the letters patent of 4 March 1790.

In truth, what constituted a province was a debatable question; and the lists of 'provinces' drawn up by different authorities at different times are by no means uniform. Fifty-eight appear in the Doisy list of 1753 and the same figure was retained by Ricci Zanoni ten years later, but even so the two lists are not identical. In 1790, fifty-two names were included in the 'table of provinces invited to nominate Commissioners' by the *Comité de division* set up by the *Assemblée Constituante*, yet eighty-seven were enumerated in the royal letters patent of 4 March 1790, which ordered the division of the provinces into departments.[18] All these hesitations are symptomatic of the heterogeneous nature of the Ancien Régime's administrative divisions, which had evolved only gradually and in the absence of any overall conceptual framework. However, the concept of the 'province' was challenged by no-one. The number of provinces listed in 1753 by Doisy and in the 1790 'table' is remarkably close to that of the *civitates* of Gaul.

What was the nature and significance of these two categories of divisions? How were they formed and how did they evolve? What did they represent and what were their functions?

The historic 'pays': from the 'pagus' to the 'pays'

The names are the first thing to study. An analysis of the names of the *pagi* is most revealing, even if the origins and meanings of a few remain unclear. Some were the names of Celtic tribes and peoples of secondary importance, which do not appear ever to have been adopted as names for towns: *Pagus Caletus* > Pays de Caux, from the name of the *Caleti*. These ethnic names provided almost all the names of the *pagi* attested in the Roman period in the Three Gauls (the *pagi* of Narbonnaise Gaul, most of which bear the obviously Latin names of particular families or deities, were probably for the most part renamed). If we then consider the whole collection of names of *pagi* known for the Carolingian period, most turn out to have been given to towns: *Pagus Vindocinus* (Vendôme) > Vendômois; *Pagus Tolonensis*

(Toulon) > Toulonnais; *Pagus Bellovacensis* (Beauvais) > Beauvaisis and so on. Whether the name of the town is derived from that of the *pagus*, particularly where the latter was the name of a Celtic people, or whether – as may have been more frequently the case – it was the other way round, this testifies to the early development, in Gaul, of a web of small towns which identified with their field of influence within the framework of a life of active interchange. Gradually, as the ethnic identities of the Gallic tribes were weakened by Romanization, these towns came to constitute the organizational basis of the surrounding territory, a basis that the subsequent devastation of the countryside at the time of the Germanic invasions had the effect of consolidating. Some of these small towns bear the names of people, large-scale landowners or influential individuals, the erectors of public buildings and the founders of urban life. Others are named after particular localities which have disappeared but some of whose names survive for particular landmarks, as in southern Mâconnais, in Burgundy, where the name of the *Pagus Tolvedunensis* is preserved in Mont Tourvéon (950 metres), formerly the site of a Celtic *oppidum*. A settlement clearly still existed here at a relatively late date, but it was then deserted and the name passed to the mountain.[19] All this testifies to an unmistakable continuity between Gaul and medieval France.

However, continuity is not the same as identity. Camille Jullian has emphasized the permanence of the *pagi* of Gaul, arguing that they became the *pays* of later periods. In some cases, he is undoubtedly right, but in others serious modification seems called for. It is true that most *pagi* were set up on sites which are geographically permanent and which provided a focus for the surrounding area, but the infinite variety of those territories reflected the instability of human societies in general: the territories changed as the societies that occupied them did.[20] The first factor to take into consideration is the historical evolution of the occupancy of the land. In the north and the east, the Germanic invasions clearly brought considerable disruption to the network of small towns. Many fell into ruins and new names appeared. Some were the names of Germanic tribes, as in Burgundy, in the ancient *cité* of Langres, where *Pagus Attuariorum* (> Atuyer) comes from the name of the *Hattuarii*, a Batavian or Frankish tribe, part of which probably came to Burgundy under the leadership of Constance Chlore, and which certainly became integrated into the Roman Empire, to the extent of becoming the focal point of the surrounding territory. Another example is provided by the ancient city of Besançon, in Franche-Comté, where the *Pagus Amavus* (= *Chamavus*) > Amous is derived from the Frankish *Chamavi*. Many of these names are applied to rivers, points of reference which perpetuate them even when the towns have disappeared. In certain cities of the north-east, names such as these multiplied and became predominant: for example, three out of six in the diocese of Metz, where we know of a *Pagus Nitensis* (of the Nied), a

Pagus Blesensis (of the Bliess), and a *Pagus Rosalinsis* (of the Rosselle), all derived from the names of these tributaries of the Sarre. Some names have an even vaguer application, denoting the more or less deserted and abandoned nature of the region, as in the case of the *Pagus Wastinensis* > Gâtinais (wasteland) to the south-east of Paris.

Next, we should take account of the upheavals that affected the very concept of administrative organization. In the Merovingian period, the *pagi* tended to be confused with the cities of the Gallo-Roman period, many of which had disappeared (see above, pp. 73–5). Many dioceses, most of them to the south of a line running between Lyons and the bottom of the bay of Normandy and Brittany, only ever consisted of a single *pagus*, that which was represented by their *chef-lieu*. This clearly indicates that many urban settlements disappeared during the turmoil brought by the invasions, leaving only the largest towns in place. In the north, in contrast, where the power was now concentrated, the *pagi* multiplied once again after the division of the kingdom, possibly partly so as to make more posts available for civil servants.[21] There is even one – exceptional – case, on the borders of Burgundy and Lorraine, where the *Pagus Basiniacus* > Bassigny was named after an individual, Count Basin, an official of the late Merovingian or early Carolingian period.[22]

If these territorial units varied in so many ways, so, *a fortiori*, did their boundaries, even where the *chef-lieu* or regional centre remained a fixed point. The *pagi* or *pays* were living entities which might either expand or contract. When the *comtés* became hereditary territories, stabilized by the feudal system, their boundaries were usually fixed according to the totally new relations of strength that existed. For example, the *comté* of Montbéliard, as established in the eleventh and twelfth centuries, presents no features of continuity with the Carolingian *pagus* of Ajoye (*Pagus Alsegaugensis*, in German: Alsegau) despite the fact that it fulfilled similar functions relating to the defence of the Gate to Burgundy. Nor do its boundaries correspond to the Germano-Burgundian ethnic frontier. This is clearly an instance of the emergence of an entirely new territorial entity.[23]

Throughout all these fluctuations, the stabilizing element was, as can be seen, essentially constituted by the town. When the town disappeared, the *pagus* lost its identity and had to be remodelled on different bases. On the other hand, once a new settlement developed as a seat of power, that of a feudal lord in the first instance, subsequently that of a royal representative, a new *pays* was created, with boundaries that corresponded to the limits of the town's sphere of influence. In the period of the absolute monarchy, in the face of the proliferation of administrative divisions of every kind, with boundaries that did not coincide, the concept of the *pays* lost its precise territorial definition and tended to become confused with the idea of the urban territory, the space dominated by the town. In the last analysis, every town worthy of the name had to have its own *pays*, which was in many cases designated by a simple suffix (*-ais*, *-ois*, *-en* and so on) tacked on to the name of the town itself. In Delamare's *Traité de la Police* (1729), southern Champagne is considered to contain as many *pays* as towns, each one being defined by its capital: the *pays* de Langres, de Chaumont, d'Epernay, de Châlons, de Vitry (or Perthois), and so on.[24] The last case provides an interesting example of transformation. The *Pagus Pertensis* (Perthes), whose name survived in that of the *pays* 'Perthois', is replaced by the eighteenth-century author by the expression '*pays de Vitry*', based on the name of the town, which had become dominant.

In the north and the east in particular, the fabric of small towns underwent profound changes after the Gallo-Roman period. Many new centres emerged in the feudal period. At first these were gathered around the castles, for a feudal lord would naturally attract a variety of artisans to supply his own needs and those of his warriors – weapons and clothing, for instance. At the same time though, others would be attracted by the protection that such proximity might afford them. Soon the nucleus of craftsmen would also be working for the peasants of the neighbourhood, who would bring their own goods to town. A weekly market or a fair would become established, perhaps initially located in the castle yard, but quite soon transferred beyond its walls. Many small towns thus sprang up around castles. They are easily recognizable from their names using Château- or Castel-: Chatillon (= little castle), Castelnaud (= new castle), Châteauroux, Castelnaudary, Château-Thierry and so on. Furthermore, in many cases, places bearing more ancient names, where a castle was added in the Middle Ages, also owed their urban development to its presence (for example, Niort, Alençon, Falaise, Dieppe, Saumur and so on). In the Middle Ages, the presence of a castle was thus the normal and principal reason for the emergence of the small, new urban centres which became incorporated into the major urban fabric left from the Gallo-Roman period. One manifestation of this phenomenon was the elaboration of a new administrative geography based on units essentially centred around points of fortification. In the course of the twelfth century, the term *châtellenie* gradually came to replace the Carolingian *vicairie* (*vicaria*). Its dimensions were roughly those of one or two present-day cantons, incorporating twenty or so villages, headed by a small town which might well acquire urban functions. This picture would not be complete without the monasteries which, throughout western Christendom, also played an important role in the development of towns. In France, this phenomenon came later than in Germany, but it proved just as vigorous. The Benedictines (Cluniacs and Cistercians) established towns of two kinds. The first type grew up around the monastery itself simply through the creation of a market. In the tenth and eleventh centuries, the revival of urban life frequently took the form of the foundation of a temporary market alongside a monastery, and with the passing of time the market would become a more or less permanent feature. With just such an intention in mind, many feudal lords granted land concessions to religious communities. The other kind of town for which the monasteries were responsible grew out of the monks' systematic colonization of the land. This was particularly common in the thirteenth century, when the Cistercians faced a crisis of recruitment. Unable to cultivate their land themselves, they set up new villages around their barns and many of these, like the *bastides* in the south-west (see above pp. 143–7), eventually assumed the functions of small urban centres.

While new small towns were emerging in this fashion, many others were disappearing or declining into modest villages. Many Carolingian *pagi* thus ceased to correspond to any real territory, although in several cases the names themselves survived, particularly in the scholarly tradition. What did remain more or less constant amid the transformations in the map of small towns which evolved over the centuries was their numerical level. In the mid-sixteenth century, Charles Estienne, in his *Guide des chemins de France*, cites 230 towns which he considered worth mentioning, in a territory two-thirds the size of Gaul and over four-fifths the size of the France of 1789. Coming as it does from a period in between that of the Carolingian *pagi* and that of the bailiwicks and seneschalsies at the eve of the Revolution, this information constitutes precious quantitative testimony regarding the nature of the basic organization of the territory, as distinct from the individual historical vicissitudes of these little towns and their surrounding spheres of influence.[25]

The provinces

From the Gallo-Roman cities to the provinces
The *pays* seem to testify to a measure of permanence at the first administrative level of the territory, despite all the changes of fortune through which these second-order urban centres passed. But as for the provinces, these were more complex spatial units, mostly established during the feudal period. The practice of periodically reassigning to various princes of the blood the privileges and duties traditionally associated with them helped to stabilize them right up to the point when they were definitively reunified with the royal domain. They were maintained as administrative divisions under the absolute monarchy, particularly in the form of the great *gouvernements* of 1789. Let us take the map of those *gouvernements* (fig. 36) as the starting point for our analysis. It shows, at a glance, that some of these units bear names of relatively recent origin (Boulonnais, Dauphiné, Dombes, Comté de Foix, Hainaut, Languedoc, Lorraine, Marche, Metz, Toul and Verdun, Roussillon) and that they came into existence as a result of certain historical

168 *Traditional organization of France*

events, in particular the episodes which determined their attachment to France in the medieval and modern periods. Most of these newer provinces are peripheral regions, for the territory was progressively expanding outwards. Others (Brittany, Burgundy, Flanders, Gascony, Normandy, Picardy) bear names of ethnic origin, testifying to the new settlements, for the most part also peripheral, established at the time of the great invasions. This category should be extended to include the name of the province of Alsace which, though undoubtedly of Celtic origin and whatever its etymological significance,[26] made its appearance only in the Merovingian period; and equally the province of Aunis, probably named after an ancient *pagus*, but which only acquired its administrative importance in the modern period.[27] The remainder of the provincial names (Anjou, Artois, Auvergne, Béarn, Berry, Limousin, Lyonnais, Maine, Nivernais, Orléanais, Poitou, Saintonge, Touraine) all stem directly from Gallo-Roman cities. This category should be extended to include Provence, the heir to the Roman *Provincia* and Guyenne, which is a deformation of the name *Aquitania* (> Aguyenne). It should also be pointed out that many of the secondary divisions within the largest *gouvernements*, which are also included in the eighteenth-century list of 'provinces', particularly those within the vast *gouvernement* of Guyenne and Gascony (70,000 square kilometres) also bear the names of Gallic cities. This testifies to a remarkable stability in the pattern of territorial divisions at this level, as well as at the level of the *pays*.

What is the explanation for that continuity? To understand the structure of these spatial units, we must take into consideration the circumstances and conditions of their formation in the feudal period. In a number of cases, good analyses have already been produced.[28] The starting point is invariably the existence of a sizeable town. In other words, the major feudal developments normally tended to take over from pre-existing urban centres wherever these had survived. A network of castles then developed, some of them initially no more than forts, and this came to determine the way in which the territory was divided up. The boundaries of the areas for which the castles provided a focus were for the most part determined by natural obstacles, usually forests or mountains, which had earlier marked off one ancient city from another. In principle, the pattern of feudal divisions is likely to coincide with the Gallo-Roman one. The differences, the new features, were introduced essentially as a result of 'major historical' political events: ethnic boundaries were shifted at the time of the invasions; the interplay of inheritances and matrimonial alliances left its mark; and finally, the struggles for influence between emerging national entities, prompted by the needs and ambitions of the modern period, all resulted in a redefinition of frontiers. Other new factors were connected with the changes in the hierarchy of major towns that had occurred since ancient times and, in particular, with the new commercial trends that emerged in the Middle Ages.

These favoured the development and prosperity of some thitherto secondary urban centres and forced others into decline. All in all, the structure of the provinces reflects a number of geographical constants affected, to a greater or lesser degree, by historical events. The expression 'historic regions', which is often used by contemporary geographers, is an apt enough description for this interaction between nature and history.

The structure of the historic regions: the model of northern France.
A productive central area with unproductive margins[29]

The most common model of such political feudal structures was, logically enough, a province pivoted on an agriculturally rich and densely populated area. The starting point for the development of these historic regions, the focus around which they crystallized, was usually provided by limestone and silty plains with well-drained soils, capable of producing regular grain harvests. From early times (see above, pp. 137–41) these had been densely populated areas and, in northern France at least, they were characterized by a concentrated settlement pattern consisting of large village clusters surrounded by open fields. In these rich agricultural plains towns developed, most of them on the sites of the *chefs-lieux* of the Gallo-Roman cities. A feudal historic region would grow out of a rich agricultural territory and its principal town. It would be surrounded by unproductive margins of ill-drained soils of clay or sand or by wasteland, still widely wooded, where frontiers long remained undefined. Many of these disputed sectors acquired the medieval legal status of 'separating marches'.[30] Their inhabitants were dependants of both the feudal lords along whose boundaries they lived and they paid half their dues to the one, half to the other. However, in consideration of the many inconveniences to which their situation exposed them and the frequent passage of armed men, they were usually exempted from the taxes and services designed to promote war efforts. Even in the modern period some of those villages did not – or at least claimed not to – know to which feudal lord they belonged. In the thirteenth and fourteenth centuries, in northern and eastern France, official enquiries were set up to determine which villages were subject to the Emperor and which to the king of France. Some certainly made the most of the situation: for instance, the nationality of Clinchamp (in the present-day department of Haute-Marne), situated on the borders of France and Barrois, had still not been decided in the eighteenth century and the village had consequently managed to dodge virtually all taxation.[31]

This model of an historic region centred on a fertile grain-producing heartland, with outlying unproductive areas given over to woodland and pastureland, possessed the additional advantage of a balanced economic structure, with other crops to complement the cereals. This seems to have been the normal type for a region in northern France, particularly in the

Paris Basin, where advanced levels of cultivation and dense population had developed relatively early on the limestone and silty plains. The model reappears in a number of places which differ from one another only as to the respective number of typical features present and their particular arrangement.

Champagne and Berry are two perfect examples, both with a central nucleus surrounded by a ring of marginal areas. The centre of Champagne consists of the kind of open chalky plain – *campania* – which in the ancient economy constituted the epitome of good grain-producing land.[32] Despite an absence of silt, light, warm soils resulted from chalk fissured by quaternary freezing on the gentle hill saddles covered by a short grassy sward (*savart*). Meanwhile, the heavier soils of the valley-beds, dotted with the occasional clump of oak trees (*garennes*), are mixed with detritic flows accumulated by solifluction dating from colder periods, and these too always produced reasonable harvests of grain. This was clearly one of the reasons why Caesar decided to winter with his legions among the *Remi*, who were allies of his.[33] All around the area topped with chalky soil, abandoned peripheral land constituted a continuous surrounding belt. To the east lay Champagne Humide, with sandy and clay soils of the Lower Cretaceous and beyond, along the horizon of hard siliceous *gaize* (friable sandstone with slight siliceous cement), the wooded hills of Argonne. To the west, beyond the Eocene *cuesta* of the Ile-de-France, lay Haute Brie Orientale, where old erosion surfaces topped with a siliceous crust intersected. This was ill-drained land, waterlogged and unproductive and scattered with ponds. To the south lay the Pays d'Othe, wooded land where the chalk is covered by a layer of flinty clay, produced by decomposition in the Lower Tertiary, and Gâtinais, which took over from it and extended westward. To the north, finally, lay the impermeable chalky marl of Thiérache. The area as a whole was structured by nature in a coherent and uncompromising fashion and the only question to be resolved was that posed by the presence of two rival ancient urban centres, both situated in a chalky strip 200 kilometres long. These had been the *chefs-lieux* of two large Gallic tribes: Rheims (*civitas Remorum*) and Troyes (*civitas Tricassium*). The feudal territorial unit adopted Troyes as its centre,[34] for Rheims, a holy and royal city, where French kings were crowned, was soon pulled into the sphere of influence of the Parisian monarchy. In Berry, the central layer of Jurassic limestone – Champagne Berrichonne – is more compact in shape, and Bourges (*Avaricum*), already one of the largest towns even in the days of independent Gaul, inevitably became the region's centre, the head of the city of the *Bituriges*, the structure of which was adopted by the feudal province. Disposed around the limestone nucleus was a less productive peripheral belt: to the north, the Cretaceous clay soils of the Pays Fort, on the edge of the Portlandian *cuesta*, which were soon covered over by the sands of Sologne; to the west, Brenne, made up of accumulated Eocene detritic strata, with a scattering of ponds; to the south, the peripheral clay-marl soils of the low-lying fringes of the Massif Central, Boischaut and Val de Germigny; the eastern borders were formed by the hills of Sancerrois, where the limestone table rises and splits into numerous cracked blocks overlooking the gulch formed by the course of the Loire.[35] The unity of this historic region did not go unchallenged. In the tenth century, the Carolingian *pagus* constituted by the city of the *Bituriges* split into two, one part looking to Aquitaine, the other to the feudal houses of Blois and Champagne. Right down to the end of the twelfth century, there continued to be a split between the high eastern Berry and the lower western Berry, the latter being attached to the estates of the Plantagenets. However, the installation of the Capetian monarchy at Bourges soon made it possible to reunite the province. The decisive factors in this process of reunification were the persistence, throughout, of a single ecclesiastical unit, with its seat at Bourges, and the major role played by this town, with which Déols-Châteauroux, the *chef-lieu* of lower feudal Berry, could not seriously compete.[36]

The principle behind the organization of all the rest of the Paris Basin is the same, although here it is less perfectly realized. The geographical structure of Picardy also approximates closely to the theoretical schema described above. The natural central focus is constituted by the productive soils of the chalky plain of Picardy, much of it topped with a layer of silt. The wooded, unproductive margins of the region are well defined: in the south-west by the incision of the anticlinal hollow constituted by the Pays de Bray, where marl and clay soils from the Lower Cretaceous come to the surface; in the north-west by the comparable area of Boulonnais, while along the coast lie the marshy plains of the Bas-Champs and Marquenterre, reclaimed from the sea and drained only in the thirteenth century, which for a long time remained uninhabited.[37] To the north-east lies the wet, Cretaceous land of Thiérache, while in the south, the Tertiary cliffs of the Ile-de-France display sandy, wooded slopes. It is only in the north that the boundary is less clearly defined, on the Artois side. The limestone hills of Artois, which constituted the core and basis of the city of the *Atrebates* and of historic Artois used to be separated from the plain of Picardy by the Arrouaise forest, which marked the border (fig. 4A).[38] But in the Middle Ages, most of this was cleared and soon nothing remained except for a few wooded strips. Not until one reaches the low-lying plain of Flanders does any truly natural change in the landscape occur. Similarly, towards the east, the boundary with Champagne is ill-defined in the Vermandois and Laon regions. The plain of Picardy, such as it was, nevertheless constituted the heart of the Gallic city of the *Ambiani* and during the Roman period it had enjoyed considerable prosperity.[39] However, the ethnic upheavals produced by the Germanic invasions made a heavy impact here, making the ancient territorial divisions pointless. The origin of the name of Picardy is not clear (possibly a Frankish tribe of *Picards* or 'pike-bearers') and the name itself did not appear until the twelfth or thirteenth century. In the Middle Ages, Picardy was characterized purely linguistically, as a domain where one of the most distinctive dialects of the *langue d'oïl* was spoken, although this language was also used beyond the borders of Picardy itself, throughout northern France in fact, as well as part of present-day Belgium. As a province, Picardy was resurrected in the fifteenth century in the form of a military *gouvernement*, as part of the strategic organization of the French monarchy, which needed just such a structure to defend this close and particularly exposed frontier. The province of Picardy incorporated not only the city of the *Ambiani* but also that of the *Veromandui* (Vermandois). The administrative borders varied considerably, being for a time extended to take in certain northern sectors of the Ile-de-France. Even today, the geographical concept of Picardy remains undefined and has little practical significance.[40]

Lorraine is another example of an historic region centred upon a similar natural core but which, for medieval political reasons, expanded beyond the boundaries of its ancient cities. Its heart consists of the jurassic limestone plateaux to the east of the Paris Basin, bounded by the impermeable Lower Cretaceous ring to the west, the hercynian Vosges mountains to the east, and the Ardenne and the schistose Rhineland Massif to the north. But the natural boundaries are much less clearly defined to the south, where a whole sector of the limestone plateaux overspills into the orbit of the plains of the Saône and the Burgundian state. Southwards the plateaux extend for a long way but are broken and divided by many physical features: not only lines of hills (*cuestas*) but also the great river valleys of the Meuse and the Moselle, so a number of natural territorial boundaries could have been viable. Several ancient cities had been set up here: those of the *Mediomatrici* in the north (Metz), the *Leuci* in the central region (Toul), and some of the *Lingoni* in the south (Langres). There were three *chefs-lieux* to control the essential intersections of overland routes and waterways. In the Gallo-Roman period, particularly at the time of the Late Empire, the most advantageously situated was Metz, commanding the confluence of the Moselle and the Seille, at the intersection of the great south–north route leading from Lyons to Trèves and Cologne, and the west–east route connecting the Channel with the Rhine by way of Boulogne, Rheims, Verdun and the Saverne pass, to the north of the Vosages. In the tenth century Metz was known as 'the head of all Lotharingia'.[41] Moreover, a series of towns were founded at points where the Moselle could be crossed, on the border of the limestone plateaux and the plains of the Vosges foothills, composed of sandstone and clay soils. These, in combination, possessed a far greater potential than the towns strung out along the Meuse, a linear feature which cut through the limestone plateaux but was joined by almost no transversal tributaries. It was here that a major political structure was to crystallize. In contrast, the Barrois, the political structure based upon the Meuse valley, which emerged in the feudal period, had a difficult life, torn between France and the Empire, and eventually returned to the orbit of Lorraine in the fifteenth century. However, when Lotharingia had been split into two separate duchies in 959, at which time only the more southern one, Haute Lorraine, retained the name created at the Treaty of Verdun, the imperial city of Metz, first a bishop's seat, then a Patrician republic, passed out of the hands of both the dukes, as did two other bishops' seats, Toul and Verdun. Lorraine was left bereft of towns. (Metz was, in any case, located too much on the edge of the new territory and too close to the ancient massifs of the north to constitute a useful capital.) In the eleventh century, the dukes of Lorraine accordingly built the castle of Nancy. They positioned it at the approximate centre of their possessions, at the foot of the Moselle hills, in a situation which offered them rich hunting grounds in the Haye forest on the other side of the ridge (*cuesta*) and at a point equidistant from the two gaps in that escarpment, namely the valleys of the Moselle and the Meurthe. To the south, however, towards Burgundy, the limits of Lorraine always remained purely artificial, expressing no more than the existing balance of power on these limestone plains.[42]

Beyond the barrier constituted by the Vosges, Alsace, for its part, was another direct and inevitable creation of the natural environment. It was not possible for the Rhine valley as a whole to constitute a single unit. The easily flooded Rhine plain, the Ried, still marshy and wooded until a relatively late period, constituted a frontier not easily crossed. The Merovingian duchy of Alsace, bounded by mountains, then virtually uninhabited, and the riverside forests, included part of the dioceses of Strasbourg and Basle and was organized around the productive terraces of the Ill valley, with its loessic surface layer, and the pebbly banks and hills of the Piedmont region. Within the framework first of Lotharingia, then of the Empire, and under the influence of the urban prosperity fostered by the presence of the Rhine waterway, a whole cluster of feudal estates and free towns developed. Meanwhile the length and narrowness of that useful waterway made it necessary to divide the area into two landgraviates, namely Haute-Alsace and Basse-Alsace (the former being a direct possession of the House of Austria). The plain of Alsace was eventually reunified when it was annexed by France according to the treaties of Westphalia (1648) and the subsequent decisions made by the *Chambres de Réunion*. Its southern boundary was, quite naturally, constituted by the outcrops of the Jura and the '*Porte de Bourgogne et d'Alsace*' (Gateway to Burgundy and Alsace)'. The northern frontier, which sliced across the Rhine valley, was dictated by a more artificial political balance and was marked out by a number of historic developments (the archbishoprics of Mainz, the Palatinate and – later on –

Hesse) on the banks of the middle reaches of the Rhine, in the region of the confluences of the Neckar and the Main.

Normandy, finally, resulted from an arbitrary ethnic division, which was prompted by the invasions of external peoples and superimposed upon the Gallo-Roman territorial units. After it became part of France, it evolved towards a bi-polar structure in part determined by its natural geography. In this coastal area, the various belts of sedimentary deposits had been levelled down by surface erosion many of which were then covered with new deposits and had remained virtually intact. These were interspersed with discontinuous patches of silt, the only major accident of relief being the Seine valley. A large number of focuses for regional polarization were thus possible. The river, which constituted an important commercial artery, was naturally the most obvious. The *Caleti* from the Pays de Caux, where the chalk was mostly covered by a layer of siliceous clay, and the *Veliocassi* from Vexin, where silt appears, had even in ancient times gravitated around capitals situated in the valley: respectively *Juliobona* (> Lillebonne), on the estuary, and *Rotomagus* (> Rouen), the first bridgehead on this tidal river. The site of the latter was determined by an unusual accident of nature constituted by a large area unaffected by flooding (a hill detached from the plateau by a network of folds and cracks) positioned on the concave bank of a bend in the river. (Convex banks, which are liable to flooding, tend not to provide urban sites.) To the south of the Seine, the *Eburovices* (> Evreux) could depend upon the good cereal-producing land of the silty Plaine du Neubourg and Plaine de Saint-André. Further to the west, on the edges of the sedimentary basin and in the ancient massif, the small Armorican tribes of the *Lexovii* (> Lisieux, Lieuvin), the *Esuvii* (or *Sagii* > Sées), the *Viducassi* (> Vieux), the *Baiocassi* (> Bayeux, Bessin), the *Abrincatui* (> Avranches), and the *Unelli* (*chef-lieu*: Constantia > Coutances, Cotentin), who had not developed many towns, had not set themselves up as truly coherent territorial units. Under the Late Empire, the cities of the *Caleti* and the *Veliocassi* had been amalgamated, creating the first major centre in eastern Normandy. It was dominated by the influence of the large town of Rouen, but that influence did not extend southwards much beyond the river, as is indicated by the limits of this diocese, which remained essentially confined to the north side of the valley. Western Normandy needed another centre of polarization. As in Lorraine, this was provided by a new creation. A number of dukes had hesitated over the choice between various locations (in particular Bayeux). But in the second half of the eleventh century, at the time of the constitution of the Anglo-Norman monarchy, William the Conqueror built the castle of Caen. Its site was an escarpment overlooking the confluence of the tidal river Orne and the river Odon; it was thus well-positioned to guard the Orne waterway at the mouth of which Ouistrehem, a dozen kilometres downstream, constituted the duchy's great western gateway to England. At the same time, it controlled the productive wheatlands of the jurassic limestone, which soon became known as the Campagne de Caen. In these conditions, it was not long before a prosperous town was flourishing. In the thirteenth century, William the Breton was already describing it as *'potens, opulata ... decora ... agrorum fertilitate'*, rating it as highly as Paris: *'Ut se Parisiis vix annuat esse minorem'*.[43] This marked the start of Lower Normandy's evolution towards administrative autonomy, which was realized in 1542, when Caen was promoted to the rank of a *généralité*. Here in the Campagne de Caen, with its impermeable margins to the east (Lieuvin, Pays d'Auge, Pays d'Ouche) and to the west (Bessin, Cotentin), the bases existed for a territorial unit made up of districts of a complementary nature, bounded to the south by the hilly Bocage Normand and Perche. However, the limestone belt was too narrow and too limited and the vast wooded area was too intermittent and too irregular for Caen's supremacy in Lower Normandy to remain unchallenged. In 1636, the promotion of Alençon to the status of a *généralité* deprived it of all the southern margins of the province. The division between Upper Normandy, centred on Rouen, and lower Normandy, centred on Caen, nevertheless became an accepted mental concept and is even reflected in the present-day pattern of regional organization.

The same principle of polarization and complementarity can be seen to be at work in the western and south-western Paris Basin and in the Loire area, but finds less complete expression here. The areas of productive land capable of constituting the focuses of organization were extremely limited as a result of increasing tertiary detritic deposits and the presence of untouched remains of ancient erosion surfaces containing deposits produced by the rocks *in situ*. In the west, the feudal political structures were organized on the basis of complementary parallel strips of land, associating a sector of the Armorican massif with part of its sedimentary margins. The ancient Gallic cities had been organized in similar fashion. Medieval opposition to the state of Brittany favoured the persistence of such a pattern. Anjou was thus defined almost exactly by the ancient limits of the city of the *Andecavi*, while Maine resulted from a fusion of the land of two related peoples, the *Aulerci Diablintes* and the *Aulerci Cenomanni*, retaining the name of the latter. In Poitou (< *Pictones*), this twofold structure is complicated by other factors. The sedimentary pass formed by the jurassic plains situated in between the Massif Armoricain and the Massif Central was extremely narrow, despite being the artery of communication whose commanding strategic value constituted the basis of this historic region. This led to the creation of a ternary structure in between the two ancient massifs, a central strip of rich cereal-producing land flanked by two crystalline margins. The elongated shape of this unit made it difficult to preserve its unity. Numismatic evidence shows that, as early as the Celtic period, a group of *Lemovices* had installed themselves at the western end, close to the Atlantic Ocean, in what is now known as the Pays de Retz.[44] The creation in 1317 of two new bishops' seats at Luçon and Maillezais, which cut into the immense territory of the bishopric of Poitiers, testified to the need for towns specially designed to provide an urban framework for the lower Poitou, whose peoples' distinctive character was to manifest itself in a violent fashion in the 1793 Vendée revolt. In the north of the Aquitaine Basin, the city of the *Santones* (> Saintonge), centred on the *'champagne'* of the surrounding cretaceous belt, presented a similar, equally fragile structure destined for eventual fragmentation: it stretched from the Massif Central in the south-east to the marshy, coastal land in the north-west. The fragmentation took place as early as the Late Empire, when the city of *Ecolisma* (> Angoulême) was split off. This was the origin of the feudal *comté* of Angoumois, which was later to retain its individuality when the *gouvernement* of 'Saintonge and Angoumois' was set up. Meanwhile, the expansion of La Rochelle, in the north-west, accounts for the creation, in the seventeenth century, of the *gouvernement* of Aunis (see above, pp. 159–63 and note 27).

Around the middle reaches of the Loire, the historic regions were smaller, centred essentially upon the river towns, important and prosperous complexes which were situated where roads crossed the waterway and profited from its traffic. These towns became centres of polarization as they more or less annexed the productive cultivated fields of the neighbouring plateaux. Touraine (< *Turoni*) thus controlled the rich land of the chalky plateau of Champeigne Tourangelle, in between the Gâtine de Touraine to the north of the river and the ponds of Brenne to the south-east. Nivernais (from *Noviodunum Aeduorum* > *Nevirnum* > Nevers), originally a city of the Late Empire which was separated from the city of the *Aedui*, depended upon its agricultural bases in the jurassic limestone plains between the Loire and Morvan, bounded to the south by the tertiary sands of Bourbonnais and to the north by the surrounding belt of Lower Cretaceous clay of Puisaye. A chance of feudal history created the vast but short-lived comté de Blois along the middle reaches of the Loire, in between the Beauce and Sologne. It stretched from Normandy to eastern Berry and caprices of dynastic inheritance from time to time extended it to include Champagne and the northern sectors of the Ile-de-France (the comté de Beauvais). However, the position of Orléans, nestling in a large bend of the Loire, which was also the junction where the cargoes of ancient navigation were transferred to the overland route to Paris,[45] was so exceptional that it was bound to become predominant once the region as a whole fell under the direct authority of the monarchy. From 1566 onwards, the Orléanais *gouvernement* extended its authority over the entire Loire region, swallowing up the Maine, Anjou, Poitou, Aunis, Berry and Nivernais sectors until, in the early seventeenth century, during the minority of Louis XIII, these were all detached from it and turned into separate *gouvernements*.

In the heart of the Massif Central, the deep tectonic rifts of the Loire and the Allier, filled with Tertiary sediments in which Oligomiocene limestone and marls were predominant, created similar possibilities. Feudal Auvergne thus took over from the large city of the *Arverni*, retaining its name. Centred on the (Limagnes) plains of the Allier, in particular the largest, most northern of them, the Grande Limagne of Clermont (which was the seat of the *gouvernement*) and Riom (the seat of the *généralité*), it proceeded to annexe vast sectors of the peripheral Hercynian and volcanic mountains in the east (Livradois, Bois Noirs, Monts du Forez), the west and the south-west (Chaîne des Puys, Mont-Dore, Artense, Cézallier, Cantal). To the north, Bourbonnais, with its accumulations of detritic sands and gravels at the confluence of the Loire and the Allier, the natural boundaries of the province of Auvergne, constituted an important feudal unit centred upon the recently built fortress of Moulins, on the banks of the Allier. At the other extremity, the volcanic mountains of the south-west, not easily linked with the Grande Limagne, from the fourteenth century on became the '*baillage des montagnes*', with a bishopric at Saint-Flour (created in 1318) and its own *Etats*. But the unity of the province as a whole, which was incorporated into the Parisian monarchy at an early date, was never really challenged. In the eastern and western parts of the massif, it did not prove possible to set up anything of this kind. In the hydrographic Loire Basin, the Forez plain (much smaller than the Grande Limagne) did not provide an adequate base for any comparable historic region. In antiquity, the *Segusiavi* had never been more than clients of the *Aedui*. Once the area was taken over by the crown, the feudal unit of Forez, for a while detached from the neighbouring large towns of Lyons, returned into the orbit of the *gouvernement* and *généralité* of that city. Finally, in the western part of the Massif Central, the only possible focal centres for the layered hercynian plateaux were the points where communication routes intersected. The great Roman roads from Bourges to Bordeaux and from Lyons to Saintes coincided for a short distance on the western slopes of the high land, forming a narrow corridor running between the Monts d'Ambazac and the Vienne. *Augustoritum*, 'the ford of Augustus', at a river crossing of the route leading to Bordeaux, enjoyed an unusual situation at the point where the overland and the river routes converged, and this quite naturally determined the centre of the city of the *Lemovices* (> Limoges). That pattern of organization was preserved in the Limousin province, the only French historic region which corresponds to a more or less totally homogeneous natural unit, with no internal areas of a complementary nature. In the northern sector, the province of Marche split off in the tenth century, in somewhat obscure circumstances which probably had something to do with certain of its own peculiar characteristics,[46] a legacy from the time of the Germanic invasions, when large Frankish settlements were established in these lowlying plateaux immediately across from the sectors of resistance in Aquitaine.

The structure of the historic regions: the ill-defined area of Aquitaine and its margins

Regional structures took quite different forms in southern France. Unlike in the Paris Basin, no comparable historic regions built around centres of polarization ever developed in the Aquitaine Basin to the south of Saintonge. A number of convergent reasons, some geographical, some human, may be adduced to account for this. The first pertains to the nature of the sedimentation in this basin, where detritic tertiary sandstone predominates and areas of limestone are rare throughout the central and southern sector. Consequently, productive land suited to cereal cultivation, which is the essential basis for structures of the kind described above, was in short supply. In this region, the good land was heavy: it was an intractable area, difficult to plough and not put under cultivation of any sort until a relatively late date. Furthermore, the Aquitaine Basin remained an area with a predominantly dispersed settlement pattern where, apart from a few exceptions, grouped villages were not introduced on a massive scale until the late Middle Ages (see above, pp. 143–7). Historical circumstances had encouraged such trends. Aquitaine, which had retained a persistent human and social individuality throughout the Merovingian period, having previously been attached to the Visigoth State, lost all its political individuality in the feudal period. For a while it was attached to Poitou, then became a distant appendage to the vast Plantagenet domain; but for years it was first and foremost a battlefield where struggles for influence were fought out. This was an unstable area, caught between the French and the English monarchies, a vast inorganic mass which, following its definitive reunification with France, survived as the gigantic *gouvernement* known as Guyenne and Gascony. The two names themselves reflect the realities of the situation. The former is derived from the word 'Aquitaine', the latter from an ethnic name reflecting the Basque expansion; and both transcend the network of the ancient cities. As we have seen, in this area that network had been largely fragmented, and the political structure of Novempopulania had already reflected the same natural conditions (see above, pp. 10–11 and 39–40).

In this vast, poorly differentiated area, the particular features of the natural topography were not of a kind to serve as a basis for administrative organization. The terms used to denote the nature of the soil, *terrefort*, *boulbène*, *mollasse*, certainly possessed a resonance in the popular consciousness, but nowhere were the natural features sufficiently continuous to rate as characteristic landscapes. Among the names of the new provinces which were designed to divide up this huge area at the time of the Revolution, only one stands out as having a natural origin, namely the Landes of Gascony, an area of morphological uniformity taking the form of an exceptionally large plain of quaternary sand. At the ill-defined heart of the Basin, the names of the rest of the small new provinces conveyed only the vague basic spheres of urban influence into which they fell: they were more like *pays* than provinces. Agenais and Bazadais had more or less preserved as their centres the two ancient cities of the *Nitiobroges* and the *Basates*, but Condomois corresponds to the territory of a small town which, despite its Celtic name (*Condomagus* > *Condomum* > Condom), only emerged during the Merovingian period. After the Basque expansion, in the valleys of Gascony, territorial frameworks had been reconstituted, under the Carolingians, on bases which were quite different from those of the small ancient cities, and the various feudal units are named after individuals, no doubt local administrators: Armagnac < *Armaniacus* < *Artmannus*; Astérac < *Asteriacus* < *Asterius*; Fézensac < *Fidentiacus* < *Fidentius*.

Only in the margins of the Aquitaine Basin was it possible to retain territorial structures based on complementary geographical units, which combined sectors of plains with mountainous areas. In the north-east, on the edge of the Massif Central, the limits of a number of ancient cities were preserved in the feudal boundaries: those of the *Petrocorii* (> Périgueux, Périgord), centred on the cretaceous plateaux; the *Cadurci* (> Cahors, Quercy), centred on the Jurassic limestone Causses de Quercy; the *Ruteni* (> Rodez, Rouergue), centred on the low Hercynian plateaux; the southern *Ruteni*, who split off from their northern fellows in the Roman period and settled in a more cramped region straddling the border of the Aquitaine Basin and the Massif (*Albiga* > Albi, Albigeois).

In the Romanized part of the fringes of the Pyrenees, Béarn (*civitas Benarnensium*), Bigorre (*Bigerriones*), Comminges (*civitas Convenarum, Pagus Convenicus* > *Commenicus*) and Couserans (< *Consorani*), all of them centred on valleys or groups of valleys, preserved the names of small ancient cities, most of which had resulted from Roman synoecisms imposed upon the mountain tribes. The comté de Foix, despite the ancient, probably Iberian, name of its main town, was the only purely feudal construction here, formed in the southern part of the territory of Toulouse and detached in the early eleventh century from the comté de Carcassonne. In the western sector of the ridge, where the Basque language continued to be spoken, the territorial divisions which cut across the mountain range were likewise a legacy from antiquity: Labourd (< *Lapurdum*) was the ancient name for Bayonne; Soule (< *Subola* or *Vallis Subolae* < *Sybillates*) came from the name of a tribe. But all these territorial units were small, many of them more like *pays* than provinces. Only the province of Béarn made something of a historical showing. Basing its powers of expansion upon the productive cereal-producing terraces of the Gaves region, in the eleventh century it spread to assimilate the vicomté of Oloron (< *civitas Iloronensium*) and between the thirteenth and the fifteenth centuries it proceeded to incorporate virtually the entire length of the Pyrenean border, all the way from Albigeois and the comté de Foix to Lower Navarre and the Atlantic coast, developing a couple of new capitals in the process: Orthez (twelfth century) and Pau (fifteenth century). After its union with France, it retained its status of a distinctive *gouvernement*.

With its ill-defined structure, this area thus seemed irremediably destined to severe fragmentation. It was, furthermore, tugged this way and that by different centrifugal forces. Both the major urban centres, Bordeaux and Toulouse, owed their existence and prosperity to external relations: Bordeaux was situated at the intersection of the estuary and the overland highway, looking outward to a wide horizon of commercial possibilities, while Toulouse was strategically placed to control the route to the Mediterranean. Neither town at any point managed to unite the Aquitaine Basin. The present-day regional division is determined by that persistent situation, forced as it is to adapt to a dualism of capitals which has no geographical basis.

The structure of the historic regions: route intersections in the Rhône valley and the Alps

In south-eastern France, in contrast, it certainly was nature that dictated the pattern of territorial divisions, for these were determined by the arteries of communication. In this compartmentalized area of dramatic contours, the prosperity of the towns – the key element in the development of the region – could only be fostered by active currents of circulation.

Lyons enjoyed the most favourable situation and it would have been reasonable to expect that, after being the major capital of Roman Gaul, it would continue to be the centre of a vast political entity. However, this was not to be. The trouble was, precisely, that Lyons was so dependent upon the ebb and flow of movements across the isthmus of Gaul and the diffusion of Roman influences that its fortunes could not withstand the strategic upheavals of the period of Germanic invasions. Under the Late Empire, it was already being supplanted by Trèves, the frontier capital, and Arles which, controlling the crossing points of the lower Rhône as it did, constituted a more natural centre for the less exposed provinces of the south. For a while, it continued to be the seat of the kingdom of Burgundy, but when this was integrated into the Merovingian kingdom, in the early Middle Ages, and circulation diminished considerably, it lost its status as a major metropolis once and for all. Like the Gallo-Roman city, the *Pagus Lugdunensis* still spread across the plateau of Dombes and the plain of Bresse in the Saône–Rhône valley, and embraced the southern end of the Jura and, to the west, the Plaine du Forez. In the feudal organization of the territory, the baronial archbishopric of Lyons was reduced to an important mountainous suburb in the Monts du Lyonnais to the west and south-west of the town. Its early absorption by the French monarchy in 1310 integrated the town within a northern state which was, nevertheless, already deeply interested in the south.

It was this that restored its importance as a focal point in the general network of communications, a centre from which French influence was beamed out towards the Mediterranean and Italy, and a town with a major administrative role. For a short spell (at the end of the sixteenth century), the Lyons *gouvernement* extended to cover Bourbonnais, Marche and Auvergne. Right up until 1789 it included the Monts du Beaujolais and the Plaine du Forez as well as Lyonnais proper. Even so, this was not much compared to the ancient domain and role of the town.

Other major political structures had meanwhile emerged in the Rhône–Saône gap, as a result of the geopolitical upheavals that accompanied the Germanic invasions and the fixing of the frontier between France and the Empire by the Treaty of Verdun. The major development took place in the north. Nominally the successor of the Burgundian kingdom, but now settled by new inhabitants, French Burgundy was the model of a state which drew its strength from the arteries of communication that passed through it, for it controlled and exploited the routes connecting the Rhône gap with the heart of the Paris Basin. It was a situation which, in antiquity, had ensured the prosperity of the city of the *Aedui*, in a territory which combined the plains of the Saône valley with the extensive limestone plateaux of the south-eastern Paris Basin. However, the borders of feudal Burgundy were drawn quite differently from those of the great city of antiquity. The province extended considerably further northwards, taking in large sectors of the ancient cities of the *Lingoni* and the *Senones*. This reflected the political attraction of the centre of gravity constituted by the French monarchy, for which Burgundy was always a frontier fiefdom facing towards the Empire and, as such, one of the major territories traditionally allotted to princes of the blood. The fact that the province which took over from the city of the *Aedui* and the kingdom of Burgundy extended much further to the north reflected the general shift of power towards the heart of the Paris Basin during the period of the Germanic invasions. In the fourteenth and fifteenth centuries, the great expansion of the dukes of Burgundy, of the House of Valois, into the Netherlands and the Meuse area simply continued the trend initiated by the very first dukes descended in direct line from the Capetians. The displacement of the capital was a manifestation of that expansion. Burgundy controlled many different routes, some of which passed through the outlying valleys of Morvan, while others cut more directly across the limestone plateaux. Autun, the ancient capital of the *Aedui*, founded when the *oppidum*, Bibracte, was shifted down to the plain, was not ideally situated from this point of view, for Burgundy needed an urban centre positioned at the foot of the plateaux escarpment and in contact with the plains of the Saône. Competition was fierce. Chalon-sur-Saône, a river port and the crossing for roads leading to Autun during the Roman period, remained the principal centre during the Burgundian and Merovingian periods. Dijon, which controlled the valley of the Ouche, which cut through the limestone plateaux, was closer to the upper Seine valley, but much further north. In ancient times, it had been a small *Lingoni* town and its importance had been no more than marginal ever since. But in the feudal period, its strategic position gave it more importance as the territory expanded northwards, and in the twelfth century it became the capital of the duchy. The monk Richer writes of a claimant to the dukedom as follows: 'He was burning to possess the citadel of Dijon, thinking that, once he controlled this place, he would be able to impose his laws on the greater part of Burgundy.'[47] But despite this political status, it was not until the nineteenth century that, thanks to its role as a railway junction, Dijon definitively established its precedence over the other towns of the Saône valley, namely Beaune, Chalon and Mâcon.

At the beginning of the fourteenth century, the comté de Bourgogne – Franche-Comté – on the eastern bank of the Saône, was for a while attached to the royal domain. In the fifteenth century it became attached to the French duchy as a result of purely personal links, but it nevertheless remained a part of the Empire until it was finally annexed by France in 1678. It preserved a distinctive individuality of its own, partly as a result of the frontier fixed by the Treaty of Verdun, but also on account of essentially geographical features, namely the obstacle represented by the river, its marshy land and the forests of the plain.[48] The borders of the city of the *Sequani* were reproduced almost exactly by the feudal organization of the territory, which incorporated the western slopes of the Jura and their surrounding plateaux and dominated the passage leading to the Rhine gap. The reason why it adopted Dole, in the upper Saône plain, as its capital while it was linked to the duchy was simply that the old capital and bishop's seat of *Vesontio* > Besançon had retained the status of an imperial town.[49]

The political separation of the two banks of the river was reflected in the geography of the area to the north of Lyons, where the width of the Saône valley made it possible for balanced territories, incorporating large sectors of the plain, to develop on either side of the river. To the south of the confluence, however, the separation was less obviously marked. For the Gallo-Roman cities, if not for the peoples of independent Gaul, the Rhône had never constituted a frontier. The spheres of influence of the prosperous towns of Vienne and Valence had extended to both sides of the river. Here, it was the frontiers fixed by the Treaty of Verdun that were the main reason for the territorial division. Only the comté of Valentinois and Diois managed to survive for a short spell, in the fourteenth and early fifteenth centuries, as a territorial unit which spread over both banks of the Rhône. No stable political unit was able to develop on the French bank, where the Massif Central rises above the river in a steep escarpment. Vivarais, which was no more than a bishopric, was soon (in 1271) swallowed up by the royal domain. On the eastern bank, the Comtat Venaissin became a papal estate in the thirteenth century and this chance development of religious history gave the area a stability that lasted right down to the Revolution. But two major political creations developed on the western incline of the Alps, where they controlled the passes and river valleys converging towards the Rhône. Savoy incorporated most of the ancient city of Geneva, together with the northern-most pre-alpine massifs (Chablais, Genevois, Bauges) and it was not until much later, at the time of the Reformation, that Geneva split off from the province. But even before that, the establishment of the feudal capital at Chambéry, situated in one of the widest cross-valleys cutting through the pre-alpine massifs, had underlined the shift of the centre of political gravity to these mountain valleys. Savoy's strength stemmed from its control over a number of important routes, as was clearly demonstrated up until 1860 by the fact that its territory extended to include both sides of the mountain ridge. Meanwhile, Dauphiné, which controlled the more southern passes, extended considerably beyond the limits of the ancient city of Grenoble, being the result of a veritable synoecism which, in the course of the feudal period, gradually spread to include all the basins of the middle and lower reaches of the Isère and the Drôme, in an area which, during the Roman period, had been extremely fragmented.

The structure of the historic regions: between the mountains and the Mediterranean
Finally, in the Mediterranean regions, the regional units combined complementary geographical features of a different nature. The climatic pattern, with its long periods of drought during the summer, made it necessary for the flocks to migrate seasonally between the plains and the mountains. Because of this transhumance and semi-nomadism, it made sense to divide the territory into a series of units extending inland from the coast between the natural boundaries formed by the relief of the terrain. The provinces of Roussillon, Provence and the comté of Nice (which split off from Provence towards the end of the fourteenth century) are all of this simple type. The establishment of the capital of Provence at Aix was in line with the historical status of the town which, under the Late Empire, had been the *chef-lieu* of the second Narbonnaise. It also reflected a constant feature of this Mediterranean region, namely that Marseilles had always been ethnically and economically marginalized. Under the Late Empire, Marseilles was attached to the province of Vienne, being far more oriented towards the Rhône valley than towards the mountainous interior, with which it was difficult for it to communicate. The latter area could be more satisfactorily organized around the internal limestone basins of Lower Provence. In contrast, Nice-Cimiez (< *Cemenelum*), which was fortunate in possessing easy means of communication with the valleys of the interior and with Piedmont, was the only possible capital for the area surrounding it, which looked outwards towards

the Mediterranean. In the Middle Ages, it had been separated from the upper valley of the Durance, to which it had been attached under the Late Empire, within the somewhat artificial structure that the Romans had organized to provide a framework for the various local small tribes. In Roussillon the competition was between the cities positioned on the major longitudinal hill-road which commanded the crossing-points of the Tech and the Têt, the two rivers from whose alluvial deposits the plain was formed. The native *Illiberis*, on the Tech, probably used to be the most important town until the Romans established a *chef-lieu* at Ruscino (> Castel-Roussillon) on the Têt. In the early Middle Ages the political and religious centre shifted back to *Illiberis*, which was renamed *Helena* (> Elne), then returned to the Têt, five miles upstream from Ruscino, where it was easier to cross the river. This was the spot where Perpignan appeared in the ninth century. We know nothing of the mechanisms of this series of shifts,[50] but at any rate they did not affect the political unity of the plain.

The situation of Languedoc, stretching from Toulouse to the Rhône, was more complex. Most of the region was taken over by the French monarchy when it began to penetrate the south following the Albigensian Crusade. To this, the former sphere of influence of the counts of Toulouse was then added in stages. In the eleventh century, this area had been centred on the eastern parts of the Aquitaine Basin, but over the eleventh and twelfth centuries it had gradually spread to include the Septimania of ancient times, bounded by the river, the sea and the Pyrenees. The political structure that emerged, which controlled the Lauraguais pass, constituted a resuscitation of the first Narbonnaise of the Late Empire. Further afield it was bounded by the original western limits of the Roman *Provincia*, with an outgrowth projecting into the Aquitaine Basin, and it also took in a slice of the south-eastern balcony of the Massif Central. The abiding principle of organization that oriented Toulouse's sphere of influence beyond the pass towards the coastal Languedoc plain thus resulted in a composite territory that was not exclusively Mediterranean, although in the eastern sector geographical complementarity certainly was achieved between the coastal plain and the hills that edged it. The difficulties of communication between the two components of this vast entity were resolved at an administrative and financial level in 1542, when it was divided into two *généralités*, the one based on Toulouse, the other on Montpellier. However, the legal organization of the province continued to be the sole responsibility of the *parlement* of Toulouse, which thus served an exceptionally vast area.

Finally, Corsica retained its unity right up until its definitive annexation by France in 1768, owing to the unusual circumstances of its historical evolution. It is not as if that unity was preordained by its natural geography. The long ridge of peaks, not easily crossed, rising as high as 2,500 metres, which structure this extremely mountainous island, certainly predisposed it

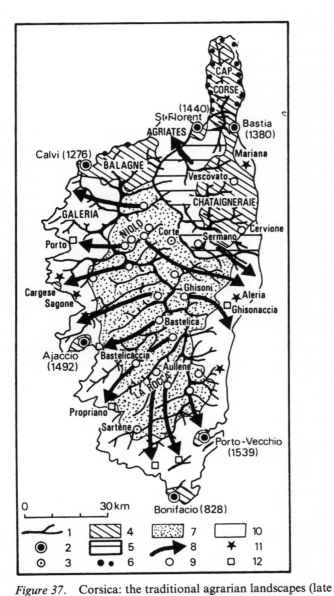

Figure 37. Corsica: the traditional agrarian landscapes (late eighteenth and early nineteenth centuries) and urban developments (after Fel, 1975, p. 195 and Kolodny, 1962, p. 39).
1. Mountain ridges. 2. Genoese towns (foundation dates are given in brackets. 3. Native towns. 4. Main sectors of tree cultivation, enclosures, terraced cultivation. 5. Zone controlled by institutionalized 'guardians of agriculture and wheat'. 6. Ports and 'marinas'. 7. Agro-pastoral mountain level. 8. Currents of semi-nomadism and transhumance. 9. Principal mountain villages. 10.'Beach' zone. 11. Attested ruins. 12. Developing beach communities.

to a system of political division into several sections extending from the coast inland, up the one or the other side of the mountains. But such a system would have required autonomous urban centres of organization of a kind that simply did not exist, for the island was dotted with very primitive inland village communities of a semi-nomadic, pastoral nature, which drove their herds down to the beaches in the winter months. The coastal urban centres of antiquity had disappeared in the early Middle Ages and further developments had long been ruled out by the danger of pirate and Arab raiders. In Corsica, the eventual appearance of towns was essentially due to external powers[51]: from the thirteenth century on Genoese colonizers founded more and more towns at fortified points, gradually establishing their control over the island (fig. 37). Except for the modest little inland towns of Corte and Sartène, which are the only native foundations in the island's interior, all the towns were creations or recreations of the Genoese, who used them to impose their unifying domination over the island. It was only after the French annexation, when the departmental organization was introduced, that the island was for a while divided into two units, one to either side of the mountain ridge: a decree promulgated by the Convention and dated 11 August 1793 created the two departments of Golo and Liamone, with Bastia and Ajaccio as their respective *chefs-lieux*, dismantling the single department of Corsica which had been created in 1790. In 1811, the island was reunified, with Ajaccio as its *chef-lieu*.[52]

The range of a jurist and the range of a horseman

What were the functions of these two levels of organization, that of the '*pays*' and that of the '*province*', in the economy and society of traditional France? How did these divisions affect daily life?

It is quite clear that neither the *pays* nor the *province* really performed any economic functions. A small town, the *chef-lieu* of a *pagus* or, later, a bailiwick may of course have attracted fairs and markets which served the outlying neighbourhood. But these would have been held at many spots, not just in the regional *chef-lieu*, and their field of attraction would in most cases not have extended to anything like the entire area of the *pays* (see below, pp. 197–206). A town would certainly consume a proportion of the produce from the surrounding countryside, but it would be a very small proportion. Pierre Chaunu believes that in the traditional economy, nine-tenths of the produce changed hands within a radius of 5 kilometres, with nine-tenths of the remaining tenth circulating within an area with a radius of between 25 and 40 kilometres,[53] and those figures seem altogether believable. But that circulation should not be imagined as a centripetal movement within a circumscribed area. Rather, merchandise would be exchanged according to a 'Brownian' pattern of diffusion, changing hands several times as it moved from one district to another. The produce used by the towns, at least, would be paid for with coins, for by the late Middle Ages the economy was already to a large extent monetarized. Exchanges in kind would take place where particular types of consumers were involved – first and foremost the large secular or religious landowners who lived in abbeys or castles in the countryside, rather than in the urban centres. A small town would draw most of its supplies from its own belt of surrounding gardens or from the immediately adjoining countryside, no more than a few kilometres away: an area far smaller than that defined by the administrative or legal boundaries of the locality. As for larger towns such as provincial capitals, in the period for which we possess accurate evidence (the eighteenth century), the area from which they drew their supplies by no means coincided with the limits of their administrative jurisdiction. Some supplies – such as grain – came from much closer in the immediate vicinity, while luxury goods in general came from further afield, beyond the province's boundaries. In more ancient times the flow of major supplies was by no means stable. In the sixteenth century, the royal court still used to move frequently from one palace to another, so as to spread the costs for its upkeep over the entire length and breadth of the kingdom, just as it had done for centuries past, under the Merovingians, the Carolingians and the early Capetians.[54] The great feudal lords did likewise within the boundaries of their own fiefdoms.

The area of the *pays*, like that of the province, was dictated by political factors and conceived in terms of power, not economics. The *pays* has often been described as an area limited by the distance of a day's march into and back from the central market[55] and such trips must certainly have been common. However, they did not represent the norm in commercial relations. For most of the people involved, the trip would be much shorter, taking two or three hours at the most (see below, p. 215). And a single day's trip to and back from the *chef-lieu* was not really possible for peasants loaded with wares or driving their flocks from the outermost reaches of the district and needing to spend several hours in the market town if they were to complete their business. The area of a *pays* should be understood in relation not to the peasants but to the city-dwellers. Now, these would be the representatives of authority, tax-collectors or policing officers, and in the traditional society they would, as such, move around on horseback. The area of a *pays* would accordingly be determined by the distances that a rider could, in normal circumstances, expect to cover in a single day, returning, after completing his business, without having to change horses en route. The dimensions of a *pays* were designed for a horseman. From the town's point of view, it was an area of intervention, not of attraction. If, from another point of view, it also constituted an area of attraction, it could only be for the *gentilhomme* or the *petit seigneur*, who would also be mounted and whose interests or relations with those in authority would take them into the town from time to time. For the mass of ordinary inhabitants, however, this did not apply.

The authorities' modes of intervention were decided by the law. In common-law *pays*, this was established and defined within a particular framework of relations. Certainly, there did exist a complex hierarchy of local customs and traditions that varied from one canton, or even commune, to the next, which the *notaires* and other practitioners of law were to classify in the nineteenth century.[56] But for all matters of importance, customs were codified in relation to the existence of an authority capable of applying the law. At the level of the *pays*, these customs could be codified as a distinct body. In the eighteenth century, there existed in the kingdom of France close on two hundred local customs and over 500 individual customs or statutes.[57] These are figures comparable to the number of bailiwicks or seneschalsies within the area occupied by *pays* possessing written laws (but particular customs also existed in *pays* with Roman laws). These local customs were sometimes given written form, generally in municipal charters.[58] But most were not recorded in detailed codes. It was essentially at the provincial level that the huge task of collection and codification was undertaken, beginning in the late twelfth century, when we find references to customs that are not local but provincial, and culminating in records of 'general customs', mostly produced in the fifteenth and sixteenth centuries.[59] 'The people of each province have the right to establish their own laws, and these are the customs', Guy Coquille wrote in the sixteenth century.[60] There were about sixty general customs in France – about as many as there were provincial units of organization or had been cities in Gaul. The province, originally a fiefdom granted to a warrior, became, in the modern period, above all the domain of jurists – at least it did in areas of common law. In southern areas, with written law, customs remained for the most part a local matter, being restricted to a particular town or the territory of a large village. But even here, a few more general customs, such as the statutes of Provence or the customs of Bordeaux, were recognized before the general return to Roman Law, which began in the sixteenth century.[61] It was essentially within the provincial framework that the most important rules applying to people and property were itemized, in particular the regulations applying to inheritances. Meanwhile, as the royal domain was progressively extended, the monarchy respected or granted a wide range of particular provincial customary practices or privileges, particularly those relating to taxation: these either affected the very principles according to which taxes were levied (which differed for *pays d'Elections* and *pays d'Etats*) or else related to the rates of taxation.

This provincial variation in legal and fiscal matters caused considerable repercussions. Over the last centuries of the Ancien Régime, the various provinces acquired a strong sense of their own particular individualities within the kingdom of France. It was frequently within provincial limits that political issues were played out and within a provincial framework that revolts took place. Often, these would erupt in one particular *pays* and then spread to others within the confines of the province, in response to the imposition of a resented tax or in defence of a particular privilege:[62] Périgord was the scene of the revolt of the *Croquants* from 1634 to 1637. The revolt of the *Nu-Pieds*, in 1639, in Lower Normandy, was connected with the threatened introduction of a salt-tax, from which the province had previously been exempted. The last major popular uprising before the Revolution, the revolt of the *Bonnets Rouges*, took place in 1675, in Brittany. There seems to be an undeniable correlation between zones of rebellion and zones with particular local customs.[63] Under the absolute monarchy, the lively consciousness of a Breton identity, in reaction to the French, was manifestly connected with the defence of Breton liberties and privileges.

To force obedience upon the provinces, the centralized state was sometimes obliged to use maximum force or at least to make a show of its prestige. The practice of the *Grands Jours*, special legal sessions held by magistrates sent from Paris to ensure respect for the law, was a typical response to the strong sense of provincial identity which sometimes found expression in anarchy and disorder. Fléchier has left us unforgettable literary evidence of this practice in his work on Auvergne, which is also a precious social document.[64] There were to be other spirited resurgences of provincial passions on the eve of the Revolution, in 1788–89, when the central power could be seen to be under threat (see Bardes, in Gras and Livet, for the case of Gascony).

Not surprisingly, these undeniably real political events have been overlaid by a cultural image. This was given particularly strong expression at the end of the Ancien Régime, when a whole geography of temperaments was elaborated. For Savinien d'Alquié, the author of *Les Délices de la France* (1670), particular provinces were characterized by particular human temperaments: 'The Angevins are subtle and capable of friendship' while the Manceaux 'are clever and dangerous'; and towns also had their own particular characters: in Brest, 'the people are gentle, frank and love an easy life', whilst at La Rochelle, 'the common people are bad-mannered and of a rather insolent humour'.[65] Nor was this 'theory of humours' limited to scholarly writers. Such character differences became proverbial: the Bretons were said to be stubborn, the Normans prudent, the Champenois somewhat simple. Some such representations even related to individual *pays* and became widely known. One remarkable example was 'the Sologne idiot only makes mistakes when it suits him', the fortunes of which have been the subject of an excellent study.[66] The saying probably originated in Champagne Berrichonne and it presented an unfavourable image of the neighbouring Sologne, a harsh, poor region where the people were said to be devious and malicious. This proverb is mentioned at least as early as the beginning of the seventeenth century (in the *Comédie des Proverbes*, probably composed in 1616) and progressively, through the seventeenth and eighteenth centuries and down to the early nineteenth, it came to be quoted throughout France, its fame being diffused via Orléans and Paris. At the beginning of the eighteenth century it was already said to be 'known throughout the kingdom'. In 1803, Dauvigny based a successful comedy, *Le Niais de Sologne*, upon it.

Alongside these stereotypes that were seized upon by outside observers, there are signs of a more internal sense of identification. People were truly patriotic about their own province, or even their own *pays*. In the modern period, to be sure, the sentiment tended to be of a self-interested nature, essentially confused with the defence of privileges that stemmed from particular local customs or statutes. But in the Middle Ages, it was more complex: in the case of knights, at least, it was fostered by a passionate pride in military caste. The *chansons de geste* are full of telling examples: *Girart de Roussillon*, which glorifies the resistance against Charles the Bald, is steeped in Burgundian patriotism; *Gaydon*, written around 1230, to encourage the Angevins to accept French rule, is – despite its purpose – also a work that eloquently glorifies the province and its knights;[67] *Huon de Bordeaux*, who was a native of Saint-Omer or at least wrote for a local readership there between 1260 and 1268, exalted the manliness, warlike courage and religious ardour of the knights of the province:

> *De Cambrésis et d'Artois, ce saciés*
> *En i ot mout de baceliers légiers*
> *Qui volentiers ferroient sour païens.*[68]
> (From Cambrésis and Artois, you should know,
> There were many brilliant bachelors
> Who were keen to strike at the heathen.)

And each province had its own warcry, by which its warriors made themselves known.[69] The sense of fraternity among these groups of knights constituted a rudimentary form of provincial solidarity but, in the absence of specific evidence, it is impossible to know whether it extended to the popular masses. However, it certainly gave rise to the sense of identity and the patriotism based on the general interests of one's own province that are detectable in the modern period. Originating in the pride of the feudal warrior and subsequently manifesting itself in the jurists' determination to foster their own provincial privileges, the provincial sense of identity continued to evolve in more subtle forms and it probably did also communicate itself to the lower strata of the population. But it did not change its principles nor did it broaden its horizons. The feeling of superiority and the pride in one's own origins was matched by an accompanying scorn for 'others'. In *Foucon de Candie*, Bovon de Commarcis asks Guillaume's messenger who was the leader of the Christians in the disastrous battle of Larchamp; and in the picture which the messenger's reply paints of Tedbald de Bourges, whose cowardice lost the day, it is hard not to read an intentional slur on his province:

> Cil de Berri, qui tant par est provez
> De coardie, honiz et vergondez[70]
> (The native of Berri, the one
> Whom this proves so guilty
> Of cowardice, shameful and despised.)

Were the relentless vendettas between the *Lorrains* and the *Bordelais* which fill the *chansons de geste* more than just lineage rivalries?[71] Did they perhaps already involve feelings of regional solidarity? Provincial hatreds, such as the enmity which set the Bretons and the Normans in opposition, were certainly deep and tenacious and Savinien d'Alquié remarks that the people of Nantes 'have such a great aversion for the Normans that they cannot bear them; and the Normans reciprocate with a deadly hatred for them'.[72] Did these violent reactions ever transcend a simple hatred of 'others' and take on nobler nuances? Probably not. We know, for instance, that the many passages which evoke Breton patriotic hostility towards both the French and the English, in La Villemarqué's *Barzaz-Breiz*, were purely and simply inventions by the author and have no historical basis at all (Gourvil, 1960, pp. 390–5).

In truth, these sentiments do not appear to have been particularly lofty, compounded as they were of narrow-minded egoism, skin-deep antipathy and intolerant belligerence. It took the genius of a poet such as Du Bellay (*Les Regrets*, XXI) to rise to nobler heights, as when he expresses his preference for his own humble home to all the splendours of Rome:

> Plus mon petit Liré que le mont Palatin,
> Et plus que l'air marin la douceur angevine.
> (Rather my own little Liré than the Palatine mount,
> And the soft air of Anjou than a sea breeze.)

Here and there, a few scholarly poets seem to manifest signs of provincial pride. Around 1100 a canon from Saint-Omer, Pierre le Peintre, hymned the praises of his Flemish homeland: '*Flandria, dulce solum, super omnes terra beata* (Flanders, whose land is so sweet, blessed above every other)'.[73] A pious Carmelite monk, Father Pierre de Saint-Louis, included in his extraordinary baroque poem dedicated to *La Madelaine au désert de la Sainte-Baume* (Lyons, 1668) a prefatory sonnet singing the praises of Provence, '*Princesse en sainteté des Provinces de France* (Sainted princess of all the provinces of France)'; but popular echoes of such sentiments seem rare.

In Du Bellay's nostalgia for his homeland paradise, however, such provincial sentiments certainly are sublimated. Can the same be said in other, less exceptional cases ? Did the area to which the feudal warriors and the jurists of the monarchy related leave its mark upon language, art and thought? Did the *pays*, the province, function as a cultural region?

The answer to these questions must definitely be 'no'. Whatever isolated points linguists may make, dialect boundaries in no way coincided with the boundaries of the administrative or religious units represented by the *pays* and the provinces. At the very most, a particular field of economic relations, which may have played a formative role in linguistic differentiation, occasionally coincided with a unit of local government. But such cases were exceptional. Pierre Bonnaud makes the point amusingly when he protests against 'the tendency to ascribe to archdeaconries or archpriests' domains mysterious virtues which determine the limits within which particular dialects are spoken – as if peasants sold their pigs to archdeacons or consulted the archpriest when it came to spreading their manure or planting their radishes'.[74] The same point could be made in relation to the province from the feudal period on. The members of its intellectual élite were cosmopolitan, expressed themselves in Latin and attended universities all over Europe, where they rubbed shoulders with scholars from a wide range of countries. Where literature develops characteristics peculiar to a particular area, it tends to be within a more restricted framework, pertaining above all to linguistic frontiers. In *langue d'oïl* literature it is possible to distinguish an Anglo-Norman area, where the *chanson de geste* appeared first and in its most elaborate form; possibly a Picardian area; and an Ile-de-France area, where the influence of Paris is manifest. But those areas were all considerably larger than the major feudal units. There may be more of a case to make out for different schools of architecture. In the vast domain of the Romanesque and Gothic arts, which flourished over more or less the whole of western Europe, local styles have been recognized, associated with particular places and called after particular regions, although specialists do not always agree as to what those regional names should be. In the Romanesque style, a number of schools have been distinguished and associated with Normandy, the Ile-de-France, the Rhineland, Poitou, Burgundy, Auvergne, Périgord and Provence. The Gothic style is represented by the schools of the Ile-de France, Normandy, Champagne, Burgundy, Languedoc and the southwest.[75] Some places certainly did spread a particular influence and in many cases they did correspond to the centres of feudal political power which encouraged the art to develop and offered it their patronage. In some cases – the Romanesque school of Auvergne, for instance – the geographic concentration of the style in its purest forms is very noticeable. It is confined within an area far smaller than the geographical province, in fact one which barely overlaps with the boundaries of the present-day Puy-de-Dôme department, which is no more than 100 kilometres across at its widest

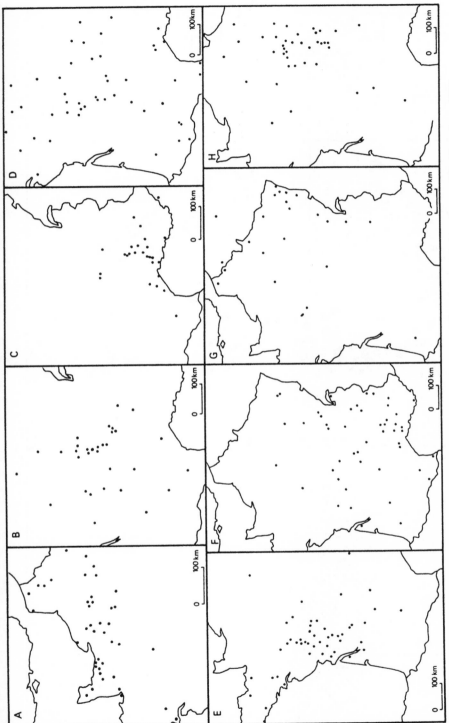

Figure 38. The range of influence of the schools of Romanesque architecture (after the maps in Brunhes and Deffontaines, 1926, pp. 640, 641, 649, 652, based on data provided by C. Enlart).
A. Normandy. B. Auvergne. C. Provence. D. Languedoc. E. Poitou. F. Lombardy. G. Rhineland. H. Burgundy.

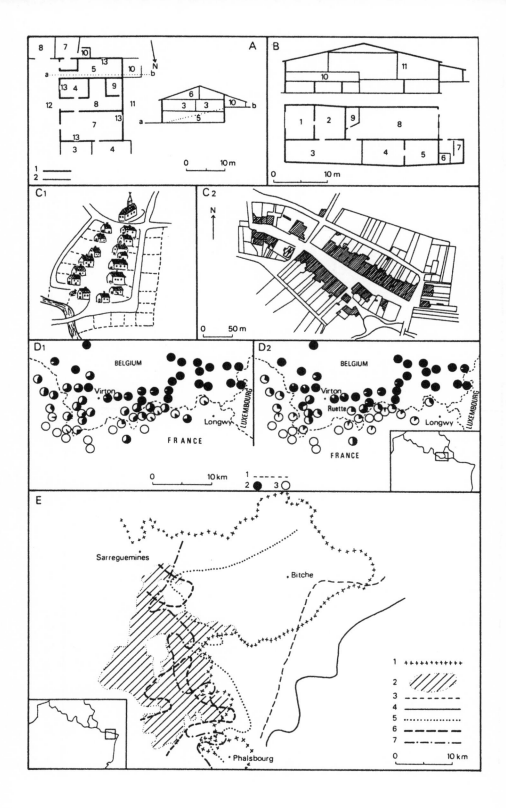

point.[76] But these schools of architecture were no more homogeneous than they were hermetically closed. They manifest many anomalies and, in particular, influences from particular regional styles may be detected at considerable distances from their centre of origin, as can be seen from the maps plotted by J. Brunhes and C. Enlart (fig. 38).[77] Artistic and technical trends, which often developed in areas of relative political stability and economic prosperity, were subsequently diffused in a fashion which took no account of territorial boundaries. Furthermore, the major centres of the initial artistic flowering were themselves unstable. Between the Romanesque period and the Gothic period, the 'schools' of Auvergne, Provence and the Rhineland all disappeared.[78]

During the modern period, *a fortiori*, the provinces lost all their cultural individuality: with the sense of French nationalism becoming stronger and

---

Figure 39. The rural settlements in Lorraine: appearance, genesis, limits.
A. House built in 1567 at Jevoncourt (Xaintois, Meurthe-et-Moselle), from Planhol, 1971, p. 76. 1. Original building. 2. Later additions. 3. Living room. 4. Kitchen. 5. Store room. 6. Granary. 7. Cowshed-stable. 8. Barn. 9. Threshing floor. 10. Shed. 11. Orchard. 12. 'Usoir' (private utility area giving on to the road). 13. Doorway. The house is wider than it is deep. It features side doors, which shows that it was detached from its neighbours. B. House built much more recently in Hammeville (Meurthe-et-Moselle) (after Demangeon, 1946, I, p. 171). 1. Living room. 2. Kitchen. 3. Barn. 4. Threshing floor. 5. Cellar. 6. Oven. 7. Pigsty. 8. Stable. 9. Store room. 10. Grain loft. 11. Hay loft. The street is on the left. The house is narrow but reaches back a long way. It is not detached from the neighbouring houses and has no side entrance. C. Sauzey (Meurthe-et-Moselle) in 1721 (C1) (from a sketch of that period), and in 1818 (C2) (from the first cadastral survey). In between the two dates the gaps between the houses were filled as a result of demographic pressure. The houses were extended at both the front and the back and side entrances disappeared (from Planhol, 1971, p. 88). D. The limits of the Lorraine rural settlement at the Belgian border (from Planhol, 1968, pp. 161–2). 1. Political frontier. 2. In D1, roofs with a slope of over 30°; in D2, slate roofs. 3. In D1, roofs with a slope of less than 30°; in D2, tiled roofs. In Ruette and the neighbouring village of Grandcourt, in the south-east, in a situation where the Belgian territory spills over into the French, the central cluster of roofs are of the Lorraine type, with shallow slopes and tiles. Slates and steep slopes appear round the edges, on houses not built until sometime between 1840 and 1900, when slates were being introduced on Belgian territory. The expansion of the Lorraine type of roof was much earlier. E. The Alsatian limits of the Lorraine rural settlement (after Planhol and Popelard, 1976). 1. The frontier of Lorraine in the mid-eighteenth century. 2. The comté de Sarrewerden, lost in 1527 and reattached to France in 1793. 3. The limits of the appearance of Alsatian houses. 4. The limits of the exclusive dominance of Alsatian houses. 5. The western limit of the exclusive dominance of roofs sloping at more than 30°. 6. The eastern limit of a 25% presence of houses not detached on either side. 7. The eastern limit of over 25% of houses adjoining their neighbours on both sides.

under the formative influence exerted by the capital, the fashions and tastes of the élites tended towards uniformity, and pressure from the Parisian centre of diffusion increased (see the conclusion to Part II and pp. 247–56). Such intellectual centres in the provinces as continued to flourish functioned essentially as points from which a more general ideology was relayed on. At this point it becomes increasingly difficult to distinguish between different schools of art. Although historians of art do sometimes speak of 'provincial schools of painting' in the last centuries of the absolute monarchy, it is only with the vaguest concept of territorial divisions. In popular art, similarly, despite this being less affected by Parisian and international European fashions, spheres of influence by no means corresponded to territorial provincial divisions. In the seventeenth and eighteenth centuries Laval altar-pieces were to be found all over western France, that is to say over one tenth of the territory of France, not counting Lower Brittany.[79] One interesting exception, Lorraine, which remained nominally independent up until 1766, despite being surrounded by French territory on virtually every side, makes the point even more tellingly. The Lorraine rural settlement pattern is extremely individual (fig. 39), characterized in particular by compact villages of terraced houses, many of them extending a long way from front to back and incorporating inner, windowless rooms. Their area of diffusion coincides to a large extent with that of shallow roofs covered by curved tiles. A study of the limits and genesis of this kind of settlement shows that processes of imitation and diffusion were certainly at work within the framework of the State of Lorraine in the sixteenth and seventeenth centuries. From a settlement pattern still, in the fifteenth and sixteenth centuries, mainly constituted by relatively loosely-knit villages of dwellings consisting of a number of separate buildings, a model was elaborated which was propagated in particular in the single-street villages founded by the dukes of Lorraine, as they systematically colonized the German-speaking sectors of their possessions and during the period of reconstruction that followed the Thirty Years War. The high demographic growth rate of the eighteenth century, which led to the spaces in the village street being filled in, encouraged the development of this model. However, nowhere do the limits of the phenomenon coincide with the historic frontiers of Lorraine. The province had a creative centre, but not all the area within the territorial framework was filled by its creations. On the other hand, the model did continue to be diffused after Lorraine became a part of France: by the early nineteenth century it had left its mark as far afield as the French borders with both Belgium and Luxembourg.[80]

As can be seen, it was essentially at a political level that the territory of France was organized into provinces, the aim being to provide a framework for different interests and relations of power. Neither in economic life with its humble everyday relationships (which were organized at a much lower

The secondary divisions 197

level), nor in the higher sphere of thought and art, which was subject to far wider trends, did the organization of the province really provide a framework or a basis for the diffusion of influences.

Perceived space
The organization of space as it was perceived by its inhabitants
Understandably enough, as a purely political unit the province did not take much of a hold on people's imaginations. It is true that the names of the major provinces have survived, even if, during the centuries of absolute monarchy, they evoked little more than a particular status and a history of rivalry and petty squabbles with neighbouring territories. But many of the names of the *pays* disappeared altogether from the spoken and even the written language and are to be found only rarely – if at all – in ancient texts. Many others, though, did survive down to the present day and do appear in the lists, catalogues and documents of the modern period and also on maps, usually tacked on to a toponym in conjunction with a preposition such as *en-*, *de-*, *aux-* and so on. But that may be as a result of scholarly efforts at reconstruction on the part of small-town jurists or clerics and also the early cartographers and naturalists of the seventeenth, eighteenth and even nineteenth centuries. Their task was made easier by the fact that, in French, it is a simple matter to create a regional name from the name of a town, by adding a suffix in *-ais*, *-ois*, *-is*, *-ien* and so on. Even at the end of the nineteenth century, geologists and geographers were still prone to resurrect the names of ancient regions in this fashion, anxious as they were to divide up the territory in accordance with their own particular preoccupations and classifications.[81]

To what extent were these names of *pagi* > *pays* a part of the popular consciousness? Did they continue to have any real significance? In general, that is a matter which it is virtually impossible to determine beyond doubt, for only very few of these names are to be found mentioned in early texts of a non-scholarly nature. But in all likelihood, the vast majority of the *pays* names that were derived from the names of towns never meant a great deal to local inhabitants. Peasants did not feel that they lived in 'Blésois' or 'Toulois', even if they were conscious of being dependent upon Blois or Toul.

Natural names
The resulting blanks in people's minds were in part filled by other kinds of names – names which described particular features of the region and of which they were well aware. 'Natural' names of *pays*, derived from certain

qualities of the geographical environment or from the way in which these struck the imagination gradually emerged alongside the names of organized regional units and, as early as the Carolingian period, some even acquired an equal official status. Alongside the *Pagus Wastinensis*, already cited, we thus find, for example, a *Pagus Waldensis* > Vaud (*'pays* of forests'). But in truth, that is a relatively exceptional case.

Unnamed space
However, it must be recognized that natural names of that kind were, on the whole, rare. Much of the territory of traditional France was certainly not 'named' at all. In common parlance, there simply were no regional names below the level of the unit constituted by a major feudal fiefdom or an historic province. The use of 'natural' regional names, to fill the gaps, was incomplete and uneven.[82]

Mountains constituted one particularly important category of places which were generally known by no traditional popular names. The names now borne by the mountainous massifs are synthetic, scholarly creations, at least so far as their overall meaning goes, for their earliest inhabitants had no concept of separate massifs, as such.[83] In some cases, the origin of a mountain's name is a common noun with a descriptive force, used as a general convenient label: Alpes, from *alp*, *alpe*, which even nowadays still denotes a high mountain pastureland (= *alpage*), formed on a pre-Indo-European base *al*, the meaning of which was probably simply 'stone';[84] Jura, from the Celtic *juris* = 'wooded hill', the form favoured by geographers from a group which has provided the language with a whole series of names (Joux, Jorasses and so on).[85] Cévennes is an example of an ancient systematization. Formed from a pre-Celtic root *keb-* (height, hill), it was Celticized as *Cebenna* (the 'back' of the mountain, with the same suffix as that used in *Arduenna* – Ardennes)[86] and was extended by Caesar to denote the whole south-eastern part of the Massif Central,[87] that is to say the highest part. The evolution of the name Ardennes itself was complex. It too comes from the Celtic and also means 'wooded hill' and in Caesar's day it was originally applied (as *Arduinna silva*[88]) to an area far vaster than that of the Ardennes of today, to wit all the major wooded areas of north-eastern Gaul. Later, it was restricted to a Merovingian *pagus* (*Pagus Arduenna, Ardennensis*) formed to the south-east of the city of Tongres (*civitas Tungrorum*).[89] But at the same time it retained a wider, literary application to all wooded hills through which the middle reaches of the Meuse pass. It is used very precisely in the thirteenth-century copies of the *Chanson des Quatre Fils Aymon*, where it appears as *'la forest d'Ardane'*,[90] *'la parfonde Ardane'*.[91] It is here that the heroes set about building their fortress of Montessor:

> *Et fermèrent chastel ...*
> *Desus Muese en Ardane, en molt haut liu et bon*[92]

(And they completed a castle ...
Above the Meuse in Ardane, in a very high, good place.)

At first, the name was probably only used by monks or in literary works, but later it passed into popular usage. Eventually, the name Ardennes acquired its modern, scholarly application to the ancient massif lying partly in France, partly in Belgium. However, other mountain names do not originate from the natural environment at all: Pyrénées is an ancient artificial creation, possibly from the name of an Iberian town (perhaps Port-Vendres) situated at the end of the ridge. Vosges was the name of a Celtic deity (*Vosegus*).[93]

The meaning and resonance of natural names
The only mountain names that really took on meaning for the ordinary people were those of smaller, precisely delimited mountains which drew the eye and stood out distinctly from the landscape: small mountain blocks rising abruptly above the surrounding sectors, as in the Massif Central where the Lévezou rises from the Ségala plateau; or the Margeride, in Auvergne; or isolated volcanic mountains such as the Aubrac and the Cantal, which bears a Celtic name (*cantes* = shining),[94] and so on. Another descriptive, Celtic name, Morvan (probably = 'black mountain'), known as early as the third century and used as a regional name in the seventh and eighth centuries,[95] has continued to apply to the extreme north-east of the Massif Central. There was also another Morvan in the heart of the Paris Basin, the *Mons Mauripensis* (Parc de Pont, to the south of Pont-sur-Seine, in the Aube department), but its name, preserved in the *Pagus Morivensis* or *Morvensis* > Morvois,[96] of the city of Troyes, has lost any connotation with mountains.

It was really in the more densely populated and intensely cultivated plains that the 'natural' regional names which entered the popular consciousness developed. The vast majority of these seem to have applied to homogeneous territories possessing distinctive vegetation or agricultural features, whether of an attractive or a discouraging nature, which gave them a character of their own and struck the imagination of their inhabitants and those of the neighbouring regions. In most cases the names reflect that distinctive character. The linguistic origin of some of them remains puzzling. For example, nobody has been able to explain the etymology of the name of the Puisaye region (*Poseia*, *Puseium* in the twelfth century), despite the fact that it constitutes a regional unit with a character all its own, determined by its impermeable sandy and clay layers from the Lower Cretaceous, to the north of the limestone plateaux of Nivernais. However, the implied meaning of a wet, wooded area cannot be in doubt, since the name reappears in toponyms with a similar context.[97] The origin and meaning of Limagne are also unknown (ancient forms: *Pagum Lemaniam* in Sidonius Apollinaris, *Arverniam Lemanem* or *Lemane* in Gregory of Tours), but *Limania* derives from a medieval reconstruction of the Provençal *lim* = mud, silt, alluvium. On the basis of a misunderstood ancient name, there emerged a popular form which assimilated it to the category of soil particularly characteristic of much of this plain,[98] and the term then acquired a regional meaning which applied to the deep tectonic rifts and fertile soil of the heart of the Massif Central. The name Perche is of equally uncertain origin. It appears in the sixth century in the forms of *Pagus Pertensis, Pagus Perticus, silva pertica* (hence the shift to *pert(ic)ensis*), and the association with wooded land, in a region of impermeable, marly chalk topped with sand, is all the more significant in that the term reappears in the Champagne Humide region in the form of Perthois (also *Pagus Pertensis*). It is clearly a term associated with the idea of a more or less wet, wooded region. The name Perche, denoting a wooded region, which from the eleventh century on became a political and administrative unit, then acquired its agricultural meaning of 'pastureland' (as opposed to the neighbouring cereal-producing land of the Beauce), after the land-clearance operations of the Middle Ages.[99]

However, the meaning of most of these names is known, for they are derived from terms which from the start were descriptive, referring in particular to idiosyncrasies of the soil or vegetation. In most cases, the meaning then evolved in an essentially agricultural direction. Thus Causse(s), from a pre-Indo-European root *kal-*, meaning 'stone',[100] subsequently, in popular parlance, came to denote a particular type of soil, suitable for wheat cultivation, as opposed to the neighbouring Ségala(s) of the southern Massif Central; and eventually it took on the sense of 'stony plateau', denoting one particular geographical region.[101] Beauce, a Celtic term meaning 'denuded plain',[102] appeared as a regional name as early as the sixth century. The Gallo-Roman grammarian, Virgile Maron, writing in that century, regarded it as a synonym of *campus* and also identified it with *gramina* – grassland, meadows: *'In belsa hoc est in campo ... Belsa enim hoc dicitur quia belsa plurima quae sunt gramina profert'*.[103] Not until the twelfth century did Suger, in his *Vie de Louis VI*, describe it as a rich wheatland. In 1217, that concept was confirmed. William the Breton, historian to Philip Augustus, compared the great host of John Lackland's soldiers to the heads of grain of the Beauce: 'The Beauce, the mother of grain, cannot display more numerous ripening heads.' A century later, however, in the *Roman du comte d'Anjou* by J. Maillart (1316), the predominant idea was still that of a bare, open, treeless plain:

> *Tant qu'il est entréz en la Biausse*
> *La li fist lo vent mal sausse*
> *Car il le fiert à descouvert*

> *Quer la Biausse est large et honnie*
> *Et si n'i a rienz qui abrie*
> *N'a quoi escouser se puisse en*

Forest, ne haie ne buisson.[104]
(Once he entered into the Beauce,
The wind played havoc with him,
For he was unprotected there

Since the Beauce is wide and rude
So that it affords no shelter,
Nowhere to take refuge
In forest, hedge or bush.)

The sense of 'rich, agricultural land' was not to become definitively established until the modern period, when the influence of large Parisian landowners became general (see below, p. 267). A similar evolution affected the concept of Brie. The word (ancient form: *Briga*) probably derives from the Gallic *briga* (= 'heights'). It would appear that, although relatively low-lying (100 to 150 metres), this region of clay soils covered with silicified travertine was a 'hill forest'. That is certainly the idea conveyed when Saint Colomban, in the seventh century, penetrated the solitudes '*intra Briagensem saltum*'. However, the term was soon to acquire the essentially pedological sense of a wet, sticky soil. That was the essential feature that Eustache Deschamps distinguished in the fourteenth century:

Sur tous pays de mortier et de boe
Ne se doit nuls à Brie comparer
Que Dieu a fait de tous biens séparer,
D'y chevauchier n'est homme qui se loe
Et en tous temps y voy gens esgarer.
(Of all the regions of mortar and of mud
None bears comparison with Brie
To which God denied every favour:
No man is glad to venture there on horseback
And in all weathers people lose their way.)

In the seventeenth century, Estienne and Liébaut painted another arresting picture of the area: 'In the Brie region, where the land is heavy and wet, ... at the point where the fields come to an end a number of quite high ridges are to be found. Here, in between the rise and the field, there is a ditch cut in the shape of a long vat, to catch the streams of water which run down in the heavy rains, for otherwise they would rot and submerge the grain; this is inconvenient for those passing through this region, which is why they are called the "grasshoppers" of Brie.'[105] As well as the regional name, the concept of the '*terre de bri*' persists in the minds of local country folk today.

Finally, many natural names speak for themselves and are immediately comprehensible. These include terms of relief or hydrography, for instance: Vallé, Vallage, Marais (the Breton and Poitevin Marais = marsh); terms relating to agricultural land: Bocage, Champagne, Campagne (Campagne de

Caen), Plaine (in the sense of an open expanse); terms describing the vegetation: Landes, Brandes (burned terrain, land freshly cleared), Gâtine, Gâtinais (land more or less uncultivated), Boischaut (the wooded land to the south of Champagne Berrichonne); terms relating to the soil: Terrefort (the marly depression round the edge of the Massif Central, below the Causses), Pays Fort, the cretaceous land to the north of Berry; and terms relating to crops: Châtaigneraie, to the south-west of the Massif Central, and Castagniccia, in Corsica; Ségala, to the south and south-west of the Massif Central and Sologne, to the south of the Paris Basin (both the latter terms come from *seigle* = rye).[106] It is clearly only in particular cases that such topographical features take on a geographical meaning and come to designate specific regions. In western France there are plenty of *landes* which have never given their name to particular geographical regions, for it is only in south-western Aquitaine that this has happened, in the case of the Landes of Gascony. The word *campania*, meaning 'wide expanse', as in the vast area of open fields in northern and north-eastern France, has only here and there provided the basis of a specific regional name: Champagne de Reims et de Troyes; Champagne Berrichonne; Champagne Tourangelle; Campagne de Caen. The same applies to the words *plaine* and *bocage*, from an early date frequently used in literature as contrasting terms and to be treated with caution when it comes to historical geography. In the *Roman de Rou*,[107] written in about 1160, Robert Wace distinguishes between two kinds of Norman peasants: '*Cil del boscage è cil del plain* (those of the bush and those of the plain)'. In this instance, it seems reasonable to follow Marc Bloch,[108] who detects a consciousness of an already existing contrast between land characterized by enclosures and land consisting of open fields. But in La Vallière's version of the manuscript of *Les Quatre Fils Aymon*, where Charlemagne and his barons, travelling to Paris by way of Monloon,[109] in Ardenne, '*Trespassent le boschage et le pais plénier*[110] (Pass through the bush and the open plain)', it is impossible to regard this as any more than a *topos*.

What were the conditions necessary for these ordinary descriptive words to be turned into the names of specific regions? The region in question had to be, on the one hand, relatively large and homogeneous, on the other markedly different from the surrounding areas. In western France, the term *bocage* is only used in individual regional names when it conveys a notion of isolation or marginality in relation to the neighbouring plains, as on the Poitevin and Norman borders of the Massif Armoricain (Bocage Vendéen, Bocage Normand, respectively). Furthermore, these were scholarly names, with no popular basis.[111] Similarly, the term *plaine* only produced regional names (Plaine de Saint-André, Plaine du Neubourg) in sectors of Upper Normandy which were already adjacent to areas characterized by enclosures. Just as toponyms are often given to places which are exceptional rather than conform to the rule,[112] regional names above all indicate distinctive features.

From the territory as officially organized to the natural territory as humanly perceived
Apart from these essentially agricultural descriptive terms, one also comes across names with a strictly historical origin among the 'natural' regional names. These may have come to apply to homogeneous regional units through gradually becoming identified with a precise geographical area. One such is the already mentioned case of the Pays de France, now restricted to the structural plain of coarse limestone to the north-east of Paris. Hurepoix (ancient forms: Hérupe, Héripoix) was an historical name initially given to more or less the entire region between the Seine and the Loire. The origin of the word is uncertain: an '*hériban*' may have been a sector that was more or less marginal in relation to the heartland of Frankish power to the north of the Seine and where royal proclamations did not apply.[113] Within this vast sector, it gradually came to be restricted to areas where no other names, either natural or popular – such as Perche or Beauce – conflicted with it. It is as a result of the scholarly work of modern geographers and cartographers that this term of probably politico-military origin has come to specify one particular well-defined natural region: the area where the limestone of the Beauce, to the south of the Seine, becomes overlaid with patches of clay and sand which turn it into uncultivated forestland, dotted with small lakes.[114] In Lorraine, to the south of Nancy, the region of Xaintois or Saintois, originally the name of a *pagus*, the meaning of which is uncertain (*Pagus Suggentensis* or *Suentensis*), in time became attached to the small natural area of shell and limestone layers (*Muschelkalk*) of Lias. On the borders of Lorraine and Burgundy, after the feudal division of the area, the region of Bassigny (derived from a personal name) (see above, p. 165) spread to include more and more land until it incorporated a number of quite distant localities; finally, however, it came to apply solely to one particular well-defined natural unit, the marly valley of the upper reaches of the Meuse.[115] The name of the Camargue region (*Camarica insula*), a recently formed alluvial plain at the mouth of the Rhône and a natural region of distinctive character if ever there was one, was originally the name of the senator Annius Camars, a great landowner from Arles.[116] In a variety of other cases, particular parts of historic regions or provinces acquired popular epithets reflecting local agricultural or scenic peculiarities: thus, Périgord Noir (which is characterized by woods located on detritic deposits on top of the chalk substrata) is distinguished from Périgord Blanc.

There was thus a good deal of interpenetration between the space as organized and the space as perceived. In naming particular regions, administrators sometimes, in the absence of significant urban points of reference, resorted to natural names; and in similar fashion, when historical names lost their significance after administrative boundaries were altered, they sometimes became attached to natural divisions of the landscape. The popular tendency has, throughout the ages, been to see the territory in terms of natural units rather than as a space defined by a theoretical framework.

Conclusion: a utilitarian concept

That is perfectly understandable. All these expressions of the way in which the concrete territory was perceived, whatever their widely ranging origins and evolutions, in the last analysis express people's consciousness of the more or less determining framework within which human beings go about their business. That is the operative concept, par excellence. The images conveyed may not be immediately comprehensible but neither are they intangible. In many cases, they have been defined, refined and transformed over many years – centuries, or even millennia. Some of these geographical names must be extremely ancient. The name of the Crau region (*Cravus campus*) is derived from the pre-Indo-European root meaning 'rock'; the region is a very distinctive geographical area, a pebbly plain in the lower Rhône valley, which is an abandoned alluvial fan laid down in the Wurmian, and this must have constituted an entity of which people were fully conscious long before the time of the Celts. Ancient authors have other names for it: Λιθῶδες (Strabo, IV, 1, 7, who shows that the region of Crau was already known to Aeschylus and Aristotle); *campi lapidei* (Pliny, III, 34; XXI, 57; this author was already aware that it was traversed by migrating herds from far away, *e longinquis regionibus*,[117] moving between their winter and their summer pastures). These ancient names reveal a deep knowledge of this inhospitable region and also of the use to which it was put – a use which has changed hardly at all since ancient times, except where modern irrigation has been introduced. Most names, however, acquired their final form and definitive meaning only in the late Middle Ages or even during the modern period, following a long development which reflects the efforts of many generations progressively trying, now successfully, now unsuccessfully, to adapt to nature.

These 'natural' regions which come to terms with the different features of the area that affect all the activities of daily life tend to be more clearly defined than the organized administrative divisions. In general, they have proved more stable. They do not disappear unless the countryside and the human uses to which it is put themselves change. Thus, in northern France, some names of forests which were in current use in the Middle Ages have now disappeared, following land clearance in the area: the Arrouaise (see fig. 4A) and the Baroeul regions of inland Flanders no longer exist except as toponyms marked on maps. But in some cases even a transformation of the landscape is not enough to wipe out the region's identity. When the nature of the substrata, for example a particular geological layering, continues to be recognizable but is used for other agricultural purposes, the concept of the region's individuality survives. Also in northern France, the Pévèle region, an ancient forest in Merovingian times, was turned into pastureland in the feudal period, yet even today it continues to be perceived as a truly individual area.[118] In organized territories, the divisions are unstable and constantly subject to revision, the only structural basis being provided by the major towns. In contrast, a territory popularly perceived as such has roots which go much deeper. Natural units, with all the activities that human beings develop within them, produce particular social typologies, particular forms of exploitation and particular 'ways of life' (*genres de vie*), as the French geographical school would put it. These determine people's attitudes to nature and the uses to which it should be put.

But that does not mean to say that they made any decisive contribution to the compartmentalization of traditional life. Only in exceptional cases did naturally determined regions coincide with specific networks of relations or constitute closed zones of circulation and interchange. The only examples of 'natural' regions with dialects of their own are situated in particularly inaccessible locations where it was difficult to move about, such as the marshlands of western France. These *marais* did produce a dialect with specific features of its own.[119] Usually however, natural regional units have been pulled this way and that by a number of different influences emanating from the towns situated along their borders or along the communication routes that traverse them, leading to neighbouring areas. There are, to be sure, a few instances of mountainous uplands, cut off from the outside world by the dominant features of the terrain, where singularly strictly circumscribed zones of intercommunication developed. In the case of the volcanic plateaux of the Aubrac region, on the south-eastern slopes of the Cantal, this resulted in the breeding of unusual strains of stock (see below, pp. 356–9). Likewise, in particularly idiosyncratic natural environments, one comes across peculiar cultural features: for example, the Breton and Poitevin Marais, the coastal marshlands of western France, produced a distinctive type of barge; equally, the *jais* (sheepfolds) of the Crau plain are typical of that particular region. But all this amounts to little more than local adaptivity at a very rudimentary level. The mountainous Aubrac region[120] seems – as we have noted – a particularly propitious area. Yet, at the level of relatively advanced technology, thorough ethnological studies have revealed nothing that really stands out from a common culture of much wider diffusion, extending at least to the Occitan area in general, far beyond the limits of this little natural sector whose peculiar identity nevertheless certainly continues to be perceived. Conversely, within these homogeneous natural sectors, plenty of separate pockets of material culture are detectable. Natural regions certainly are perceived as providing a particular framework for life. People are aware that they live '*en* Beauce' or '*en* Limagne'. But there has never existed any 'culture' peculiar to either the Beauce or the Limagne areas. And within those regions, the framework of ordinary everyday life was infinitely more restricted.

The spatial framework of daily life

If the real horizons of relations in traditional society were not created by either the limits of administrative territorial units or by people's representations of the separate regions of France, what did determine them?

The framework of rural life

In this world of a bygone age, there were tramps and vagabonds and many others who were nomads either by profession or by accident: soldiers, craftsmen, salesmen, scholars and pilgrims. These were sometimes joined by casual agricultural labourers or peasant families driven by hunger.[121] Numerically, however, these were exceptions. The vast majority of the rural population was stable and these people lived out their lives within an extremely limited framework.

An accurate idea of the dimensions of this framework of daily life is provided by the sites and spheres of attraction of the fairs and markets which constituted seasonal or weekly meeting points for the peasants. At the beginning of the nineteenth century, forty-five weekly markets were held in the department of Calvados (5521 square kilometres). In the Pays d'Auge, already a region with a strong market economy, which was a large consumer of cereals and a large producer of milk products, the distribution of markets was even denser. In the largely self-sufficient Bocage it was less so.[122] On average, a weekly market would serve an area of just over 120 square kilometres, that is to say with a radius of roughly 6 kilometres, which corresponded to the distance which could be covered on foot in an hour and a half or two hours at the most for a slow cart or a heavily laden individual. Fairs constituted more important occasions and were held less frequently, on fixed dates of the year. Their main function was the sale of livestock and they attracted custom from a slightly wider area. For the small fairs of western Brittany, the radius would be about 10 to 20 kilometres, the distance that a herd of animals could cover in two or three hours.[123] In the mountains, the distances involved might be slightly greater. On the eve of the 1914–18 War, the catchment area of the Goncelin fair in the Sillon Alpin, not far from Grenoble, varied between a radius of 10 and 25 kilometres, depending on the direction.[124] That seems to have been the maximum. One could always find a fair that was within striking distance. At the beginning of the nineteenth century, 192 fairs were held in the department of Calvados, 180 in that of

Manche.[125] For peasants, economic life was limited to a circle with an average radius of under 10 kilometres, which could be covered in two hours at the most: that is to say a far smaller area than any constituted by an historic region or an administrative unit.

The same could be said of the horizon of social relations. A sensible way to gauge this is by studying matrimonial relations. The vast majority of marriages would be contracted within the framework of a single parish or one which included its immediate neighbours: 70 per cent of the husbands and 97 per cent of the wives married in Joué-sur-Erdre, between Nantes and Châteaubriant, between 1700 and 1789. Barely 2 per cent of the marriage partners came from more than 25 kilometres away. Those percentages were even higher in certain cut-off places, such as islands (on the Ile d'Yeu, under Louis XIV, 95 per cent of the marriages were between natives of the island) or marshes (in Brière, at Crossac, 89 per cent of the inhabitants married fellow-villagers). Endogamy and consanguinity are evident from the high incidence of particular patronyms in particular villages.[126] A careful region-by-region analysis of early nineteenth-century Alsace[127] has made it possible to define and pinpoint the basic areas of matrimonial relations, grouping on average three to five villages together, that is to say about 3,000 inhabitants, all living within five kilometres of one another.[128] The normal horizon of social relations was determined by the distance that one could walk in one hour. That was as far afield as people would be prepared to go in order to take part in one of the evening gatherings which played such an important role in traditional social life. The distances might be slightly greater in the case of religious minorities, whose members would be prepared to travel further in order to find a marriage partner. But they would be even smaller in areas where the natural geography made travelling difficult. These spheres of restricted relations were in general clearly defined, bounded by clear limits, and characterized by a strong internal cohesion. Neighbouring social units would meanwhile be scorned or shunned or even subject to positive taboos.[129] Communities of this kind tended to display remarkably homogeneous patterns of demographic behaviour, social structures (the organization of hierarchies and types of property), economic structures (the nature of agricultural and artisan production) and modes of communication with the outside world. It is thus possible to define particular cells of homogeneous living, some of them centred upon a head-village, and these were certainly what constituted the most meaningful aspect of traditional life.

That elementary rural living space was certainly not structured by official patterns of organization. It was much smaller than a *pays*, considerably larger than a parish – the smallest unit of religious life, which definitely counted in some day-to-day relationships of solidarity but which, in others, was frequently irrelevant. Usually, it was also larger than the area served by the lowest-level court of law, presided over by the local feudal lord, who was the most immediate representative of authority. Admittedly, the geography of feudal courts was extremely complex, the product of a long series of struggles, alliances and usurpations; and some courts, which were associated with particularly large feudal domains, served an area which incorporated dozens of parishes. Overall, however, jurisdictional courts clearly outnumbered parishes. (At the beginning of the eighteenth century, there were 3,000 of these courts in Brittany, almost twice as many as there were parishes.)[130] The framework of a peasant's daily life was not determined by the Church nor by the limits of his feudal lord's authority. Lying somewhere in between the space organized by the king and his representatives and that presided over by the parish priest or the feudal lord, it was the purely spontaneous product of human relations, owing nothing to God or to any master.

However, the traditional structures of France included horizons vaster than the social framework of a peasant's life. The range of matrimonial and human relations expanded the higher one moved up the social scale. The social circle of the local rural nobility would always include such neighbouring small towns as were administrative centres and would even incorporate relatives and friends from distant areas. The area within which the daily life of this social category was lived could be defined as a space suited to a man mounted on horseback, whose more or less regular visits to the local lord's château and the manor houses of the gentry might be contained by a circle with a radius of 20 to 30 kilometres.[131] Only in sectors where there was little social differentiation and where communication with the outside world was difficult would the dimensions of the framework of daily life be identical for all members of the community. In most cases, different spheres emerged in relation to social status. Some humble folk, such as domestic servants, farm-hands and apprentice craftsmen, would normally live out most of their lives within a restricted framework but might sometimes be called upon to move considerably outside it. In the early nineteenth century, the framework of life for the leisured classes, wealthy landowners and members of the liberal professions, was relatively unconfined and fragmented, incorporating large areas of the countryside as well as towns of various sizes. It was the middle classes of country society, agricultural labourers who owned their own plot of land and craftsmen and merchants who were their own masters, who were the most restricted, living within a limited framework determined by economic factors.[132]

The framework of life for city-dwellers. Relations between town and countryside

How did urban influences affect the countryside which was fragmented into these small sectors? How far did the catchment area of a town extend and what was the range of its relations? There are various ways of discovering, at least so far as the last periods of traditional society are concerned, that is to say the eighteenth and early nineteenth centuries.

The distribution of property owned by urban landlords
This has been the subject of a number of careful studies, most of them based on cadastral registers for the first half of the nineteenth century, some of which also include earlier records. Studies carried out in Upper Normandy[133] (fig. 40) and in Languedoc[134] reveal a broadly similar picture. In the early nineteenth century, at the lowest level, landowners tended to be the inhabitants of small or average-sized towns, whose estates might be quite large but were always located in the countryside immediately adjacent to their native town, usually within a radius of about 20 to 30 kilometres at the most. Examples of such towns in Upper Normandy were Dieppe, Yvetot, Fécamp, Louviers, Elbeuf, Pont-Audemer and also Le Havre which, though an important port, was a town of relatively recent development which exerted relatively little attraction upon the interior; in Languedoc, these include the small towns of the region's mountainous margins: Lodève, Uzès, Pézenas, Le Vigan, Ganges, and so on, which owned about 4,000 to 9,000 hectares, and also Béziers, situated in the plain with 12,000 hectares of property. Even Nîmes, a much larger town which owned 34,000 hectares, also belonged to this category. The estates owned by its inhabitants were located strictly in the immediate neighbourhood, in a territory stretching across the plain from the mountains to the coast. The next category up from these towns with a purely local sphere of influence comprised the provincial capitals, which controlled more land. It was distributed over a far greater area throughout the province and overlapped with the spheres of influence of smaller towns: Rouen in Normandy, Montpellier in Languedoc (72,000 hectares). The towns were the seats of *parlements*, whose local nobility managed to accumulate particularly large estates (ninety Montpellier families owned 62,000 hectares). Finally, over and above both these levels, the estates of Parisian landowners were equally dispersed throughout more or less the entire territory (accounting for 40,000 hectares in Languedoc), for the capital had by this stage already attracted members of the upper classes from every corner of the kingdom (see below, chapter 7).

The schema as a whole bore the mark of the official divisions of the territory of France. The extent of the land owned by the civil servants, lawyers and bourgeois of small towns was approximately bounded by the limits of the historic *pays*; that of land owned by large towns such as Rouen and Montpellier was approximately bounded by the limits of the 'province' or the zone served by higher jurisdictional courts. The landownership of the upper classes related to the organized territory of France as a whole and, like that organized territory, reflected past or present relations of power. In traditional society, city dwellers who owned land elsewhere would tend to

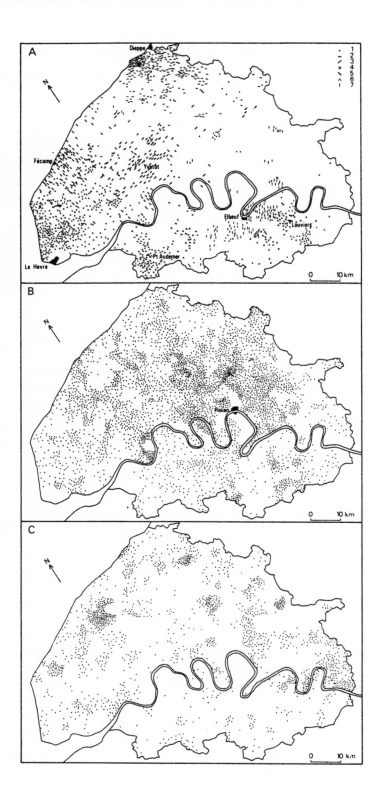

divide their time between their two residences, spending the winter in their town mansions, their summers in their country châteaux or manors, on their own land. This seems to have been the general pattern throughout the kingdom, at least so far as the plains were concerned, for city dwellers seem in general to have been deterred from acquiring property in mountainous areas. According to the same source as that used for Rouen and Montpellier, in 1841 much of the landed property of the inhabitants of Grenoble, also an old *parlement* seat, was located in the Sillon Alpin and the Isère cross-valley and also along the borders of Dauphiné (Voiron), with very little in the pre-alpine massifs despite their proximity to the town. A similar situation obtained in Chambéry in 1725. Its inhabitants owned property in the cross-valley and the Sillon Alpin but practically none in the mountain massifs.[135] Mountain property remained in the hands of the peasants (fig. 41).

The basins of demographic attraction

A study of the origins of urban populations leads to other conclusions. Surveys for the first half of the nineteenth century show that there existed at this time two distinct geographical sectors.[136] In northern France, the range of attraction was much wider. This difference reflects, first and foremost, the contrast between the rural environments peculiar to the two areas. Southern France includes many poor mountainous regions from which the inhabitants had already long been emigrating to distant destinations. In the more prosperous north, such emigration as took place tended to be limited to within the normal social spheres of influence of the towns, except in times of famine or other unusual kinds of distress. There were, of course, some anomalies: Paris, naturally; Brest, a military port which exerted a pull on immigrants from Brittany, which was relatively poor; Nancy, which attracted immigrants from not only the Vosges but also the Alps; and Mulhouse, already an industrial town offering employment for a large workforce. The contrast between the attraction exerted by Mulhouse and that exerted by Strasbourg was remarkable.[137] Strasbourg was a typical town of northern France. Of its population, 47 per cent was drawn from within the normal migratory limits, 32 per cent from occasional, discontinuous spurts of influx from the vicinity, and only 20 per cent from further afield. In Mulhouse, the normal migratory limits accounted for no more than 13 per cent of the population; the zone devoted to the textile industry (the Vosges valleys),

---

Figure 40. City landowners in Upper Normandy in 1825–35 (after Elhaï, 1965, pp. 54–5). Each dot represents 20 hectares owned by a city-dweller. A. 1. Le Havre. 2. Dieppe. 3. Yvetôt. 4. Louviers. 5. Fécamp. 6. Pont-Audemer. 7. Elbeuf. B. Property owned by landlords from Rouen. C. Property owned by Parisian landlords.

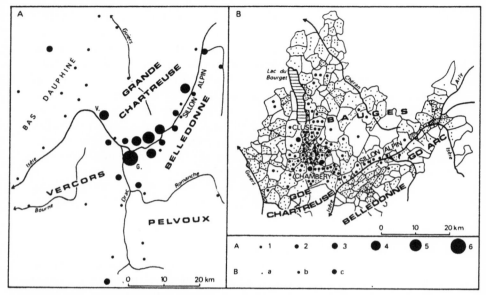

Figure 41. Traditional areas of property in the Alps (after Armand, 1974, pp. 123 and 126).
C. = Chambéry. G. = Grenoble. V. = Voreppe. A. Sums paid as land tax by the Grenoble upper classes in 1841 and places of payment (in francs). 1. Under 1,000. 2. 1,000–2,000. 3. 2,000–5,000. 4. 5,000–10,000. 5. 10,000–15,000.
6. 40,000–50,000. B. Places where land taxes were paid by the bourgeois of Chambéry in 1725, located by commune. a. One quota. b. Ten quotas. c. Sixty quotas.

which cultivated business interests in the town, for 10 to 15 per cent; while the flow of emigrants from distressed areas such as over-populated rural regions not included within the normal catchment area (for example, the Outre-Forêt region of northern Alsace, beyond Strasbourg) accounted for over 50 per cent and migration from even further afield for over 15 per cent. But in northern France such cases were exceptional. Overall, the contrast between the poverty-stricken south and the wealthy north is clearly reflected in the map of urban catchment areas.

When did that contrast become established? Two assessments fifty years apart would seem to indicate that it developed mainly during the eighteenth century. The Bordeaux map for 1737–42 shows that immigrants were already flooding in from all over the Aquitaine Basin at this time but that the percentage fell progressively the further out one moves from the town and its immediate neighbourhood. The map for 1782–86 shows, in the first place, that greater numbers were coming from elsewhere, but also that they were coming from much further away. Above all, although the percentage

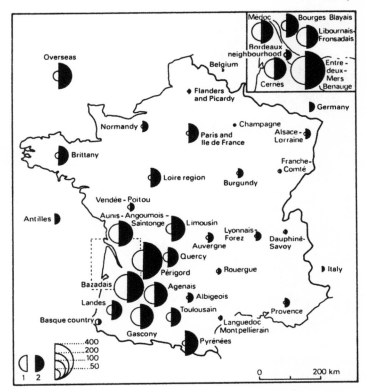

Figure 42. The development of the catchment area of Bordeaux in the eighteenth century. Immigration to Bordeaux in 1737–42 (1) and 1782–86 (2), after *Histoire de Bordeaux*, V, and Clout, 1977d, p. 488.

regularly decreases outwards, there is an increase in the percentage of immigrants from the poor mountainous environment of the valleys of the Pyrenees and Limousin, beyond what could strictly speaking be called the Aquitaine Basin. And many other immigrants were now coming from extremely far away: from virtually every corner of Brittany, the Loire region, Paris and even eastern and south-eastern France. Between those two periods, Bordeaux's catchment area seems to have opened up dramatically, particularly in mountainous regions such as Limousin and the Pyrenees, for it was at this point that migration from these areas to Bordeaux began (fig. 42).[138]

That does not mean to say that in some, special, cases, the model of migration from afar did not become established at a much earlier date. Between 1529 and 1563, almost two-thirds of immigrants to Lyons came

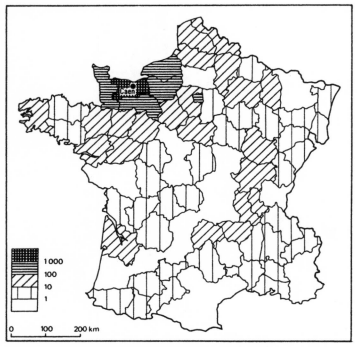

Figure 43. Travellers putting up in Caen in 1790–91, expressed department by department (after Perrot, 1974, p. 648).

from the surrounding countryside, particularly the poor mountainous neighbourhood: Savoy, Dauphiné, Monts du Lyonnais, Beaujolais, Forez, Bresse. But a steady stream was also arriving from as far afield as Paris and the towns of the lower Seine.[139] The picture presented is that of a town situated at a cross-roads, which had always depended heavily on its distant contacts. We have no statistical evidence of comparable assessments for the Middle Ages, but significant clues abound, particularly thanks to anthroponymy. In the thirteenth century, the names of people living in Provins, one of the major locations for fairs in Champagne, reveal a catchment area stretching as far as Anjou, the northern edge of the Massif Central and southern Burgundy.[140] In the same period, the names of the inhabitants of La Rochelle demonstrate the cosmopolitan character of the population of this port and also the high proportion of immigrants from Normandy and Eastern Brittany.[141] There were thus always – or at least had been ever since the Middle Ages – two types of city: one type drew its population from a limited surrounding region which barely extended beyond the range of the town's social influences; the other maintained distant commercial relations and dominated an extremely extended catchment area.

Commercial fields of influence and sources of supply

In truth, all towns of any importance maintained distant commercial relations. But their essential food supplies – cereals, meat and the customary beverages – would come from the immediate environment. In the eighteenth century, in Caen for example, cereals came from the neighbouring Campagne de Caen, meat and cider (the drink of the popular classes) from wooded areas slightly, but not much, further away: the Orne valley to the south, the Pays d'Auge to the east, Bessin and the Bocage Normand to the west.[142] However, for many materials used by craftsmen and industry and, obviously, for food products of an exotic nature or of colonial origin, the range of sources of supply extended to include most of the markets of Europe, as a map of human relations shows. A map showing the origins of travellers staying in Caen at the beginning of the Revolution, in 1790–91[143] (fig. 43) indicates many arrivals from beyond Normandy, that is to say from not only western and northern France generally but also from the Rhône valley, which clearly played a major role in the town's commercial relations. *A fortiori*, the same applies to an important sea-port such as Bordeaux, whose commercial and human relations extended beyond its immediately dependent region of the Aquitaine Basin to more or less the whole of western Europe.[144] Supply and demand for luxury or semi-luxury goods, particularly for the upper classes, resulted in a network of commercial – and, consequently, human – relations between virtually all the towns of Europe. Inventories recording the possessions of deceased persons, even small-town bourgeois, list many objects from a wide range of distant places.[145]

The loose control exercised by the towns over the countryside

That network was, however, loosely knit. In this same period, most peasant homes would contain only objects made locally or, at the most, two or three leagues away, in the case of pottery or metal products.[146] Below the level of the tenuous network of inter-urban commercial relations, vast expanses of rural France were virtually self-sufficient. Much of the territory did not even maintain regular relations with any kind of town. In the early nineteenth century, in Alsace, the map showing the regions from which trips were made into the various towns on a regular basis leaves many areas totally unaffected. Many sectors remained altogether marginal, responding to distressful circumstances with migrations only at times of agricultural catastrophe. In rural areas such as these, only in the most exceptional circumstances would merchandise or services be required of any town. Haberdashery would be obtained within an average range of 6 kilometres.[147] The legal services of a notary, which would be required on no more than a few occasions in a single lifetime, could be procured close at hand, not more than 8 kilometres away. The few shops and services used by country folk would be provided locally, in the nearest village-centre. The worlds of city dwellers and country folk may have overlapped or touched from time to time, but frequently they were unaware of one another.

In traditional society, the territory was by no means integrated. *A fortiori*, no hierarchical network of towns had been organized. As we have seen, in the early thirteenth century the respective zones of influence of urban centres, such as they were, left many areas unaffected: they were certainly not part of an interlocking system. Even a large town such as Strasbourg presided over no subsidiary network of relations in the surrounding countryside. Although many towns, both large and small, had sprung up, they were as yet not organized into an hierarchical network.

Given these conditions, it is not surprising to find that this imperfectly structured traditional territory did not, apart from in a few exceptional spots, evolve to any degree of regional economic specialization, in the service of a single, national market. One region was still largely undifferentiated from all the rest.

The beginnings of agricultural specialization

Traditional France was essentially rural, and the agricultural land remained for the most part undifferentiated. Apart from wine, agricultural produce was sold more or less on the spot, or at any rate was not moved far. Each town lived mainly on produce from the neighbouring countryside. Agriculture was fundamentally a subsistence economy, devoted to the production of basic foodstuffs, first and foremost cereals. Everywhere, the first priority was to produce sufficient grain for local consumption.

Monoculture and polyculture

In circumstances such as these, differences in the agricultural landscape appear to have been dictated first and foremost by the constraints of the local environment, in particular climatic factors. There was thus a contrast between the landscape of the north and that of the south. The north was characterized by the almost exclusive cultivation of cereals and by land left fallow in the vast plains, and the very limited cultivation of textiles, vegetables and fruits was essentially restricted to the belts of gardens surrounding the villages. The landscape of the south, on the other hand, was characterized by the cultivation of a variety of crops, including fruit and olive trees planted among the cereals or – particularly in the Aquitaine Basin – arranged in ordered rows between one field and the next. Here, an economy of polyculture was a response to the more temperate climate, which particularly favoured trees and fruits, and also to the uneven pattern of rainfall, particularly in the Mediterranean zones, which made it necessary to diversify production so as to minimize risks.[148]

In ancient Gaul, the landscapes characterized by southern polyculture had reflected the impact of Graeco-Latin influences (see above, pp. 32, 53–7). The pattern was then repeated in the north, so far as the climate allowed, spreading so gradually that it is possible for us to distinguish its very last stages. Thus, the limestone plateaux of Périgord and, further north, the Loire region (Touraine, Berry, Nivernais) were organized as open fields under cereal cultivation, dotted with oil-producing walnut trees, the presence of which testifies to the lack of animal-fat products in the absence of natural meadows suitable for pasturing cattle, in this area of permeable soils.[149] These 'open fields with walnut trees' were still being rapidly developed in the fifteenth and sixteenth centuries, following the Hundred Years War, at which time leasing regulations often stipulated that a certain number of these trees should be planted. In north-western France, it was also from this period that much of the territory was organized into enclosed pastures dotted with cider-apple trees. Cider had, to be sure, been produced here since considerably more ancient times, for it is mentioned in Brittany and also in the Frankish world, as early as the ninth century.[150] William the Breton, probably writing in 1219, mentions the abundance of apple trees along the boundary between the Pays de Caux and Picardy, in the Bresle valley. To give some idea of the large numbers of John Lackland's soldiers, he remarks: 'The *pays d'Eu* boasts fewer of those apple trees from which the Neustrians are in the habit of making an agreeable beverage.' But the general cultivation of apple trees dates from later, probably not before the fifteenth or sixteenth century, as is proved by the insults which the French, wine-drinkers one and all, hurled at their Norman adversaries on the eve of the Battle of Agincourt in 1415, when the latter were still, it would seem, drinking beer:

> *Entre vous Anglais et Normands*
> *Retournez à la cervoise*
> *De quoi vous êtes tous nourris*[151]
> (As for you, English and Normans,
> Get back to the beer
> On which you are all nourished).

The furthest that this pattern of fields planted with fruit trees was to spread was to the plateaux of Lorraine where, following the ravages of the Thirty Years War, a leafy landscape was developed, featuring fields liberally planted with fruit trees (mostly pear trees).[152]

Areas devoted to staple crops

The earliest estates under cereal cultivation: wheat, rye and buckwheat

However, the most fundamental contrast between one region and another depended upon the type of cereals cultivated, the main opposition being between the regions that cultivated rye and those that cultivated wheat. Rye, less easily digested than wheat, had been much less popular with French consumers (unlike those of central and eastern Europe), particularly since the Middle Ages, when progressive milling techniques had made it possible to produce very fine white bread from wheat. However, rye could be grown in acid, flinty soils requiring much less nitrogen, phosphoric acid and potassium than wheat did and it adapted well to the granitic sands and gravelly soils of ancient mountain massifs and to the sandy soils of sedimentary basins. Rye is, moreover, a cereal that tolerates the cold and grows fast, adapting well to mountain conditions. Rye (*seigle*) was accordingly the cereal for regions with poor soils and for mountainous areas, particularly the high rocky zones of the Massif Central, where the term *segala* came to be applied to some areas. It was also the cereal for certain sandy plains such as those of the Sologne region, whose name stems from the same origin.

But rye is not the only inferior cereal. In the low-lying rocky land of western France, it was replaced by black wheat, or buckwheat, which had found its way to France in obscure circumstances (see above pp. 83–4), but which began to spread in the fifteenth century. This cereal was used in soups and pancakes, but was not very well suited to bread-making. It suffered both from the frost and from summer drought, but flourished on low-lying rocky land and in the Atlantic climate (Brittany, Bocage Normand and the low-lying Limousin plateaux), where rye suffered from the summer rains. In the Landes of Gascony, finally, where buckwheat never thrived on account of the heat in summer, the inferior summer cereal was millet, which produced rather bitter-tasting soups. In some sectors, on relatively well-drained valley slopes, where cultivation was possible, the main crop was a mixture of rye and millet: the rye was cultivated in winter on the ridges between the furrows, the millet in the summer, in the furrows between the ridges, on soils which were very mediocre but could be kept under continuous cultivation without being left fallow, thanks to natural composts produced in the surrounding Landes.[153]

The maize 'revolution'

At the other end of the spectrum of regional differentiations in the traditional subsistence agriculture was a whole favoured area where maize, a cereal of American origin, had been introduced after first appearing in Bayonne at the end of the sixteenth century. This had made it possible to produce two cereal crops in quick succession without letting the land lie fallow, over most of the region. It was in the Aquitaine Basin that this 'wheat–maize' cycle was developed, in the course of the seventeenth and eighteenth centuries. Maize, a plant hostile to weeds, purifies the soil. This made it finally possible to colonize the mollassic heavy loams of the interior Aquitaine Basin, where the soils were rich but produced a profusion of weeds. Here, the pattern of wheat alternating with maize now took over from the old agricultural system which, however, continued to be followed in the neighbouring Landes, where relatively small pockets of continuous cultivation depended upon the vast adjacent tracts of natural vegetation which produced the organic components that were indispensable for this. The Aquitaine Basin was the only region where such a 'maize revolution' was possible.[154] Further north, the climate was not warm enough for the grain to ripen, while in the Mediterranean regions, the summer rainfall was inadequate. The combination of wheat and maize set the countryside of Aquitaine in a league of its own in the traditional economy and also made its fortune. It is this that explains its relatively astonishing prosperity up until the Agricultural Revolution of the late eighteenth century, which transformed the countryside of northern France by replacing fallow land with fodder crops. Arthur Young paints an idyllic picture of the Aquitaine countryside: 'Between Calais and Cressenac, in the Quercy, you never emerge from land lying fallow, but as soon as you reach the climate for maize fallow land is abandoned, except on the poorest of soils ... The maize frontier marks the dividing line between the good rural economy of the south and the bad rural economy of the northern part of the kingdom. Until you get as far as the maize you see extremely rich land left fallow, but after that never.'[155] Complemented, as the wheat-maize combination was, by intensively cultivated kitchen gardens, many draught animals, fed mostly on maize straw, and an abundance of poultry the produce from which formed a regular part of the daily diet, it produced pockets of development, particularly in the Béarn region, which stood out as being quite exceptional in this generally ill-nourished south-western area of France (see above, pp. 153–4).

The chestnut groves
Finally this traditional economy also included particularly unproductive areas where the local cereal crops did not suffice for subsistence throughout the year. Here, it proved necessary to turn to a different type of agricultural economy, which made it possible for large settlements to develop in mountain zones, giving these localities a special regional character all their own. This economy was based upon plantations of chestnut trees,[156] silicicolous trees which were well adapted to the medium-sized mountains surrounding the Mediterranean basin, where the rainfall was relatively heavy, and also to the Atlantic-facing slopes where the winters were still relatively mild. In terms of calories, the product from a well-kept orchard was approximately three times as valuable as that of a traditional cereal crop cultivated in the same ground. The chestnut tree economy became a veritable single-crop system of cultivation, totally dominating the countryside in two regions: to the south-east of the Massif Central (Cévennes and Vivarais); and in the Corsican Castagniccia, in the north-east of the island. In the latter area, chestnut plantations covered close on 90 per cent of the cultivated ground at

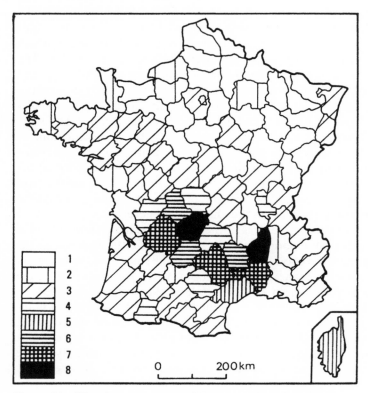

Figure 44. The chestnut groves of France at the time of the old cadastral survey (first half of the nineteenth century, after Fel, 1977, p. 222).
Percentage of the total area. 1. Negligible. 2. Less than 0.01%. 3. 0.01–1%. 4. 1–2.5%. 5. 2.5–5%. 6. 5–7.5%. 7. 7.5–10%. 8. Over 10%.

the time of the map of land distribution made between 1770 and 1790. In the mid-sixteenth century, the chestnut groves of the Cévennes were producing a crop whose value was three and a half times as great as that of the cereals in the area.[157] In other regions, essentially to the west of the Massif Central, one can speak of *pré-châtaigneraies*, the chestnut trees being less omnipresent but still numerous: the *châtaigneraie* of Cantal accounted for 18 per cent of the cultivated area in the late eighteenth century; in lower Limousin, chestnuts covered on average 10 per cent of the cultivated land, although in certain communes that percentage rose to 50 per cent.

The presence of chestnut plantations accounts for the high population rates in these mountainous areas: about a hundred inhabitants per square

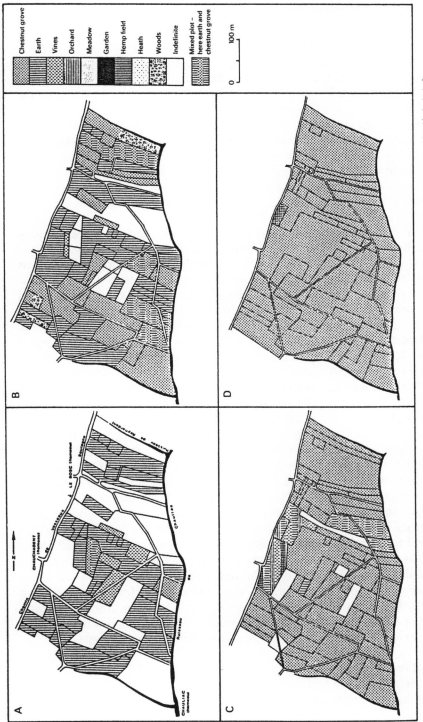

Figure 45. The evolution of the rural landscape of the area known as Les Chaussadenches, commune of Vesseaux (Ardèche) (after J.-R. Pitte, 1986, pp. 384–5 or 1978, pp. 170–1.
A. 1390. B. 1534. C. 1658. D. 1772.

kilometre in the eighteenth-century Corsican Castagniccia; approximately forty per square kilometre in the Limousin *pré-châtaigneraies*. These communities were established relatively late. In Corsica, it was the pressure from Genoa, bent on colonization, that was the decisive factor: in the second half of the sixteenth century and in the seventeenth, Genoa was encouraging the establishment of plantations of chestnut trees as a matter of policy, one that the local village communities were later also to adopt.[158] In the Cévennes and Vivarais, chestnut cultivation developed spontaneously in precisely the same period, between the sixteenth and the eighteenth centuries (fig. 45).[159] To the west of the Massif Central, the *pré-châtaigneraies* were established in the Cantal and Rouergue regions at about the same time, but their presence in lower Limousin may date from earlier, in fact from before the Hundred Years War.[160]

It is not easy to determine why these developments took place precisely where they did. There are many mountainous regions in the southern Massif Central where no large groves were developed yet which appear to possess equally propitious qualities. Basically, the answer seems to lie in the existence of earlier demographic concentrations which started off the process and whose presence, in its turn, may be explained by particular historical circumstances – possibly fewer ravages suffered from the Black Death or from the Hundred Years War, as a result of which a more substantial demographic basis existed even before the population growth of the modern period. In the case of the Corsican Castagniccia, where the communes are particularly numerous and communal territories are much smaller than in the rest of the island, the answer may lie in the fact that the area provided a mountain refuge high above the coast which, in turn, was ravaged by pirates and malaria. But other factors must also have played their part, in particular the extremely intense Genoese influence throughout the north-eastern part of the island, the traditional base for all the most socially and economically advanced sectors of the population (see below, p. 225 and fig. 37). Genoese propaganda promoting chestnut plantations was also directed, albeit without comparable success, beyond the mountains where, admittedly, there was probably much less demographic pressure and where, consequently, a pastoral economy seemed more attractive. The absence of organized chestnut groves in other naturally propitious areas may be explained by the availability there of alternative solutions. For instance, there was an active and early phase of emigration on the part of the workforce in the western and south-western parts of the Massif Central. Workers from this region began at an early date to gravitate towards Paris (masons from Limousin) and also to commercial centres in Spain (workers from Auvergne).

The potato
This might well have had as important a nutritional role as the chestnut in sectors where cereal production was inadequate, particularly in rocky,

mountainous areas. However, it was not introduced until relatively late.[161] By the late sixteenth century, it was certainly known and by the mid-eighteenth century it constituted an important crop in certain mountainous localities such as the Pyrenees of the Ariège district.[162] But it was only in the 1780s that it became widely distributed throughout France, after which, within twenty short years in the first half of the nineteenth century, it became a definitive part of the daily French diet. The cultivation of potatoes did not result from the conscious adaptation of a terrain of a particular ecological nature, but was essentially a social phenomenon. With hard work, large crops could be produced on quite small plots and with a minimal outlay by poor peasants whose assets amounted to little more than the strength of their own arms. Potatoes provided them with an assured source of nourishment and also offered the possibility of raising a pig or two. In the early decades of the nineteenth century, before the epidemics of the 1840s, the potato became an essential part of the French diet, but at this stage it was still cultivated mainly by small-scale farmers with restricted areas of land at their disposal. By the beginning of the century, it was already being widely cultivated not only in the rocky uplands of the Massif Central, but throughout eastern France, from Lorraine to the Alps, in the Pyrenees, in coastal Brittany and in Maine and Anjou.[163]

Sectors of specialized agriculture

The necessary conditions for the introduction of specialized crops

In this traditional economy, there were not many regions where the main products were not foodstuffs. For a different pattern to emerge, it was necessary for a number of physical and human factors to coincide in a combination in which it is the latter that seem to have played the crucial part: natural advantages or constraints which prompted developments, linked with the existence of markets in towns or on feudal estates, in conjunction with the presence of active craftsmen and an enterprising bourgeoisie in the prosperous towns.

Flanders

The earliest of such regional developments and the one from which most of the later agricultural progress of western Europe evolved took place in the plain of Flanders, spurred by the presence of the many industrial and commercial towns which had been developing there ever since the Middle Ages.[164] It began between the tenth and the fourteenth centuries, in an economic situation which was of central importance to the whole of Europe at this period; for Flanders was situated at the point where the overland routes of the Rhône and the Rhine valleys, having passed through the Escaut valley, met at the North Sea. This urban development soon led to over-population; and meanwhile, because the clay or sandy soils of the plain were either unproductive or hard to work, they soon became the object of persistent efforts to improve them.[165] Animal and human manure was collected, as were many other kinds of urban waste, and this eventually resulted in a veritable recreation of the land which made it possible, as early as the fourteenth century, to switch to continuous cultivation: fallow land was replaced by artificially constituted meadows and fields where turnips were cultivated; and meanwhile products destined for industrial purposes were increasing. By the seventeenth century, records testify to the appearance of crops grown in between the cereal harvest and the new autumn crops. On the eve of the Revolution, Arthur Young tells us: 'Here, wheat is not the only resource; flax and rape predominate and broad beans, carrots, turnips and a variety of other products are all grown, so the land remains under cultivation year in year out.'[166] Long before the Agricultural Revolution, this was the most intensely cultivated region in the country.

Alsace

The polyculture of Alsace shares a number of common features with that of the Flanders plain. This too was a rich agricultural region put under intense and painstaking cultivation at an early date and farmed by small family units. The plain was certainly an open area of cereal cultivation, but also constituted a magnificent and impressive market garden which was already attracting the admiration of the French administration under Louis XIV.[167] Industrial crops always held an important place here: first hemp, oil-yielding crops, madder for dye, and tobacco; later on, also hops (the large-scale production of which was introduced sometime between 1850 and 1875). The particular nature of the agriculture of Alsace was largely explained by the climatic conditions: unusually high summer temperatures, with early, warm springs occasioned by the embanking of the plain explain the presence even here of delicate crops of a Mediterranean character; meanwhile, the irregularity of the heavy inland summer rains encouraged the production of a wide range of different crops, just as it did in more Mediterranean regions. Over and above these climatic factors, however, another essential explanation for the agricultural situation in Alsace was the proximity of the urban markets along the commercial thoroughfare constituted by the Rhine.

The Loire valley

A similar combination of favourable climatic conditions and human factors is to be found in the Loire valley. First and foremost, the hot, bright summers of this region result from the protection afforded by the Atlantic anticyclone, creating an enclave of relative dryness, affected by far fewer outbreaks of cyclonic weather than the rest of the north-western Paris Basin.[168] Furthermore, downstream from Blois, the entire right bank of the river provides a series of sheltered spots, exposed to the sun, at the foot of the

Cretaceous cliffs. In the Anjou valley, palms and pomegranate trees flourish well inland. Then, a number of human factors facilitated the exploitation of the natural possibilities. In the first place, the more or less continuous presence of the royal court, from the Renaissance onwards, created important outlets for horticulture and for the products of the market gardens and fruit tree plantations which some of the French monarchs encouraged. Addressing Catherine de Medici, Ronsard sang the praises of:

> ... *les arbres fruitiers*
> *De vostre Anjou, ou les fruits que Touraine,*
> *Plantez de rangs en vos jardins ameine*[169]
> (... the fruit trees of
> Your Anjou, and the fruits which Touraine
> Planted with rows of trees brings to your gardens.)

Secondly, the great waterway of the Loire created outlets towards Paris by way of Orléans and, from the eighteenth century on, through Nantes towards overseas ports and the Antilles. As a result, a continuous pattern of cultivation soon developed, with no fallow periods, for permanent communal pasturelands for the flocks were available in the plain, which was subject to flooding. In the eighteenth century, the Loire valley seems to have constituted an uninterrupted strip of specialized crops: the vineyards of the Orléans–Blois area supplied Paris (see below, pp. 227, 263); Touraine produced vegetables and fruit; Anjou produced flax and hemp for processing into linen cloths in Cholet and Angers.[170] No comparable developments took place in the Rhône and Garonne valleys, despite their favourable natural conditions and the possibilities of circulation that they offered, for they lacked the attraction of such large local markets. Most of the prunes which were produced as early as the late seventeenth century in the Rouergue and the mid-Garonne areas were exported abroad from Bordeaux; the tobacco, also cultivated from that period on, in Agenais, was sold in Italy, via Marseilles.[171]

Urban agricultural regions
Essentially, the development of specialized crops such as these was influenced by the towns and limited to their orbits, for it was the towns that provided market outlets for these products and that dominated trade. From the Middle Ages on, every town of any importance was served by a belt of gardens and orchards either situated within the town precinct or around it. It was in such areas as these that industrial crops, destined for the textile or dye trades, were also grown. Between the mid-fifteenth and the mid-sixteenth centuries, the countryside around Toulouse produced woad on a regular basis, funded by the natives of that city or else by foreigners (mostly Spanish) who had settled there. Other centres too, such as Albi, played an influential part in developing the production of woad on a commercial scale.[172]

226 *Traditional organization of France*

From this point of view, the case of Corsica is particularly remarkable. Colonized by Genoa from as early as the fourteenth century onwards, but particularly from the mid-sixteenth, it was regarded as an agricultural dependency of that large city, a large importer of cereals, oil and wine. The Genoese not only embarked upon a number of systematic development schemes, but the Republic imposed upon the Corsicans, thitherto little inclined towards tree-cultivation, obligations to start up plantations. As a result, in the north and the north-east of the island, where the Genoese influence was naturally strongest, a whole series of specialized agricultural ventures developed, the products of which were destined for exportation. By the eighteenth century, these were well established: the Balagne plain was producing cereals, while Nebbio and Cap Corse were producing wine. However, developments such as these were limited and localized, for the culture of the Genoese city-state, founded upon commerce and regimentation, was basically incompatible with that of the native Corsicans, which was founded upon domestic independence.[173]

The evolution of regions devoted to viticulture

Among these specialized types of agriculture a place of particular importance goes to viticulture, for in the ancient economy this was by far the most important, and also the only one to develop into a truly commercial proposition.

The early Middle Ages: the vineyards survived but trade declined or disappeared

The great vineyards established far to the north, well beyond the Mediterranean regions, during the Gallo-Roman period, survived the collapse of the Roman Empire, despite the general regression in population numbers and in civilization generally that followed the Germanic invasions. Their survival was due to the social prestige of viticulture and the fundamental role that wine played not only in daily life but also in festivals, ceremonies and hospitality generally. Vineyards of high quality continued to be cultivated all over the territory around the residences of princes and, above all, bishops, who played a major role in the creation of vineyards throughout the Middle Ages. Many of the latter must have earned the title of *pater vinearum*, 'father of vines', that was conferred around 800 AD upon Theodulf, the bishop of Orléans.[174] In the Middle Ages, the suburban vineyard was thus a major feature of the countryside throughout the land. Equally, monasteries, which fulfilled an essentially hospitable function, tended to surround themselves with vineyards or sought out, for their foundation, sites which were suitable for viticulture,[175] even in quite northern regions. Vineyards thus became a common feature of the countryside in the Middle Ages. However, even as vineyards were becoming established generally, the wine trade for its part declined. Now vine cultivators had to produce their wine for local consumption, near to their own residences. There were no great commercial vineyards at this period. Up until the Norman invasions traces of wine exportation from the Loire and the Gironde to England and Ireland can still be found, but even these disappear in the ninth century[176] not to reappear for close on a hundred years.

The revival of south–north trade in the late Middle Ages: the first high-quality vineyards
From the early eleventh century on, a revival of the wine-trade was occasioned by the startling prosperity of northern towns situated around the North Sea, in a climatic zone where it was impossible for vines to flourish. Most of its traffic passed along the major river waterways that crossed the continent. This was to be the decisive factor in mapping the major regions of commercial viticulture which progressively in the late Middle Ages, from the eleventh to twelfth centuries on, spread southwards. In order to supply these northern regions where vines could not be cultivated, the movement of expansion now turned in the opposite direction from the spread of viticulture in ancient times. The new vineyards were essentially suburban, funded by rich citizens profiting from grants for planting, and their commercial reputation rested upon the discipline and dependability guaranteed by certain communal institutions which were set up during this period.[177]

Along the northernmost limits of possible viticulture, the Middle Ages thus saw the establishment of a whole series of quality vineyards, whose products were largely destined for export: the vineyards on the south-facing slopes of the northern Ile-de-France hills and around Laon, Soissons and Beauvais, situated little more than 100 kilometres away from the upper valley of the Escaut, across the plateaux of Picardy, where transport was no problem; the vineyards of the Rhine (Alsace) and the Moselle; the vineyards of the Seine valley, chief among them those which produced what were known as *vins de France* (that is to say, essentially the wines of Paris and its surrounding communes), the most reputable being the vineyard of Argenteuil, whose wines were exported to England, Flanders, Artois and Normandy; the Vernon vineyard, along the frontier between Normandy and France, thanks to which Normandy was able to produce its own wines from the eleventh to the thirteenth centuries; and finally the vineyards of Rheims, Epernay and Châlons, along the eastern border of the Ile-de-France, whose wines were not as yet known as 'Champagne'.[178]

However, these northern vineyards were handicapped by the serious unreliability of their harvests, occasioned by climatic problems. Further south, central provinces, less well situated from the point of view of communications, benefited from the hotter summers which diminished risks and

228 *Traditional organization of France*

raised the alcoholic content of the wine. Here, a number of large commercial vineyards accordingly developed: in Auxerre, Tonnerre and Chablis sur l'Yonne – the wine of Auxerre was particularly prized in the twelfth century, when it was described as *pretiosissimum*;[179] the vineyards of Orléans, with easy transport to Paris across the Beauce; the vineyards of Sancerre and Nevers, on the Loire, and Saint-Pourçain on the Sioule, situated at the highest point upstream that was accessible to barges; and the vineyards of Anjou on the lower Loire, which exported their products to England and the borders of Lower Normandy and Brittany. Apart from a few exceptional cases, this expansion was in general circumscribed by the upstream limits of river traffic in the Loire and Seine basins. Clamecy, upstream from the regular traffic of the Yonne, and Le Blanc, 45 kilometres upstream from the limit of river traffic on the Creuse, both developed highly rated vineyards which, however, could not be maintained in the face of the competition from vineyards better served by communication routes. On the other hand, a concern for quality and the development of overland communications resulted, later, in success for a number of vineyards seemingly less well-placed for trade. Thus, it was only in the thirteenth century that wine from Beaune began to be supplied to the dukes of Burgundy and, thanks to its quality, sold particularly well in the markets of northern France, being shipped from Cravant on the Yonne. It was meanwhile also transported southwards by way of the Saône and the Rhône, to the papal court in Avignon. Similarly, the Arbois vineyard, on the edge of the Jura region, originally developed to serve the plateaux of the upper Jura and Switzerland, won the favour of first the dukes of Burgundy, then Charles V and, as a result, penetrated the Flanders market.[180]

The maritime expansion of northern sources of supply: the Atlantic seaboard and the development of the vineyards of Bordeaux.
Meanwhile, in the twelfth century, another area of distribution developed along the Atlantic seaboard of Aquitaine, favoured by the facilities of maritime transport afforded by the region. The initial spur must have come from the evolution of a new type of ship which made its appearance at this time, on this north-western coast of Europe, producing a great increase in maritime traffic. These vessels were rounder, bigger and capable of carrying much larger cargoes than the deckless Viking long-boats. They put in here chiefly in order to load up with salt from the salt marshes of Aunis and Saintonge, the most northern of the salt-producing areas. The vineyards in the Charente region were developed in liaison with this traffic, whose importance received tangible recognition with the foundation of La Rochelle by Count William of Poitiers in the 1130s. In the thirteenth century, Bordeaux, which until then had been left out of these developments (in 1199 the English court was familiar only with the wine from Poitou), also became

involved, thanks to its loyalty to the king of England, for which it was rewarded by the establishment of communal institutions in 1224, precisely at the point when La Rochelle became subject to the king of France. The export trade to England now spread rapidly throughout the Aquitaine Basin. But Bordeaux soon gained a decisive advantage thanks to a ruling known as 'the privilege of Bordeaux',[181] initially passed on its own authority but ratified by the king of England and subsequently, in the fifteenth century, also by the French monarchy. This privilege gave Bordeaux a monopoly over all sales before Christmas, prohibiting the distribution of wine from upstream before that date and even the shipment, downstream from Bordeaux, of wines which were not destined for the Bordeaux market. Only Bergerac, on the Dordogne, which was particularly loyal to the king of England and the last town of Aquitaine to remain in English possession, had been exempted from that prohibition, a fact which made possible the early development of a vineyard of quality which still exists there today. Dispensations were also granted by the king of France to the vineyard of Gaillac in Languedoc, from which wines could be delivered as early as 11 November, provided none were sold to the English before Christmas. The privilege enjoyed by Bordeaux had considerable geographical effects, for it wiped out all the quality production beyond the area of the town's direct influence (which extended no further than Saint-Macaire, between Langon and La Réole, on the river) and at the same time ensured the development of the great Bordeaux vineyards, despite a damp, foggy climate which is extremely unfavourable to vines, making a veritable geographical paradox of the area. The only other vineyard that survived, apart from those of Bergerac and Gaillac, was at Cahors, a bishop's seat whose wine was already famous in the Middle Ages. This town was favoured by its situation at the highest point of navigation on the Lot, which provided it with outlets for its wines in the non-wine-producing mountainous regions of the Massif Central; it also enjoyed very benign climatic conditions and was surrounded by land suitably disposed for the cultivation of vines. In the rest of the area, however, commercial viticulture virtually disappeared.[182]

The absence of commercial viticulture in Mediterranean regions up until the modern period
In the Middle Ages, the availability of suitable sea and river outlets for the distribution of commercial wines thus limited the trade to an area well to the north of the Mediterranean region, whose climatic suitability for viticulture was meantime hardly exploited at all. In the Aquitaine region there were virtually no vineyards upstream from Bordeaux. In the interior, the southern limit of exports to the north coincided with the political frontiers of the duchy of Burgundy, in Mâconnais. At this period, the wines of lower Languedoc were purely for local consumption. The first wines to be exported

from this area, in the sixteenth century, were fortified and treated with additives: the Frontignan muscat wine, for example, which had probably been developed in the Middle Ages to cater for the northern ships which were drawn to the area by the presence of the salt marshes.[183] Serious exportation to northern France began only after the severe winter of 1709, which ruined the vineyards throughout much of the north. In the Rhône valley, exports in the Middle Ages had been aimed at the south, in particular the papal court at Avignon, and even Italy. Not until the seventeenth century did the establishment of a good overland route between Condrieu (on the Rhône, downstream from Vienne) and Saint-Rambert in Forez, on the Loire, make it possible for wine from l'Hermitage (opposite Tournon) to reach Paris (where it was much in demand) and for the vineyards of Vienne to expand (Côte Rotie). Subsequently the construction of the Canal de Briare (1642) and the Canal du Loing (1723), connecting the Loire and the Seine, increased and extended this traffic.[184]

The decline of the northern vineyards in the seventeenth and eighteenth centuries

It was in this period that, as a result of new factors, the vineyards that had been developed since the Middle Ages along the northern limits of viticulture began to decline. As yet, however, those factors did not include competition from the Mediterranean south. The first destabilizing factor stemmed from the progress made by Atlantic shipping, which had already brought the vineyards of Aquitaine to the notice of the countries of northern Europe. In the seventeenth century, Holland owned three quarters of all the merchant ships of Europe; this was the time when Dutch commerce, then at its peak, began to impinge upon the wine trade. The sale of mediocre wines with a low alcohol content, destined for distillation or for fortification by the addition of spirits, now increased considerably. The area affected by these sales covered virtually every part of the Aquitaine Basin that was accessible by river routes. Here, mediocre wines now began to be produced on a massive scale. Meanwhile in the basin of the upper rivers, beyond the reach of river shipping, spirits began to be produced in considerable quantities, as a means of increasing profits in this area where transport presented problems, for distillation reduced the volume of liquid to be moved. This was the origin of the spirits of the Armagnac region (produced in the upper basins of the Adour and its tributaries, the Baïse and the Gers) and those of the Cognac region (produced in the upper basin of the Charente). The mediocre wines which were not used for distillation now began to reach northern markets, where they competed with the quality products of the northern vineyards for the custom of the popular classes of the northern regions.

But meanwhile, similar developments were taking place in northern France itself. From the end of the sixteenth century on, approximately in the reign of Henri IV, the consumption of wine increased considerably amongst the lower classes of the towns, where the population was growing. A new class of independent workers was emerging, largely composed of former servants who had acquired a taste for wine from their masters and were not prepared to give it up. Regulations affecting taverns and wine-shops were relaxed. To cope with the new demand, the quality vineyards of the north now all switched to the production of larger quantities of mediocre wine which was not for export but was sold on the spot. Conflicting interests developed between the bourgeois, who were keen to safeguard the quality of the suburban vineyards, and the vine-growers who, thanks to the mechanism of leases, according to which vine-growers received part of the vineyard once production started, had now become landowners and were producing low-quality wines on their own plots of land. Because of the climatic conditions, which disadvantaged higher-quality production in these northern regions, it was the lower-quality wines that won the day in this competition. The last phase in the life of these northern vineyards was thus characterized by a shift to the mass production of low-quality wines until, in the nineteenth century, faced with competition from wines produced in southern France, they disappeared entirely. The modern period thus saw the eclipse of many of the famous wines of the Middle Ages: those of Auxerre, Orléans, Beauvais, Laon and Paris (on Paris, see also below, pp. 261–3).[185] Only one survived, the vineyard of Beaune, but at an entirely different level – producing wines of the most prestigious quality: that was the only way that these exposed vineyards of northern France could hope to survive.

The origin of the greatest wines
The last episode in this long evolution was to be the development of the greatest wines of France.[186] Up until the end of the seventeenth century there was no equivalent of the great vineyards of today. The art of nurturing great wines and subsequently allowing them to mature was not yet known. Most wines that were exported had been produced that same year and none were ever more than three or four years old. Quality wines did exist, but only in the form of what we should nowadays call 'new wines', pleasant to drink and fairly fruity once fermentation was over. In 1750, the price of the best wines of Burgundy was still only one and a half times that of ordinary wines.

From the end of the seventeenth century on, a number of circumstances combined to encourage the development of high-quality wine production, initially in the Bordeaux and Champagne regions. The first factor was the revolution brought about by the introduction of new types of beverages which entered into competition with the ordinary wines. At the end of the seventeenth century, the Dutch invented gin (made from a basis of juniper berries), which soon became popular in England, making it possible to bypass the importation of French wines that had been distilled. At about the same time, they perfected the technique for making strong beers; these were at first developed in Germany, to take the place of the viticulture ruined by the ravages of the Thirty Years War. Meanwhile, the countries of northern Europe were being introduced to new drinks of an exotic nature: first chocolate and coffee (in the second half of the seventeenth century) and a little later, tea, and the effect of this was to diminish the demand for wines for everyday consumption. Furthermore, Louis XIV's wars led to several blockades, during which England stopped buying French wines, turning instead to Portugal. The blockade of the League of Augsburg War (1688–97), followed by the Treaty of Methuen (1703) sparked off the massive English importation of Port wines. The only option left to Bordeaux, if it was to remain in the running, was to move into the production of high-quality wines. The way was led by the members of the Bordeaux Parlement who owned castles in the neighbourhood of the town and in Médoc. The movement began in Haut Brion in the 1670s, becoming firmly established over the first two decades of the eighteenth century at Lafite, Latour and Margaux. These produced the 'New French Clarets', whose success was closely linked with the prosperity of England and the sophistication of its eighteenth-century society. The wine was now sold in bottles, no longer in casks, and techniques continued to be refined throughout the century.[187] The last episode in this evolution was the creation of high-quality dessert wines, such as the Sauternes modelled on Tokay, in the second half of the eighteenth century.

England was also behind the success of the wines of Champagne. At the end of the seventeenth century, there were still relatively few vineyards in this area: quite a number in the Marne valley, but very few around Rheims. The first sparkling champagnes were prepared between 1665 and 1685, in London, where they were bottled with various spices and sugar added, and left to ferment once again. In view of their success, it naturally occurred to French wine producers to prepare them in their own country of origin, which Dom Pérignon, the bailiff of the Abbey of Hautvilliers, proceeded to do, as did wholesalers in the Ay region, from 1695 onwards. By 1710–20 the Champagne wines were an established success in England and also on the Paris market, so that the future of Champagne was assured well before the mid-eighteenth century. This was the period which saw the establishment of the major wine-trading companies. The high-quality wines of Bordeaux made their appearance in Paris a little later, between 1760 and 1780. The last to be recognized were the great wines of Burgundy, which did not appear on the markets of Paris and London until the 1780s. Up until this period, the only wines of good quality available to connoisseurs had been new ones, such as the Chablis and Beaujolais wines of today. The great wines of the Côtes de Beaune and Côtes de Nuits came later.

Finally, it was also in the eighteenth century that high-quality spirits made their appearance. The technique of ageing in casks of oak was attested in

Cognac as early as 1702 and continued to be perfected throughout the century in this area, with the encouragement of trading companies of partly English origin. Charente thus became the first region to set itself up as a specialized producer.[188]

The establishment of top-quality vineyards for particular wines and spirits was a relatively late development but it constituted the first sign of definite regional specialization. By the end of the eighteenth century a number of top-quality French vineyards had already switched to the production of mediocre wines, a move which led to their eventual eclipse. But a few new prestigious regions were beginning to emerge. The transformation was completed in the nineteenth century, when transportation was revolutionized.

Dispersed industry[189]

Spontaneous industry and organized industry

While certain types of agricultural specialization had already emerged in the traditional economy in connection with particular products or in certain particularly favoured geographical sectors, nothing of the kind had really manifested itself in the geography of traditional industry. Up until the advent of the steam engine and the economic integration on a national scale which followed this, the picture of industrial France continued to be characterized by spatial dispersion, the small scale of its businesses, no more than local integration with other sectors of the economy, and the use solely of raw materials and sources of energy to be found essentially in the immediate neighbourhood.

Industrial production was principally the province of the artisans or craftsmen. These were widely dispersed throughout the territory. In the last centuries of the absolute monarchy, organized corporations operated only in the larger towns. Crafts and free craftsmen were to be found throughout the land, in all the small towns and villages. In 1769 at Larrazet, a commune in what is now the department of Tarn-et-Garonne, with a population of about 1,000 inhabitants, there were 24 different kinds of craftsmen engaged in non-agricultural occupations.[190] In many regions, craftsmen were as numerous as peasants, in fact many of them were peasants seeking an extra income from various forms of industrial work. From the sixteenth to the eighteenth centuries, industrial development was thus mainly limited to regions where the growth of a landless or virtually landless population created a pool of available labour. This kind of industry, for the most part spontaneous and unorganized, was mainly devoted to the extraction and processing of local raw materials. On the other hand, in places where it was necessary to import raw materials a much more highly organized form of industry developed, mainly under the auspices of urban entrepreneurs. For instance, as early as the sixteenth century such entrepreneurs were distributing work for textile production around the towns. The silk industry of Lyons was the first major example where labour was distributed within an urban orbit, under the influence of the silk merchants. Finally, a third category was that of industries introduced by the state in order to diminish national dependence upon imported products, particularly luxury goods which involved the export of large quantities of currency: factories for the manufacture of tapestries (previously imported from the Netherlands); carpets; glass in the Venetian style; and so on. The key period for developments of this kind was that of Henri IV. Of the forty-eight factories of this type in operation in 1610, forty had been created during his reign. Most were situated around Paris, in the Ile-de-France and on the banks of the Loire. Further afield, around Lyons and in a number of other sectors, similar developments followed later, in the second half of the seventeenth century, with the encouragement of Colbert.

The distribution of the principal branches of industry

It is only at the end of the eighteenth century, with the introduction of statistics, that a clear national picture of the distribution of the various industries begins to emerge.

In 1788–89, the iron industry[191] was the subject of an enquiry set up by the Bureau de Commerce, which listed all the establishments involved, ranging from simple forges burning charcoal to large, already complex furnaces. This industry used a wide variety of sources of energy: wind, water, and above all charcoal (despite the fact that, as a result of deforestation, charcoal was beginning to be in short supply in some regions) and it was very widely dispersed, essentially in districts where shallow seams of iron ore were available: the tertiary deposits, rich in iron, of Périgord; the oolitic iron of the jurassic limestone plateaux of the eastern Paris Basin and Nivernais; the seams in the ancient Vosges, Ardennes and Normandy massifs; and the mountainous regions of the eastern Pyrenees and the Alps (see fig. 75). Other mineral deposits, of lead and copper, were used by the industry in the Beaujolais and Lyonnais regions. The manufacture of weapons and knives flourished in Thiers, in Auvergne, and also in Saint-Etienne, employing iron which came from Berry. Locksmiths were dispersed throughout the Vimeu countryside, in Picardy. Scattered glassworks were to be found in wooded regions rich in sand, such as the sandstone foothills of the Vosges, in Lorraine.[192] Paper-mills were also widely distributed, mostly located where the water supply was plentiful, for this was essential for the washing of rags.

The textile industry was also widely scattered, particularly that of hemp, since the raw material was produced ubiquitously; likewise the wool industry, which was served by the sheep raised all over France, but in particular on the limestone plateaux of the sedimentary areas encircling the Paris Basin (Berry, Nivernais, Champagne and Picardy), in Languedoc and in the Landes. Raw materials were also imported from Spain, Portugal, the Levant and even Germany and the Baltic States. Virtually every region boasted numerous weaving looms, just as more or less every village had its own flock of sheep and its own hemp field. So many towns engaged in the weaving of cloth that it is hard to pick out a few larger centres such as Amiens, Beauvais, Rheims, Rouen, Troyes and Tours. By the beginning of the eighteenth century, however, three principal areas do emerge: Picardy–Normandy, Champagne and Languedoc. Languedoc accounted for about 30 per cent of production in terms of both value and volume, while northern France (within the quadrangle formed by Caen, Lille, Sedan and Châlons), with 50 per cent of the volume of all production, accounted for 56 per cent of the total value.[193] However, heavy cloth was produced all over the country, as were linen materials, although the latter industry was rather more concentrated in the north-west (Brittany, Normandy, Picardy and Nord) by reason of the climatic factors here, which suited the steeping of flax.

A more noticeable pattern of organization and concentration was detectable for the silk industry, dependent as it was upon the merchants who received and distributed the raw material. At the end of the eighteenth century, the silk industry was well established throughout Touraine, lower Languedoc, Vivarais, Provence and, principally, in Lyons and its sphere of influence. But it was the cotton industry, par excellence, served by raw materials imported from the Levant, America and above all the 'islands' (Guadeloupe and Santo Domingo) that had the air of a relatively highly organized industry. It had grown up mainly in the vicinity of ports such as Rouen, Nantes, Bordeaux, in the towns of Provence and Languedoc and in Lyons. Alsace was also important, because techniques of printing and industrial dyeing, developed in Switzerland, became known here at a time when the machines involved were banned in France, as a result of the opposition of the corporations (the ban was not lifted until 1775). This explains the early establishment of a large cotton factory in Mulhouse, an independent town which did not become a part of France until the Revolution. At the end of the eighteenth century, the cotton industry spread widely through the countryside of upper Normandy (in particular in the Pays de Caux, in the area dominated by Rouen) and Picardy, and it also became established in Paris, which was to remain a major centre up until 1815, at which point the high cost of land and labour ruined the textile industry here. Overall though, at the end of the eighteenth century, it was still hard to distinguish the principal industrial regions. Industry was flourishing on all sides and the siting of factories was idiosyncratic. Every town had its own specialities, its craftsmen particularly skilled in one field or another.

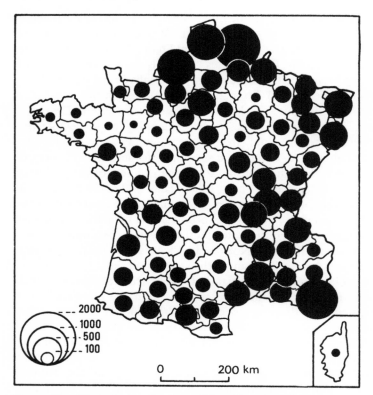

Figure 46. 'Factories' at the time of the old cadastral survey (first half of the nineteenth century), after Clout, 1977c, p. 460.

The situation changed little over the first half of the nineteenth century. At the time of the old cadastral survey – between 1810 and 1840 – traditional sources of energy were still dominant more or less everywhere, reflecting the pattern of rural-based workshops. Mills were to be found all over France: windmills particularly along the western coasts; water-mills more generally throughout the Massif Central. The map of mill locations[194] is complementary to the map of forests, which were another fundamental source of energy. The nineteenth-century map of forges and furnaces, compared to that of the late eighteenth century, reveals the progress of many large ports: Bordeaux, La Rochelle, Brest, Le Havre and Dunkirk, all of which needed supplies of metal for the ship-building yards. The general map of factories and manufacturing plants, where workers were employed on site, unlike the craftsmen working in their dispersed workshops, shows considerably larger concentrations in the north – Paris and the Seine valley – and in the east – Lyons and the Mediterranean coast (fig. 46). It is a map that calls to mind the map of literacy, with its empty 'Atlantic triangle' (see above, pp. 151–7) in the west and the Massif Central, broken only by a zone of activity centred on Bordeaux. In the first half of the nineteenth century, the map of industrial activity still essentially reflects the degree of initiative and sense of enterprise of different populations and their level of cultural development, rather than basic economic conditions.[195]

The industrial scene in the regions

It would be interesting to achieve a more detailed level of analysis than these macroscopic impressions. However, few exhaustive regional studies exist and even fewer cartographic diagrams. The most detailed that we possess are the maps that were made of Lorraine on the basis of information provided by its various communities in 1708 (fig. 47).[196] Mills appear everywhere, but other types of craftsmanship are sparsely scattered. In the Vosges forests, there is a thin sprinkling of saw-mills and paper factories, situated close to flowing water. Tileworks, which depend on outcrops of clay, are more concentrated, as are glassworks, in locations that combine the presence of wood and sand. Metal industries are concentrated into even more close-knit groups. Forges are to be found in the valleys of rivers which provided a source of energy, whilst complementary furnaces fuelled by wood are grouped on the surrounding plateaux. But natural factors alone cannot explain why they are, in the main, distributed along the edges of the duchy. Also to be taken into account are the dues that were payable for 'iron marks', which were lower on the French side of the border. In other respects, the industrial picture remains very incomplete. There is virtually no sign of a textile industry. At this period, textile work was still exclusively domestic in Lorraine. The material for clothes was spun and woven at home. There were no industrial urban centres. Industries were still essentially rural, their location being dictated by the distribution of raw materials and the availability of local sources of energy. The density of industrial establishments was greatest in the Vosges mountains. Each of 163 communities declared a number of industrial workshops of various kinds. Apart from the western Barrois region, where metalworks were to be found, they were located essentially in the Vosges or along the edge of the mountain range.

Small industries were, to be sure, widely scattered in many poverty-stricken plains that were already quite densely populated: metal-works in the upper Saône, textiles in lower Normandy, among many other examples. But the mountain regions held pride of place in the expansion of traditional industry. There are a number of reasons for this: in this sphere, as in that of cereal production, the isolation of the communities; their self-sufficiency; and their need to produce everything on the spot in those days before modern communications. But in the mountains, other factors also came into

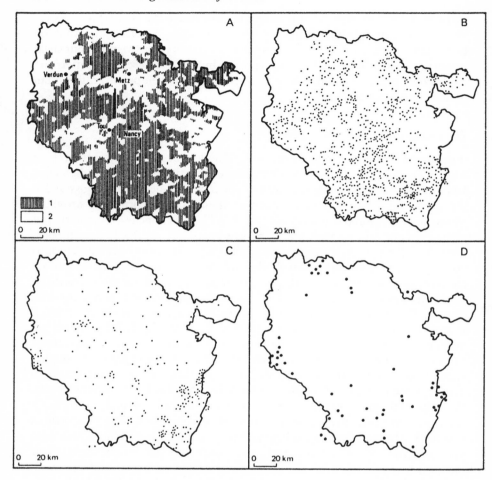

Figure 47. The industries of the duchy of Lorraine, according to the declarations made by its communities in 1708 (after Peltre, 1978, pp. 155, 156, 157, 160).
A. 1. Sectors covered by the census. 2. Sectors not covered by the census.
B. Mills. C. Mines and industries of various kinds. D. Metal mines and metallurgical industries.

play: the availability of abundant energy (provided by water or wood); the presence of raw materials, especially minerals, whose seams, revealed by erosion, were discovered at an early date to contain quantities of ore which, though small, were quite sufficient for the purposes of traditional industry; the wool and leather provided by the large flocks; and a large workforce with nothing else to do during the winter, when it was impossible to move about outside with the flocks. Such a workforce was available in particular at the lower and middle altitudes, where people were cooped up for between two and four months, a period of time not long enough to justify moving far away, as the inhabitants of the upper valleys sometimes did. The Préalpes and the Jura were thus particularly active in the traditional industries. In the Ariège Pyrenees, quite clear distinctions between different altitudes were established: mostly mining and metallurgy in the upper valleys, where activities were purely rural; many small industries scattered throughout the sub-Pyrenean zone; and the textile industry established in many small, already urban centres.[197] In the traditional economy, the Alps as a whole presented a remarkably complete industrial picture,[198] dominated by three main branches – textiles, metallurgy and leather – with woodwork and pottery on a smaller scale.

These mountain industries, born of local possibilities and needs, were soon aiming to supply the outside world. This could only be managed by concentrating on the manufacture of light-weight objects which were easy to transport, especially if some of the necessary raw materials had to be brought in from outside. A number of characteristic features thus soon emerged. Light industry and, in particular, precision work developed, producing artefacts of high value and drawing upon a tradition of skilled craftsmanship. This pattern of industry was particularly strong in the Jura where, in the wake of the woodwork of Saint-Claude, the origins of which went back to the Middle Ages, clock-making and gem-work became widely established on the plateaux during the seventeenth and eighteenth centuries, as did the making of spectacles and optical instruments at the end of the eighteenth.[199] In the Préalpes, metallurgy soon developed into the manufacture of nails in the Bauges (with wood used as the source of energy) and the production of clock parts in the Arve valley, which fell under the administration of Geneva.

In this picture of intense industrial activity in the mountains, within the framework of the traditional economy, a relative exception must be made for the Massif Central, which was much less industrialized than other mountain areas. The mineral seams of ore here had been exploited at a very early date, during the Celtic period when the area was quite densely populated, and they had soon been exhausted. Metallurgy, paralysed particularly by the dearth of wood in these mountains which had been subjected to deforestation at an early date, was virtually non-existent, except in Thiers, which constituted a notable exception. Only the textile industry was widely established: the hemp industry was very widely distributed in the high, rocky regions; the wool industry predominated in the south, all around the Causses and the Cévennes, and heavy cloth such as caddis and serge was manufactured in the small towns of the Cévennes and Gévaudan; cloth of fine quality was produced to the south of the Causses, the centre of this industry being Lodève, to which Cardinal Fleury, Louis XV's prime minister, who was a native of the town, had granted a monopoly in the manufacture of cloth for the army, thereby providing employment for a workforce of up to 4,000 in the eighteenth century; Velay was the home of lace production, born from the pilgrimage to Le Puy and the demand for lace to adorn reliquaries and shrines; finally, the factories of Lyons, to the east of the Massif, produced silk. At the end of the eighteenth century, glove production developed in Millau, using sheepskins from the Causses. Overall, however, the textile industries aside, industrial activity in the Massif Central was not nearly as great as in other mountainous areas.

Traditional industry and the countryside

To sum up, it is fair to say that, despite its unequal distribution, industry was present all over traditional France, although density certainly varied from one region to another. And industry left its mark upon the countryside far more obtrusively than it does today. Industrial activity was manifest everywhere, first and foremost in the shape of countless mills. Most of these (the flour-mills, for example) were of course necessary in food production, but there were also many other hydraulically powered machines, such as sawmills, tilt-hammers, fulling-mills and forge-bellows. Such machines were installed at intervals in virtually uninterrupted series running the length of even the smallest river valleys. Many of these were dammed to create chains of pools arranged in such a way as to produce small waterfalls. The countryside of today displays many traces of such technology – embankments, apparently purposeless bends in the roads, hedges and tow-paths.[200]

The countryside of the past was a noisy place, resounding with the din produced by all these installations. It was also, without a doubt, much more polluted than it is today. The steeping of flax and hemp, for example, turned the smallest streams into rivers of ink flowing with black, stinking water, whilst the stench of the hemp spread out to dry in the fields hung heavy in the air.[201] The interpenetration of industry and rural life created many types of pollution which it has been possible to circumscribe and thus diminish only by dint of concentrating industry.

Conclusion to Part II: The beginnings of a territory with a spirit of its own?

Over and above all these divisions, over and above the basic compartments within which life was lived and over and above *pays* and provinces, did France exist as an entity, in the framework of this traditional society, before the great highways were developed and the Industrial Revolution took off? Was France any more or anything other than a purely political concept? Did the land known as France have any functional meaning?

The first point to note is that the France of the Ancien Régime was not an economic entity. Trade was impeded by innumerable internal barriers and customs dues. Even the movement of grain from one province to another was in many cases prohibited. The first liberating measures were not taken until 1763 and, more importantly, 1774 (by Turgot). It is true that there were a number of regulations designed to prevent certain commodities from being exported outside the kingdom, cereals in particular (these were allowed to be sold abroad in 1764); and other regulations to prevent the importation of other goods (luxury products in particular but also, throughout much of the eighteenth century, painted linen). However, these restrictions, inspired by exclusively monetarist preoccupations, were not matched by any attempts to organize a single internal market. On the eve of the Revolution, only the territory known as the Cinq Grosses Fermes, where Colbert had set up a customs union in 1664 and which more or less covered the northern half of France (excluding Brittany, Artois, Flanders, Lorraine, Alsace and Franche-Comté), constituted an organized market run as a single unit. The only gestures towards general economic legislation were made in uncoordinated spheres: forests (the ordinance of 1669); trade (the ordinance of 1673); shipping (the ordinance of 1681). The introduction of a unified system of measures seemed an inaccessible dream. A few attempts at this were made as early as the sixteenth century (the edict of 13 April 1540, establishing the royal ell, and the general ordinance of 13 November in that same year, which could not be applied), but these were not followed up.[1]

This compartmentalized, divided territory, lacking any unified system for levying taxes and customs dues, in no sense constituted a general framework of activity. Merchants' trading relations reached into no more than a fraction of the territory of France but, on the other hand, frequently extended beyond its frontiers into neighbouring countries. The horizon of a small-scale cloth merchant in Amiens, at the end of the sixteenth century, was on the one hand limited to Paris, Rouen, Rheims and Troyes while on the other hand it reached as far as Antwerp, Brussels, Bruges and Flushing. For precious goods, the field of commerce included virtually the whole of western Europe. Woad from Toulouse, in its heyday, namely the fifteenth and sixteenth centuries, was sold in London, Antwerp, San Sebastián, Pisa, Leghorn, Florence, Venice, Geneva and, by way of Strasbourg, as far afield as Germany.[2] France does not seem to have constituted a preferred market for any type of merchandise. At the end of the eighteenth century, with a few particular exceptions, there were still no clearly discernible homogeneous, national economic zones. For Paris, the area of supply for cereals was vast but limited to northern France, stretching from Flanders to the channel ports and the Beauce, in the west, and from Barrois to the Langres plateau in the east. Wood supplies came from no further afield than the lower Burgundian plateaux. Wines came essentially from northern France, as did iron and ironmongery products. Only luxury goods came from far afield, from well beyond the frontiers of the kingdom. Similarly, Bordeaux maintained commercial relations with much of north-western Europe, even the Baltic ports, but within the national territory, apart from the Aquitaine Basin, its relations covered only the western and north-western regions and were virtually non-existent in the eastern half of the country. Only where the textile industry was established did there already exist at least an emerging national economic area. Paris played an increasingly dominant role and cloth factories in northern France made use of it as an intermediary, to redistribute their products throughout much of the country to a network of clients strung out along the main arteries of communication. Textile warehouses were to be found in even the remotest sectors of the kingdom. The Lyons silk trade was much more active within the 'French isthmus' as a whole and, in similar fashion, from Rouen and Paris to the Mediterranean seaboard, passing by way of the Rhône valley, the manufacturers of the Lyons silk industry and the wholesalers of that city made their presence felt through their representatives in every town of any importance. France as a whole did thus provide the framework for a real 'national urban market' but it was, in truth, but one element within the far more extensive field of trade relations which Lyons maintained with the Iberian peninsula, northern Italy and the greater part of Germanic Europe. Anyway, the textile industry was an exception in this respect. And even within the textile industry, the pattern of commercial relations was extremely uneven over France, with many areas remaining completely unaffected. The Alps (except for Grenoble), much of the southern Massif Central and the Aquitaine Basin, and many inland sectors of western Poitou, Brittany and Normandy remained beyond the influence of the Parisian trade.[3]

A similar picture is suggested by the pattern of people's movements. Temporary migrations from the mountainous massifs often involved great distances and destinations beyond the national frontiers. It is true that, by the late fifteenth century, long before Savoy was a part of France, Savoyards were to be found in Paris (see below, pp. 255–8), but people migrating from the northern Alps at the beginning of the modern period tended, rather, to make for the Rhineland and other German areas.[4] As for the people of the southern Alps, in the seventeenth and eighteenth centuries most were still migrating along eastbound routes leading on the one hand to lower Dauphiné and Languedoc, and on the other to Piedmont.[5] In the Massif Central, similarly, while many Limousin masons were, already in the seventeenth century, migrating to Paris, others were moving to Spain; and from the south-western slopes of the Massif Central, particularly Cantal, craftsmen and merchants were emigrating to the Iberian peninsula as early as the fifteenth century and were still doing so in large numbers in the eighteenth century.[6]

In other domains, however, there were early signs of a developing consciousness of France as an entity. At the time of the *chansons de geste*, the north and the south constituted two totally separate literary regions (see above, pp. 131–2). But already by the end of the twelfth century, the scene was changing. Between 1150 and 1160, the courtly romance, a specifically French creation, was already reflecting an interaction between the south and the north and testifying to the fact that southern literary themes and love psychology were making an impact in the provinces of the north.[7] The poetic frontier was the first to collapse. In figurative art the moment of unification came later. But by the fifteenth century, there was no longer an exclusive division between north and south. There were no more provincial schools of architecture. The flamboyant Gothic style, a national product, now spread uniformly throughout the kingdom.[8] At this point France, as a cultural entity, already existed.

It probably existed not only for the monks and the master craftsmen but also at a deeper, more popular level of consciousness. A specifically French mythology had emerged.[9] Figures such as Gargantua, the giant, Mélusine, the snake-like fairy, ogres, magic animals such as Bayart the horse, and various incarnations of processional dragons, such as the Gargoyle of Rouen (and of Gothic art) and the Tarasque of the lower Rhône valley[10] only acquired their particular forms and names within the French culture, however ancient their origins and however universal the myths with which they are connected: Gargantua in the early Carolingian Middle Ages; Mélusine originally in the Plantagenet circles of Poitou;[11] Bayart, the magic horse,[12] also in the *chansons de geste* of the Carolingian period (*Maugis d'Aigremont* and *Les Quatre Fils Aymon*). It was, furthermore, essentially within the framework of the French language (with one or two overlaps into the neighbouring linguistic areas of Walloon and the dialect of Vaud) that these figures became popular. Those that passed into the Germanic tradition (as Bayart the horse did, into Flemish mythology) did so only at a relatively late date and within a framework of literacy. Gargantua, for example,[13] never became part of the Flemish tradition of giants, which was very distinctive and already developing along quite separate lines in the late Middle Ages.[14]

The monarchy then took certain steps which were superimposed upon these popular trends. They took the form of a linguistic intervention. The ordinance of Villers-Cotterêts (1539) prescribed the use of French as a universal language in preference to Latin, wherever particular dialects were in use, and thereby created an extremely powerful instrument of cultural unity. Measures taken in the legal domain were also particularly important. It is true that a general and single legal codification, though pressed for and planned on several occasions,[15] was not achieved before the Revolution; and that even in the fourteenth and fifteenth centuries the French monarchy did not take systematic action designed to promote centralization (Guenée,

1968); nevertheless, the great royal edicts of the fifteenth and sixteenth centuries (1498 and 1539) and the seventeenth century (1667 and 1670) did normalize the procedures of civil and criminal law.[16] Above the level of provincial legislation, the wider framework of the royal jurists took shape.

This national framework of laws and regulations was already reflected in the movement of individuals employed in professions which were subject to strict corporative organization or constricting technical regulations. Their training more or less had to be acquired within the framework of the kingdom of France. Two apothecaries of Amiens, who qualified in 1582, had thus studied and served their apprenticeships in Paris, Mantes, Rouen, Orléans, Tours, Bordeaux and Lyons.[17] Their horizon of contacts and relations was thus quite different from that of the above-mentioned cloth merchant of the same town. For one thing, it did not extend, as his did, to the Spanish Netherlands.

Such a cultural unity was exceptional in the Europe of the sixteenth century and, from the Renaissance on, foreigners were very much aware of it. The Venetian ambassador Marino Cavalli wrote in 1546: 'There are some countries that are more fertile and richer, such as Hungary and Italy; and there are others that are larger and more powerful, such as Germany and Spain. But none is so united and easy to manage as France. That, to my mind, is its strength: unity and obedience.'[18] However, as can be seen, all this pertained to a higher sphere, the sphere of tastes and intellect, sentiments and passions. France certainly existed by the end of the Middle Ages. It was alive, personified and loved, as were its kings. Major literary, artistic and religious movements took place, unimpeded, within its frontiers and assumed specific forms there. But very little of all this was reflected in the material culture or in day-to-day relations. To its inhabitants, the concept of France was a distant one. The immediate realities and framework of their lives were enclosed and infinitely fragmented. France existed as an imagined concept before it did so as a lived reality. It was already part of people's imaginary representations of the world, a part of their dreams, beliefs and faith, long before it became a practical reality in daily life. The France of the modern period was already an Idea, but no more than that. Down through the centuries, it had been deliberately created, mentally constructed within the frontiers established by the Treaty of Verdun. It existed in the souls of its children long before it began to shape their lives.

PART III
The centralization and diversification of the French space

7

Paris and the Parisian centralization

The traditional geography of France was dominated by contrasts between the north and the south, which were a legacy of the Germanic and other invasions, and between the north-east and the south-west, which resulted from the socio-economic evolution of the Middle Ages. But this pattern was to be deeply upset by the emergence of a major centre of polarization, a development which took place much earlier in France than in the other countries of western Europe: this was the major urban centre constituted by Paris. The geographical effects of this development were considerable. Paris remodelled the countryside all around it. Above all, it refocussed and completely reoriented the network of communications, for routes were now designed in a fashion that no longer had much to do with the deeper natural tendencies of the isthmus. All the regional differentiations connected with the contemporary economic revolution made their appearance within a country that was no longer divided by major contrasts but was, instead, dominated by a single headquarters.

The development of Paris

Why Paris? Why a capital with a relatively northern situation, the choice of which definitively established the centre of gravity of France in the north of the country?

The geographical conditions

The gap routes[1]

Paris appears, at first sight, to be favoured by an exceptional convergence of waterways: the valleys of the Seine, the Marne, the Yonne, the Oise, the Ourcq and the Loing, to name only the principal ones. Their convergence results from a series of specific morphological factors: the bending of the Oligo-Miocene erosion surface (the 'Paris hollow'[2]) at the time of the renewed

248 *Centralization and diversification*

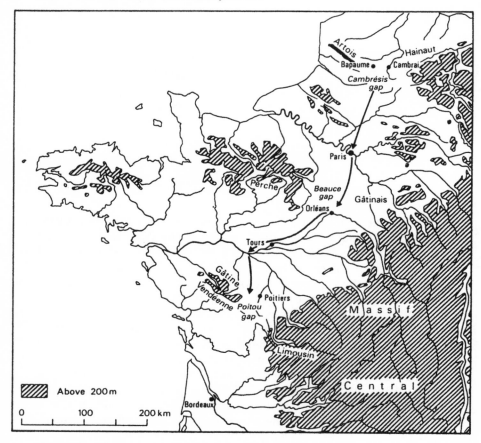

Figure 48. The 'gap route' and the position of Paris.

tectonic activity of the Upper Miocene, producing an incurvation which attracted the water courses at the point when a concentrated drainage network was being created. But in fact, even if a corporation of boatmen was active in Lutetia as early as the Gallo-Roman period,[3] and even if the presence of river boatmen was recognized as early as the Middle Ages in the heraldic arms and motto of the town ('*Fluctuat nec mergitur*'), the advantages of this situation were only appreciated relatively late, when the existence of the convenient supply routes to the large town became important. That was not the basic cause for the growth of Paris, however. The city in truth owes its existence to a major overland route, not easily discernible at first glance, which passed across the plains and plateaux of northern France.

R. Dion has dubbed it the *route des seuils* (gap route). It leads from the north-east to the south-west, from the North Sea to the Pyrenees, across

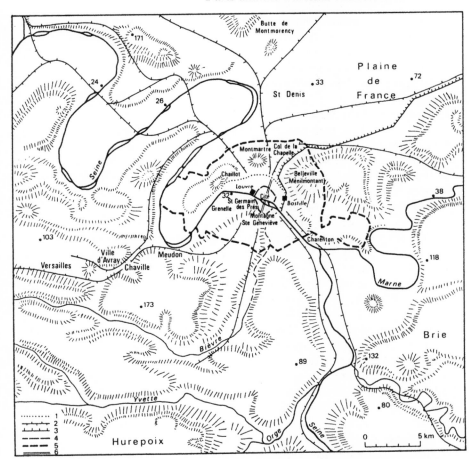

Figure 49. The site of Paris. 1. Streamlet (now disappeared) from Ménilmontant to Chaillot. 2. Railways. 3. Canal de l'Ourcq. 4. The precinct established by Philip Augustus. 5. The present-day communal boundary. 6. The rue Saint-Martin/rue Saint-Jacques axis.

northern and central France: in the north, the Cambrésis or Bapaume gap, between upper Artois to the west and Hainaut to the east; in the centre the Beauce gap, between the Perche hills to the west and the Gâtinais hills to the east; in the south-west, the Poitou gap, between Gâtine Vendéenne to the west and the Limousin plateaux to the east (fig. 48). The role played by these gaps in attracting traffic is due not so much to the difference of altitude, which is negligible (no more than 50 metres below the level of the surrounding area in the case of the Cambrésis gap; just over one hundred metres in the Beauce), but rather to their value in the task of delivering supplies for both inhabitants and armies. The gaps cut across plains of low-lying limestone plateaux which are surrounded by less fertile hills where the lime has disappeared. These are rich agricultural plains, clearings which at an early date provided communication routes, lying as they did in between less domesticated and less fertile areas.

The route across the Seine

The line of the three gaps marked out the orientation of the great route. But as it crosses the extremely fragmented tertiary plateaux of the Ile-de-France, it runs into considerable obstacles in a landscape characterized by juxaposed valleys and wooded heights. The normal itinerary runs from Senlis to Etampes, reaching the Seine valley at the site of Paris itself. After it has crossed the coarse limestone plateaux of Valois and the little plain of north-eastern France, the point at which to cross the river is clearly indicated by the natural features of the countryside (fig. 49). It has to be somewhere between Charenton and Grenelle. To the east of Charenton, the confluence of the Marne presents an extra obstacle; to the west of Grenelle, immediately after crossing the river one encounters a high wooded massif (the woods of Meudon, Chaville and Ville d'Avray), formed by the extremity of the Hurepoix plateau, which is topped with a silicified calcareous crust. Lying precisely along the axis of the most favourable direction, an opportunely placed natural feature facilitates the descent into the valley. This is the La Chapelle pass, a gap between the hills of Montmartre and Belleville which constitutes a feature of major importance, cutting through the amphitheatre wall formed by the northern hills. This is the passageway taken by the railways of today which lead to the north and the east. On the other side of the Seine, the spur of Lutetian limestone constituted by the Montagne Sainte-Geneviève provides a short cut across the alluvial plain and affords easy access to the Beauce platform. At precisely this point, islands are to be found in the river-bed, which is of relatively recent origin, created by the cutting of a meander whose ancient course was marked, up until the modern period, by a streamlet running from Ménilmontant to Chaillot: even today it affects the distribution of zones liable to flooding. Above this ancient alluvial plain, along the right bank, there survive numerous vestiges of an ancient terrace seven metres high, which usually remained unaffected when the river level rose. This provided the axis of the rue Saint-Martin/rue Saint-Jacques, which was the course that the gap route was bound to take to cross the Seine. On this route, the Ile de la Cité, despite the risk of flooding (for the island certainly was liable to occasional submersion), offered an excellent defensive site. Situated between the two arms of the river, it was in a good position to control the route and block the path of any army advancing from the north.

Stages in the city's political destiny

Paris before the Germanic invasions

We do not know in what historical circumstances the Celtic *Parisii* tribe set up its *oppidum*, but we might understand how it happened if, following R. Dion,[4] we reflect upon the conditions created by the influx of the *Belgae*, the last in a long line of Celtic invaders. The territory of this ethnic group spread as far as the neighbourhood of Senlis and bordered upon that of the *Parisii*, who did not belong to the *Belgae* people.

In the Roman period, the Gallic town of Lutetia expanded considerably.[5] Active corporations, such as that of the Seine boatmen mentioned above, already existed. The town spread mainly southwards, to the slopes of the Montagne Sainte-Geneviève, where water supplies for the public baths (from the Bièvre) were more easily organized. The architectural achievements of the town were remarkable. Apart from Paris, only four other large towns, Lyons, Vienne, Nîmes and Arles, all of them in the Rhône valley, could boast three separate places of public entertainment.[6] However, Paris was not situated on any of the main thoroughfares, chief amongst them that which led in the direction of Britain by way of Rheims, Soissons, Amiens and Boulogne. The Seine was, to be sure, an important waterway, but the two most important routes across the isthmus seem to have been those which led on the one hand to the Pas de Calais, on the other to the Gulf of Gascony.[7] Paris was certainly a big town, but it did not rank amongst the most important of the cities of Gaul. Strabo[8] placed it on the same level as Meaux and Lisieux, far below Rheims, which served as a residence for the Roman leaders: 'δέχεται τοὺς τῖον ‛Ρωμαίων ἡγεμόνας'; and Ausonius did not consider it worth mentioning at all in his *ordo urbium nobilium*, in which he did refer to Trèves, Toulouse and Bordeaux.[9]

After the Germanic invasions: the emergence of a political role

By the fourth century however, the barbarian threat had revealed the full strategic importance of Paris as a stronghold in the Roman defences, which commanded one of the essential invasion routes. The emperors Julian and Valentinian both settled down in Paris for longish periods in the second half of the century. All the same, Ammianus Marcellinus refers to it simply as a *castellum*, not a *civitas*. By the fifth century the town had shrunk to within the confines of the island and was considerably smaller than it had been in Roman times, but throughout the Merovingian period it remained one of the most important centres of the Frankish kingdom, one of the capitals between which the Merovingians oscillated in this northern sector of France which from now on was to be the centre of gravity of the country both in terms of population and power. Clovis had already made it his royal seat, *cathedra regni* (Gregory of Tours, *History of the Franks*, II, XXVIII, 38), and during the brief periods when one or other of the Merovingian leaders gathered the whole of the Frankish kingdom beneath his sceptre, it was in Paris that he would install himself. So at this time it was certainly regarded as a capital (*sedes*). In 567, at the death of Caribert, who had received Paris in his share of the patrimony after Clotaire's death in 561, the town remained an undivided entity, the common property of Caribert's three surviving brothers. This mark of its importance meant that the town could be entered by none of the three without the agreement of the other two; it was a privilege that Paris shared with two of its neighbours, Senlis and Chartres. The necropolis of the Merovingian kings was established nearby, in the basilica of Saint-Denis, the saint who was adopted as patron of the realm in the ninth century, in the reign of Charles the Bald. Despite competition from Rheims which, promoted by its archbishop, Hincmar, was selected as the place for royal coronations, the christening of Clovis having earlier taken place there, the Church thus bestowed its spiritual blessing upon what was at the time still no more than a tentative political tradition. This is no doubt why it was that Abbot Abbon, of Saint-Germain-des-Prés, writing in this period, before 898, praised Lutetia above all other towns in Gaul, in his account of a Norman siege to which it had been subjected.[10] Clearly, Paris' destiny as a metropolis was already to a large extent prefigured in the Merovingian and Carolingian periods. Maçoudi, an Arab geographer and compiler writing in 947, reported that *Barîza* was the capital of the confederation of the Franks and was a 'very large town'.[11]

In truth, that picture of Paris was already out of date. Even as Maçoudi wrote, the reality was somewhat different, for Paris' political role seemed deeply compromised. At the end of the seventh century, after the reigns of Clotaire and Dagobert, who had still resided at Clichy, close to Paris, the royal court became itinerant and the Merovingian administration disintegrated. The town slipped into a dormant state which lasted for close on four centuries. Charlemagne did not install himself in either Paris or Saint-Denis, while for Charles the Bald, Paris represented but one of many palaces. The threat of Norman attack was another factor that kept monarchs away from the place.

The Capetian capital
When the Capetian monarchy emerged at the end of the tenth century, Paris' role as a capital, within the boundaries of the very modest personal possessions of the king, thus remained extremely controversial. The first three Capetian kings maintained a palace there, but preferred to reside at Orléans, probably not so much because of the more central position that it occupied in their kingdom, but rather so as to be better placed to keep an eye on one of their most powerful vassals and most redoubtable enemies, the count of Blois. In the last third of the eleventh century, the danger point shifted, with

the constitution of the Anglo-Norman power and it was this that prompted the Capetians to concentrate their main defences in the Seine valley and make Paris their principal residence. The royal administration now acquired a new lease of life and began to establish itself as a permanent structure, tentatively under Philip I, then more resolutely under Louis VI. This monarch relied heavily upon the monastery of Saint-Denis, designated as *caput regni* in an act of 1124, and on the burghers of Paris, as he confronted his lesser vassals and the Anglo-Norman threat. With Louis VII's accession to the throne, developments speeded up. The post of royal chancellor was now regularly filled by Parisian clerics and, when the king set off on a crusade, he entrusted the regency to Suger, the abbot of Saint-Denis, who ruled the whole kingdom from his abbey. Under Philip Augustus, Paris was the uncontested capital of the country. The Louvre (1190–1202), built at the westernmost point of the town, facing the fortress of Gisors constructed by the Plantagenet kings, was its most spectacular edifice. In 1196, Paris itself was for the first time called the *caput regni*, by William the Breton. Now Paris was the definitive capital of France, whose moral personality was already vigorously asserting itself (see above pp. 105–6). Coincidentally, the territorial conquests of Philip Augustus, the true founder of the French monarchy, assured Paris of a wider sphere of influence. By the twelfth century, the royal town had already spread over both banks of the Seine. The *cité* on the island, together with a new commercial and residential quarter on the right bank and a burgeoning university quarter on the left bank now constituted the *ville, cité et université de Paris*. In the mid-thirteenth century, secular and religious authorities from all over France converged on Paris to set up residences there. At the end of the century, Pierre de Belleperche described it as *communior et excellentior civitas in regno Franciae*.[12]

It was, on the whole, by political chance that Paris, rather than Orléans, became the capital of France. Certainly these were the only two possible choices, in the *Francia occidentalis* mapped out by the Treaty of Verdun, to which the Germanic invasions had bequeathed a tradition of northern-based power. With Rouen – situated in an extremely advantageous position comparable to that of London – ruled out since the formation of the State of Normandy, Orléans would have been a conceivable capital for France. Strategically, it was just as well situated as Paris and it was definitely more central. It is interesting to imagine what France would have been like with its capital on the Loire, protected behind the barrier constituted by the Seine. It would doubtless have been much less expansionist northwards, more open to influences from the southern provinces and sooner affected by them. The establishment of a relatively excentric and exposed capital at Paris was to result in deep repercussions upon the history and the geographical structure of the nation.

The heart of France

It was not long before the new situation was reflected in literary representations.[13] The Paris pictured in the historical and legendary texts of the Carolingian period was still, by and large, the modest town of the Gallo-Roman period. In the space of two centuries a deep transformation took place, as the poets were fully aware. *'Paris estoit à cel jour mout petite'* wrote the author of *Moniage Guillaume* in about 1180,[14] referring to the legendary siege of Paris in the reign of King Louis, the son of Charlemagne, which was conducted by the Saracens under the leadership of the giant Isoré (an event possibly documented by a real text dating from 978). In the epic texts of the twelfth century, in contrast, Paris is constantly the subject of paeons of praise. It was extolled for its monuments, its riches and the festivals that the presence of the royal court encouraged: *'De joiaus, de richesces toute Paris resplent*[15] (The whole of Paris is resplendant with jewels and riches)'. It certainly was *la mirable cité* (the admirable city) lauded by the *Chanson des Narbonnais* (line 1171). The same text, which dates from the early years of the thirteenth century, devotes comical sections to the influx of visitors attracted when the king held court there and the overcrowding that ensued.[16] Prodigious quantities of precious metals appear to have been amassed in Paris: hence the declaration, still sometimes used today: 'I would not do it for all the gold in Paris.'[17] In this idealized representation of Paris, the Abbey of Saint-Denis continued to play a role of particular importance, alongside the royal residence. 'The ideas that burgeoned within the Abbey and the influences coming from the awakening town complemented one another and interacted. Saint-Denis provided the neighbouring town with the traditions that it lacked. ... Saint Denis is portrayed as the protector of France in over 300 passages from epics of a very wide variety of origins and periods'.[18] From 1124, the 'oriflamme', the banner of Saint-Denis, was borne into battle as the French army's faithful companion. The name of Saint-Denis became the army's warcry. In epic, the emperors and kings of France are frequently (for the first time, in the *Couronnement de Louis*, lines 1461 and 2521) referred to as *emperere* or *rois de Saint-Denis*.[19] Suger, the minister of Louis VI and Louis VII and abbot of Saint-Denis, transferred the responsibility for celebrating the glory of the king of France from Saint-Benoît-sur-Loire to the Abbey of Saint-Denis and entrusted to its monks the task of writing up the chronicles of the kingdom. Around the most favourable of the real facts a mythical geography was soon constructed. A town so noble and so rich could only be situated at the centre of the kingdom: *'Locum utpote medium, vel quasi, tenet urbs inclita Parisius'*, wrote Jean de Jandun in 1323, in his *Tractatus de laudibus Parisius*,[20] and two centuries later that ideal geography was reflected in Ariosto's lines:

> *Siede Parigi in una gran pianura*
> *Nell'ombilico a Francia, anzi nel core*[21]
> (Paris lies in a great plain
> At the navel of France, or even at its heart.)

Situated at the heart of France, Paris was its teacher and its sovereign; and the entire poetic and literary tradition of the late Middle Ages and Renaissance was to embroider upon this theme. Its cultural supremacy was plain to see. In the second half of the twelfth century, a letter written by Guy de Bazoches, having painted a picture of the bustling river with its ports and the nobility and power of the palace, proceeds to sing the praises of the island, which had long been a centre for philosophy and the liberal arts.[22] In 1461, Villon, who was himself a Parisian and accordingly not altogether impartial,[23] declared '*Il n'est bon bec que de Paris* (Paris tongues can vanquish all)'. Rabelais, however, was a native of Touraine, yet, having had his Pantagruel visit a series of universities, each one satirized in some way, even he has him settle down in Paris, and his purpose in doing so is clearly to exalt the capital.[24] His view of Paris was still shared in the sixteenth century by poets who were natives of the Loire valley, when the court resided here, having abandoned its seat in the capital after the Hundred Years War. In an elegy published in 1567,[25] Ronsard wrote:

> ...*Paris, admirable cité*
> ...
> *Mère d'un peuple abondant et puissant*
> (...Paris, admirable city
> ...
> Mother of an abundant and powerful people.)

and Du Bellay reached new heights in his extravagant eulogy of 1575:

> ...*semblable à la grand' mer*
> *Est de Paris sans pair, où l'on voit abysmer*
> *Tout ce qui là dedans de toutes parts abonde.*
> *Paris est en sçavoir une Grèce féconde,*
> *Une Rome en grandeur Paris on peult nommer,*
> *Une Asie en richesse on le peult estimer,*
> *En rares nouveautés une Afrique seconde*[26]
> (... like the great sea
> Is peerless Paris, where all who flock to it
> From every side are swallowed up.
> In knowledge Paris is a productive Greece,
> In grandeur Paris may be named a Rome,
> In riches it may be thought an Asia,
> In rare novelties a second Africa.)

The prestige of Paris was equally high for writers from even further afield, as can be seen from a fine text published in 1588 by Montaigne, who was

Mayor of Bordeaux: 'I will not forget this, that I can never mutinie so much against France, but I must needes looke on Paris with a favourable eye: it hath my hart from my infancy ... I am no perfect Frenchman, but by this great matchlesse citie'.[27] By now the image of Paris was inseparable from that of France. The capital was identified with the nation. Henri IV, for whom 'Paris was certainly worth a mass', did not feel he was truly king of France until he had overcome it and installed the monarchy there.

Paris' sphere of influence: the formation of its population

In this respect too the imaginary representation of Paris had anticipated the reality. In the sixteenth century, the capital, which the entire French people idealized, was in truth a centre of attraction only for a few northern provinces. For a long time its sphere of influence by no means extended to the entire national territory. It is true that even in the Middle Ages Paris attracted small groups – mainly students – from all over western Christendom, including representatives of the 'nations' of Normandy, Picardy, England and Germany (with the last two groups accounting for most of the foreigners) who mingled there with the French: merchants and preachers from Italy; temporary migrants from the mountains, such as the '*basteleurs traynans marmottes* (showmen and their performing Alpine marmots)' from Savoy mentioned by Villon in 1461.[28] But statistically speaking, these arrivals from distant parts counted for little. In the fifteenth century, most of the merchants who visited Paris came from the Seine valley and the large industrial and commercial towns of the north. The radius of regular commercial relations reached as far as Coutances, Dunkirk and Tournai, Liège and Cologne, Châlons and Langres, Beaune and Dijon. But to the south it barely extended beyond the Loire. Visitors from Orléans were to be found in Paris but very few from the Tours, Berri and Anjou districts and virtually none from Poitou and Auvergne.[29] The Seine and its tributaries and the routes to Flanders and Orléans were the essential features in the Parisian sphere of influence at this period.

Under the absolute monarchy, this picture remained virtually unchanged. The development of the large consumer market constituted by Versailles and Paris certainly attracted temporary migrants in great numbers, chief amongst them Limousin masons, who worked throughout the summer on construction sites. Once the summer was over, the Savoyards arrived. In the eighteenth century, these were organized into 'a kind of confederation'[30] and, at this time of the year, they made a major contribution to the amusements and activities of Parisian street life: street porters, fiddlers, the owners of Alpine marmots and, above all, little chimney sweeps such as those described by Voltaire:

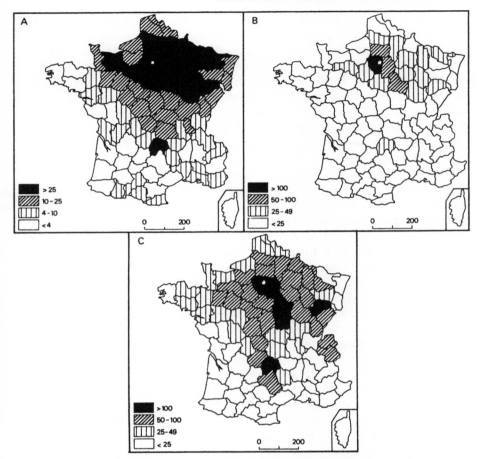

Figure 50. The origins of the population of Paris (late eighteenth to late nineteenth centuries).
A. Parisians per 1,000 inhabitants of each department in 1791 (from the information provided by Bergeron *et al.*, 1970 and the map by Le Bras, 1986, p. 82 (d)). B. and C. Birthplaces of non-natives of the Seine department who died in Paris in 1833 (B) and 1891 (C) (after L. Chevalier, 1950).

> ...*des honnêtes enfants*
> *Qui de Savoie arrivent tous les ans*
> *Et dont la main légèrement essuie*
> *Ces longs canaux engorgés par la suie*[31]
> (... Those worthy children
> Who arrive each year from Savoy,
> Whose hands deftly sweep
> Our long flues blocked with soot.)

258 Centralization and diversification

However, these visitors left little mark upon the permanent population. Apart from the nobility and their retainers, who flocked to Versailles from every corner of France and also maintained residences in Paris, the essential bulk of the population was recruited locally or from the surrounding Parisian region. A study of the sectional registers of the revolutionary period provides a certain amount of information (fig. 50A), but the first accurate statistics date from a survey made in 1833,[32] showing the distribution, department by department, of the birthplaces of 10,246 French inhabitants who were not from the Seine department but who died in Paris (fig. 50B). It shows that the area of recruitment was essentially limited to the Seine basin, Champagne and Picardy, with a clear bias to the east of the Paris Basin over the west. Of the more distant regions, only the Cantal department was already making a significant contribution to permanent emigration to the capital, prefiguring what was later to be known as the 'Auvergnat colonization' of Paris. With that one exception, Paris, in the first third of the nineteenth century, remained, as it always had been, a northern town and in particular a centre of attraction for people from the Seine basin.

It was only as the nineteenth century proceeded that a real transformation took place. The 1891 map shows appreciable differences from that of 1833.[33] The capital was exerting a greater pull and its geographical sphere of influence had increased. Irregularities still existed however. High percentages were emigrating from the Yonne basin (the Yonne and Nièvre departments) which had naturally always been oriented towards Paris; likewise from the southern Massif Central, where the tradition of Auvergnat emigration had been consolidated; and from the Jura and Savoy mountains – a similar but more recent phenomenon that had gained in strength since the incorporation of Savoy into France. But in northern France, at least, the Parisian sphere of influence tended to display a concentric pattern, with a regular decrease in the rate of emigration to Paris towards the periphery. Above all, the northern and western margins of the Paris Basin were now incorporated into the area supplying immigrants to Paris, as was most of the Armorican peninsula; and the asymmetry between east and west had disappeared. The areas where resistance to the attraction of Paris persisted are significant: south-western Poitou and Aquitaine; the south-eastern Mediterranean, Dauphiné and the Rhône valley and also non-francophone lower Brittany. At the end of the nineteenth century, Paris was still not exerting its attraction upon the whole of the French population, but the capital had certainly extended its sphere of influence considerably, particularly in the southern Massif Central, and had thereby become a partially southern town. Its domination over northern France was total.

At the same time, it had largely remodelled the face of the territory of France.

The geographical impact of the growth of Paris

The urban phenomenon of Paris was for a long time unparalleled in western Christendom. With roughly 200,000 inhabitants at the beginning of the fourteenth century (there were 61,098 hearths in the city in 1328), Paris was already by far the largest town in western Europe, followed by Florence, Venice and Milan, each with 100,000 inhabitants, and London, with between 40,000 and 50,000. In France, Bordeaux, Toulouse and Rouen, each with 30,000 to 40,000 inhabitants, were far outstripped. Paris was to remain the first town of Christian Europe until the late seventeenth century, when it was overtaken by London. In the eighteenth century, its inhabitants numbered between 500,000 and 600,000 (calculated at 509,000 in 1719, 576,000 in 1763 and 647,000 in 1794, on the basis of the bread tickets issued. The first census, carried out in 1801, when the revolutionary troubles were over, produced the figure of 546,858 inhabitants).[34] At this time it was four or five times as large as Lyons, which was France's second city. In 1851, it exceeded the one million mark (1,053,000; 1,400,000 including the suburbs) and continued to develop rapidly under the Second Empire, totalling 1,851,000 inhabitants in 1872 (2,500,000 including the surrounding built-up areas). By the end of the nineteenth century, Paris together with its suburbs could claim over 3,500,000 inhabitants.

The presence of such a large urban population in a still largely rural France soon produced considerable geographical effects. In particular, a distinctive agricultural region, designed to cater for the needs of Paris, began to stand out against the background of the traditional rural world.

The Parisian agricultural sphere of influence

A fruit- and vegetable-producing belt[35]

The first effects were felt at the level of the immediate environment. Lavoisier[36] calculates the value of bread consumed per year in Paris at the end of the eighteenth century as 20,600,000 French *livres*, that of meat as 40,000,000 *livres*, that of wine as 32,500,000 *livres* and that of fruit and vegetables as already 12,500,000 *livres* (compared with 20 million *livres* for food products of other kinds). The first mark that Paris made upon the agricultural landscape was the early development of a large area devoted to the production of fruit and vegetables – a phenomenon no different from that which we have already noted (p. 225) in connection with other large towns, but more intense in the case of Paris, on account of the exceptional size of the market here. In the Middle Ages, cultivation of this kind was confined to within the town walls, particularly in the 'Marais' quarter of the north-east of the town. But it soon expanded to the point where, by the end of the eighteenth century, a virtually continuous belt of market gardens

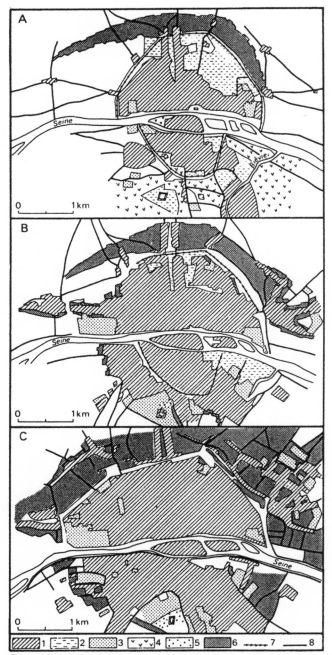

Figure 51. The development of the belt of market gardens around Paris (after Philipponneau, 1956, pp. 45, 46, 47).
1. Built-up zone. 2. Urban vegetable gardens and cottage gardens. 3. Aristocratic and ecclesiastical gardens. 4. Vines. 5. Tree nurseries. 6. Market gardens.
7. Precinct. 8. Roads. A. End of the fourteenth century. B. Middle of the seventeenth century. C. Beginning of the eighteenth century.

surrounded the capital (fig. 51). Forced cultivation, by means of cloches and forcing frames, which was introduced in the royal kitchen gardens in the seventeenth century spread to these urban, marshy plots in the eighteenth century.[37] Other specialized forms of cultivation were also introduced: flowers had been grown ever since the thirteenth century; tree nurseries and fruit shrubs and trees appeared in the fifteenth century, as did vines; mushroom cultivation sprang up in the sixteenth century, using urban fertilizers, initially outdoors or in cellars, then, from the seventeenth century on, on a large scale in abandoned underground quarries which had provided building materials for the construction of the town. Beyond the immediate suburbs, likewise, in the valleys and hills of the Ile-de-France, fruit and vegetables were increasingly grown for the urban market (fig. 52).

This rural suburban life reached its peak in the nineteenth century, after which it declined under the combined effects of rapid transport and refrigeration. But the decline was slow in coming and only really got under way in the twentieth century. In 1892 still, the Paris region was growing 20 per cent (in value) of all the vegetables produced in France, 23 per cent of the cultivated flowers, 26 per cent of the nursery products and 5 per cent of all fruit. Supplies far exceeded the demands of the city and were sold well beyond the Parisian region. Paris was also surrounded by a belt of suburban dairy farms (in 1892 there were 490 nutritive dairies in Paris itself and a further 500 in the suburbs) and pig-farms, which made use of the swill discarded by the town.

The evolution of the vineyards

The decline of the Parisian vineyards, which had enjoyed a period of massive production, came much earlier, for wine travelled better and was transported much earlier than vegetables and fruit. The evolution of the ancient Parisian quality vineyards went through phases of decline similar to, if somewhat more acute than, those of the commercial vineyards of northern France generally (see above, pp. 230–1 and below, pp. 350–3), and the consequences for the region as a whole were even greater.[38] The Parisian vineyard began to spread in area at a relatively early date, just as the quality of its wine began to decline – and this despite many efforts and a series of regulations designed to prevent wines of inferior quality from entering Paris: these included the institution of *jurés vendeurs de vins* (accredited wine purveyors) (1351); a ban imposed on local wines within a radius of 20 leagues, prohibiting their supply to the drinking houses of Paris (1577); the imposition of severe restrictions on the introduction of ordinary wines into the capital, in the form of an entrance due levied on all wines entering Paris, relating to the quantity, not the quality (which meant that inferior wines were taxed much more heavily than the rest) (1529); the erection of fiscal barriers around the town limits, the latest of these being the *mur des Fermiers Généraux*, set up on the eve of the Revolution (1784–87), an obstacle which

262 *Centralization and diversification*

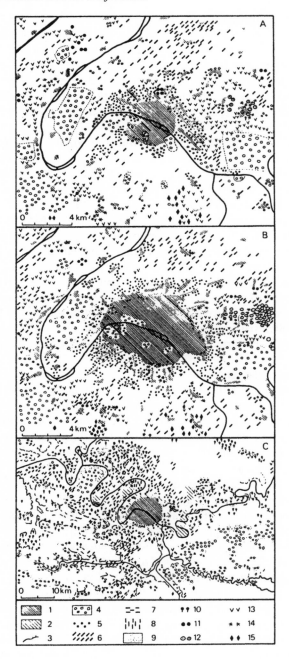

the people of Paris proceeded to get around by going to drink in the suburban establishments which were proliferating rapidly in the last quarter of the eighteenth century. But all these measures proved powerless to halt the serious degeneration which began at this period. By the mid-eighteenth century, the wine of Argenteuil, in the north-western suburbs, which had been much sought-after in the Middle Ages, had completely forfeited its reputation. The regulation pertaining to the 20-league limit also had serious geographical consequences. Wines from so far away, encumbered furthermore with such heavy entrance dues, could only be an economical proposition if the least expensive means of transport were used, in this instance river transport. Wine production was accordingly now concentrated in the larger valleys, close to the navigable rivers which converged upon Paris, and it was here that the quality of wine declined most rapidly. In the Yonne basin, quality wine cultivation survived only at Chablis, which was beyond the reach of the commercial barges. The deterioration set in particularly early in the Orléans region, which was linked to Paris by good highways across the Beauce, even before the construction of the Canal de Briare (1642) and the Canal d'Orléans (1692), which provided waterway links between the Loire and the Seine. By the beginning of the seventeenth century, Orléans wine was already banned from the royal table and it lost the last shreds of its reputation over the decades that followed. Finally, the establishment of an overland link between the Saône and the Loire (running from Belleville-sur-Saône, south of Mâcon, to Pouilly-sous-Charlieu, north of Roanne), followed by the construction of the Canal du Loing (1723), which greatly facilitated transport to the city of Paris, explains how it was that the capital's supply zone for ordinary wines now extended all the way to the Beaujolais area, whence ordinary wines were arriving in bulk in Paris by the mid-eighteenth century.[39] It was now that the Beaujolais production began to be geared to ordinary wines, whereas the prestige of the wines of neighbouring Burgundy was maintained and confirmed.

The market for dairy produce: butter and cheeses
It was not solely the immediately surrounding area that was affected by the existence of this huge market for food products. It also played a decisive role in structuring the nationwide map of semi-perishable milk products. In the

---

Figure 52. Suburban cultivation around Paris from the late eighteenth to the late nineteenth century (after Philipponneau, 1956, pp. 48, 81, 101).
1. Dense built-up areas. 2. Sparsely built-up areas. 3. Relief contours. 4. Woods and parks. 5. Market gardens. 6. Field-grown vegetables. 7. Watered vegetable cultivation. 8. Irrigated vegetable cultivation. 9. Sewage works under construction. 10. Vertical fruit trees. 11. Intensive orchards. 12. Espalier fruit trees. 13. Vines. 14. Cultivated flowers. 15. Nurseries.

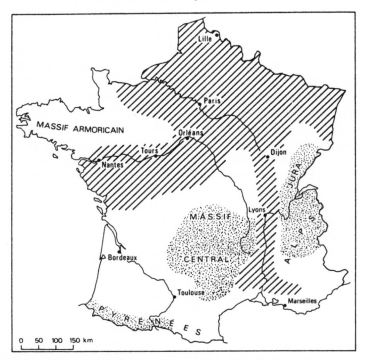

Figure 53. The traditional areas of cheese production (after Pitte, 1985).
Hatched area: regions long urbanized, producing fresh cheeses with powdery or soft rinds, mostly of small dimensions. Shaded areas: wet, mountainous regions producing large pressed, cooked or blue-veined cheeses. Blank areas: 'cheese deserts' with no produce sold anywhere except locally.

eighteenth century, the most distant sectors of Normandy, Isigny and Carentan were still sending their salted butter to Paris, as was Brittany.[40] Then, closer to Paris, the production of unsalted butter (which does not keep for so long) gradually developed and eventually spread to the whole of Normandy, whereas Brittany, which was further away, remained faithful to the salted variety which could be transported over greater distances. In similar fashion, the influence of Paris modelled the map of the different types of cheese products (fig. 53).[41] Cheeses with soft rinds, which cannot be kept for long (Brie, the cheeses of Normandy and so on), were produced essentially within the immediate commercial orbit of the capital: the Paris Basin, the Loire valley; but also in regions where communications were relatively easy and a dense network of roads had been created at a relatively early date, running from the north to the north-east (see above, pp. 153–7 and fig. 53); and also along the great communication route constituted by the Rhône corridor. The more remote plains and low-lying regions remained 'cheese deserts', at least so far as commercialized products were concerned. In mountainous areas, where the presence of large flocks led to the production of cheese on a commercial scale (Massif Central, Alps, Jura, Pyrenees), these areas remained faithful to cooked, pressed or blue-veined cheeses, which would keep longer and could be transported over greater distances.

The large agricultural units of the Ile-de-France
The influence of the great conurbation of Paris also made itself felt, this time indirectly, in the area of, not agricultural products, but the structures of cultivation and the landscape. The silty plains and plateaux of the Tertiary region of the heart of the Basin, around Paris, are even today still characterized by an agricultural structure quite different from the strip fields of Champagne and Picardy. The cultivated land is mostly divided into extremely large fields, more or less square or trapezium-shaped, with sides several hundred metres in length. This is what is sometimes called an 'openfield mosaic'.[42] Recent reorganization is in no way responsible for this pattern, which it has simply regularized. This fragmentation is not dictated by landownership. It is a division of the land according to the pattern of cultivation that results from the presence of large farms which incorporate land from a number of properties; the cadastral survey produces a very different picture. Today, large farms such as these are extremely modern and totally mechanized, but they came into being long before mechanization. Essentially, their formation dates from the eighteenth and nineteenth centuries. In the eighteenth century, there were already many large farms, but few were organized in a logical fashion. Large feudal farms with a scattering of plots of land around the village still presented a contrast to large isolated farms where the land was cultivated in a more concentrated fashion (fig. 54). The agricultural reorganization which was completed in the nineteenth century involved an exchange of plots between farms and the concentration of cultivation around centres of agricultural activity that already existed at the end of the eighteenth century. This ferment of reorganization was sparked off by the high profits realized as a result of the Agricultural Revolution.

What part did Paris play in this evolution? Its direct role seems not to have amounted to very much, for at this time the capital of landowners remained largely unproductive. The concentration of fields under cultivation was for the most part not the work of Parisians. However, the indirect role played by the metropolis was much more important, for it stimulated technical progress, supplied urban fertilizers and, above all, constituted a large market for food products, thereby generating a commercialized agricultural economy. This was the geographical area of France where grain was transported the most widely, at the earliest date and in the greatest quantities.

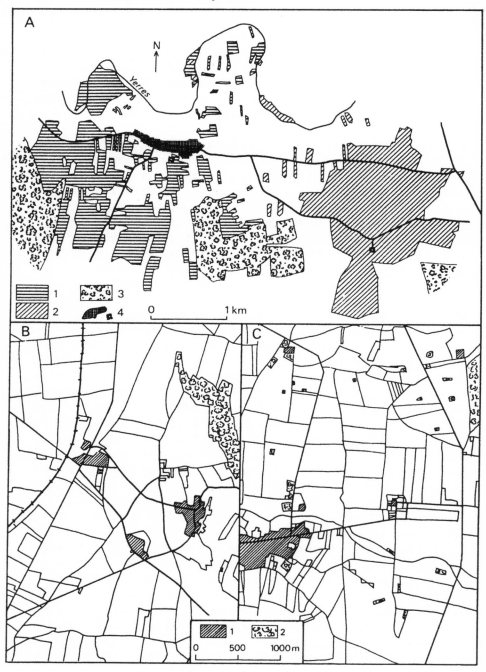

The role played by the metropolis seems to have been much more important in the immediately preceding periods, when it accounted for the very creation of the large farms. In the seventeenth and eighteenth centuries, it was the presence of Paris that mostly determined the development of big properties by absentee landlords (members of the regional *parlements*, Parisian bourgeois or members of the nobility living at court). The origin of these large properties is complex (some had started as Gallo-Roman *villae* situated on the silty plateaux; others had resulted from land clearance effected by the monasteries, on mediocre soils round the edges of the forests). But it was clearly the presence of Paris that, from the seventeenth century onwards, prompted the general switch to these large-scale farms with non-resident landlords. This was a period when the need to feed a large capital with a growing population prompted investments which turned the surrounding countryside into a large agricultural region composed of farms which, even if their bases were modest, were eager to make the most of improving technology. In the fifteenth century, the Beauce, devastated by the Hundred Years War, seems to have been an uncultivated wasteland. In the early sixteenth century, it was still represented as a region of hardship. As Rabelais put it, in 1532,[43] 'the gentlemen of Beauce do ... break their fast with gaping, and do spit the better for it'. Not until the mid-sixteenth century did the Beauce begin to be lauded as a rich, wheat-producing region. Charles Estienne's *Guide* runs as follows: 'The true Beauce is a flat region, fertile in grain, like a storehouse for France, just as Sicily was for Rome';[44] and not long after, Ronsard was writing:

Figure 54. The division of land in the Ile-de-France.
A. The land belonging to a large village farm and an outlying farm at Combs-la-Ville (Seine-et-Marne) in 1790 (after P. Brunet, 1960, p. 103). 1. Feudal farm in Combs-la-Ville (162 hectares) with parcels of land still widely dispersed. 2. The Egrenay farm (190 hectares), already much more condensed. 3. Woods and thickets. 4. Habitation. B. The Plain of France. Land belonging to Villeron (Seine-et-Oise). Sketch based on an aerial photograph published in Brunet, 1960, p. 23. A type of openfield mosaic. Huge tracts of land of between 20 and 40 hectares, many of them irregularly shaped. Three groups of buildings: to the south, the Vollerand farm (520 hectares), the only agricultural property in the commune; to the north-west, the Fantuzzi sugar refinery, with several buildings clustered close to the road and the railway; to the east, the village of Villeron, hardly more than living quarters for agricultural labourers. C. Western Brie, close to Moissy-Cramayel (Seine-et-Marne). Sketch based on an aerial photograph (Brunet, 1960, p. 27). Plateau of crushed limestone covered with silt. Large plots of land of, on average, about 10 hectares. A geometric chequer-board resulting from recent reorganization. Legend common to B and C: 1. Habitation. 2. Woods and thickets.

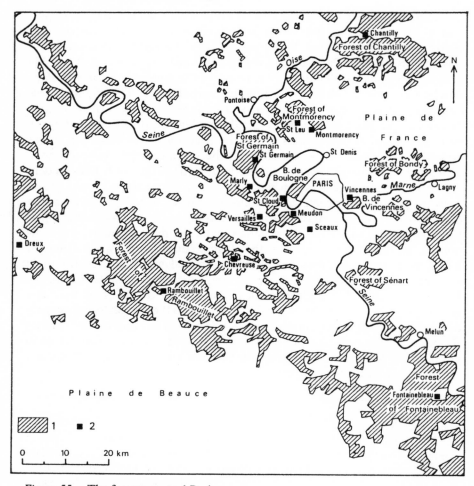

Figure 55. The forests around Paris.
1. Forests. 2. Principal royal and princely castles.

> *Que dirons-nous d'Auvergne ...*
> *... de Champagne et de Beauce?*
> *L'une riche en troupeaux, les deux autres en blé*[45]
> (What shall we say of Auvergne ...
> ... of Champagne and of Beauce?
> The one so rich in flocks, the other two in wheat.)

But it was not until the seventeenth century that the expression *'ferme en Beauce'* came to signify copious abundance. The influence projected by the court and Paris was the basis of that wealth.

The large wooded massifs of the Parisian region

A combination of social factors and the need of supplies for the urban market determined the permanence of one of the major features of the landscape in the neighbourhood of Paris: the large expanses of forest which are so characteristic of the Ile-de-France (fig. 55). They are, it is true, situated on poor land: more or less podzolized sandy and clay soils produced by the silicified calcareous crusts which cover the Beauce limestone in the Hurepoix region and which crown the hilltops (the forests of Montmorency and Saint-Germain); or sand and sandstone (the forest of Fontainebleau). These areas had never been settled in medieval times. But the presence of the monarch's court and the aristocracy which surrounded it proved decisive in their preservation and development as hunting zones within reach of the royal and princely residences in the area (the royal forests of Saint-Germain and Fontainebleau and the forest around Chantilly, which belonged to the Princes of Condé, the first princes of the blood). The distribution of these large wooded areas was the result of a progressive redefinition of their boundaries carried out by secular authorities in the modern period. At the same time, the proximity of the capital and the large number of navigable waterways leading to it made it possible to put to good use the timber provided by these large woods preserved as hunting grounds, for Paris had always needed large amounts for building. Meanwhile, the supplies of game produced in the course of hunting could be sold at a profit in the markets of Paris. A combination of social and economic factors thus produced the landscape of forests and castles that typifies the Ile-de-France of today.

The centralization of communication routes

The effects of the organization of the countryside in relation to the needs of the capital were clearly manifest in the heart of the Paris Basin and, where certain semi-perishable products were concerned, also further afield in much of northern France. Another essential consequence of the expansion of Paris produced much wider effects: this was the development of a radial network of major communication routes, first roads, then railways. Of all the countries of Europe, France was the only one where all these routes converged upon the capital, a circumstance which further confirmed the predominance of Paris and made for increasing centralization. This situation came about as a result of the fact that the major modern routes of communication in France had initially been developed by an already strongly centralized monarchy which had definitively fixed its seat in Paris. It is a manifestation of the fact that the capital was already well established before the communication routes were developed and it reflects the early growth and predominance of that capital. Apart from the 'gap route', none of the communication routes in this design were dictated by the general distribution of the major masses of mountainous terrain.

270 *Centralization and diversification*

Figure 56. The postal routes in 1632 (after the map by Melchior Tavernier, Paris, 1632, and Cavaillès, 1946, pp. 40–1).

The roads[46]
The function of transcontinental relations in the isthmus of Gaul had been reflected in an entirely different network of communication routes, the Roman network, with its major Rhône–Saône axis linked to the Rhine, on which transversal routes were grafted, running from east and west and leading to the major ports and towns of the Atlantic seaboard: Bordeaux, Saintes, Nantes, Rouen, Boulogne (see above, pp. 37–9 and fig. 9). After the Roman period, no new network of roads was organized until the end of the fifteenth century, when Louis XI set up the royal postal service, establishing[47] relay points all along the major routes. The sixteenth century saw

the development of a completed network, known to us chiefly through the *Etat des postes* ... of 1584. It consisted of roads radiating out from Paris to every frontier of the realm. It is represented on the *Carte géographique des postes qui traversent la France* produced by Samson d'Abbeville and printed in Paris in 1632 (fig. 56).[48] The only important routes that did not radiate out from Paris were those leading from Lyons to Limoges (through Clermont) and from Lyons to Toulouse (through Rodez), across the Massif Central; the road from Bordeaux to Narbonne, through the valley of the Garonne and along the pass of Lauraguais; and in the north, the road from Calais to Blois, through Rouen and Chartres. Such was the basic design of the great royal routes. It was improved considerably by successive efforts on the part of a number of officials: Sully, for whom the post of chief inspector of roads was created in 1599; Colbert, who in 1669 created the *Ponts et Chaussées* bureaucracy; d'Orry, *contrôleur général des Finances* from 1730 to 1745 and the author of the fundamental *Instruction* of 13 June 1738, which laid down that major routes must be repaired before any transverse roads could be built[49] and thereby gave absolute priority to the major royal roads radiating out from Paris; and finally Trudaine, the councillor of state in charge of *Ponts et Chaussées* from 1743 on, who was the major organizer of gangs of road-maintenance workers. This system, in operation from 1738 to 1776, made the construction and upkeep of roads the responsibility of the inhabitants of the areas through which the roads passed, upon ministerial request. It resulted in a network of roads which at the time was unrivalled in the whole of Europe and which remained fundamentally centred upon Paris. It was only in northern France, the highly developed and prosperous France of large concentrated villages and open fields (see above, pp. 153–7 and fig. 35), that a dense network of secondary, transversal roads of any importance existed by the end of the eighteenth century.

The railways
The railway network (fig. 57) was, naturally enough, conceived with the same general perspective.[50] It is true that it was in the mining regions that railways first appeared, on the initiative of mining engineers and essentially for the transport, over relatively short distances, of heavy materials, in particular coal: the railway from Saint-Etienne to Andrézieux which carried coal from the Saint-Etienne basin to the Loire (1827), extended in 1834 from Andrézieux to Roanne; the railway from Saint-Etienne to Lyons, passing through Givors (1832); the railway from Epinac to the Canal of Burgundy (1830)), for the transport of coal from the Epinac basin; the track from Alès to Beaucaire on the Rhône, completed in 1841, for the transport of coal from the Alès basin to Marseilles. The first rail links were thus organized in a slightly confused fashion, essentially to serve the emerging industrial regions. At this stage,[51] the mining engineers were dreaming of a network

272 *Centralization and diversification*

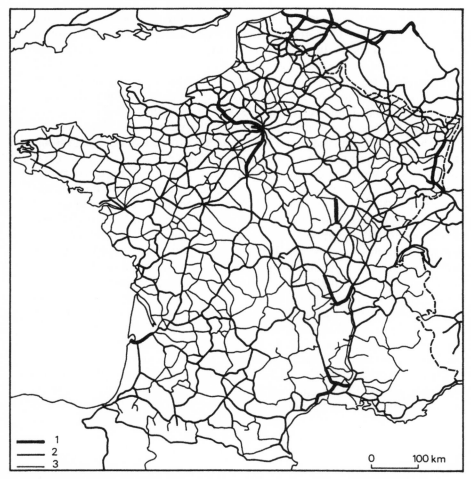

Figure 57. The development of the railways (after Brunhes and Deffontaines, 1926, p. 154, and Demangeon, 1946, I, p. 411).
1. Railways built between 1832 and 1846. 2. Built from 1846 to 1870. 3. Built after 1870.

which would link Boulogne to the textile and steel industries of Champagne and thence proceed to the Saône and Rhône valley, thereby reconstituting – with the aid of the waterways – the major routes of the isthmus of France, on to which they hoped then to graft a transversal link between Alsace and the lower Seine, passing by way of northern Champagne. However, the engineers of the *Ponts et Chaussées* were quick to impose their own national plan of 1842. This restored the upper hand to the politically motivated project for a railway system radiating outwards from the capital and established the major arteries of a network centred on Paris. By 1860, railway links had been created with all the frontiers of France. By 1870, 17,490 kilometres of track had been opened up and the network already possessed all its major features, making Paris the centre for converging routes not only from the rest of France but from the whole of western Europe. The imposition of breaks in traffic (as a result of discontinuous tracks and separate stations for different regions) in Paris and only in Paris (despite Lyons' efforts to secure the same privilege) confirmed the pre-eminence of the capital as an *entrepôt*[52] and speeded its further growth. It was not until the end of the nineteenth century, with the Freycinet plan, promulgated by the law of 17 July 1879, which affected over 10,000 kilometres of new track, that serious steps were taken to provide more transversal links and fill the considerable gaps that remained in the west, the Massif Central and the Alps, by building what were known as 'lines of local interest'. As for the Lille–Strasbourg line, a major artery running parallel to the north-east frontier, which created French competition for Belgo-German traffic and a link of major industrial interest between the coal-producing basin of Nord–Pas-de-Calais and the steelworks of Lorraine, this was only built bit by bit, between 1855 and 1890, and did not assume its full economic role until the end of the First World War, after the recovery of Alsace-Lorraine.

Paris, the cultural capital of France

Finally, Paris ruled people's minds, developed their sensibilities and shaped their passions. That determining role evolved only in stages and not without competition and rivalry from other major towns. During the Renaissance, under Henri II, important intellectual centres such as Lyons, Poitiers and Bordeaux still existed in the provinces, in competition with the capital.[53] But Paris and the court there were already becoming the centre which produced the major influences. The foundation there of the Collège Royal (the future Collège de France) by François I, in 1530, established a new centre of intellectual stimulus. Under the last Valois monarchs, an academy was set up for the first time (1570–87). It was at first known as the Académie Francaise, then in 1576 took the name Académie du Palais. But it was only in the seventeenth century that the ultimate decisive step was taken: the permanent and definitive installation of the French court in the Parisian region turned it into the centre of taste and fashion, soon uniform all over the country among the upper classes, and a powerful cradle of artistic talent. In 1631, the first newspapers were produced in Paris and it thus became the centre from which news was disseminated throughout the national territory. The foundation of the Académie Française in 1635, the Académie des Inscriptions et Belles-Lettres (initially known as the 'Petite Académie') in 1663, the Académie des Sciences in 1666 and many other artistic academies testifies to the progressive institutionalization of intellectual life in Paris, under royal authority. Around 1660, with the triumph of Classicism and the edification of an ordered humanism with pretensions to universality, the intellectual supremacy of the capital became overwhelming.[54] Literary production in regional dialects, still worthy of note in the first half of the seventeenth century, now collapsed.[55] The dissemination of ideas now moved exclusively from the centre outwards to the periphery and the Parisian sense of superiority was consolidated. Mockery of provincialism became a common theme of comedy in the theatre, where laughs were easily raised at the expense of not only the dialects, fashions and customs of the *langue d'oc* regions but also those of the rural north. Molière's comtesse d'Escarbagnas comes from Angoulême and Monsieur de Pourceaugnac is a *gentilhomme limosin*; on the other hand, in Racine's *Les Plaideurs*, it was the cavilling temperament of the Normans that the author lampooned.

However, some intellectual centres did still survive in the provinces[56] and it is perhaps significant that in 1656–57 Pascal chose to present his ideas in the form of *Letters to a Provincial*. In the eighteenth century, provincial France, which had been eclipsed for a time, made a come-back. The diffusion of culture and the widening band of social strata with access to it provided the basis for a vigorous renaissance, the principal manifestation of which was a proliferation of provincial academies (twenty-eight large, well-established ones and at least a dozen that were shorter lived), some of which were to play a by no means negligible national or even international role, particularly through the international competitions that they launched.[57] Jean-Jacques Rousseau thus first won fame when he was awarded a prize by the Académie de Dijon. At this point, an appreciable current of intellectual influences does seem to have been moving from the periphery to the centre, unlike in the seventeenth century. All the same, Montesquieu, a native of Bordeaux who spent part of the year in Paris, wrote in 1740 of a 'France in which only Paris and the most distant provinces are of any account, for those are the ones that Paris has not yet had a chance to swallow up'.[58] And in the second half of the century, that tendency speeded up. As Tocqueville correctly observed: 'At the time of the Fronde, Paris was simply the largest town in France. By 1787, it was France itself.'[59] At the time of the Revolution, Arthur Young, who left Paris shortly before the storming of the Bastille, had found nothing but noise and turmoil in the capital. In the provinces, he found nothing but inertia and silence. In every town, he asked the inhabitants what they were going to do and, he relates, 'the answer was everywhere the same: "We are only a provincial town; we must wait and see what they do in Paris"'; and Young comments: 'These people do not even dare to have an opinion until they know what is thought in Paris.'[60] Thanks to the centralizing measures of first the Jacobins, then the Empire, Paris eventually won the day definitively, as the failure of Girondin federalism strikingly demonstrates. All future major political movements would henceforth begin in the capital. By the nineteenth century, the intellectual predominance of Paris was beyond question. Provincial centres presumed to do no more than relay its influence in the direction of the countryside. In 1897, the situation was such that Maurice Barrès commented: 'God is no longer heard in the departments.'[61] Paris' peculiar achievement was to combine the capital's political role with cultural predominance, to be both Rome and Athens; to have managed to cultivate two very different, opposed civilizations, the Roman and the Greek.[62] In western Europe, where political capitals tended to acquire their cultural supremacy only somewhat late in the day, if at all, Paris proved to be an exception. It had responded to the deepest aspirations of France itself, whose imperial ambitions were matched by a pronounced sense of its function as a civilizing influence. Paris thus became identified with the very essence of French civilization.

Conclusion

The French territory in this way acquired a framework and a pattern of communication arteries which had very little in common with the initial international tendencies of the isthmus of Gaul. Gaul was a territory conceived as a recipient and relayer of outside influences and commodities, essentially a transit area. France, in contrast, was a centralized, albeit definitely assymetrical territory, 'inward-looking', organized according to its own needs, whose international relations depended solely on the attraction exerted by its capital. And it must be recognized that, in the eighteenth century, when French culture dominated the whole of Europe,[63] the influence of that capital beamed out far beyond the frontiers of France, drawing visitors, curious travellers and intellectuals from far and wide.[64] This organization of the territory in the interests of France itself and France alone led in the nineteenth century, despite the French colonial adventure, to the formation of somewhat narrow, nationalistic attitudes. At an economic level, these were reflected in a persistent protectionism, a deeply introspective 'agrarian' mentality[65] and, in the last analysis, a certain backwardness and dogged conservatism. However, it also resulted in an energetic reorganization and elaboration of currents of interchange, poles of attraction and frameworks of communication. The effects of all this are manifest in the culture, economy and social relations that found expression in the urban networks of France.

8
Cultural action and reaction: unity and diversity

This territory of France, now dominated by Paris, had to be broken in to the rules and the culture elaborated in the capital. With its political power, its intellectual prestige and its radiating complex of highways and railways, Paris was a wonderful agent of uniformity. However, uniformization did not proceed without difficulties, setbacks and failures. Fusion within a single mould was constantly challenged by the reaffirmation of local and regional individuality, the emergence or resurrection of local peculiarities and a succession of movements of resistance or even revolt. It seemed that the ineluctably triumphant pressure from the capital in itself aroused local suspicions, pride and resentment. From this painful dialectic there emerged the complex structure of 'French passions'[1] which without doubt constitute the most basic framework within which the contemporary history of France has developed.

The achievement of cultural unity

Songs and stories

All along the great royal roads, elementary forms of popular oral culture took shape. Songs were the first to do so and were certainly the most widely diffused. The French, just like other peoples, or possibly even more so, loved to sing. In France, as the saying goes, '*tout finit par des chansons*'. A living, complex, ephemeral body of songs became established. The compilations that were made, essentially in the nineteenth century, preserved only a small proportion of them, as it were an 'archaeological residue'.[2] The careful analyses that this material has inspired[3] nevertheless give us a clear idea of when and how these songs came into being. The process of their elaboration falls into two phases:[4] the antecedents and the folklorization. In the first phase, the influence of the upper classes was negligible, that of the popular urban strata far more important. These were the strata from which most street singers emerged, technicians with 'more talent than genius',[5] a few of them literate or semi-literate, but the great majority uneducated. These were the performers who provided the prototypes on the basis of which rural singers proceded to produce many variants, thereby ensuring the incorporation of these songs into folklore. Songs born in the towns became folklore in the countryside.[6] In the course of this process much was lost, in particular many of the countless political songs. 'In France, under our kings, songs for many years constituted the only possible opposition; in those days, the government could be described as an absolute monarchy tempered by songs.'[7] However, many of those political songs, ephemeral by nature, soon disappeared. Similarly, oral traditions preserved very few of the popular songs that could be called 'historical' (in the widest sense of the term), although in every period these were equally numerous. Only a handful were preserved as stereotypes in the popular traditions.[8] However, a statistical analysis has been made of those that took root, from which it is possible to draw certain conclusions.[9] Nearly all were produced between the fifteenth and the nineteenth centuries, particularly towards the end of this period, that is to say during the eighteenth century. Of the 400 prototypes or archetypes of folkloric songs which can be dated, just over one fifth belong to the seventeenth century, rather more than three-fifths to the eighteenth. After the revolutionary period, in the early nineteenth century, popular urban culture ceased to be productive. Once Romanticism took over, the rural folkloric stock ceased to grow. Essentially, the French popular song truly was born along the 'royal roads' of the eighteenth century.

Stories, as opposed to songs, represent a 'closed' form of popular culture, stemming for the most part from a universal repertory which probably goes right back to Paleolithic times and which seldom accommodates reworked themes, except in the form of variations of an altogether superficial nature. French stories were no exception to that general rule. After the late seventeenth century, when the literary fashion for fairy stories, started by Charles Perrault's *Contes* (1697), had resulted in the publication of many of these tales in popular editions, only those which had been lifted from the peasant tradition without undergoing serious deformation could subsequently be restored to it.[10] The influence of higher literature upon the French folkloric tale was by no means negligible as regards forms, atmosphere and even structure; and this certainly did not go unnoticed by foreign observers who, in the mid-nineteenth century, were commenting that French stories 'had a smell of eau de Cologne and orris-root about them'.[11] But perhaps that judgement is a little too cursory. It is not really the case that French stories are contaminated by the 'courtly air' of the classical period; rather, they are marked by a specific cultural evolution whose distinguishing features have been well analysed and contrasted in particular with the neighbouring domains of German and Celtic folk tales.[12] The magical content of French stories is pared away, disciplined and simplified – made to seem almost reasonable in comparison with German fantasy and the enhanced, extravagant world of the Celtic tales. The trials to be overcome belong to a more recognizable, familiar and everyday world than the dark forests where the German stories unfold and the seas and distant islands of the Celtic universe. The fantastical creatures who crowd the German tales are far less numerous in the French versions, pretty well limited to fairies and ogres, with a scattering of dwarves. The French tendency towards simplification does away with many fantastical creatures, replacing them by human beings, and includes far fewer magical objects, emphasizing the familiar aspect of those that do appear (such as Cinderella's slipper). Equally, the plots of French tales depend less upon magic and more upon a dramatic development based upon purely human factors. This is very noticeable in Perrault's *Contes*. The fantastic plays virtually no part in his *Bluebeard*, whereas German versions of the story depend heavily upon it. The setting for his *Little Red Ridinghood* is quite banal and the more dramatic on that account. In fact, Perrault's stories follow a deeply French tendency, detectable in virtually all French popular tales. Within the sphere of French culture popular tales have a gentle air about them (cruelty, though evoked, is seldom described) and a definite humanity. The social elements are no less characteristic. Most stories unfold in a peasant setting, where kings, if they appear at all, live like other men. Finally, the structure of French stories is usually simple, the style is plain and direct and the ternary grouping of episodes is more strictly observed than in the stories of neighbouring cultures. The picture presented is characterized by a simplification of magical elements, a taste for rationality and a profound humanity and sense of the realities of everyday life: the very intellectual and psychological characteristics that are generally recognized to reflect the essence of the French genius. That process of rationalization and humanization certainly began before there was any question of literary contamination, but it was clearly speeded up under literary influences, in the eighteenth and nineteenth centuries, as can be seen wherever it is possible to compare the written version of these stories with the oral ones. Even before the end of the seventeenth century, peasant culture could no longer be completely autonomous. It was already, to a large extent, 'dependent'.[13] In the cultural sphere, in France, mental attitudes became increasingly uniform in the modern period and have continued to do so in contemporary times.

Over and above these elementary forms of oral expression, since the invention of printing and the progress of literacy, the countryside of France had also been affected by a written popular culture. Particularly in the seventeenth and eighteenth centuries, this was disseminated by a wide range of rather uncouth publications, many of them extremely simple and naïve, such as the pamphlets distributed by pedlars, the most famous and the most widely available being those which belonged to the 'Bibliothèque Bleue', published in Troyes. A number of careful analyses of this popular culture have appeared:[14] it included works on daily life and handbooks full of practical advice; pamphlets on piety and conviviality; fairy stories and historical legends; funny stories and comical accounts of social *mores*. This literature of escapism, as it has sometimes been described, was never particularly concerned with social inequalities, pre-revolutionary discontent or mythical emancipation.[15] It was characterized by a taste for emotion very alien to the rationalism of the classical period and the eighteenth-century *philosophes*[16] and it perhaps expressed a peasant reaction to the pressure of that all-powerful trend. It was a culture of resignation. Such as it was, transmitted by the already dense strata of peasants who could read and who, even in areas where dialects were spoken, possessed the rudiments of a common French language, it constituted the principal manifestation of the first steps towards cultural unification.

Enlightenment and language

That peasant culture was overlaid by a superior culture already widely diffused by the eighteenth century, which was shared by all the élites. That is how it was that Tocqueville could declare that 'in truth, the whole of France had but one soul. The same ideas were current the length and breadth of the kingdom. The same customs prevailed[17] ... Men from different classes, at least all those above the most popular levels, were becoming more and more alike.'[18] But of course this was only true of the upper social strata: it was no more than a superficial unity reserved for the educated minority who lived in the cities.

In the first place, true cultural unification of the popular masses was inseparable from the diffusion of the French language which, at the end of the eighteenth century, was still only spoken by a small proportion of the population. The great majority expressed themselves in *patois* which, throughout the *langue d'oc* region, was more or less incomprehensible to any French speaker. Then, there were also the areas where truly foreign languages were spoken: Breton, Basque and Germanic languages. In truth, cultural unity for the peasant masses was conditional upon linguistic unity. The problem was seriously tackled for the first time only during the Revolution. Up until then, under the absolute monarchy, no attempts had been made to impose the use of French forcibly upon the entire population. The advent of popular sovereignty, with the Revolution, changed the status of all the languages of France. Grégoire's questionnaire, mandated by the Constituent Assembly (13 October 1790), declared its revolutionary intention to communicate the knowledge of the state language to the masses and at the same time assigned to all the various *patois* the status of 'spoken' languages.[19] The Revolution called upon all citizens to use the national language. Grégoire's Report, presented to the Convention on 16 Prairial of Year II (4 June 1794), demanded 'the sole and invariable use of the language of liberty'.[20]

These efforts were premature. In the mid-nineteenth century, statistics still indicated 40 per cent illiteracy; and in 1863 it was established that one quarter of the population still did not speak French.[21] Furthermore, the culture of the masses in the countryside remained the popular culture of the Ancien Régime. At this period, this was still being diffused throughout France by Pyrenean pedlars from the upper Comminges, as they carried their wares all over the territory of France apart from the easternmost regions (Lorraine, Alsace, Franche-Comté), which appeared to have been already won over to a superior type of culture and where the popular culture no longer found a clientèle.[22] The cultural conquest of the masses was, however, to be definitely launched under the July Monarchy, as a result of two factors: the Guizot law of 1833 on primary education, which stipulated that every commune should maintain a school; and the appearance of cheap newspapers (1836), which were soon being distributed to the small towns, which in their turn relayed them into the countryside. The date at which more developed forms of city culture began to reach the countryside can be determined relatively accurately. By and large, it was during the Second Empire. The peddling of books and pamphlets fell sharply throughout the following decade.[23] The decline was no doubt hastened by particular regulations affecting the legal status of the pedlars' activities, but there was a deeper historical cause for it too, namely the advance of civilization.[24] The simple, passive culture of resignation, which was essentially what was purveyed by the pedlars, had clearly lost its hold over the rural population which, through primary schooling and newspapers, was becoming more receptive to the culture of the cities.

The cultural unification of France was brought to completion by the laws of 1880 which, by making primary schooling compulsory, made the national language available to all. Five years later, all French children were receiving some schooling and, as they acquired proficiency in the French language, they could be said to be undergoing an 'apprenticeship of France'.[25] Everyone now became aware of the Republic, of an ideal France. 'The country that I come from thus turned out to be a language that I had had to learn, my birthplace a school which I had had to attend.'[26] But this change brought about by the universal moulding affected by schooling was not felt as a rupture. There was nothing traumatic about the disappearance of the *patois* dialects, the decline of regional languages and the diffusion of common values. It was not a painful break, rather an initiation. 'The switch from dialect belonged to another, more exciting adventure ... It was as if we were living in town.'[27] France, which for years had existed as a remote concept, was now a concrete and familiar reality. In the smallest villages, it took shape on the classroom maps. It had already been an aspiration. Now it became a representation. The Idea had at last given birth to an Image.

Rationalized space: the organization of the departments[28]

It was an image powerfully structured with a strong fundamental framework. Almost a century before the eventual cultural unification took place, the Enlightenment had left its mark on the territory of France, which had at this point been divided up in a fashion designed to be as rational and egalitarian as possible. The system organized by the Constituent Assembly was foreshadowed by a long pre-history. Plans had proliferated as early as the beginning of the eighteenth century.[29] Initially, they envisaged only large divisions: Fénelon suggested about twenty, in the *Plans de gouvernement* which he produced in 1711 for his pupil the duke of Burgundy, proposing to 'divide the kingdom into parts as equal in wealth as possible'; Saint-Simon suggested twelve divisions. In 1757, in the article 'Généralités' in the *Encyclopédie*, Diderot and d'Alembert recommended dividing the territory into *généralités* of more or less uniform dimensions. In 1765, the marquis d'Argenson, in his *Considérations sur le gouvernement ancien et présent de la France*, suggested *départements* rather smaller than the *généralités* and each containing about two hundred parishes. In the last third of the century, more detailed proposals were put forward, suggesting a hierarchy of divisions and subdivisions. Some of these were abstract utopias, bearing no relation to reality and modelled on the large, regular squaring maps then being used as the basis for divisions in North America and also in Scandinavia. One such was the geometrical dream of Robert de Hasseln, the king's topographer, who in 1780 imagined dividing the kingdom into nine squares produced by the intersection of two north–south meridians and two parallel east–west lines, each of which was then divided in the same fashion into nine counties, each in turn divided into nine districts, themselves split into nine territories. But many plans were more serious, taking into account not only the two major levels of organization that already existed and were accordingly familiar to the theoreticians in the administration, but also the perceived need to create basic units which corresponded to the spheres of relations of the peasants. Thus Letrone, in 1779, in his *De l'administration provinciale et de la réforme de l'impôt*, proposed 25 *généralités*, 250 districts and 4,500 *arrondissements*. Condorcet, in his *Essai sur la constitution et les fonctions des assemblées provinciales* (1788), defined the smallest administrative unit desirable, the basic *communauté*, as a territory of such dimensions that 'within the space of a day, the citizens furthest from the centre could travel to the *chef-lieu*, spend several hours conducting their business there and return home; three leagues thus seems to be the distance which should determine

282 Centralization and diversification

Figure 58. From the provinces to the departments.
The heavy lines indicate the provincial boundaries (after Mirot, 1947, II, p. 417).

the limits of the territory'. That was pretty well exactly the catchment area surrounding a rural fair (see above, p. 206–8). A combination of several communities would form a district, all accessible within a short day's march, and this would constitute the lowest level in the organized territory. Above all these districts would come the provinces, already a major reality in eighteenth-century customary and administrative practice.

The system ratified by the Constituent Assembly was a compromise between these various plans. It was based upon the ideal of an egalitarian and artificial division which was put forward on 7 September 1789 by Sieyès, whose plan for 81 separate departments was largely inspired by Robert de Hasseln's model. On 29 September, a report drawn up by Thouret proposed, in its first article, that France should be divided 'into units each of 324 square leagues, that is to say 18 by 18, so far as possible, taking Paris as the central point and spreading out from there in every direction as far as the frontiers of the kingdom'. But even at this stage, Thouret envisaged adapting this plan to the existing realities, for he went on to say: 'Nearly everywhere, the division can observe the local conventions, above all with regard to the boundaries of the provinces.'[30] Other plans were based upon demography: Mirabeau's proposed to create 120 departments, each of which would contain an identical fraction of the overall population, an unrealistic idea which would have involved constant readjustments to take account of demographic variations; Malouet and Vandreuil's plan proposed working on the basis of the population of the Paris region (700,000 inhabitants) and forming 40 provinces of equal size, so as to create a balance.[31] Eventually, in its first vote, on 22 December, the Assembly decided that there should be between 75 and 85 departments. On 15 January 1790, the number was fixed at 83, including Paris and Corsica. The area of a department should be at least 302 square leagues and not exceed 342. These departments were divided into 544 districts, which corresponded to the lower level already in existence, and into 4,710 cantons, which approximated to the average catchment area of rural markets. Discussions on the size of these units, in which local representatives were allowed the greatest say, resulted in a pattern of division which to a large extent respected the existing provincial boundaries. Brittany, Normandy and Provence, and others too were in effect divided into departments which accounted for the entire area and did not overlap with neighbouring provinces. Other departments were formed by combining historic *pays*, which were constituted by separate but neighbouring provincial units that were left intact: for example, the Ain department was composed of Bresse, Bugey, Valromey and Dombes; Pas de Calais was composed of Artois, Boulonnais and Calaisis; Haute-Loire, in which the (Auvergnat) *pays* of Brioude was added to the (Languedoc) Velay, and so on. Regions over which there was any doubt were mostly allocated to the departments which were more powerful by reason of their population, wealth and the political clout of their representation in the Assembly. Thus Châteaumeillant, on the road to Bourges, close to Marche, was given to the Cher rather than the Creuse department; the representatives of Haute-Garonne managed to deny the department of Ariège several communes close to Pamiers but partially oriented towards Toulouse.[32] But some inextricable problems remained, defying reasonable solution and leading to monstrous hybrids, such as the Yonne department, made up of the left-overs from five provinces,

284 Centralization and diversification

four dioceses (Sens, Auxerre, Langres, Autun), twelve *élections* and as many bailiwicks; but such confusions were the exception rather than the rule. On the whole, the departmental map respected previously existing divisions. Where it made alterations, for the most part these constituted a definite improvement in that, for example, they adapted to natural boundaries or ancient historic frontiers which transcended *généralités* and *intendances*. Thus, the Bourges *généralité* included sectors on the right bank of the Loire (La Charité and several parishes in the Donziois region) whose communications with Bourges were unsatisfactory. These were removed from the Cher department, which was now bounded by the Loire, the traditional historic frontier between the Berry region and the diocese of Bourges. All in all, it was a process of rationalization, not revolution.

But this continuity with, or even resurrection of, the past, albeit sometimes in a new guise, passed unnoticed. It was the change that struck people. Europe was horrified, appalled by it. Such an upheaval was without precedent. It could only be a sign of barbarity. Such, certainly, was Edmund Burke's reaction in 1790. He declared: 'The present French power is the very first body of citizens who, having obtained full authority to do with their country what they pleased, have chosen to dissever it in this barbarous fashion. It is impossible not to observe that, in the spirit of this geometrical distribution and arithmetical arrangement,.. [they] ... treat France exactly like a country of conquest. Acting as conquerors, they have imitated the policy of the hardest of that harsh race. The policy of such barbarous rulers, who contemn a subdued people and insult their feelings, has ever been, as much as in them lay, to destroy all vestiges of the antient country,.. to confound all territorial limits... When the members who compose these new bodies of cantons, communes and departments, arrangements purposely produced through the medium of confusion, begin to act, they will find themselves, in great measure, strangers to one another. The electors and the elected throughout, especially in the rural *cantons*, will be frequently without any civil habitudes or connexions or any of that natural discipline which is the soul of a true republic. Magistrates and collectors of revenue are now no longer acquainted with their districts ... The new colonies of the rights of men bear a strong resemblance to that sort of military colonies which Tacitus has observed upon in the declining policy of Rome ... When all the good arts had fallen into ruin [the Romans] proceeded, as your Assembly does, upon the equality of men, and with as little judgement and as little care for those things which make a republic tolerable or durable.'[33] Burke, the theorist of natural liberties, considered the dissolution of the ancient territorial boundaries to constitute an indefensible break with the past; while Burke the empiricist, convinced that 'the science of politics is experimental',[34] regarded this logical reconstruction as the worst possible manifestation of a destructive urge which he saw as the very essence of the revolutionary psychology.[35]

In truth, for a long time the existence of the departments was precarious. They were not easily accepted. Under the Convention, at the time of the elaboration of the constitution of Year II, a number of plans for a revised territorial distribution were put forward: one reduced the number of departments (which had meantime swelled to 88 as a result of splits in some of them or the extension of the national territory) to 78; another, significantly, regrouped the departments into 29 provinces (the re-emergence of this term is worth noting), divided into 10 districts (that is to say a total of 464), divided into 10 cantons (a total of 7,484).[36] None of these plans were adopted. However, the second did at least recognize the persistence of the '*grands Gouvernements*' of the Ancien Régime to be the most efficient administrative division and also the one most deeply anchored in imaginative representations and customs. Only when the legislative unification introduced by Napoleon's Civil Code wiped out the sources of provincial individuality did the department really come into its own as an administrative unit. It proceeded to develop throughout the nineteenth century and, in contemporary times, even in some cases became a truly economic and cultural unit.[37] For a long time, however, the parts only existed by virtue of the greater whole, with no real life of their own.

The Tour of France

Over and above these artificial divisions into fragments with no soul of their own, France was already creating itself a real identity. An image of the territory as a united whole was emerging. From being a purely territorial concept, France was becoming, if not the framework of daily life, at least – for some – a territory worth visiting.

The national territory was, at an early date, seen as a place to be visited, in the first place by its kings. For these princes from the north, it was above all the provinces of the south that constituted the major focus of interest and attraction. The crown desired at all costs to preserve its links with those provinces separated from the centre of political gravity in Paris or the Loire valley by the obstacle of the Massif Central, most of which had been integrated into the royal domain at the time of the painful crisis of the Albigensian Crusade. The 'visit to the South' was one of the royal duties. At the beginning of 1336, Philip VI, on his way to Avignon to visit Pope Benoît XII, travelled slowly 'for interest and pleasure' and 'to learn to know his cities, towns and castles and the nobility of his kingdom'. From Avignon he proceeded to Marseilles, to inspect his ships, with a possible crusade in mind.[38] Following the episode of the kingdom of Bourges which, during the Hundred Years War, had drawn its capacity for resistance, its focus and its supplies from the southern parts of the kingdom, these distant possessions came to be regarded by the French monarchy as an object of constant solicitude and unfailing vigilance. For the French princes, touring France initially meant discovering the south.

That was certainly the model created by the first systematic royal tour of France, which Charles IX made between 24 January 1564 and 1 May 1566, 'to get to know his good and loyal subjects and to let them get to know him', at the instigation of his mother, Catherine de Medici, who was of Italian origin and consequently perhaps particularly sensitive to the need to promote such wide contacts, at a time when the progress of the Reformation was threatening national cohesion.[39] The tour of Charles IX (fig. 59A) even incorporated a number of diplomatic visits beyond the frontiers of his kingdom. He thus met with his two sisters: Claude de France, married to the duke of Lorraine, whose first-born child he held in his arms at the christening held at Bar-le-Duc on 7 May 1564; and Elisabeth, the queen of Spain, for whom he laid on sumptuous entertainments, punctuated by negotiations, in Bayonne (14 June–15 July 1565). His tour incorporated incursions into the heart of the country, as far afield as Auvergne, although the initial itinerary had been planned to circumvent the Massif Central. However, he left out Normandy and Picardy, the provinces situated to the north of Paris, which were too close to be worth including in the circuit. A similar pattern was followed in the early eighteenth century for the Tour of France made by the dukes of Burgundy and Berry (fig. 59B), between December 1700 and April 1701. They were accompanying the duc d'Anjou, the future Philip V of Spain, on his way to his new kingdom. After passing through Orléans, Poitiers and Bordeaux to Bayonne, they returned by way of Toulouse, Marseilles, Toulon, Grenoble and Lyons, this time with no detours to the provinces of the east or the west and no visit to the north either.[40]

This model of a circular journey, taking in southern France, can clearly be accounted for by fundamental geographical constraints, in particular the location of mountainous areas. Basically, the 'Tour' followed a route which circumvented the Massif Central. Having reached the south from one side or the other of the Massif Central, travellers tended to make their return journey northwards by way of the other side. Men of letters, driven by curiosity, were no exception. In September 1656, Chapelle and Bachaumont, going to take the waters at Encausse, at the foot of the Pyrenees, passed through Blois, Bordeaux, and Agen. On their return journey, they passed through Toulouse and Marseilles and then pushed on as far as Toulon, Hyères and the Sainte-Baume mountain, before returning northwards along the main highway from Aix to Avignon and Lyons (fig. 59C). The itinerary of Parisian companies of actors, taking entertainment to southern France, seems to have followed a similar pattern, to judge by the list of towns that Molière is known or thought to have visited, also in the mid-seventeenth century (fig. 59D).[41] The northern provinces were not included. The tour did

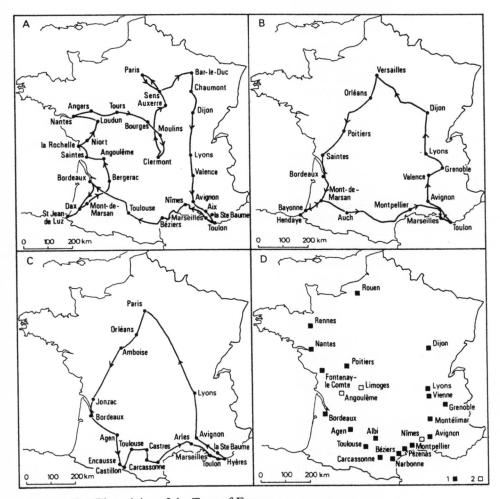

Figure 59. The origins of the Tour of France.
A. Charles IX's Tour of France (1564–66). B. The travels of the dukes of Burgundy and Berry (1700–01). C. The itinerary of Chapelle and Bachaumont (1656). D. Places which Molière's company of actors, touring between 1643 and 1658, visited or may have visited. 1. Visits attested. 2. Probably visited (according to the information provided by Mongrédien, 1965).

not encompass the whole of France. The same certainly went for the voyages of countless humble, anonymous folk who left no precise records of their travels, particularly itinerant workers. For it was the tradespeople, more than the princes and the actors, who established the tour as a regular, and soon codified, practice. As early as the Middle Ages, these *compagnons* would move from one working site to the next, working with a succession of masters in order to perfect their skills, before eventually settling down. It is not known when these voyages began to be organized according to a planned itinerary, with a sequence of predetermined halts where a welcome and a lodging would be assured and the chances of employment good. The first attested use of the expression *Tour de France* amongst these *compagnons* dates from 1731, in the *Livre des règles des Jolis Compagnons Tourneurs de Bordeaux*,[42] but by this date the institution already seems to have been well established, with a body of precise rules and regulations which testify to a long tradition. All the essential elements of the system seem to have been organized at least as early as the second half of the seventeenth century. But we even know of a 'mother' who was taking in *compagnons* in Dijon in 1541,[43] and it is perfectly possible that, as the best historian of these *compagnonnages* (or trade guilds) believes,[44] the practice of the 'Tour', at least in a rudimentary form, went back to the late Middle Ages. It is impossible to say whether the *compagnonnages* emerged as a result of the Tour of France or vice versa.[45] Whatever the case may be, it is not until the end of the eighteenth century that the phenomenon can be studied with any degree of accuracy, that is, at least enough to make it possible to understand its importance.

The first, and essential, point to make is that this was typically and exclusively a French practice. It is quite natural that it should be in the most ancient nation of the European continent that an institution so closely linked with the perception of a national territory should appear and develop. Nothing of the kind existed in neighbouring countries despite the fact that they were more or less familiar with such *compagnonnages*. In England, working men do not appear to have embarked on such migrations. In Germany they did, but the *Wanderschaft* that took place on the other side of the Rhine was quite different: there was no 'Tour of Germany' or 'Tour of the Empire', but instead complex, non-structured, international itineraries which, between the seventeenth and the nineteenth centuries, extended from southern Germany, Switzerland and Austria all the way to Denmark and the Baltic States, with Hamburg providing a centre of attraction, and also to Franche-Comté which by then was already part of France. Meanwhile the Tour of France also attracted men from beyond the national frontiers, a fact that testifies to its longer standing and firm establishment. *Compagnons* from the Spanish Netherlands, now Belgium, where guilds, offshoots from the French ones, are attested from 1650 on, would undertake the Tour of France, as too would workers from the Swiss canton of Vaud.[46]

What is remarkable is that this specifically French phenomenon, which attracted foreigners from neighbouring countries, never, until contemporary times, extended to take in the whole of the national territory. Until well into the modern period, no touring guilds appear to have existed north of the Seine, in Normandy, Brittany, Aunis, or in Guyenne or Provence, nor – *a*

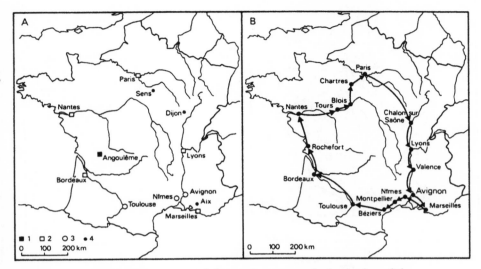

Figure 60. The Tour of France of the *compagnons* at the beginning of the nineteenth century.
A. The organization of the Tour of France for cobblers in 1808 (from the data provided by Coornaert, 1966, p. 191). 1. *'Ville de Fondation'*. 2. *'Villes de boîte'*. 3. *'Villes bâtardes'*. 4. Towns providing only *'chambres d'aspirants'*. B. The Tour of France of Agricol Perdiguier, alias 'Avignonnais-la-Vertu' (1824–27) (from the data provided by Coornaert, 1966, p. 190).

fortiori – in French-speaking regions not yet incorporated into France, such as Savoy, Franche-Comté and Lorraine. Their absence, equally, from the Massif Central is easily explained by the archaism of techniques and methods of production here and the inadequate means of communication. But in the busy northern regions it is harder to understand. The Tour of France did not take in northern France until relatively late and even then proved difficult to establish there. In the eighteenth and nineteenth centuries no staging posts had yet been organized (fig. 60). The extreme limit was Rouen, where an attempt to set up a *cayenne* (inn) in 1824 proved abortive.[47] The *cayennes* of Strasbourg, Lille and Rheims were not founded until after the Second World War. It must be said that the essential function of the Tour of France seems to have been to put the Paris region and the Loire valley, which constituted the heart of the French monarchy, in touch with southern provinces as far distant as the Mediterranean seaboard. Toulon and the Sainte-Baume mountain were two of the earliest established halts there. The Tour of France of the *compagnons*, like the tours of French princes and men of letters, began as a discovery of the south. It was an early symbol of national unity, which may first have appeared, in a primitive form,

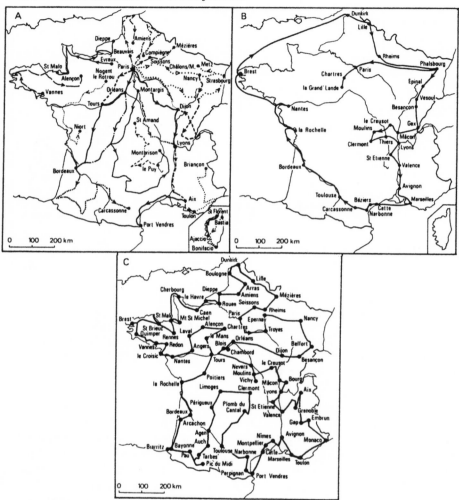

Figure 61. Didactic itineraries of the nineteenth century.
Mme Amable Tastu's *Voyage en France* (1846). The towns shown are departure and arrival points. The different routes correspond to the different chapters of the book. B. *Le Tour de la France par deux enfants*, by G. Bruno (1877). C. *Voyage en zigzags de deux jeunes français en France*, by Gaston Bonnefont (1881).

in the aftermath of the Albigensian Crusade and France's annexation of Languedoc. It was a means, conscious or unconscious, of establishing closer links with the provinces which, however distant, were rapidly conceived to be an integral part of France. The northern regions, so much closer to Paris, did not hold such interest. However, men from the north, who lacked any such organized structures on their home-ground, did also embark – as did the Belgians – on the Tour of France. A Prudence-le-Cambrésis is known to have stopped in Nantes in 1736; in 1760 a Laurent-de-Nancy, a Fleming, a native of Arras, and a Lillois-le-Sérieux all passed through Chalon-sur-Saône.[48] The Tour of France was an exceptional expression of the links that existed between the Loire–Paris heartland and the provinces of the south and it was already attracting the working population of the entire country as well as that of neighbouring French-speaking districts.

In the nineteenth century, the circuit of the *compagnons* took on a new function and underwent a decisive transformation, entering into the consciousness of the whole nation. At the same time, its itinerary was extended to take in the national territory in its entirety. The beginning of its fame should probably be traced to George Sand's novel, *Le Compagnon du Tour de France* (Paris, 1841), which popularized and disseminated the theme; and meanwhile the publication of the collected pamphlets of Agricol Perdiguier, under the title *Le Livre des compagnonnages* (Paris, 1839), brought the practice and purposes of the Tour of France to general attention. Significantly enough, Perdiguier was passionately in favour of the eradication of dialects.[49] In the atmosphere of idealized unification of the Second Empire, when the city culture was spreading rapidly throughout the country, the concept of the Tour of France became practically useful, for it answered a need in a society in the process of becoming conscious of its definitive national framework. The idea was now taken up by didactic literature, particularly in books aimed at children. It marked the emergence of a new attitude towards the teaching of children. At the beginning of the century, knowledge of France was first acquired through an understanding of the system of departmental organization which had just been introduced and with which many people were still unfamiliar. In Constant Taillard's *Les Jeunes Voyageurs* (6 vols., Paris, 1821) ('letters on France, presenting a general map of France, particular maps of the departments, agricultural products, curious natural phenomena, the names of famous men, etc.') the descriptions of the various departments took the form of a series of letters. Mme Amable Tastu's book entitled *Voyage en France* (which first appeared in 1846 but was to run to several more editions by 1869) was deliberately designed to publicize the by now successful Parisian centralization of the road and railway networks. The chapters were organized to cover a series of itineraries, all starting from Paris and following the great highways which radiated out to the natural frontiers. Only a few (seven out of twenty-one) introduced complementary itineraries and even these were essentially no more than extensions of the major routes (fig. 61A). But in the last quarter of the century, the Tour of France became a literary *topos* which was to enjoy a huge success. *Le Tour de la France par deux enfants* (Paris, 1877), by G. Bruno (the pseudonym of Mme Alfred Fouillée, née Augustine Tuillerie), was a collection of reading passages and lessons designed for elementary schools, which enjoyed a phenomenal success (6 million copies sold between 1877 and 1901).[50] The story of André and Julien, two young natives of Phalsbourg in Lorraine, who fled their annexed province determined to learn to know and love their true country, appeared at exactly the right moment, at the point when the laws of Jules Ferry made schooling compulsory and the 1870–71 defeat and the secret hopes of revenge which it engendered set the seal upon the national ideology. It was a work of high-minded simplicity, crammed with earnest sentiments and Kantian moral philosophy which, at the time, was a source of inspiration for the state schools and the Université de France. But countless other reworkings of the theme, of varying success, soon appeared amongst the books on offer for children: *Le Tour de France d'un petit parisien*, by Constant Améro (Paris, 1885), an account of travels between 1870 and 1883; *En Route avec l'oncle Mistral*, by Paul de Courselles and Sixte Delorme (Paris, 1900), a gastronomic tour of France, and many others. Since a purely circular itinerary only provided a partial acquaintance with the territory, efforts were made to complement it. In 1889, Gaston Bonnefont produced *Le Voyage en zigzags de deux jeunes français en France*, in which carefully researched and complex detours ensured that an exhaustive itinerary was followed (fig. 61C). From the end of the nineteenth century onwards, major sporting events were organized on the basis of the itineraries evolved for the Tour of France, the most popular being the *Tour de France* for cyclists (started in 1903). Such institutions testified to the grip that the image had established upon the imagination of the entire French people.

But the Tour of France was not a circle without a beginning or an end. Certain points in the circuit were particularly important, first and foremost Paris, whose domination over the territory was constantly reaffirmed. The great cyclists' *Tour de France* of today always finishes in Paris. In the works cited above, Constant Améro's *Petit Parisien* and likewise the young companions of 'l'oncle Mistral' and Gaston Bonnefort's students from the Ecole Normale and the Ecole Polytechnique all set out from Paris and also returned to it. The itinerary of the two children in G. Bruno's story (fig. 61B) returns them, after their great adventure, to Phalsbourg; they then pay a visit to Paris, after which they eventually settle in another place which is full of significance, at the Ferme de la Grand' Lande, situated 'in the most mountainous part of the Orléans region,.. in the part of the Beauce which is close to Perche':[51] this spot is not far from the *Carnutes*' meeting place, where the annual assembly of the Druids of Gaul used to be held. This was their Journey's End, at the focal point for the circumscribed world of their travels. Truly, it was a case of France closing in on itself and contemplating its own navel.

Friedrich Sieburg was later to comment: 'Everything that has occurred in the development of this country has helped to attribute to the Franks a definitive perfection, which in this day and age is liable to become a danger to it.'[52] This was a formal, aesthetic and moral vision of France which took over from and perpetuated the image of the *jardin de France* and the *pré carré*. It encapsulated and crowned all the secular efforts to centralize the monarchy, the Revolution, the Empire and the Republic.

But it was a unity which incorporated plenty of diversity. *Le Tour de la France par deux enfants* presents a didactic description of the disparate elements which together make up the whole: 'The most beautiful garden is the one which contains the most beautiful and the most numerous species of flowers. Well, my little one, France is a garden. Its provinces are like flowers of every kind, between which it is difficult to choose but which in combination produce the most beautiful of countries, the sweetest to live in, our beloved land.' Père Guillaume's touching words to André and Julien[53] expressed many realities; but many tensions, passions and dramas underlay them.

The differences

The cultural unification engendered reactions – defensive reactions favouring diversification and individuality. Faced with pressure from the centre, local characteristics, thitherto dormant, and new expressions of individuality rose to the surface.

Variations on a theme: the arts, costumes and dances

It is easy enough to see how this came about. The main elements of the national model certainly were adopted and assimilated by the provinces. But their progressive infiltration into a deeply traditional way of life in the long run produced a proliferation of idiosyncrasies which encouraged certain rapidly shrinking groups to reaffirm their identity and distinguish themselves from their neighbours. This process evolved in a series of phases spanning several centuries.

Popular art

The earliest manifestation of the tendency was the appearance of what may loosely be called 'popular art'. It first became apparent mainly in the field of furniture but was also reflected in a wide range of useful objects and ornaments: household implements, pottery, and pious or comical images.[54] Under the Absolute Monarchy it had become increasingly difficult to differentiate between regional schools in the major arts (painting, sculpture, architecture). Now, in contrast, regional styles gradually emerged and asserted themselves, influenced by the general trends of European art but incorporating a multitude of variants. The catalyst for this development was the invention of the printing press which, from the Renaissance on, made it possible to circulate prototypes, patterns and compositional schemata which provided basic models for popular art. Those prototypes were then given infinitely diverse forms, according to the inspiration and imagination of the local artists. The local styles were then reproduced within spheres of influence of varying dimensions. Religious art, in particular, produced prototypes universally recognized, whose motifs frequently reappeared in artefacts designed for daily use. This evolution went through a sequence of phases as various objects became familiar to an increasingly wide public among the popular strata. Thus the moulds for *pain d'épices* and *pain d'anis*,[55] which first made their appearance in the Middle Ages when such cakes were all the rage, were originally earthenware. But by the sixteenth century they were being made from wood, sculpted into a variety of designs, and in the seventeenth century, when they became larger and rectangular in shape, they were produced in an amazing proliferation of forms, ornamented by either heraldic or religious motifs. They finally disappeared in the late nineteenth century, following the introduction of industrialization.

Beds had been produced in a wide variety of styles since the mid-seventeenth century. Thus, in the case of the enclosed Breton bed, which was gradually detached from its wooden frame, five successive stages of development have been distinguished.[56] The oldest beds, dating from 1659–67, had only one door; in the eighteenth century, the semi-solid sides opened up and sliding doors appeared, whilst in the Rennes region a remarkable variant developed: the wheeled bed, with two levels and four doors. From 1750 on, semi-enclosed beds with doors appeared, with a central opening which became progressively wider during the nineteenth century. Finally, four-poster beds became common, of rectangular shape and with a column at each corner. These were to be found in the neighbourhood of Rennes and Nantes and in the Guérande and Brière regions. It was clearly in the larger French-speaking towns of Brittany that new styles were introduced and developed.

For earthenware, regional differentiation came later. Originally for the exclusive use of the aristocracy, it did not become available to the popular classes until the eighteenth century.[57] At this point, workshops were set up in every province and different decorative styles began to develop. By and large, the triumphant period for popular art, when a multitude of different forms of expression developed, was the late eighteenth century and the first third of the nineteenth. For example, many cribs were carved during the last quarter of the eighteenth century and the first quarter of the nineteenth.[58] For many particularly elaborate religious objects, such as statuettes of saints and fonts, the most productive period came later, around the mid-nineteenth century.

In the domain of furniture, at least, it has been possible to produce a

Table. *The spread of furniture styles*

Style	Renaissance	Louis XIII	Louis XIV	Louis XV	Louis XVI	Empire
Accepted date of its appearance	1500	1620	1680	1740	1775	1805
Earliest date when reflected in popular furniture	1626	1720	1726	1740	1779	1809
Latest date when reflected in popular furniture	1768	1817	1840	1930	1903	1849
Life-span of the style in the field of popular arts (years)	125–270	100–200	50–160	0–190	0–120	0–40
Number of examples	7	15	25	187	38	3

statistical survey of the length of time that it took for a central model to be transmitted and become a source for regional popular styles. It is based upon a total of 275 findings.[59] They are set out in the table.

As can be seen, the lapse of time that it took for a central style to influence popular culture becomes progressively shorter as time passes. It starts by being very long (over a century), but by the mid-eighteenth century has shrunk to virtually nothing. By this date, influences were being rapidly diffused by way of the great royal highways. It was now that imitations of central models began to multiply. Similarly, the influence of a particular style upon popular art lasted longest for the Louis XV style, quite long for the Louis XVI style, but for a very much shorter time for the Empire style. That is because regional styles, expressing regional reactions, were by now more firmly established and blocked the diffusion of new central models. The style which exerted the greatest influence of all was without doubt the Louis XV style. It was at this period that pressure from the centre, received as innovatory, was at its strongest and most productive. The great flowering of popular regional styles was founded on imitation of the Louis XV style and

296 *Centralization and diversification*

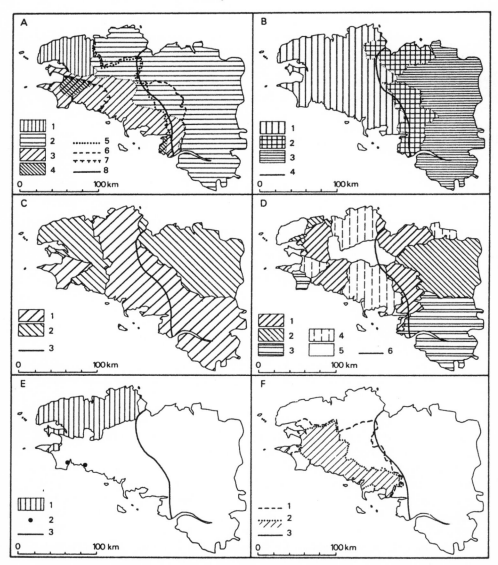

also, though less dramatically, the Louis XVI style. The process ceased to be productive in the second third of the nineteenth century.

Costumes

Geographical differentiation was a later phenomenon but one of great importance, which frequently led to clear-cut territorial divisions, particularly where costume was concerned: a visual field if ever there was one, which forcefully reflected the essentially human urge to express one's identity and which was strongly affected by the more or less universal availability of manufactured textiles. Local costumes made their appearance in the nineteenth century, as a symbol par excellence of local idiosyncrasies and a colourful and spectacular expression of group individualism.

Such a remarkable phenomenon was bound to provoke curiosity and surprise. At the time, all kinds of fantastical and speculative interpretations of its origins were produced. In the Romantic period, the Breton peasant *coiffes* were declared to be inspired by the 'capulet' (hood) of Anne, duchess of Brittany.[60] A number of 'anthropological' theories attributed the origin of local costumes to foreign components of the population: for instance, an Asiatic origin was postulated for the monumental, spire-like headgear of the women of Pont-l'Abbé (in the *bigouden* region) which, it was claimed, was based on a Mongolian pattern. Meanwhile, a theory favouring a 'spontaneous' origin claimed, on the contrary, that this same *coiffe* could be explained by the events of the Papier Timbré revolt of 1675. The women of Pont-l'Abbé were supposed to have set these towering *coiffes* upon their heads as a sign of protest at the destruction of the bell tower of Lambour. Other theories imagined connections between local costumes and hypothetical Breton 'clans' which in truth never existed. (There is, in particular, no connection at all between the local costumes of Brittany and those of the Celtic peoples in Britain from whom the Bretons are descended.) An exemplary monograph[61] has now established beyond doubt that the origin of the costumes of Brittany, like that of the regional styles in other domains, should be sought in the fashions disseminated from the centre, that is to say Paris and the court, essentially from the seventeenth century on: those of Louis XIII's period, for example, which was the time when fashions first began to be diffused throughout the country. But the process of differentiation was

---

Figure 62. The differentiation of costumes in Brittany: general features (from the facts and figures provided by Creston, 1974).
A. Masculine costumes. 1. Jacket in the Breton style of about 1900. 2. Coat in the French style (*c.* 1900). 3. A short jacket derived from 2. (*c.* 1900). 4. Baggy breeches. 5. The limit of the Breton hat. 6. The limit of the open waistcoat (in the west) and the waistcoat as worn with city costume (in the east), in about 1939.
7. The limit of the fastened waistcoat at the same date. 8. The linguistic boundary.
B. Loose working over-shirts (*blouses de travail*) (men). 1. Short. 2. Semi-long.
3. Long. 4. Linguistic boundary. C. Major types of feminine *coiffes* in about 1850.
1. Upswept wings. 2. Downswept wings. 3. Linguistic boundary. D. The later evolution of *coiffes*' wings. 1. Upswept. 2. Drooping. 3. Spire-shaped.
4. Atrophied. 5. Suppressed. 6. Linguistic boundary. E. Women's hennins and mob-caps. 1. Continuous presence. 2. Isolated fashions. 3. Linguistic boundary.
F. Features of feminine costumes. 1. Bodices and surcoats (*surcots*). 2. Collars and collarettes. 3. Linguistic boundary.

very gradual. True, as early as the seventeenth and eighteenth centuries, or even before, there are few indications of things to come. The hat worn by the women of Bresse in the nineteenth century, round, with a pointed cone in the middle, is represented in an illustration dating from 1567. Contemporary Norman *coiffes* are uncannily like various forms of the hennins of the fourteenth and fifteenth centuries. It would accordingly seem exaggerated to attribute the elaboration of local varieties of dress exclusively to the Revolutionary and post-Revolutionary periods.[62] Nevertheless, those were certainly the years during which the main creative explosion took place. At the outbreak of the Revolution, there were as yet no distinctive local costumes in Brittany. A masculine costume in the Louis XIII fashion, reminiscent of the neutral-toned suits worn by seventeenth-century valets, appears to have been common and is referred to, in legal statements of the time, somewhat vaguely as 'peasant clothing'. Feminine costumes were even more archaic, completely plain and strictly utilitarian. They had undergone no change since the sixteenth century. Distinctive *coiffes* had not yet been introduced.[63] It was after the Revolution that peasants all over France adopted costumes imitated from those of the ruling classes.

Not surprisingly, then, masculine costumes were derived essentially from the 'French-style' or 'Court' costumes which, under the Empire and the Restoration, became the uniform of the bourgeoisie. Meanwhile, around this time but perhaps slightly later, peasant working clothes also became fixed. In Berry, from 1830 on,[64] the long, linen tail-coat, known as the *biaude* or *dômaie*, the daily wear of a peasant at the end of the Ancien Régime, gave way to the loose shirt, which may have evolved from the over-shirts worn by Parisian workers. This was the model generally adopted over most of the national territory. Only a few outlying, old-fashioned regions, such as parts of lower Brittany, remained unaffected by this new trend. By the end of the eighteenth century, three major areas were emerging. In eastern Brittany, where French was spoken, the Gallic region adopted the central model of dress, albeit somewhat late in the day. In the north-west, beyond a line running along the river of Morlaix, the woods of Huelgoat and the course of the Aulne, an old-fashioned type of jacket persisted, modelled on the Louis XIII costume, which remained, par excellence, the Breton fashion. In the south-west, southern Cornouaille, western and north-western Vannetais and the neighbourhood of the Brière, a short jacket made its appearance, sometimes tricked out with basques and tucks: this was the *kramailhon* (derived from '*carmagnole*') of the Vannes region. Its appearance marked the beginning of the evolution towards a particularly typical form of Breton coat, derived from the jacket in the 'French style'.[65] These were the bases for the three major groups of coat styles that were well defined at the end of the nineteenth century (fig. 62A). Variations in other items of clothing did not necessarily fall into quite the same divisions. In the south-western sector, in the Vannetais and Rosporden areas, French-style breeches were worn, with a matching coat. But the Quimper region on the one hand, and the Guérande and Brière regions on the other, remained faithful to baggy breeches, a left-over from the seventeenth-century fashion. As to headgear, the south-western and north-western areas stuck to the old-style round Breton peasant hat, although the limits of those areas did not quite coincide with those that applied to coats, while the limits for the city-type waistcoat and the peasant waistcoat did more or less coincide with the limits for different styles of coats. As for peasant waistcoats, a zone of fastened waistcoats appeared around Quimper – a fact that is explained by the spread of influence from the town into southern Cornouaille. For working clothes, the divide was a more straightforward north–south one, coinciding quite closely with the linguistic frontier, separating the longer over-shirts of the east (also to be found throughout France generally in the nineteenth century) from the short over-shirts of the west, with a transitional zone between the two (fig. 62B).

The evolution of feminine clothing started later and followed different principles. In Berry, there were virtually no changes from 1780 to 1835. In Brittany, autonomous developments took place, which owed far more to phenomena of local differentiation *in situ* than to reactions to the diffusion of Parisian styles. The sphere of feminine relations, more introverted than the masculine one, was reflected in a number of specific ways. It is true that some of the major regional divisions noted in masculine fashions reappear: for example, the north-western sector, in particular, is characterized by festival hennins and mob-caps (fig. 62E); the south-west by bodices and surcoats (*surcots*). However, the evolution of *coiffes* is totally idiosyncratic (fig. 62C). Early on, for no obvious reason, extended separate areas adopted *coiffes* with upswept wings and these areas later spread and joined up. However, they also produced a number of extra features, including various simplifications which in some places led to the disappearance of *coiffes* altogether. Throughout the nineteenth century, the major stylistic areas underwent a process of fragmentation which produced countless detailed variants, leading to a stylistic map of prodigious complexity. In the field of women's fashions alone, 66 major styles have been distinguished in Brittany, subdivided into close on 1,200 variants. Some 55 basic types of peasant *coiffes* have been identified and a further 19 types of craft headgear.[66] The greatest number of variations are to be found in Breton-speaking regions, but they are unevenly distributed. In Cornouaille, the areas of distribution are clear: 13 basic groups are associated with territories that are, for the most part, clearly defined (fig. 63A). The limits of these territories are usually traced by major geographical features such as the contours of the Montagnes Noires or sections of river courses. Particular styles tend to develop in peninsular areas: the Plougastel–Daoulas headland, with its very archaic style; Cap Sizun and the island of Sein, with a style known as *kepenn*; the

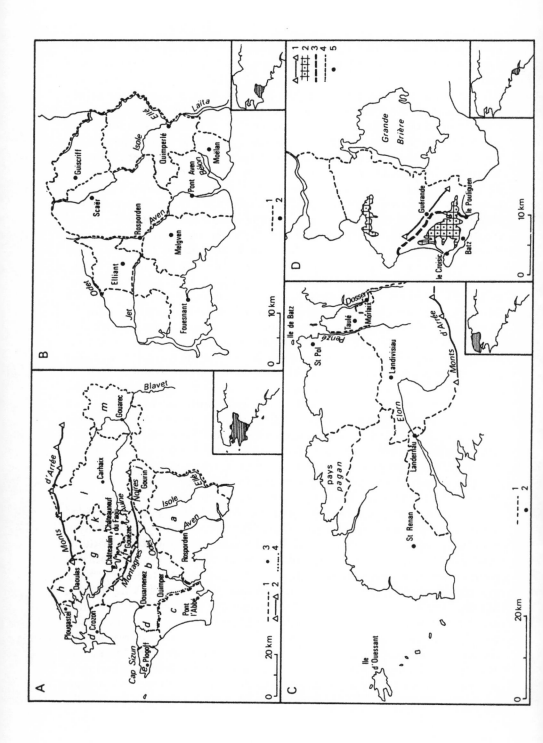

Douarnenez–Crozon region around the Douarnenez Bay; the *bigouden* region around Pont-l'Abbé. These Cornouaille styles reflect a strong sense of regional identity and are often accompanied by regional nicknames which are applied to the various human groups and are usually inspired by the dominant colour of their costume: *glazig* ('little blue') in the Quimper region; *rouzig* ('reddish') in the Châteaulin region; *menelik* ('lemon yellow') in the Elliant region, which is a subdivision of the Rosporden region, and so on. In northern Brittany, Léon and Trégor, in the Saint-Brieuc region, the compartmentalization is much less complex and the individual areas are generally larger. The Léon region, for example, falls into only nine areas, but they are, admittedly, less naturally determined than those of Cornouaille. The Trégor region has even fewer subdivisions and this pattern continues as one moves eastwards, although a number of quite small subdivisions reappear around Saint-Brieuc. But above all, the boundary lines are vaguer and regional nicknames disappear. The fragmentation is certainly much less accentuated in French-speaking Brittany, where the pressure for a uniform adoption of the central model seems to have been far stronger and individual reactions much more pronounced. But there is also an intermediary zone where the virtually uniform style adopted from the Channel right down to the mouth of the Loire in the first half of the nineteenth century later became diversified, particularly along the road between Rennes and Carhaix.

It is not difficult to interpret this geography. The maximum and most noticeable diversification of fashions occurred in Cornouaille, where the traditional French-style coat gained acceptance in a Breton-speaking region. This diversification expressed a defensive cultural reaction within a framework receptive to a number of very different trends. Diversification was much less a feature of the north-west, that is to say Léon and Trégor, which were particularly traditionalist, introverted regions that had remained faithful to the old peasant costume derived from the Louis XIII style. It was even weaker in the French-speaking area, where acculturation went far deeper

---

Figure 63. The differentiation of costumes in Brittany: local aspects (from the facts and figures provided by Creston, 1974).
A. The principal styles of Cornouaille. 1. Boundaries between different styles.
a. Rosporden. b. Quimper (*glazig* style). c. Pont-l'Abbé. d. Douarnenez and Crozon. d. Plogoff. f. Gouézec. g. Châteaulin. h. Daoulas. i. Plougastel.
j. Gourin. k. Châteauneuf-du-Faou. l. Carhaix. m. Gouarec. 2. Relief axes.
3. Towns and large villages that were stylistic centres. 4. The old western limit of the *glazig* style in the Douarnenez region. B. Local feminine styles of Rosporden, around 1880. 1. Limits of styles. 2. Towns and large villages, centres of styles.
C. Principal styles of the Léon region. 1. Limits of styles. 2. Towns and large villages, centres of styles. D. Styles of the Guérande region. 1. Guérande hills.
2. Salt marshes. 3. (Inland) limit of the marshland styles. 4. Other limits of styles (known as 'sharecropper' or *métayères* styles). 5. Towns and large villages.

and the cultural domination of the capital was virtually unchallenged. Perhaps the model of Brittany may serve as the basis for an analysis to help us to understand the geography of regional costumes throughout the territory of France. The studies that have been made of the phenomenon in other regions are not so full as those relating to Brittany, but it has been shown that differentiation in styles of dress was greatest in isolated mountainous sectors such as Savoy, where distinctive costumes sometimes vary even at village level,[67] whereas contrasts are much less sharp in the open plains, where changes are very gradual and come about slowly.

The origin of these detailed local differences and the dates when they first appeared are for the most part hard to pin down. But in some cases, such as Brittany, it is possible to be relatively precise. The *glazig* style of the Quimper region originated after the wars of the First Empire,[68] when stocks of blue cloth for uniforms from the military stores in Brest were sold off in the markets of the town and the surrounding countryside. The Douarnenez–Crozon style began as a style of dress for workers, which emerged when the tinned sardine industry was developed with a labour-force of women drawn in from the Brittany headlands. (The *coiffe* which these women wore was known as a *penn sardin*.) But the style of dress favoured in Douarnenez itself was derived from the costume of the peasant commune of Beuzec.[69] In many cases, the origin of a particular style is traceable to ornamental motifs from the general fashions of the rest of nineteenth-century France, spread by fashion magazines to the little towns of Brittany and copied by embroiderers in the towns or countryside. The major divisions were for the most part established around the middle of the century. In the '*pourlet*' region of Guémené-sur-Scorff, in the Vannetais area, there was no regional costume during the first thirty years of the century, when individuals still dressed as they chose. Here, the regional group consciousness only emerged in the course of the second third of the century. The fashion for the short, white jacket of the Pontivy region developed between 1850 and 1870. The *coiffes* of the *bigouden* area did not make their appearance until after 1850. Differentiation, particularly in feminine *coiffes*, continued throughout the second half of the century and into the early twentieth century. But already, spheres of influence were shrinking, particularly those of the towns, which had been the first to become firmly established at the beginning of the century. Now they shrank and broke up and this favoured the appearance of autonomous rural styles. Thus, the area of the *glazig* style (around Quimper) reached much further out towards Rosporden in the nineteenth century than it did in the twentieth; the sphere of influence of Cap Sizun–Ile de Sein was shrinking from the nineteenth century onwards.[70] But other regions remained surprisingly stable, one being that of the five 'marshland' styles, all well-defined in the Guérande area, which was neatly separated from the interior by a continuous ridge of hills: here, those five styles remained totally unaltered from 1840 to 1914.[71]

Dances

These constituted another model of diffusion,[72] but in this case it took longer for city forms to penetrate rural areas and for new varieties and local diversifications to be elaborated. That is because there already existed a rich collection of deeply-rooted rustic traditions of dance and resistance to innovations was accordingly hard to overcome. In matters of dress which, for them, had formerly been essentially utilitarian and lacking in frills, country folk had been remarkably open to city and court influences, quite ready to imitate new styles and receptive to the dreams that came with them. But where other traditional peasant dances were concerned, resistance was far more stubborn, on account of the social functions that they fulfilled. Traditional dances were a major force in maintaining the cohesion of the group, for they strengthened ties within it and manifested its unity. The fundamental form of dance, to be found throughout France, was the *branle*, a collective chain dance, either open-ended or in a closed circle (*ronde*), in which a series of steps was repeated over and over again in identical fashion, inducing automatism and abandon in the dancers. But in many local versions (such as the Auvergne *bourrée* and the dances of the Trégor and Léon regions to the north of lower Brittany), the dancers faced one another in two rows, performing a set of basic figures, and here the division between actors and spectators was carefully organized and deliberately controlled. These dances seem to have possessed an extraordinary stability, reflecting that of traditional society. In lower Brittany, a form noted in 1588 survived virtually unchanged down to almost contemporary times.[73] The prestigious and ceremonial dances of the court, solemn and noble exercises in deportment and gesture, which transmitted the idealized image of life that predominated until the end of the eighteenth century, for a long time influenced the popular entertainments of rural France not at all. Throughout the Middle Ages and most of the modern period, there is no evidence of any interaction or contamination between the two styles. It was left to the *contredanse*, whose spontaneity and vivacity lent it obvious affinities to popular dances, to exert a certain influence upon the latter. The *contredanse* was a joyous affair, an expression of liberation and familiarity and a form of social entertainment far more dynamic than decorative. In the *contredanse*, the steps and the movement mattered more than the gestures, and couples, rather than the group as a whole, were the centre of attraction. It is known to have developed from the country dances of Britain[74] which, in its confusion of social classes, was already combining popular forms with forms from the higher social culture. The *contredanse* made its first definite appearance in France in Paris, in 1684, and by the eighteenth century it was enjoying a triumphant success in all city circles, where it was favoured as an expression of the more liberated manners of the period, as opposed to the traditional aesthetics which survived in the minuet. Its eventual infiltration into rural areas was fortunately recorded in many accounts, so we are able to follow the phases of its development and consequences in quite a detailed fashion.

The *contredanse* did not really take a hold until the very last years of the eighteenth century. In 1755, in Auxerre, it had already won over the popular strata amongst town-dwellers, but countryfolk were still dancing their *rondes*. In 1781, in the Rouergue region, the *contredanse* was already all the rage in Millau, but in the surrounding countryside only the *branles* and *farandoles* were known. In Poitiers, just before the Revolution, the repertoire of city dances was the same as in Paris, but even as late as the first quarter of the nineteenth century, the people of the surrounding countryside continued dancing only *rondes* or in procession. In 1807, in the Pyrenees, the spa-towns of Bagnères, Saint-Sauveur and Cauterets were dancing the *contredanse*, the *monférines* and the waltz, but nothing of the kind was yet known in the neighbouring countryside.[75] The *contredanse*'s conquest of the rural areas began, significantly enough, in the open countryside scattered with large grouped villages in northern and north-eastern France. The first mention of it outside a town setting is in 1781, in the villages of Picardy. In 1810 it was danced at country weddings in the *arrondissement* of Remiremont, in the Vosges. At village festivals in the Ardennes, dancers were skilfully executing its complicated steps by 1827, so they must already have been familiar with it for some time. As we have seen, these were areas already criss-crossed by a dense network of carriage roads and consequently very much open to external influences. But above all, the large rural communities in these parts of France were prosperous and forward-looking, with an active network of human relations, and the function of social cohesion served by collective dancing was no doubt less crucial than in the dispersed settlements of southern and western France, where such amusements sometimes provided the sole opportunity for strengthening tenuous links. In the south, the *contredanse* was known by 1805 in a few villages in the immediate neighbourhood of Marseilles, which served as a powerful relaying point for the influence of Parisian models; but it is also interesting to note how early (between 1820 and 1830) the new forms also became implanted in the village festivals of Labourd and Béarn and in the valleys of the Pyrenean Gaves, where communities were large and tightly knit and the rate of literacy was high – areas which constituted pockets of early development and receptiveness to outside influences (see above pp. 152–3). In most places, the *contredanse* did not make its appearance in peasant circles until between 1830 and 1850, when new communicating roads were built. In the Drôme region, it was known in the Rhône valley by 1835, but not yet in the mountainous villages. It appeared in Bresse and Berry between 1840 and 1850. Around the middle of the century, it became established in the countryside of Poitou, Vendée, Aunis and Saintonge, but records show that in 1866 the most widely known dance was still the local *branle*. In 1868, the *contredanse* and the polka were known in Beauvoir-sur-Mer, on the Vendée coast of the Marais Breton, but the surrounding countryside was still loyal to the *ronde*. In the Landes of Gascony, there are still no signs of the new dances in the period between 1870 and 1880. But it was inland Brittany that proved to be the most inaccessible region of all. Between the two World Wars, the traditional dances were still popular in the Montagne d'Arrée, the Montagne Noire, part of Vannetais and the mountainous part of Cornouaille, whereas the Trégor region, lower Cornouaille and lower León had opened up much earlier to the new trends that spread from the towns, the coast and the linguistic frontier, by way of two major axes: along the Channel coast towards Morlaix; and from Nantes and the lower Loire, through Vannetais to lower Cornouaille and lower León.[76]

As the *contredanse* spread, new elaborations and transformations soon appeared, usually through the city model being first simplified, then contaminated by the old traditional local dances, so that different steps and figures, or remodelled ones, became incorporated. The *bourrée* of Berry, itself a late extension of Auvergne and Bourbonnais forms (for the ancient dances of Berry were *branles*) was to borrow a number of the steps used in its figures from the *contredanse*. But that was a late and complex development. At the end of the nineteenth century, in the western part of Berry, on the borders of Touraine and Sologne, the *bourrée* was still considered an Auvergne dance. It had arrived by way of the Cher, moving from east to west, and only acquired its widespread popularity here between 1880 and 1930. As it did so, it was strongly influenced by the 'salon *bourrées*' favoured by the bourgeois of the small local towns, and also by the *contredanse*.[77] The phenomenon has been carefully analysed in lower Brittany, where it developed over the second half of the nineteenth century, assuming different forms in different areas. Thus *jabadaos*, *rondes* performed by various permutations of dancers, or *rondes* danced 'forwards and backwards', which made their appearance in about 1850 around Brest, Quimper and the little towns of lower Cornouaille, became generally popular in the region stretching from Quimperlé to Audierne and Crozon, where they assumed a number of different forms the distribution of which is extremely significant. The direct influence of the various urban centres of diffusion (Quimper, Châteaulin and Quimperlé) is marked within, on average, a radius of 20 to 30 kilometres, but so too is the influence from the coast. Meanwhile, the cultural character of the headlands which always – even if they were large ones – seemed to be at the very ends of the earth, remained highly individualistic (the Crozon headland, the *bigouden* region, the part of Cornouaille to the north of the Aulne, and the Aven region). Quite different developments seem to have taken place in marginal situations, far away from the principal centres: the *ronde* with different permutations of dancers, in a form in which Part B is danced by the boys and the girls alternatively (fig. 64B) was, curiously enough, diffused on either side

306 *Centralization and diversification*

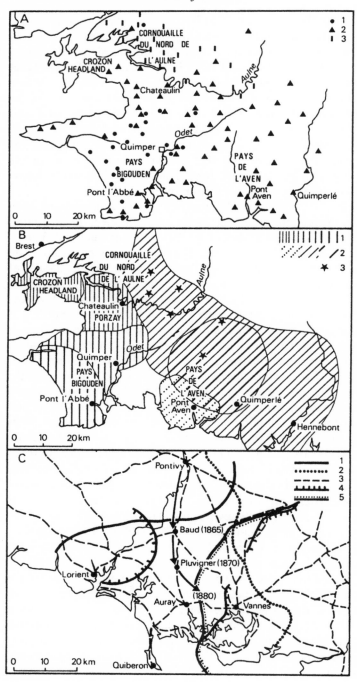

of the Aulne valley, which was relatively unaffected by urban influences. Different currents of influences intersected and overlapped and sometimes led to a discontinuous distribution of particular dances. Thus the *laridés*, which developed in the last third of the nineteenth century, between 1880 and 1914, and enjoyed great popularity in Vannetais, at the expense of the traditional dance *en dro* ('in a circle'), seem to have spread in two separate stages (fig. 64C). The first version (in 6-beat time) spread from the east to the west, starting in Vannes, passing along the coastline and eventually reaching the borders of the *en dro* region. Then, between 1860 and 1880, a second version in 8-beat time, emerged from the important inland town of Pontivy and moved south towards Auray and Quiberon, cutting the earlier area into two distinct zones, the one centred upon the sphere of influence of Vannes, the other upon that of Lorient. The 6-beat *laridé* then resumed its expansion, moving westwards and taking over the corridor between the two. Naturally enough, none of these areas is altogether continuous: interpenetrations are frequent, as they are where linguistic regions are concerned. It is also worth noting that in the vast majority of cases, these zones of local dances did not necessarily coincide in detail – or even in general – with the zones that we have distinguished for costumes, although some urban centres were, of course, trend-setters for both clothes and dances and natural geographical features did apply in some cases to both domains (for example, the headlands of Breton Finistère: compare figs. 63A and B, 64A and B). Local costumes and popular dances represented specific reactions to the cultural unification of France at two different periods: the development of local dances took place, on average, half a century later than the development of local costumes.

From regional cultures to regionalism

Local cultures and regional cultures

What was the outcome of this extraordinary effervescence and activity? Did they result in the creation of a culturally coherent geography, in which it is possible to distinguish clearly defined zones?

Figure 64. Aspects of the differentiation of dances in Lower Brittany (after J.-M. Guilcher, 1963, maps V, VIII, IX).
A. Backwards and forwards *rondes*. 1. Part B danced in couples. 2. Part B danced in an undivided circle. 3. Two-beat forms. B. *Rondes* with different permutations of dancers. Zones where new versions/types were developed: 1. Western sector. 2. Eastern sector. 3. Part B danced alternately by girls and by boys.
C. Development and evolution of the *laridés* in the *en dro* region. 1. Limit of the *en dro* region. 2. Linguistic frontier. 3. Main roads. 4. Limits of the western sector (Lorient) and the eastern sector (Vannes) of the 6-beat phrase, after the 8-beat phrase had spread (from Pontivy) southwards. The dates given are when the eight-beat phrase appeared. 5. New limit of the western region of the 6-beat phrase.

By and large, the answer must be 'no'. Only very exceptionally do the boundaries of individualistic zones for the various categories of objects, customs and beliefs coincide, although just occasionally they do. Sometimes beliefs and customs were determined by particular techniques as, for instance, where these were linked with local patron saints.[78] But such coincidences are hardly more than curiosities. Even within a single, well-defined sector such as that of viticulture, it has been shown that the penetration of new techniques, diffused along the major historical axes of circulation, was by no means uniform.[79] Different types of equipment made their marks more or less rapidly in different regions: heavy equipment (buildings, presses); light equipment (bill-hooks, containers for harvested grapes and so on); varieties of vines. Various aspects of technology were inextricably interwoven. The widely diversified cultural domain of the nineteenth century, like that of pre-industrial France (which, however, was much more homogeneous), remained ill-defined spatially. Its very complexity precluded it from being highly structured.

To what extent can one nevertheless pick out separate areas with local and regional cultural features of their own? The concept of local culture is certainly not misplaced in this context, provided – of course – that it is understood as no more than the product of the superposition of a series of cultural features which, for the most part, are equally characteristic of neighbouring regions: different for each place as for each individual, but only minimally so. The number of those small, individual particularities would clearly increase the more isolated the locality (for example, mountain villages or apparent 'world's ends', constituted by remote headlands) and the more discontinuous its activities. Thus, the small vineyards of central France, mentioned above,[80] which were topographically quite separate from one another, gradually, through absorbing various influences to varying degrees, came to constitute quite distinct complexes, each with its own particular features, at least in fields such as technology, but also in social life, festivals, beliefs and so on – features that more or less depended upon their own particular activity. In this nineteenth-century France in the grip of constant diversification, it probably is indeed possible to pick out the individual local, cultural worlds of certain groups of vine growers or fishermen, and particular villages of specialized craftsmen – worlds which manifest quite specific differences from the overall national culture.

Did any true regional cultures emerge from this increasingly accentuated process of diversification? It would seem not, apart from a few exceptional cases. Between the central model on the one hand, and the countless local variants on the other, there seems to have been hardly room for overall regional patterns to develop. The legislative unification instituted under Napoleon led to the disappearance of the provincial feelings of patriotism that had been based on provincial privileges, but as regards ideas, arts and objects, no regional identity had really emerged in between the national level and that of purely local relations. It is true that in the field of the various popular arts, definite regional styles did develop. However, they have as yet never been mapped and any attempt to do so would run into serious difficulties. The fact is that the emergence of a number of separate, homogeneous geographical areas was a chancy matter that depended upon the situation of particular kinds of workshops. Some, producing certain manufactured goods, were quite rare and their products circulated widely in some areas, less so in others. However, it is fair to say that, from certain points of view, relatively vast separate regional areas did exist, each with its own particular characteristics, and in some cases they coincided with the provinces of the Ancien Régime.

Distinctive regional styles of furniture certainly developed in Normandy, Brittany, Provence, Lorraine and even Berry. Many provinces certainly constituted zones within which a special network of communications operated around and radiated out from a *chef-lieu* where the central models were intensively promoted. At a lower level of compartmentalization, peculiarities sometimes manifested themselves within the sphere of influence of a small town or even within a naturally homogeneous area of settlements. A style of furniture characteristic of the region of Arles is thus to be found throughout that city's sphere of influence in the plains of the lower Rhône. Bresse also produced characteristic furniture in the Louis XV style, using two tones of wood. The geographic area where this style prevailed coincides remarkably closely with the tertiary clay plain bounded by the Jura and the Beaujolais hills, within the quadrangle formed by Tournus, Louhans, Bourg and Mâcon and it is not hard to understand the presence of such a distinctive style here.[81] Bourg and Mâcon were centres from which many new ideas emanated, where workshops supplying furniture to the local bourgeoisie evolved this particular variant of the Parisian model. The Bressan style emerged rapidly at the end of the eighteenth century, just at the point when this region, thitherto extremely isolated and not at all easy to move about in, was opening up to the external world and beginning to enjoy a growing prosperity which allowed even peasant homes to equip themselves with furniture. Many architectural features and individualistic styles of clothing appear to be peculiar to exactly the same zone, so up to a point perhaps it is quite reasonable to speak of a 'Bressan culture'. Its individuality is explained by a combination of factors: a very distinctive neighbourhood, which remained cut off until a relatively late date, together with a relatively sudden economic prosperity, which created a heavy demand that was met by intense creative activity within a limited area. Such cases must have been exceptional, however. 'Closed' cultures of this kind stood in definitive contrast to 'open' cultures, which seem to have been the more normal type. The new cultural developments of the nineteenth century may have created new forms in certain well-defined areas, but mostly these eventually became integrated with the general run of much less specific cultural features.

So it was with certain forms of rural housing. Their emergence, at a relatively late date, testifies to the establishment of certain regional stereotypes of agricultural exploitation and socio-economic structures. In Livradois, for example,[82] a complex kind of house emerged during the second half of the nineteenth century. It was composed of three elements (living quarters and a farm shed divided into two sections) and its emergence corresponded to new levels of local wealth and prosperity among farms devoted to cattle raising, within a social context already marked by the beginnings of a rural exodus and a reorganization of the available land. These fine 'Livradois houses' resulted from the amalgamation of already existing buildings into a structure that was now organically integrated. From 1840 onwards, essentially, new living quarters were erected alongside older buildings, introducing a certain functional and stylistic clash between the new living quarters and the older buildings. The new developments came about as a result of many abandoned properties being taken over by neighbouring farms at a time when the increasing prosperity of these cattle farmers was making them more demanding about their homes. This model of an enlarged farmhouse was widely reproduced, becoming characteristic of the region. Copies of it were even to be found beyond the limits of the mountainous Livradois massif. But the fact that there was a 'Livradois house' by no means made the region a Livradois cultural area, for striking though it was in the regional human landscape, the Livradois house appears to have been a totally isolated cultural phenomenon.

The birth and development of peripheral cultures
In truth, it was only in the non-French-speaking linguistic areas, around the edges of the national territory, which were much less accessible to central pressures, that distinctive elements multiplied to the point of producing all the appearances of a real regional culture. Beyond the barrier of the Vosges mountains, Alsace manifested its originality in its popular costumes, arts and architecture and in numerous objects. Although the zones in which they appeared do not coincide completely with the area of the province itself, they nevertheless certainly conferred upon it an overall individuality.[83] The German annexation of 1871, which came at a moment when this Alsatian culture was reaching the peak of its individuality, encouraged it to crystallize symbolically in a particularly vigorous manner. Thus the large bow in which women caught their hair on the nape of the neck and which was first introduced only in about 1860,[84] became after the war an aesthetic and sentimental symbol of the lost province. Meanwhile, during the first half of the nineteenth century, at the opposite end of the national territory, the particular character of lower Brittany was also gradually asserting itself.

This region functioned more and more as a residual area, where many practices and customs characteristic of the French-speaking territory penetrated, if at all, only very gradually. It is worth noting one feature in the series of items of dress noted above (pp. 297–302) which similarly became symbolic: namely, the Breton hat worn by men. While the French-speaking territory rapidly adopted the three-cornered hat in the eighteenth century and the stove-pipe top-hat in the nineteenth, lower Brittany remained overwhelmingly faithful to the round Breton hat (fig. 62A), an old peasant fashion that had been common to the whole of Europe. In a popular French ditty of today this hat, though gently mocked, is represented as the very epitome of the province:

> *Ils ont tous des chapeaux ronds*
> *Vive la Bretagne*
> *Ils ont tous des chapeaux ronds*
> *Vive les Bretons*
> (All of them wear round hats there
> Long live Brittany
> All of them wear round hats there
> Long live the Bretons.)

At this period, it is much harder to distinguish a real Occitan culture, as opposed to the national model. However, compared to most of the rest of France, this region could be considered as an area of retarded development, where many archaisms persisted. It is certainly remarkable that it was in the Occitan area, in the widest sense of the expression, that collective dances of the *farandole* type were preserved much better than elsewhere.

However, these marginal cultures should not be regarded simply as relics in areas where the past was preserved. For many of their components were of recent origin. In the nineteenth century, Alsace was a powerful centre of specific developments in all sorts of popular arts, particularly in the sphere of the dwelling and its ornamentation. Perhaps these areas round the edges of France were particularly receptive to innovations from outside which never became integrated into the central model. It has been shown that in the Basque country, the collective dances acquired far more complicated forms than those of the neighbouring Béarn and Bigorre regions, where the dance forms preserved were very old-fashioned. In the French Basque country, in contrast, the dance forms were much more elaborate – a fact which must be explained by the special contacts maintained with the Spanish Basque country, which was a major centre of differentiation and evolution where dances were concerned.[85]

In fact, the culture of the French Basque country presents a remarkable example of progressive enrichment and helps us to understand how the secular elaboration of these peripheral cultures proceeded, grafting new developments on to archaic substrata. The introduction of many features can be dated precisely. The discoid tombstones which are nowadays considered typical of the Basque country, were originally a pan-European phenomenon. They are to be found all over the place – in England, Sweden and elsewhere too. The fact that they are much more numerous in the Basque country accounts for their being considered as a regional characteristic. Yet in truth they owe nothing to regional traditions or to hypothetical Iberian models and are in no way connected with the Vascons. They made their appearance in the French Basque country at about the same time as the tombs outside churches constructed for passing wanderers whose graves had to be marked by some sign, originally of an anthropomorphic nature. The most ancient of these tombstones dates from 1501. In the course of the sixteenth century, they multiplied in the context of a period of general prosperity and a proliferation of new houses. In similar fashion, the 'comma-cross', sometimes known as the 'Basque cross', the oldest of which dates from 1690, stemmed from a generalized decorative motif which had nothing particularly Basque about it.[86] The Basque beret, which was worn over a vast area, was initially regarded as a sign of backwardness and poverty. At the beginning of the seventeenth century, it was 'all the poor and wretched people beyond the frontiers of Gascony and Béarn who wore on their heads little white toques, as round and as flat as a plate'. In 1787, Arthur Young, travelling in the Toulouse area, in the foothills of the Pyrenees, met mountain people wearing 'round, flat bonnets'. Between Pau and Monein, in 1789, in the Toulousain, at the foot of the Pyrenees, Arthur Young met

'mountain folk ... They wear flat round caps' ... 'The men are all dressed with red caps, like the highlanders of Scotland'.

Some fifty years later, Michelet, attending the Tarbes fairs, remarked: 'You often find, in one and the same place, the white bonnet of Bigorre, the brown one of Foix, the red one of Roussillon ... The Basque carter ... wears the beret of Béarn.'[87] Later on, the Basques adopted the beret as their own, making it a symbol of their identity and changing both the material used and the colour. Then, at the end of the nineteenth century, industrialization and fashion popularized the beret throughout the territory of France. Other developments are of even more recent date. The Basque game of *pelota*, now considered an essential element in the regional culture, was invented relatively recently, as a variant of the *jeu de paume*, which the Basques played up until the Revolution just as it was played in the rest of Europe. The new variant of the game began to take shape when, towards the end of the first half of the nineteenth century, someone had the idea of replacing the traditional ball stuffed with tow by a nugget of rubber bound by cotton threads and covered in dog-skin. As the unpredictable bounces of the new ball were impossible to control when the players stood facing one another, the ball was now hurled against a wall. In 1862, a few years after the invention of the facing wall, the introduction of the wicker glove (*chistera*) completed the basic process of elaboration.[88]

All in all, these peripheral regional cultures are an intriguing mixture the various components of which – arrested archaisms on the one hand, specific new developments on the other – need to be carefully analysed in each separate case. The importance of the new developments should not be exaggerated. Features shared in common with other parts of France outweigh local differences. As we have noted, in even the most conservative of all domains, that of traditional dances, French forms of collective dances – the *rondes* of the French-speaking interior – penetrated lower Brittany, crossing the linguistic frontier as they passed from one rural community to another.[89] The formation of these peripheral cultural complexes was, furthermore, a slow, hesitant process, spanning several centuries. It started in the Renaissance, when the invention of the printing press promoted the diffusion of a national model. In the nineteenth century, this clearly speeded up, as relations developed all along the network of main roads and pressure from the centre built up; yet still, in areas where other languages were spoken, it encountered particularly entrenched resistance.

The appearance of cultural regionalism
Understandably then, a positive awareness of these cultures took a long time to develop. For many years their development was a subterranean, instinctive process, which had to take place before a clear sense of identity could emerge. Those centuries of gestation, under the Ancien Régime, were centuries of obscurity and silence.

They were marked by the quasi-disappearance or extreme mediocrity of regional literature. Only in Alsace, in a territory where two major cultures came face to face, did the literature of both flourish alongside each other in the eighteenth and nineteenth centuries.[90] The French language was spoken only by the upper classes in the eighteenth century, but in the nineteenth century it made great progress among the bourgeoisie and, thanks to schooling and military service, bilingualism soon appeared among the peasantry. German remained the essential language, especially of scientific thought, up until 1830, by which time most literary and periodical journals were published in French. Between 1830 and 1850, an even balance prevailed between the two languages. Thereafter French definitely began to gain the upper hand, but poetry continued to be written almost exclusively in German.

Everywhere else had seen a major retreat of literature in regional languages. Particularly striking, during the centuries under the Absolute Monarchy, was the collapse of the Occitan literature, despite its brilliance[91] up until about 1660. The identity of Languedoc had continued to be keenly felt even after its annexation to the royal domain in the thirteenth century.

The term *Occitania* appeared in 1290[92] and Aquitaine also declared itself to be Occitan. The French language certainly made rapid progress within the political orbit of the French monarchy. Right from the start, in 1444, the registers of the Toulouse *parlement* were written in French. In 1521, municipal deliberations were recorded in French and the town's corporations took to recording their decisions in French during the first decades of the sixteenth century. The last inventory of the Cathedral of Saint-Sernin to be recorded in the *langue d'oc* dates from 1514.[93] The *Gai Savoir* poetry competition, which had been established in 1323, from 1513 on[94] crowned only poets writing in the French language; and the competition was held entirely in French from 1694 on,[95] the date when the Académie de Jeux Floraux took over from the Collège de Rhétorique. However, in the sixteenth century, this social and administrative advance of the French language was, thanks to the printing press, matched by the parallel progress of Occitan in popular and theological literature, for religious propaganda used the Occitan language so as to be understood by the common people. The centres from which the French culture was diffused were Auvergne, Bordeaux, Toulouse, the court of Marguerite de Valois at Nérac, and Aix-en-Provence. In opposition to these, a new Occitan literature was developed from the mid-sixteenth century on. It reached its peak at the same time as the French culture did in Occitania. The literary language of Gascony originated with Pey de Garros (about 1525/30–1581). It was the official language in Béarn, which was still independent, and remained the religious language of the Reformation in Gascony even under a French-speaking administration. In 1578, on the occasion of the entry of Queen Catherine and Queen Marguerite at Nérac, Salluste du Bartas, Henry of Navarre's Protestant secretary, composed a *Dialogue de Nymphes* in which three nymphs (one Latin, one French and one Gascon) argue as to who shall be the one to welcome the visitors. The Gascon nymph declares her pre-eminence and her pride:

> Carot Nymphe besée et tu Nymphe Romane
> N'anes de tous grans mots ma Princesse eichanta[96]
> (Hold your peace, neighbour nymph, and you too Roman one,
> Don't go frightening my Princess with your big words).

At the beginning of the seventeenth century, the Gascons set about besieging Toulouse, even though their language was not spoken there. In 1611, Toulouse saw the publication of Jean de Garros' *Pastourade Gascone*, written to mark the death of Henri IV, the peasant-king, shepherd and protector and himself a native of Gascony, and also Guillaume Ader's *Lou Gentilome Gascoun*, dedicated to the duc d'Epernon. In the latter work, the author presented his fellow Gascons with the idea of a collective super-ego, a courageous and courteous figure fired by all the qualities of feudal nobility and a chivalric righter of wrongs. 'The South was dazzled by itself at the very moment when the North was ridiculing it' with the creation of the stereotype of the cowardly Gascon braggart.[97] Meanwhile, pure Occitan poetry composed by Godolin (1579–1649) was triumphing in Toulouse. In the sixteenth century,[98] Provence had also enjoyed a similar evolution. The French language had begun to make inroads here at an early date, thanks to the presence of the papacy in Avignon and the Rhône, which acted as a major artery of communication. At the end of the fifteenth century, the people of Provence learned French in order to communicate with their king, René d'Anjou. After Provence's annexation by France in 1523, the decrees passed by the Etats de Provence were issued in French and, between 1523 and 1536, members of the Aix *parlement* rallied to the language of the king. The French language was definitively established at the time of the wars against Charles V and his invasion, when Provence asserted its French patriotism. The only literary use of the Provençal language was now in the songs of the students of Aix.[99] However, Provençal too was to enjoy a renaissance in the second half of the century, with the idealization of a mythical Provence in the work of Jean de Nostredame and, above all, in Bellaud de la Bellaudière's *Don-don Infernau* (1583 or 1584), which celebrated a Provence of games and laughter which the author was loath to leave when he had to return to the mists of the north:

> *You mouory de regret pensant a la partenso*
> *Que fayre siou contrench au terren franchiman*
> *Car senty que mous ueils eila mourran de fan*
> *Luench de mon beou souleou qu'esclaro la Prouvenso*[100]
> (I die of regret at the thought of the departure
> That constrains me to return to the land of the French
> For I sense that up there my eyes will die of hunger
> Far from my beautiful sun which shines upon Provence).

And Provençal was the language adopted by the ephemeral Republic of Marseilles set up by Casault (1591–92), a Catholic who, with the support of César de Nostredame, Pierre Paul and Robert Ruffi, rebelled against the Protestant king.

It is worth pausing to reflect upon the surge of Occitan culture at the end of the sixteenth century. Should we really follow Robert Lafont,[101] who regards it as a reaction against the first wave of French acculturation in Occitania, a movement of peripheral defiance, a reflex of protest on the part of *langue d'oc* literature, which was only allowed to bloom provided French literature was seen to be enjoying a parallel success? Was Occitania a kind of 'Arcadia', an erotic-pastoral refuge from the tensions at the age?[102] Was this truly a first manifestation of cultural regionalism that was part of the construction of the French nation and quite the opposite of provincialism?[103] In truth, it was not so much a 'surge', rather a swan-song lent artificial prominence by the appearance of the printing press and the troubles of the Reformation and the Wars of Religion. So much seems to be confirmed by the fact that it was no more than a fleeting flowering which was soon to wither away under the Absolute Monarchy, precisely when pressure from the central cultural model increased. Provençal literature was the first to die in Occitania. In 1610, Pierre Deimier, a native of Avignon and a disciple of Malherbe, excluded it from his *Académie de l'art poétique*.[104] In Toulouse, Occitan poetry died with Godolin. From 1660 to 1789, *langue d'oc* literature took refuge in religious Catholic propaganda aimed at the common people, in burlesque, in (often coarse) comedies of manners and in Christmas carols. It never reflected any Protestant or Jansenist ideas. The great issues of the day were foreign to it. The élite classes had deserted to French in droves. The Occitan language survived only among '*pastre e gènt di mas*' (shepherds and cottagers).[105]

Similar situations evolved in Brittany and the Basque country. Albeit in a less flamboyant fashion, lower Brittany had also seen the beginnings of a Breton literature, as printing presses began to spread in the sixteenth century. (They were first introduced in Brittany in 1484.[106]) Ever since the Middle Ages, the aristocracy of independent Brittany had been exclusively French-speaking and only French was used by the administration. But a written Breton literature did emerge at the end of the fifteenth century. It consisted chiefly of accounts of saints' lives and texts for the theatre such as Mystery Plays, composed as imitations of the French models, in regions close to the linguistic frontier (the dioceses of Tréguier and Vannes).[107] Three of these Mysteries were published in the sixteenth century, as were books of religious and moral poetry.[108] But in the seventeenth and eighteenth centuries these activities came to a halt. The general ban imposed on Mystery Plays by the Paris *parlement* in 1548 and the clergy's hostility to these plays, which they regarded as impious, put an end to their production. Breton literature was now reduced to a few collections of canticles and psalms and bilingual dictionaries for practical use. A certain curiosity about the past of Brittany and its language lingered on, but only within the general framework of French interest in its Celtic origins, not on account of the regional culture. In similar fashion, the Basque country, where the first work published in the local language appeared in 1545, went through a modest literary renaissance which lasted until the middle of the seventeenth century. This was also the period of the dramatic genre of tragic 'pastorals', which was developed particularly in the province of Soule, following a model of Christian drama of religious and warrior inspiration, directly imitated from the French Mystery Plays of the Middle Ages. These plays owed their genre name of 'pastorals' to the dramatic pastoral poems which had invaded the region of Béarn and Gascony in the sixteenth century. The term was so popular that it was used as a synonym for theatrical representations of all kinds. The subjects of these plays were almost all derived from the popular pamphlets written in the French language and purveyed by pedlars. The tradition was preserved into the nineteenth century, but unlike in Brittany, the texts in the local language were never printed until then.[109] In the case of the Basque country, there was, moreover, no Celtic past to attract the erudite curiosity of Gallic enthusiasts; and the eighteenth century in this region was particularly obscure.

It was only in the nineteenth century, in the great European current of Romanticism and the rebirth of nationalities, that this tendency was reversed and cultural regionalism began to emerge. Here again, Alsace constitutes an exception to the general rule. The first two thirds of the century had been marked by linguistic competition between the two major cultures of the region and the progressive predominance of the national language, which was reflected in the Alsatians' unquestionable desire to belong to and remain with France. At this period, no true cultural regionalism existed in Alsace. The deliberate choice of the French culture, rather than the German intellectual culture which had still been largely predominant at the beginning of the century, left no room for the use of a specifically Alsatian German dialect. This was only to appear after the annexation and as a reaction against it. A similar situation obtained in the tiny geographical sector of French Flanders bordering on Belgium where, since the affranchisement of 1830, French had been becoming increasingly predominant over the Flemish language.

The rest of the regional cultures, in Occitania, Brittany, the Basque country and Corsica, all produced written texts, of varying degrees of brilliance, during the nineteenth century. What kinds of texts were these and how should we interpret this Renaissance? The texts produced were exclusively the work of members of the intellectual and social élites, all deeply marked by a French education but who nevertheless reacted against the triumphant uniformity of France, and prized their origins and their own marginal cultures. We are bound to recognize that the impulse for the resurgence of these regional cultures came from a general movement of ideas that was quite external to the regions in question. Corsica constitutes an extreme case and a particularly revealing one. Up until the end of the eighteenth century, the dominant culture here had been exclusively Italian[110] and poems written in dialect by educated Corsicans were rare curiosities.[111] The popular culture survived in *nanne* (cradle songs), *canzoni* (songs mainly of a satirical or professional nature), a few serenades and *canzoni d'amore* and, above all, *lamenti* or *voceri*, funeral dirges, the foremost mode of emotional expression for the people of this island, constantly haunted by the spectre of Death. (At the beginning of the eighteenth century, on average 1,000 violent deaths occurred[112] each year in vendettas and clan warfare.) This culture was purely oral. The printing press was not set up on the island until 1750, in Bastia, and in the early nineteenth century in Ajaccio.[113] Between 1825 and 1830, this Corsican culture was 'discovered' by visitors from the mainland, all full of curiosity and in love with exoticism. The first text to be printed in Corsican dialect, in 1817, was a serenade inserted into Book IV of *La Dionimachia*, an Italian poem written by Salvatore Viale and published in London.[114] But the first systematic publications of popular songs were collected by Benson, an Englishman, and reproduced in his *Sketches of Corsica* (1825). He was followed by a number of Frenchmen, such as Valéry (*Voyage en Corse*, 1837), Pastoret (1838), and Mérimée (*Notes d'un Voyage en Corse*, Paris, 1840) and the Italian Niccolo Tommaseo, who published the first large collection of *Canti popolari corsi* (Venice, 1841). It was Mérimée's novel *Colomba* (1840) which first stirred up considerable interest in the *mores* and culture of the island. In 1843, when publishing the first volume of *Canti popolari* collected by a Corsican, Salvatore Viale, the editor was careful to point out in his preface that he offered them to the public not as a model of fine poetry, but as a document on the sentiments and customs of the Corsican people.[115] The perspective adopted was thus purely ethnological. Sandwiched, as it was, in between the Italian culture which was still very influential, and the French, which was being adopted increasingly rapidly by élite circles, the situation of Corsican culture was somewhat akin to that of Alsatian culture during the very same period. A truly regionalist and autochthonous cultural renaissance did not develop until the very end of the century, when cultural societies such as A Tramontana (1896) and Cirnea (1905) were set up.[116]

Similar phases and outside influences may be distinguished even in cultural areas where the renaissance was the work of natives of the region. The Occitan movement[117] was foreshadowed by a wave of historical feeling. The impulse came from a national fashion which started at the end of the eighteenth century, namely resurrecting the Middle Ages. It resulted in the 'troubadour' genre of poetry which flourished from the time of Tressan (1705–83) down to the Romantics.[118] The second stage came with the reconstruction of the historical dignity of the south, through the works of Augustin Thierry (1795–1856), who analysed the confrontation between the Franks and the Gallo-Roman world in his *Lettres sur l'Histoire de France* (pre-publication, 1820; published as a single volume in 1827) and his *Considérations sur l'histoire de France*, which formed Part I of his *Récits des Temps Mérovingiens* (1840). Similarly influential were the works of Fauriel (1772–1844): *Histoire de la Gaule méridionale sous les conquérants germains* (1836) and *Histoire de la poésie provençale* (published posthumously in 1846). Against this background, a large scattering of centres of Occitan literature developed around 1830–40. Initially, they were based in Béarn, Bordeaux, Toulouse, Nice and Agen, where the poet Jasmin was particularly active, but later they were concentrated mainly in the area between Béziers,

Marseilles and Avignon. 1854 saw the official birth of the Félibrige Movement, which had decided what to call itself the previous year. Mistral (1830–1914) was to produce its greatest works, starting with *Mireille* (1859) and ending with the *Poème du Rhône* (1897). In 1867, affiliation with the Catalan movement (the Floral Games of Barcelona had been revived in 1859), through the relations established between Balaguer and Mistral, completed the structure set up by the Félibrige and confirmed the maturity of the movement. In 1876, it adopted its definitive statutes and this was followed by a triumphal expansion. The Mistralian Provençal ideology was thus born from the connections established between the Occitan culture of the nineteenth century and the medieval splendours of the land of *oc*.[119] Mistral himself was a scholar who possessed a double, perfectly controlled culture, part urban, part rural, and his ideology reflected an admiring vision of an Occitan peasant world still bursting with vitality, which could only be fully reached through poetry. '*Il faut être provençal si l'on n'est pâtre* (Unless one is a shepherd, one must be Provençal)'.[120] As for the origins of the Breton movement, these may be traced to two intermingled currents of thought. One was the persistent interest in the Celtic origins of France, which we have already noted. In the course of the second half of the eighteenth century, this crystallized into a 'Celtomania' so intense that in 1807 Napoleon himself founded a short-lived Académie Celtique. The other influential current which then converged with the French Celtic movement came from the rebirth of Celtic studies in Britain and the publication of the poems attributed to Ossian. These scored a resounding and lasting success in France[121] and played a vital role in the formation of the Romantic sensibility. Such was the prevailing atmosphere of the days which saw the appearance, from 1836 on, of the *banquets bretons*, centred upon Jean-François-Marie Le Gonidec; the publication of the first *Barzaz-Breiz* in 1839, by Théodore Hersart de la Villemarqué; the foundation of the *Revue d'Armorique* in 1842, by Aurélien de Courson; and the creation, in Vannes, of the Association Bretonne, in 1843.[122] Finally, for the Basque country, the opportunity came in 1833, with the Carlist Wars: these generated intense regionalist excitement among Spanish Basques and great curiosity in educated circles. Up until this point, more or less the only interest shown in the Basques had been of a philological nature. Augustin Chaho's *Voyage en Navarre pendant l'insurrection des Basques, 1830–1835* (Paris, 1836) prompted a whole movement of ethnological and historical investigation into the Basques, which culminated with Francisque Michel's *Le Pays Basque, sa population, sa langue, ses moeurs* (Paris, 1857).

The nature of the movements of cultural regionalism in nineteenth-century France seems clear enough. At no point did they express the popular aspirations for differentiation which developed in many parts of France in reaction to the pressure exerted by the central cultural model. Rather, they were responses to a natural, even pan-European movement of ideas which discovered a means of creative expression in this distant and little-known material. These were not defensive movements. They came into being at different dates but in every case well before the languages in which they expressed themselves were really threatened by the inroads being made by French in rural areas. The progress of the French language only became truly decisive in the French countryside when schooling became compulsory in the 1880s. The flowering of the movements of cultural regionalism took place roughly fifty years earlier. In truth, each one was an imitation of the central model and was historically independent of the development and destiny of the peripheral regional culture in which it found its roots. The definitive elaboration of the French nation through the propagation of its language and the diffusion of its ideals had now engendered superior forms of expression in the thitherto spontaneous and confused substrata of those popular cultures. However unconcerned with the elaboration of the French nation those forms of expression may appear to be, they were in truth, in a sense, born from it and cannot be separated from it.

The late emergence of political regionalism and its significance
A fortiori, the same applies at a political level. The peculiarity of the French national model and, needless to say, its merit was that, up until the developments of our own day, those regional cultures were never opposed to the French nation and never sought to dissociate themselves from it. They were its daughters, sisters, even emulators, never its enemies, as was made quite plain as early as the Occitan Renaissance of the sixteenth century. Henry of Navarre's accession to the throne of France had set the seal upon the union of France with Gascony. Guillaume Ader celebrated that union:

> ... *Que la regine mute*
> *Sie la noste Gascougne, è sie la hlou den lis*
> *Empentade eus laurés d'aquet noble païs*[123]
> (... Let the royal pack of hounds
> Be our Gascony, and let the lily
> Be grafted on to the laurels of this noble land).

And when, two and a half centuries later, Mistral put into the mouths of the Saintes Maries words that painted an unforgettable picture of Provençal history, that history also achieved its goal and found fulfilment in the union with France:

> *La Prouvènço cantavo, e lou tèms courreguè;*
> *E coume au Rose la Durènço*
> *Perd à la fin soun escourrènço*
> *Lou gai reiaume de Prouvènço*
> *Dins lou sen de la Franço à la fin's s'amaguè.*

– Franço, emé tu meno ta sorre.
Diguè soun darrié rèi, iéu more.
Gaudissès-vous ensèn alin vers l'aveni.
Au grand prefa que vous appello ...
Tu sies la forto, elo es la bello:
Veirés fugi la nive rebello
Davans la resplendour de vòsti front uni.[124]
(Provence was singing and time was passing by
And, as at last the Durance is lost in the Rhône,
The gay kingdom of Provence
In the bosom of France finally came to rest.
– France, lead your sister along with you,
Said its last king, for I am dying.
Set forth together into the future
Towards the task that beckons you ...
You are the strong one, she the beautiful:
You will see the rebellious night flee
Before the twofold splendour of your united brows).

In his *Français du Nord et Français du Midi* (1867), Eugène Garcin accused the author of *Mireille* and *Calendal* of separatism. On 13 October 1878, Jules Claretie similarly denounced the Félibres.[125] In his reply to Garcin, Mistral pointed out that, by awakening the south, 'he was producing patriots for France'.[126] For him, fervent attachment to the greater country was part and parcel of his love for his homeland. Provence's attachment to France created a union of equals.[127] His Provençal idealism was resolutely 'non-historical'. In 1907, he declined to carry a banner for the vine-growers of Languedoc, who had risen in revolt.[128] Occitania could never be any more than an idea, a museum piece, protected by an 'initiatory optimism'.[129] What the Félibres had to say was a 'touristic declaration'.[130] For Mistral, it was essential that France should continue to create itself. And the federating role of the republican liberties was essential in this union. Only through them could the Provençal 'nation' exist, holding its own banner high. The spirit of 1848 was projected on to the twelfth century.[131] And, conversely, the Occitan culture had always been integrated into the soul of France, in a position of the highest honour. In the reign of Louis-Philippe, Jasmin, the barber-poet of Agen, was all the rage in the salons of Paris and was made a Chevalier de la Légion d'Honneur. By the end of the century, Mistral was also considered a glory to the nation.

In Alsace, the general picture was similar. For a long time Alsace manifested a measure of reticence towards France and after the union of 1681 and throughout the eighteenth century it was royal policy to recognize that fact and to display remarkable prudence where Alsace was concerned. A Controller General of Finance of the period scribbled on the back of an envelope addressed to the maréchal d'Uxelles,[132] 'It is important not to interfere in the affairs of Alsace.' But in 1871, protest against the annexation of the province was unanimous. At the time of the great liberation of 1789, Alsace had given itself to France with spontaneous unanimity, as Fustel de Coulanges had rightly noted.[133] Though its tongue was German, its heart was French, as Napoleon understood: he saw no harm in allowing the Alsatians to speak their own language, 'provided they wielded their sabres as Frenchmen'.[134]

In Corsica, the beginnings of a national movement had manifested themselves with the first anti-Genoese uprisings in 1729 and, until the victory of the French troops in 1769, these had been partially followed up by Paoli's attempts to gain independence. But already a large proportion of the island's élites were positively seeking external protection, first from Spain, then from France, and the French party had been particularly active throughout Paoli's régime.[135] The Napoleonic adventure set the seal on reunion with France, just as southern France had annexed the rest of the French territory, two centuries earlier, when a Gascon king came to the French throne.

Only in the west, in the Basque country and Brittany, can divergent (or apparently divergent) tendencies be glimpsed in the nineteenth century. But it is important to try to understand the real implications of the manifestations of opposition. Above all, they reflected a rejection of modernity, not a desire for a truly autonomous region. In the Basque country, where Deputy Garat's proposal of 1790 to create a Basque department had never been followed up, there are no signs of a truly political regional movement in the nineteenth century. However, Basque society as a whole remained on its guard against the new France that had been modelled by the Revolution. While the foremost figures of Basque society found that they were able to preserve their social pre-eminence by lending their support to a succession of régimes, the clergy jealously preserved their grip on the common masses, erecting defences against innovations that they considered suspect or impious and thereby protecting the regional identity. But this was a purely passive resistance which found expression in the retention of persistently old-fashioned ways: in particular, literacy gained ground extremely slowly here, for French schooling was considered to promote new ideas. Evasion, in various forms, was another symptom of the Basque attitude: the evasion of military service (the highest proportion of service-dodgers in France came from the Basque country);[136] and trans-Atlantic emigration to South America, Mexico and California, which was extremely active in the second half of the century. Similar behaviour characterized Brittany, but under the influence of numerous intellectual members of the élite, all affected by years of French education, a far clearer impulse of dynamism made itself felt. At the Breton banquet of 1837, the following verses by la Villemarqué were declaimed:

Joyeux jadis étaient nos pères,
Et nous, maintenant, nous pleurons,
Mais le bonheur revient, mes frères,
Nous sommes encore bretons.

Ils étaient libres, nous aux chaînes,
Mais nos fers, nous les briserons,
Leur sang coule encore dans nos veines,
Nous sommes encore bretons.

Oui, nous reverrons nos hermines
Reflotter sur nos bataillons, etc.[137]
(In the old days, our fathers were happy
Yet we, who are here, all are weeping,
But joy is returning, o my brothers,
For we are still Bretons.

Then they were free, yet we are in chains,
But our shackles shall soon be smashed,
For their blood flows still in our veins
And we are still Bretons.

Yes, our ermines once again
Shall flutter over our battalions, etc.).

These lines were judged 'remarkable' by the Brest newspaper, *L'Armoricain*,[138] and remarkable they certainly were. But in truth all that these bellicose sentiments manifested was traditionalism and a rejection of the Revolution, linked with devotion to the province of Brittany and its ancestral liberties. Initially, the origins of the Breton movement were above all Catholic and legitimist; it was a struggle against the despotism of the Revolution and the Napoleonic period.[139] Breton culture provided a means to preserve the Catholic faith and tradition. In 1845, La Villemarqué, again, wrote as follows: 'They cling to this language as to the guardian of their religion, their morality, their customs, their *mores*, their traditions, their spirit, their intelligence ... In a word, everything which today constitutes their nationality, something dearer to them than life itself.' He went on to invoke 'the oath, sworn by our forefathers fourteen hundred years ago: "For as long as the sea beats upon their shores, the Bretons will keep their language, the Bretons will bless their God".'[140] Like Mistral, however, La Villemarqué denied that he harboured any autonomist or separatist inclinations. For him, the renaissance of Breton culture and liberties was simply an enrichment for the whole land of France. As early as 1837, in his reply to the strictures of *L'Armoricain*, he wrote: 'In truth, I see, not only for Brittany but for the whole of France, a happier, more prosperous and freer state in the future ... That is how my rhymes should be interpreted.'[141] In his preface to *Barzas-Breiz* (p. viii) in 1867, he repeated those sentiments: 'Here then, what matters is not a purely local interest, but a French interest; ... France, in its heart, is as Celtic as today's Armorica is French. They share a common flag.' And in 1876, at the congress of Vitré, Arthur de la Borderie made the same point eloquently, in his closing speech: 'When the Breton Association studies and exalts Brittany and its heroes, it is serving France well; for by restoring the great figures of the past to the light of day, it is telling those of today: "These were your fathers; children, be worthy of them. Just as they loved France and Brittany, you yourselves must continue to love them both; as they served them, serve them yourselves".'[142] This was soon after the defeat of 1871, which had engendered in Brittany, as in all the peripheral provinces, a huge swell of patriotism. Michel Bréal has pointed out that the only good poetry on national themes produced in France during the 1870–71 War was written in Provençal, Breton or the dialect of Alsace.[143] Perhaps those who expressed themselves in these other languages felt the greatest need to bewail the misfortunes of France, their motherland. At no point did the construction of the French nation give rise to regional political movements. The French model was that of an Empire which integrated many nationalities, as it was to be overseas in the period of colonial expansion. It was a model that possessed remarkable persuasive powers and freely won universal support even when faced with political ideas of the most diverse kind, whether these were inspired by revolutionary passion or by nostalgia for the Ancien Régime.

The emergence of political regionalism came very much later and was quite different. Again, it was fostered by a movement which developed centrally and was in line with a long tradition of national political reflection. There was a whole school of thought which had never accepted the division of France into departments. Admittedly, it had never occurred to the Girondins to call themselves Occitan and they had condemned provincialism just as much as the Jacobins did.[144] But a persistent current of ideas, mainly monarchist and traditionalist but shared by many Republicans, had always remained attached to the provincial liberties and the territorial divisions that went with them. Auguste Comte reflected such ideas in his *Système de politique positive* (1851–54), in which he proposed dividing France into seventeen *intendances* or regions, set up as 'positivist republics'.[145] Equally, they are to be found amongst socialists such as Proudhon, who also divided the territory into regions (ten seemed to him the ideal number, although he sometimes speaks of twenty). They inspired the liberalism of men such as Tocqueville and Lamennais, as they did, at the end of the century, the nationalism of Barrès and Maurras. In practical politics, however, the general notion made slow progress. It is true that an initial charter of regionalist inspiration was adopted in Nancy as early as 1865. The policy that it proposed was to 'strengthen the commune, create the canton, do away with the *arrondissement* and emancipate the department'. It was chiefly the brainchild of conservatives anxious to provide the provinces with a social structure capable of resisting the revolutions of Paris (O. Voilliard, p. 299, in Gras and Livet, 1977). This was followed by a steady stream of plans for a new regional division of France. No less than twenty-nine private projects were put forward between 1851 and 1947, as well as fifteen parliamentary bills between 1871 and 1963.[146] The word 'regionalism' itself was invented in 1874 by a Provençal Félibrige poet, Berluc-Perussas, and was used by Joseph Reinach in the *Les Débats* issue of 6 October 1875, in connection with Italy (P. Vigier, p. 162, in Gras and Livet, 1977). But the idea of regionalism only began to be diffused widely in 1892, the year of the declaration produced by the Félibre federalists, and the word itself remained a technical, scholarly and little used term until it was popularized by J. Charles-Brun (1870–1946), a native of Languedoc, from Montpellier. In 1900, he set up the Fédération Régionaliste Française and in 1911 he published his major work, *Le Régionalisme*. It was only at this point that local regionalist movements began to emerge, most of them in Brittany.[147] In 1898, the marquis d'Estourbeillon, the royalist deputy for Vannes, founded the Union Régionaliste Bretonne, which claimed to be inspired by the ideas of J. Charles-Brun.[148] It was now that the attack on the departmental division of France became more specific, led by geographers. P. Foncin, in his pamphlet *Régions et pays* (Toulouse, 1903), showed that the ideas defended by Target at the time of the creation of the departments had by now become anachronistic. (Target had declared: 'Our desire was that it should be possible to reach the centre of administration within a day's journey, from any point in the department.') Foncin now analysed the structure of the eighty-nine departments and came to the conclusion that, while thirty of them were more or less homogeneous, the other fifty-nine were all, to varying degrees, incoherent, with neither natural unity nor even, in most cases, historical unity. In the *Revue de Paris* of 15 December 1910, Vidal de La Blache published an article on the French regions, proposing a new division into seventeen regions. In the years between the Hovelacque bill of 29 May 1890, which proposed eighteen regions, and the Sarrieu plan (for sixteen regions) of July 1907, a whole succession of proposals were put forward in Parliament.[149]

The purpose of the regionalist movement's rejection of the departmental division was to achieve decentralization and devolution. The movement was sympathetic towards federalism, but always refused to be won over to autonomism. The 'Pacte National Breton' of 1911 mustered no more than a tiny group of intellectuals (only thirteen in 1914).[150] It was not until the inter-war period that autonomist ideas gained wider support in Brittany, particularly under the influence of the Irish independence movement and not without a certain contamination by national-socialist ideology. Meanwhile, an Alsatian autonomist movement was also developing, provoked essentially by the clumsiness of the French administration after the reunion of

1918 and its intensive application of legislation within the province. By and large, the historical implications of the regionalist movement are clear. Again, it was not a matter of a reaction against the unity of the national territory; rather, it reflected an increasingly clear recognition of the need for an intermediate level of organization, below the national level and above the department. The unification of France had been achieved in the 1880s, when the railway network was completed and compulsory schooling was introduced. But in the early twentieth century, communications were speeding up, a hierarchy of towns was developing (see below, pp. 405–14) and the web of social relations was becoming increasingly complex. The regionalist movement expressed new needs and sought to provide for them. In order to do so, it naturally enough drew upon a political tradition that was still very much alive. But what was involved was certainly not purely and simply a return to the province, for that form of organization was by now clearly superannuated. The size and the number of the new divisions proposed varied considerably from one project to another (from a maximum of thirty-four regions to a minimum of seven).[151] But there was one idea which was to win general acceptance from all regionalists. In his article of 1910, Vidal de La Blache argued that the new regions ought to be organized around large urban complexes and according to the radius of their fields of influence. The regionalist movement helped France to move forward into the 'era of big organizations', although this happened slowly (see chapter 11). The regional organization of France came about with the creation of twenty-one '*régions de programme*' (planning regions) in 1956 and the decrees of 1964, which institutionalized the system. All this had certainly never involved anything more than a desire for good and efficient administration. In the long process towards centralization and unification, France as a nation had never seriously been challenged either at a political or at a cultural level.

The emergence of different regional and political attitudes

However, this common attachment to national unity did not rule out the appearance of internal oppositions, clashes of belief and ideas and conflicts frequently of a bitter and violent nature.

The map of religious practices[152]
This was the earliest one to emerge clearly, at least in its broad lines, displaying geographical features which were to reveal an astonishing stability. The great crisis of the Revolution, at the end of the eighteenth century, suddenly laid bare deep regional differences in the distribution of religious attitudes. They were differences which had until then been masked by the strong social pressures which, under the Ancien Régime, had imposed quasi-universal religious obedience and practice, at least of a formal kind,

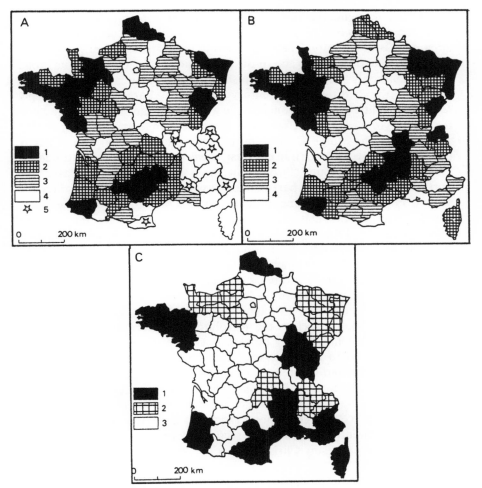

Figure 65. The origin and stability of religious attitudes.
A. Percentage of refractory parish priests in 1791 (after C. Langlois and T. Tackett in F. Lebrun, *Histoire des catholiques de France*, Paris, 1985, and Le Bras, 1986, p. 78a). 1. Over 75%. 2. 55–75%. 3. 35–55%. 4. Under 35%. 5. No data.
B. Rural religious practice (percentage of weekly communicants) between 1960 and 1970 (from F.-A. Isambert and J.-P. Terrenoire, *Atlas de la pratique religieuse des catholiques français*, reproduced in Le Bras, 1986, p. 78b). 1. Over 42%. 2. 28–42%. 3. 15–28%. 4. Under 15%. C. *Pays* organized as *Etats* and *pays* organized as *Elections* under the Ancien Régime (partially from Le Bras, 1986, p. 80b). 1. *Etats* surviving in the eighteenth century. 2. *Etats* suppressed at various dates since the fifteenth century. 3. *Pays d'Elections*.

manifested by attendance at weekly mass and Easter communion. As early as 1791, the *Constitution civile du clergé* revealed the artificial nature of this apparent Catholic unity. The map showing the distribution of priests who refused to take the oath (to accept the Constitution) (fig. 65A) is, in its broad lines, remarkably similar to the map showing the religious practice of rural adults in the mid-twentieth century (fig. 65B). Large blocs of deeply religious localities are to be found in the west, in Brittany and Normandy; in the north; in Lorraine, Alsace and Franche-Comté in the east; in the south-east of the Massif Central, in the Basque country and also in Savoy (which was not yet French). In contrast, areas with much cooler attitudes to religion are to be found throughout the heart of the Paris Basin, the north-western (Limousin) part of the Massif Central and the countryside around Poitiers and Saintonges, the mountains of Béarn and Ariège, Corbières, Provence and Dauphiné, and Bourbonnais. Between the dates to which maps A and B of figure 65 relate, de-Christianization certainly makes progress in the countryside around Bordeaux, but the general pattern remains unchanged. Even when one looks into the situation in more detail, at the level of individual cantons or even communes (as a number of studies have done),[153] the general stability is striking. Contrasts between communes remained largely identical throughout the nineteenth century. It is even possible to go back to before the Revolution: the records kept by seventeenth- and eighteenth-century missions lead one to suspect that the same differences existed to a large extent even then, marked by the façade of apparently generalized religious practice. Both in the southern Vendée and around the middle reaches of the Garonne (in the region of Montauban), it has been shown that the sectors which the missionaries' reports describe as difficult and unreceptive correspond exactly to those that are indifferent to religion in the twentieth century. The sectors where 'good missions' operated are those where practising Christians are active today.[154]

The map of political attitudes
This cannot be grasped in sufficient detail until later, after the definitive adoption of universal suffrage at the time of the 1848 Revolution or, even better, from 1871 on, for the 1851 *coup d'état* and the practice of official candidacies had the effect of complicating the documentation for the Second Empire.[155] This map displays a number of obvious correlations with the map showing the distribution of religious attitudes and, equally, testifies to a remarkable degree of stability in certain regions. Such is the case of the conservative bloc of western France, Brittany and its margins, and that of the south-eastern part of the Massif Central – all areas which remain stable throughout the Third Republic, in opposition to other regions with equally stable leftist loyalties: the northern and north-western parts of the Massif Central from Bourbonnais to Limousin; the Provençal south-east and lower

Dauphiné. In other sectors, on the contrary, the map shows clear signs of instability. The eastern part of the Paris Basin and Lorraine voted for the Left at the beginning of the Third Republic and moved progressively towards the Right from 1890 to 1900 onwards, as did Franche-Comté after the First World War. In the south, Languedoc was won over to the Left only from the 1880s onwards. In 1871, it was still voting against Gambetta, faithful to a long royalist tradition which had been particularly noticeable in 1815, at the time of the White Terror. The southern margins of the Paris Basin, Nivernais and Berry, also moved into the leftist bloc only in the 1880s. This seems a somewhat contradictory, or at any rate confusing, picture.

A tentative interpretation
What were the origins of these contrasts and what is their significance? The picture is relatively plain so far as religious practice goes. Clearly, it may in part be seen as a centre–periphery opposition. To a large extent, the process of de-Christianization spread from Paris. It reflected the philosophical propaganda which developed there in the eighteenth century as part of the Enlightenment movement. But it also reflected the slackening of social constraints which began in the great city and the consequent spread of a certain laxity to the countryside. A similar pattern seems likely in other large centres (such as Marseilles and Bordeaux, perhaps). At a local level, similar developments are noticeable along the communication arteries which disseminated the examples and models of the towns – not only the great roads but also waterways such as the Garonne and the Tarn.[156] Clamecy, the highest point for floating down logs from Morvan, on the navigable route from the Yonne to Paris, was remarkable in the eighteenth century for its low percentage of Easter communicants (less than half the population).[157] Generally speaking, regions of open countryside and concentrated populations were more receptive to these influences. The Béarn of the lower valleys, a literate region scattered with large villages, was an area of religious indifference as compared to the neighbouring Basque country, where religious fervour was preserved by the dispersed settlement pattern and the language barrier. On the edges of the Armorican peninsula, the limestone plains of Poitou and Saumurois were pervaded by religious indifference long before the granite areas characterized by dispersed settlements and large estates.[158]

In the mountainous regions where religion was on the whole more generally practised, the areas where religion was in retreat can, obviously, be explained by relatively early relations with the outside world, mostly established during the people's temporary migrations in search of work in the towns. A similar explanation no doubt accounts for the massive contemporary religious indifference of Limousin, whose masons had long been in the habit of migrating to Paris (see above, p. 256). In this case, a granite region did lose its religious faith early on.[159] A similar interpretation could probably be applied to much of Dauphiné. In this context, it has recently been suggested that another, analogous though slightly different mechanism may also have been at work. It is noticeable[160] that a major correlation exists between the map showing which parish priests swore the oath and the map showing which regions were organized into *Etats* and which into *Elections* (fig. 65C). In the regions still with *Etats* at the time of the Revolution or where they had been established only shortly before, the clergy played an important part in providing a religious framework for the population. Encouraged by the support of the faithful around them, these priests were in a better position to manifest their rejection of the oath, thereby expressing a kind of irredentism, a revolt against the reforms imposed by the central authorities. We are thus confronted yet again by a centre–periphery opposition, for the map of the *Etats* and the *Elections* in the eighteenth century reflects the degree to which the kingdom was really united and the extension of the practical authority of the central power. Of course, exceptions such as Provence and Dauphiné still need to be explained. But at any rate, the religious crisis of the Revolution, with all the persecutions that it brought in its wake, mustered certain local populations in support of their parish priests who had refused the oath and this strengthened their faith. In these regions, this had the effect of crystallizing religious behaviour and it became so deeply rooted that it remained essentially unchanged for close on two centuries. The rapid de-christianization of the urban and industrial classes in the nineteenth and twentieth centuries had no effect at all upon the basic framework of the rural areas.

However, this somewhat simplistic schema fails to account for all the facts. Some regions characterized in the eighteenth century by highly cultivated open plains and a high level of literacy, such as the north and Alsace, did remain overwhelmingly religious. Contemporary industrialization has not affected the solid bloc of practising Christians in the Nord and Pas de Calais departments. The mountains of Savoy, where large temporary migrations had been taking place for centuries, do not seem to have become contaminated by irreligiosity and preserved a solid Christian faith. Admittedly, this region was only incorporated into France at a late date, but Paris had been acting as a powerful magnet upon its population ever since the eighteenth century. And why did the Pays de Caux, close to and dependent upon large towns such as Rouen and Le Havre, produce such a high proportion of priests who would not take the oath; and why is it even today an area of impressive religious practice, whereas the neighbouring Pays de Bray, with more dispersed settlements of small isolated villages, appears to have manifested a much cooler attitude towards religion throughout the nineteenth century?[161] The presence of large landowners' estates in Caux does not seem a sufficient explanation. The basic schema was no doubt modified by plenty of local factors which only a detailed analysis could uncover. These probably included the crystallization of oppositions between Catholics and Protestants. In many places where the latter were numerous, such as Alsace, this led the Catholics to assert their faith with extra determination. Elsewhere, individual historical episodes may have left their mark. In the interior of Brittany, the Montagne d'Arrée produced a high proportion of priests who swore the oath and it has retained secular inclinations ever since, whereas the Montagne Noire and Léon manifested deeply religious attitudes. The explanation may lie in a persistent heritage from the *jacquerie* of the *Bonnets Rouges* of 1675 and the repression which followed, for it was the local squires and the clergy who benefited and thereafter they were distrusted by the common people.[162]

Simplistic explanations seem even more inadequate when it comes to the interpretation of political attitudes; and in this domain many areas remain shadowy. As we have seen, the religious element certainly played a major part here. The correlations are frequently impressive, even down to small details, and in many cases they indicate great stability, as for example in western France[163] and in the south-east of the Massif Central. The raftsmen of Clamecy, disaffected from Christianity in the eighteenth century, were on the extreme Left in the nineteenth century and in the forefront of resistance to the *coup d'état* of 2 December 1851. In general, in regions of mixed religious affiliations, the Catholic sectors appear to be further to the Right than the Protestant ones, as in Alsace,[164] for example, and the Diois region of the southern Alps.[165] On a macroscopic scale, however, a point of major importance should be recognized. Over the years, relations between religion and political attitudes may vary. In the elections at the beginning of the Third Republic, the correlation between religious practice and the conservative vote was still by no means absolute (fig. 66A, B and C). Only in the twentieth century did it begin to emerge increasingly clearly, so that by 1910 it was virtually absolute and remained so up to and including the elections of 1936 (fig. 66E and F). This was clearly a consequence of the progressive crystallization of the political debate over the religious question throughout the Third Republic and particularly after the law of 1905, which separated the Church and the state. After the Second World War, however, such correlations again became less inevitable and a large area of northern France, which was disaffected from Christianity, in particular the whole of the central Paris Basin, swung massively back to the Right. The major problem that then arises is why some regions retained stable political attitudes throughout the period that it is possible to analyse, while the behaviour of others changed and why the role of religious fervour was crucial in some places but not in others. Some of the deepest anthropological structures of French society may be involved here, namely the systems of family organization. A recent study[166] has produced a general interpretation

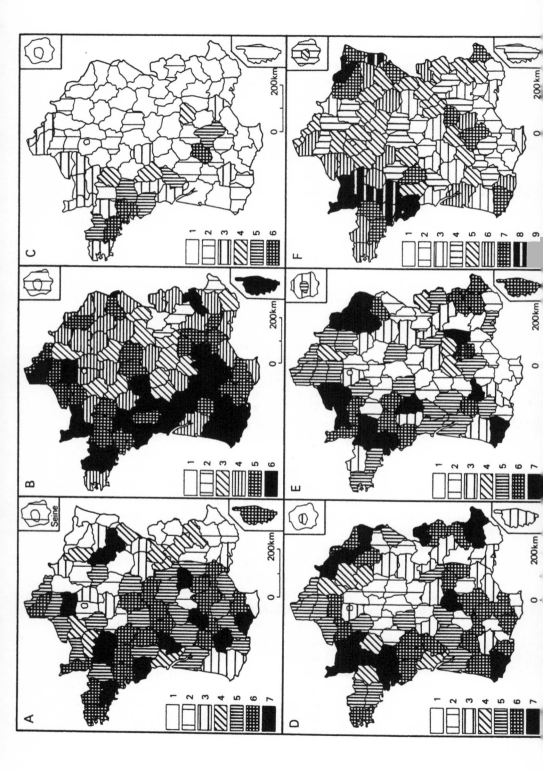

of the pattern of political behaviour, based upon those family systems. Three major types of family structures may be distinguished among the French population. A nuclear family comprising only parents and unmarried children predominates in Normandy, the interior of western Brittany, Maine and Anjou and throughout the Paris Basin, Burgundy and Franche-Comté. A complex 'community-type' family (what Le Play calls a 'patriarchal family') predominates in the south-western quarter of the country (except in the Basque country), Provence, lower Languedoc and the Nord department. In this system, marriages are not subject to strict family control and several generations and several couples may live together under the same roof; the régime is fairly lax and couples tend to marry young. The third type is a complex family of another kind, the 'stem family' (Le Play's *famille-souche*). This too combines several generations, but is more authoritarian; marriages are strictly controlled and only the heir has the right to live with the parents once he is married, whilst the other children must choose between on the one hand celibacy and a home under the family roof, on the other moving away. This pattern predominates in Brittany, the Basque country, the southern Massif Central, Savoy and Alsace. This third type of family unit, which produces a stable society within a firm framework, seems in particular to have fostered an enduring conservative ideology and religious

---

Figure 66. The development of political attitudes.
A. The elections of 8 February 1871 (after Goguel, 1951, p. 15). 1. No conservatives elected. 2. Only one conservative elected. 3. A minority of conservatives elected. 4. 50% conservatives elected. 5. A majority of conservatives elected. 6. All but one conservatives elected. 7. All conservatives elected. B. The elections of 14 October 1877 (Goguel, p. 19). The percentage of right-wing votes: 1. 7.5–15%. 2. 15–22.5%. 3. 22.5–30%. 4. 30–37.5%. 5. 37.5–45%. 6. 45–52.5%. C. The elections of 22 August–September 1881 (Goguel, p. 21). 1. No conservative elected. 2. Only one conservative elected. 3. A minority of conservatives elected. 4. 50% conservatives elected. 5. A majority of conservatives elected. 6. All but one conservatives elected. 7. All conservatives elected. D. The elections of 20 August–3 September 1893 (Goguel, p. 31). 1. No conservatives or moderates elected. 2. Only one conservative or moderate elected. 3. A minority of conservatives and moderates elected. 4. 50% conservatives and moderates elected. 5. A majority of conservatives and moderates elected. 6. All but one conservatives and moderates elected. 7. All conservatives and moderates elected. E. The elections of 24 April–8 May 1910. 1. No right-wing or centre candidates elected. 2. Only one right-wing or centre candidate elected. 3. A minority of right-wing or centre candidates elected. 4. 50% right-wing or centre candidates elected. 5. A majority of right-wing and centre candidates elected. 6. All but one right-wing or centre candidates elected. 7. All right-wing or centre candidates elected. F. The elections of 3 May 1936. The percentage of right-wing votes: 1. Under 7.5%. 2. 7.5–15%. 3. 15–22.5%. 4. 22.5–30%. 5. 30–37.5%. 6. 37.5–45%. 7. 45–52.5%. 8. 52.5–60%. 9. Over 60%.

faith. The clergy managed to maintain and even increase its influence in such societies throughout the nineteenth century by reason, in particular, of the large number of younger siblings committed to celibacy who embraced a religious vocation. Essentially, this type of family was to be found in the regions which remained continuously faithful to both religious practice and the conservative vote.[167] The regions where community-type families predominated tend, in contrast, to the egalitarian and democratic. Kinship constitutes a veritable social force that the authorities are bound to take into consideration. These families are characteristic mainly of the south, where feudalism never completely took root, or at least had to respect titular rights that were registered in written law.[168] Even in the elections of 1974, a clear correlation was detectable between the regions where community-type families predominated and those which gave the majority to the left-wing candidate, François Mitterrand. With the disintegration of the community-family in recent years, communism was in many areas welcomed as a substitute for the failing solidarity of the family group. Up until 1978, communism was stronger than other left-wing parties in Provence, Roussillon, Limousin, Berry, Bourbonnais and in Picardy and the Nord department. At that date, it was still making progress in Aquitaine, in the south-west, where the dislocation of the community-family, which set in later here than in other areas, was still taking place.[169] Finally, regions where the nuclear family predominates and society is accordingly much less structured, tend to be politically unstable.[170] Many of these, within the orbit of Paris, became disenchanted with religion at a relatively early date; in the nineteenth century, their sympathy with the philosophy of the Enlightenment made them Republican and they were faithful to the Left so long as it was represented by opportunists or radicals in the liberal tradition or who were attached to the idea of private property. But when collectivist parties began to represent the Left, these areas swung over to the Right. The rhythm of the political evolution in these regions has been much affected by local circumstances and the influence of social leaders and other strong personalities who managed to create clientèles that were dependent upon them. This is an extremely interesting analysis. However, it by no means accounts for all the facts. For example, the western interior and the borders of Brittany, where nuclear families predominated, was an area of remarkable political stability, which remained faithful to both the conservative Right and to Catholicism. The dominant factor here was religion and the family structure seems to have been irrelevant. In the south, Languedoc, an area of community-families, remained right-wing until around 1880, just as Provence (with the exception of the pocket around Marseilles) had been at the beginning of the century, during the Revolution and up until the White Terror and the Restoration, at which point, in the mid-nineteenth century, it switched to support the Republic. In any event, the problem of the origin and significance of these family structures has yet to be resolved. Apart from a few surveys, retrospective studies are not yet far enough advanced to be able to put forward any serious hypotheses as to their origins and the circumstances in which these family traditions became established. All we know is that the nuclear family is of relatively ancient date (earlier than the seventeenth century) in some northern regions such as Normandy and that the complex, multiple family used to predominate throughout a considerably larger area. Between the seventeenth century and the nineteenth, that area shrank considerably in northern Dordogne and since then has become very dislocated in Provence, in the north-western Massif Central, Périgord and Nivernais.[171] But can the predominance of the community-family and the left-wing politics of the north really be traced to the presence of people of Frankish origins?[172] Certainly, where the stem family (*famille-souche*) is concerned, there seems to be some confusion between the effect and the cause. Its distribution today – in Brittany, Alsace, the Basque country and the mountainous south-east of the Massif Central (a region traditionally unaffected by temporary migrations to the towns) – can only be explained as a phenomenon of backwardness and inaccessibility (either geographical or linguistic), typical of such areas. This is a picture that has only crystallized relatively recently, under the influence of the inflexible and strict Catholicism of which this family structure is to a large extent the social expression. In any case, the two factors – the family structure and the deeply rooted religion – have certainly interacted strongly and this is no doubt the most likely explanation for their remarkable coincidence. As for the political changes which took place in the nineteenth century in southern France, in Languedoc and Provence, other factors than family structures may have been at work here. For instance, it has – not unreasonably – been suggested that the Occitan affiliation may be relevant.[173] It is certainly remarkable that the Gavacherie de Monségur, a pocket of *langue d'oïl* on the margins of the Dordogne and the Lot-et-Garonne departments, has always voted more to the Right than the neighbouring Occitan districts.[174] The conservative vote is clearly deeply rooted here and the only left-wing successes must be accounted for by outside influences propagated in the Garonne and Dordogne valleys. The south's late swing to the Left may be understood in just such a perspective. At the time when regional political attitudes began to emerge, during the Revolution, the south had no particular reason to give the changes an enthusiastic welcome. Feudal rights were on the whole rigorously defined; most towns adhered to extremely liberal charters; there were many small freeholdings in the countryside and many ancient privileges had been preserved there. The region could see no reason to support changes whose principal purpose was to render all citizens equal before the law. At this period, community-families and democratic ideals were by no means synonymous with Republicanism – rather with attachment to a distant monarchy perceived as the guarantor of ancestral liberties and privileges. In 1848, it was still only in Marseilles and the southernmost part of the Rhône valley, influenced at an early date by the central models, that Republican ideas triumphed. The south did not swing to the Left until the end of the nineteenth century, when the motivating force of the age ceased to be the centralist Jacobinism of Gambetta, Jules Ferry and Edouard Herriot and was replaced by international socialism which, furthermore, was converted to the notion of decentralization in about 1960. The south's swing to the Left expressed a somewhat confused sense of alienation. It was the reaction of people who felt that they had been left out of the economic race and, having lost their ancient privileges, had lost their prosperity.[175] This political swing above all affected the countryside, a domain of small-scale agriculture which in southern France always votes more to the Left than the towns, which are inhabited mainly by bourgeois and retired people. In the north, the situation is reversed, for there the industrialized towns vote more to the Left than the countryside, where large-scale, prosperous agriculture predominates.[176] But left-wing political sympathies do not rule out other attitudes. In the second half of the twentieth century, alongside them (fig. 67A) one comes across abundant, seemingly contradictory signs of extreme right-wing reactions (fig. 67B and C). All in all, the south seems characterized predominantly by attitudes that are somewhat defiant, whatever their political hue; attitudes which no doubt reflect a certain more or less conscious frustration. That southern defiance also finds other ways to express itself, for instance through its fidelity to the southern accent, the last linguistic resistance, and its mocking scorn for the 'sharp' accent of Parisians (S. and C. Gras, 1982, p. 30).

As can be seen, the interpretation of political attitudes is, on the whole, a remarkably delicate business. The various analyses conducted so far have revealed indisputable correlations, but none of them are of truly general significance. They are so inextricably confused that it is impossible to construct any overall interpretation. The contemporary history of France has fostered a variety of changing oppositions. For a long time they certainly were dominated by the religious conflict; but some may have stemmed from other deeply rooted structures, foremost among them, indubitably, different types of family organization, but also cultural cleavages of a latent or more or less unsuspected nature. And the whole scene was, of course, further affected by an infinite variety of local circumstances and environments.

In the last analysis, the influence of Paris and the models that emerged from it, together with the resistance that these models provoked, cannot account for the entire cultural, religious and political geography of France. Nevertheless, the centre–periphery opposition is easily the major explanatory element.

Cultural action and reaction   337

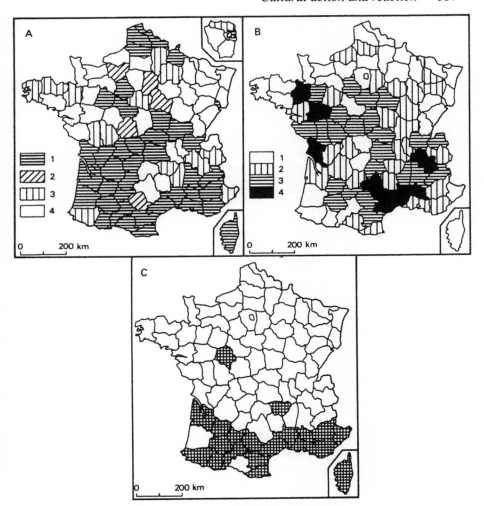

Figure 67. Aspects of the meridional political contest.
A. The cantonal elections of September–October 1967. 1. A left-wing (Fédération de la Gauche Démocrate et Socialiste, Parti Socialiste Unifié, Parti Communiste) President of the general council. 2. A Centre–Left President. 3. A Centre–Democratic President. 4. A Union Nationale Républicaine et Ve République President (after Grosclaude, p. 185 in Lafont, 1971a). B. Map of Poujadism at the legislative elections of 2 January 1956. 1. Under 10% of the votes. 2. 10–15%. 3. 15–20%. 4. Over 20% (Grosclaude, p. 188). C. Votes obtained by J.-L. Tixier-Vignancour in the presidential election of 5 December 1965. Hatched: departmental percentages over 20% (Grosclaude, p. 188).

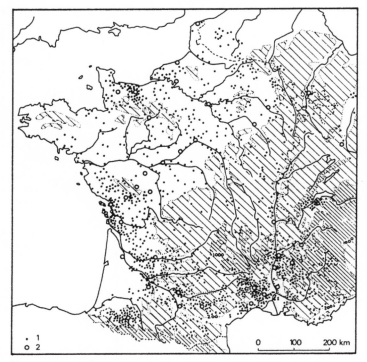

Figure 68. Reformed churches 'set up' in 1562 (after S. Mours, 1957 and Mandrou, 1961, p. 177).
1. Reformed church. 2. Reformed church with several pastors.

Rejections

Over and above these contrasts and oppositions, there were positive rejections. Their localization is not explained solely by the geographical milieu, but it probably was largely instrumental.

Protestantism in the Cévennes

In the late seventeenth century, after the revocation of the Edict of Nantes (1685), an early manifestation of this kind of rejection came with the Protestant resistance in the Cévennes. It took the form of the war of the Camisards, a rebellion which continued for close on ten years (1702–10).[177] The fact that the centre of the revolt was in the mountains reflected the difficulties of establishing communications and royal authority in this intractable region on the south-eastern slopes of the Massif Central. There were many Protestants here, in fact it was the only part of the Massif where, close to the currents of influence which had spread through the plain of Languedoc

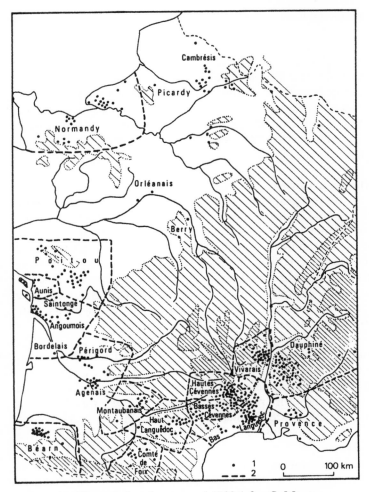

Figure 69. 'Desert' churches around 1788 (after S. Mours, 1957, p. 143, reproduced in Lafont, 1971a, p. 166).
1. Churches. 2. The limits of synodal provinces.

and along the Rhône valley, the Reformation had made considerable inroads in these central mountain areas, the rest of which remained unaffected by French Protestantism (figs. 68 and 69). Protestantism took root here in the Cévennes thanks, in particular, to the presence, within the Alès-Mende-Ganges triangle, of a large body of leather-workers, a group which, here as elsewhere, was among the first to rally to the new ideas and give them a welcome as early as 1530–60, passing them on to a prosperous, sedentary peasant community whose own commitment manifested itself between 1570

340 *Centralization and diversification*

and 1590.[178] At regional level, however, the Protestants were not really in the majority, their numbers and influence being much the same as they were in many other sectors, for instance in the plain of lower Languedoc, the Garonne valley, the Béarn region, the Atlantic seaboard between the Gironde and the Loire, the Rhône valley and lower Normandy. It was the extreme harshness of the natural setting that here, and only here, gave them the power to mount a prolonged resistance. In the lower Alps of Dauphiné, an equally propitious area from the point of view of resistance, the density of the Protestant population was not great enough to sustain a revolt of comparable force.

The revolts in the west: Vendée
The influence of the geographical setting is equally clear in the *chouan* uprisings of the west and in the Vendée revolt of 1793. Here, it was a case not of mountainous country but of a much more human and agricultural landscape, the *bocage* (meaning 'wooded, bushy district'): a region with a dispersed settlement pattern, divided up by thick hedgerows and cut across only by deep paths, often converted into quagmires and virtually inaccessible to any kind of speedy traffic and to all large-scale movement. The area was difficult for the central authorities to control, resistant to new ideas and well-suited to partisan warfare.

Yet on their own these factors do not offer an adequate explanation, for they recur throughout more or less the whole of inland western France. They do make it possible to understand the phenomenon of *chouannerie* in Maine and Brittany. But the Vendée revolt was a different matter. Here, and only here, it involved the entire population. *Vendée militaire* was defined as an area of about 10,000 square kilometres. To the north, this was bounded by the stretch of the Loire between Saint-Nazaire and Ponts-de-Cé; to the east by a virtually straight line running from Ponts-de-Cé to Parthenay; to the south by a more sinuous line which reached the Atlantic coast at Saint-Gilles-Croix-de-Vie. The area comprised 700 parishes belonging to three different provinces (Anjou, Brittany and Poitou) and four departments (Maine-et-Loire, Loire Inférieure, Vendée and Deux-Sèvres). The concept of 'Vendée', the name given to the movement, to give it a common identity, was an invention of the central government (produced on 17 March 1793) and it seems definitely reductivist. But at any rate, the area is clearly defined. It was to be defined even more clearly by the repression carried out by the 'infernal columns' of 1794, which led to the massacre of about 15 per cent of the population of 800,000 inhabitants and the destruction of nearly 20 per cent of their property.[179]

Why did the Vendée revolt happen and why did it happen here? The most recent historian of the revolt,[180] to whom we are indebted for the above first statistics of the repression, reckons that to be an 'ill-conceived question'. At this time, the spring and summer of 1793, the whole of Normandy and Brittany was in a state of unrest, as was the whole of the eastern and south-eastern Massif Central; over the national territory as a whole, about thirty departments were no longer obeying the central government. Its authority was, in truth, unchallenged only in the north-east and the Paris Basin.[181] Only the extreme vigour of the government's reaction to the rebellion and the overwhelming repressive action taken by the authorities prevented open revolt from spreading throughout the country. Vendée happened to be an area in which it could be contained. But this is still merely to sidestep the problem. Why was it that, in geographical conditions which did not differ significantly from those of the rest of inland western France, it was here, and only here, that the great revolt could not be blocked? The question is still far from resolved: the Vendée episode remains to a large extent an enigma.

However, we are one step closer to an explanation, thanks to an improving understanding of the phenomenon. In Vendée, the rebellion was not just massive but total. It involved every social category, the town-dwellers as well as the peasants, who placed at their head a sometimes rather unwilling aristocracy. Here, particular groups of craftsmen, who elsewhere were generally open to new ideas, do not stand out from the rest of the population. In the neighbouring region of Sarthe, the weavers remained loyal to the Revolution and *chouannerie* was an essentially peasant phenomenon. In Vendée, no correlations can be found between people's professional activities and their political views. The weavers here were as royalist as everyone else.[182] The unity of the region seems to have been 'preformed'. The revolt itself, however, was complex. In some ways it harked back to the Ancien Régime. It broke out at parish level, in rural communities which had defied conscription in March 1793, by which time the unrest had already been simmering for some time, ever since the eviction, two years before, of all priests who refused to take the oath. It was above all a movement to defend rights, a struggle against an oppressive central authority which, having evicted the 'good priests', was now demanding the blood of the young men. It has been described as a movement 'more religious and autonomist than political'.[183] From that point of view, it certainly was the last revolt of the Ancien Régime. On the other hand, its legitimacy was confirmed by the very principles of the Revolution, for it was in tune with the Declaration of Human Rights. Thus, paradoxical though it may seem, it was perfectly understandable that, speaking of Vendée before the Nantes Assizes of October 1832,[184] Captain Galleran should declare: 'These people are basically Republican'. Besides the geographical framework had nothing to do with the Ancien Régime. The Vendée uprising may thus be seen as the first regional movement of rebellion of a resolutely modern nature, long before the Paris Commune in 1871 and the twentieth-century revolts of the

Languedoc vine-growers and the miners of the north, which were of a trade-unionist and political inspiration.

Having got so far, it is possible to advance explanations for the geographical localization of the movement. The most convincing working hypothesis is that which emphasizes the role played by certain missions promoting a Catholic renewal, which were launched in the seventeenth and eighteenth centuries, particularly those of the Monfortains (started by Grignion de Montfort).[188] These were particularly active in Vendée (seventy-one missions and retreats between 1740 and 1779 in the territory of the present Vendée department alone). By welding the people to their priests, they rendered them particularly sensitive to the persecution of those who refused to swear the oath. By playing upon the emotions of the people and the 'charisma' of the priests and through their methods of bringing the people together, the missions no doubt developed a feeling of social cohesion among the dispersed populations of this wooded region, producing favourable conditions for an explosion which, though totally uncoordinated, soon discovered a common purpose and common ideals. Before the Revolution, religious practice and faith certainly do not appear to have been unusually strong in the Vendée countryside, as compared with the rest of inland western France. It was the Revolution and the repression that strengthened them. But in 1789, there was certainly a deep sympathy between the parish priests and the grievances of the common people.[189] The ground was prepared for a collective explosion.

But in truth that is simply to pose the problem in a different form and push it further into the past. For the Montfortain missions spread throughout the whole of western France – Aunis, Poitou, Anjou and Brittany; so why were they particularly successful here? It is certainly remarkable to find the missionaries' records contrasting the welcome that they received in some regions, such as these wooded areas, with the more reserved attitude of others, such as the plain of Poitou, the river valleys and the vine-growing districts.[187] The latter indeed turned out to be the areas which did not join in the revolt and later, in the nineteenth century, were to produce a left-wing vote. This turns the Vendée revolt into a problem of collective psychology and religious sociology. It seems reasonable to suppose that the missions made relatively little impact upon areas that were already open to outside influences or accustomed to community life by reason of their grouped settlements (the river valleys, the plain of Poitou). But among the scattered populations of the wooded areas of Vendée, which in 1793 still was only crossed by three royal highways (one of which was unfinished), the missions must have answered a deep need to find social expression and to affirm the identity of the group. In the last analysis, the geographical explanation does seem to be the fundamental one: this was a region which was still particularly marginal and isolated. However, this is overlaid by an historical circumstance, namely the massive intervention of a missionary campaign which opened up perspectives of solidarity and transcendence. The Vendée seems to have been the product both of the region's inaccessibility as an isolated wooded region and of the work of a large number of Montfortain missions, a combination which does not reappear anywhere else in anything like the same intensity.

However that may be, the moral unity of this region was there to stay. The concept of Vendée grew stronger as it was put to the test and was sustained by its own past. It has been said that 'whereas all the provinces became departments, Vendée is the only department that turned into a province'.[188] The individuality of the region was to be a lasting one.[189] Already at the end of the nineteenth century, historians and political theorists were pondering the matter, still unable fully to understand the essence of this phenomenon. Basically, it remained a mystery to them. In 1889, Emile Beaussire acknowledged that fact and devoted a lengthy analysis to 'the spontaneous formation of a province at the end of the eighteenth century: Vendée'.[190] The political character of the region was indeed to remain stable. Up until the Second World War, Vendée was overwhelmingly anti-Republican, essentially legitimist and monarchist, but occasionally rallying to the Bonapartism of the plebiscite, which had brought the region amnesty and religious peace in 1799.[191] The consequences of the marginalization of Vendée were to be spectacular. Seemingly paradoxically, it adapted remarkably early to economic progress. In this pious area, where the birthrate was high, the demographic pressure was considerable. The first response to this was emigration. This took the particularly dynamic and original form of a rural recolonization of the abandoned land in the Aquitaine Basin, which was going through a phase of depopulation, and above all in the neighbouring Charente department. This movement got under way in the 1880s, following the phylloxera crisis, when the people of Charente, ruined and incapable of cultivating anything other than vines, sold up their land cheaply and moved to the towns. It is estimated that of the 105,000 people who left Vendée between 1870 and 1950, 40 per cent settled on the borders of Poitou and Saintonge (half of them in the department of Charente, one fifth in the department of Vienne, 15 per cent in Charente Maritime and a few in Deux-Sèvres), where the populations of many villages thus became 50–80 per cent Vendéen. By 1895, the Vendée emigrants were reaching Périgord, the Gironde and the Garonne and Adour areas, where Vendéen agricultural workers took a relatively upward step on the social ladder by becoming farmers paying rent in kind, although these distant regions were less attractive to them than those nearer home.[192]

Within a framework organized by an extremely active clergy, and with the benefit of state aid provided for resettlement, the success of the rural expansion of the Vendée people underlined the Vendéen courage and spirit

344 Centralization and diversification

of enterprise that had been forged in adversity. In particular, the emigrants from Vendée made a decisive contribution to the development of dairy farming and the production of butter in the Charente area at the end of the nineteenth century. But Vendée itself was also receptive to development in every form. Abandoned and spurned by the Republic, or at any rate believing itself to be so, Vendée sought its salvation in its own strengths. It became a remarkable centre of innovation not only in the agricultural sphere, with a strong framework of trade-unionism and under the aegis of local traditional leaders and the Church, but also in the industrial sector: many factories now sprang up in this rural area, particularly in Choletais, founded on the old and widely established textile industry and the traditional skills of its labour force. The result of all this activity is an amazing socio-political scene, where an extremely high proportion of workers in the population is combined with a massive right-wing vote: for example, in 1981, the canton of Saint-Fulgent, close to Cholet, with 40 per cent of its population employed in industry, was one of the constituencies which produced the most decisive results in favour of V. Giscard d'Estaing.[193] The ideal of community solidarity and the exaltation of human effort are somewhat reminiscent of certain manifestations of the Protestant economic ethos, although the latter places far more emphasis upon the individual. Political conservatism and social dynamism, religious traditions and economic progressiveness thus all came together in the astonishing synthesis of a Vendée faithful unto itself and at the same time bubbling with life.

9
The economic differentiation of space

The gigantic growth of Paris and its suburbs and the general movement towards urbanization (see chapter 10) created huge consumer markets which needed to be supplied by the countryside. Meanwhile, the development of communication networks in the shape of first roads, then railways, was creating the conditions that would make it possible for these markets to be supplied from distant parts of the territory.

The new conditions were created, essentially, in the nineteenth century in two phases. In the first half of the century, the transport of freight in bulk was ensured; then, in the second, with the development of the railways, means of transport were speeded up. The eighteenth century had established a network of major highways which made efficient centralization possible. But communications were slow, despite the improvements introduced in the last third of the century. (Between 1765 and 1780, the duration of the journey between Paris and Marseilles was cut by one third, from twelve days to eight, and that of the journey between Paris and Bordeaux by a half, from twelve days to under six.)[1] But above all, although many new links were built between the major provincial towns, the network of royal highways was limited to roads between only the largest of towns, except in north-eastern France (see above, pp. 153–7). After the revolutionary and imperial periods, when attention was concentrated upon routes of military or strategic importance, the great improvement in communications began under the Restoration. Some 29,000 kilometres of national and departmental roads were built in the thirty years following 1815.[2] The law of 21 May 1836, which obliged the communes to maintain local roads, in effect, over the years that followed, made virtually the whole of rural France accessible by road. As for rapid transport, from the middle of the century on, an explosion of activity speedily developed the railway network: by 1851, 3,554 kilometres of track had been laid; by 1870, 17,440; and by 1900, 36,800. Between 1870 and 1880, a national market economy came into being throughout the land.

A new map of land use: the development of agricultural specializations

This new infrastructure made it possible, in the first place, to develop the agricultural possibilities peculiar to each natural environment and progressively to diversify a countryside which, up until then, had remained largely homogeneous, or had been differentiated only by major contrasts of climate (see above, pp. 216–23).[3]

Transformations in the agriculture of southern France

It was in the south that the new tendencies first made a definite impact, when it became possible for its produce to be transported to distant destinations. The polyculture of Mediterranean France had, it is true, begun to evolve long ago, in order to serve a number of local markets. Thus, as early as the beginning of the nineteenth century, lower Provence was beginning to diversify and specialize, particularly within the orbit of the Marseilles market. Some areas concentrated upon wheat (for example, the Arc basin and the plateau of the upper Touloubre); others on viticulture (the Beausset basin); others on the cultivation of olives for oil (the lower Touloubre and the banks of the Etang de Berre). Upper Crau (around Arles) had been receiving irrigation from the Durance and the Canal de Craponne ever since the sixteenth century (from about 1550). In the late eighteenth century and the early nineteenth, the Canal de Boisgelin and the Canal des Alpilles brought it new supplies of water. This led to the creation of many artificial meadows, alongside the cereal fields, and to the massive production of Crau hay, highly valued and sold throughout southern France as fodder for horses. Finally, in the Marseilles basin, all around the town, polyculture became increasingly diversified, incorporating large market gardens for vegetables and also pig farms. The traditional form of agriculture persisted only in the Var department of lower Provence.[4]

However, the most spectacular developments were those designed to keep distant markets supplied. Large agricultural areas were devoted to specialized produce destined for other parts of the country. The appearance of these was a direct consequence of the organization of rapid transport for the delivery of perishable fruits and vegetables. From the mid-nineteenth century onward, complexes of market gardens and orchards thus sprang up all over southern France. Most developed out of plots in the neighbourhood of large towns, which had been irrigated in the past. Until now, such plots had been limited strictly to peripheral belts around the towns and had been designed to supply purely local needs, but with the arrival of the railways, they were now greatly expanded. Thus, in Roussillon, where a belt of old market gardens surrounded Perpignan and the *hortolas* ('gardeners') had always played an important role in the life of the city, the great expansion dates from the arrival of the railway in 1858. The area of the Saint-Georges gardens increased from 186 to 913 hectares in the space of thirty-six years. This was the starting point for the great Roussillon fruit- and vegetable-producing region of the *regatiu*, the irrigated land along the valleys of the Têt and the Tech rivers. The development of the market gardens of the Comtat Venaissin dates from the same period. At the end of the eighteenth century there had been horticultural suburbs around all the most prosperous towns, such as Avignon, Carpentras and Cavaillon, and gardening techniques had been perfected at an early date. But there had been no really large-scale commercialization. Irrigation techniques in the papal possessions were very old-fashioned, way behind the economic developments in neighbouring Provence, where the waters of the Durance were abundantly used at this period. Essentially, the economic leap forward dates from the introduction of railway transport in the decade between 1865 and 1875, which was complemented by the organization of a network of irrigation canals in the last third of the nineteenth century. On the Côte d'Azur, the cultivation of flowers also took off from 1860 onwards, catering for external markets. It grew out of the local sale of flowers to the aristocratic clientèle of the very first wave of tourism, when visitors would settle on the Côte d'Azur for the winter.[5] Finally, at the end of the nineteenth century, the development of fruit and vegetable cultivation in the mid-Garonne area was also linked with the availability of railway transport. In the initial stages, it supplanted the industrial crops of the old system of polyculture (mostly plants producing dyes and textiles). The only crop that survived from the old days was tobacco, which had been introduced in the seventeenth century along the communication route constituted by the Garonne and had been revived at the beginning of the nineteenth century. From 1850 onwards, vegetable cultivation was introduced in many different forms, constantly adapting to market demands (tomatoes from Marmande; peas and green beans from the region of Villeneuve-sur-Lot; onions, asparagus and cauliflowers from Agenais; artichokes and melons from the lower Tarn). The cultivation of fruit began a little later on. The starting point for the evolution towards specialized orchards was provided by an old local speciality, plums. Plum trees had long been planted on the 'greenhouse' slopes of Agenais and the Lot valley (around Villeneuve) and sent off to be sold commercially, by way of the road along the Garonne valley, in the form of dried 'Agen prunes', which kept well even before the introduction of rapid transportation. The cultivation of plums came into strong competition with Californian production around the end of the century, but a wide variety of orchards (cherries from lower Quercy, dessert grapes from Moissac) were multiplying rapidly between 1880 and 1900, taking the place of the crops of the traditional system of polyculture, in particular the vines which had been abandoned after the phylloxera disaster.

348 *Centralization and diversification*

The development of the vineyards of Languedoc

The establishment of large-scale viticulture in the Languedoc Plain was a much more complex historical process,[6] in which the revolution in transportation methods through the creation of the railway network was but one component, if an essential one. It is possible to distinguish a sequence of phases.

Before the end of the seventeenth century: hillside vineyards designed to supply local needs

In the Middle Ages and at the beginning of the modern period, the vines of Languedoc were, as we have seen (pp. 229–30) reserved for local consumption. The vines grew on the hillsides: at Lansargues, in the plain between Lunel and Montpellier, a survey of 1653 shows 130 hectares of vines compared with 109 hectares of pasture-land and 3,000 hectares of ploughed fields. The vines of Languedoc, planted in amongst olive groves, were relegated to the slopes of stony hills. This was small-scale cultivation, just one element in the Mediterranean system of polyculture. In the plain itself, vineyards took up no more than a few stony plots. Although Languedoc wines occasionally made their way to the fairs of Champagne or even as far afield as England,[7] exports were virtually non-existent. The trade of Montpellier was principally directed at the Arab countries, and these were not interested in buying wine.

The first extension of the vineyards: the beginnings of large-scale trade

The first stage in the development of the Languedoc wine trade began at the end of the seventeenth century, when the port of Sète was opened (1666) and the Canal du Languedoc was built, linking Sète with Toulouse (1681). Some years later, the harsh winter of 1709 caused many wine merchants to look to Languedoc for supplies. The expansion of the vineyards began on the shores of the Etang de Thau, whence it spread to the stony hillsides of the plain. Around Saint-Drezy, to the north-east of Montpellier, the area devoted to vines increased from 17 hectares in 1678 to 175 hectares in 1744 and 270 hectares in 1789. But the vineyards were still mostly confined to the hillsides. The plain produced a few mediocre wines which in bad years were distilled and turned into spirits, for which there was considerable commercial demand. This first expansion of the Languedoc vineyards in fact seems to have been essentially linked to the production of spirits or even of *trois-six* (spirits which undergo a second distillation in order to reduce transport costs). In 1791, the port of Sète was exporting 41,000 hectolitres of ordinary spirits and 164,000 of *trois-six*, a quantity which corresponds to about one million hectolitres of wine.[8]

After the Revolution, a new phase in the expansion of the vineyards began, at the expense of communal land, which was now divided up. In the Hérault department, for example, the area devoted to vineyards increased from 63,000 hectares in 1788 to 103,000 in 1834. This expansion involved a reduction in wasteland but the essential features of the Mediterranean polyculture underwent little change. The area devoted to cereal production likewise increased. In 1838, Hérault was still only the third largest wine-producing department, coming behind Gironde and Charente-Inférieure. It was followed by Dordogne and Charente, with Gard in sixth place. In 1830, Narbonne was still a wheat-market.

The appearance of vineyards on a massive scale (1830–75)
The decisive turning point came between 1830 and 1850 and the new trend was accentuated under the Second Empire, when the railways came into use. It was during this period that Languedoc definitely turned to the mass production of ordinary table wines for the popular urban clientèles, which were growing. Many high-quality vines (such as Terret noir) were replaced by higher yielders (Aramon in the plains, Carignan on the hillsides) and hybrids which gave a large yield and produced very dark wine (Alicante-Bouschet) were also developed. The yield increased from 15 to 20 hectolitres per hectare in the eighteenth century, to 45 hectolitres per hectare. The area covered by vineyards in Languedoc increased from 270,000 hectares in 1838 to 380,000 hectares in 1863, at which point their expansion was more or less completed. But although production had increased enormously, the nature of these vineyards had still not totally changed. Two-thirds of the vines still grew on the hillsides and lower Languedoc still produced a high percentage of good-quality wines. What was new was that more and more of the poorer wines, which used to be distilled, were now sold for daily consumption and vineyards of mediocre quality had come into being in the plain, alongside the older hillside vineyards. Cheek by jowl with the small, traditional properties, more and more large properties were being developed, as the bourgeois from the local towns resolutely turned their backs on the industrial investments of the old days (such as textiles) and on commerce, in order to devote themselves essentially to viticulture. Throughout this period, however, traditional elements still survived alongside the new.

The phylloxera crisis and the definitive establishment of mass vineyards
The final agent of change was the phylloxera crisis, which resulted in total ruin for the region between 1874 and 1886 (the vineyards of Hérault shrank from 222,000 hectares in 1874 to 47,000 in 1883). The battle against the insect involved flooding, which subsequently led to the reclamation of the marshland of the Petite Camargue and the lower Aude plain, and to planting in sandy soils, which brought the coastal strips into use. The vineyards were then reconstituted by grafting on to slips from America, starting with the

riparia. This was ill-adapted to limestone soils and intolerant of the thin soils of the stony hillsides, but took more kindly to the deep, fertile soils of the plains. Only later did the *rupestris* come into use, together with hybrids adapted to stony soils. This adaptation of American slips, which produced a large yield in the low-lying plains, brought about a definitive transformation in the appearance of areas under vine cultivation and also in the social structures of the region, since the major undertaking of adapting the marshy plains to viticulture could only be tackled by large-scale landowners or limited companies with sufficient funds for the organization of drainage and irrigation. At the beginning of the twentieth century, the reconstituted Languedoc vineyards had attained their maximum extent of 440,000 hectares, that is to say they covered 15 per cent more land than they did just before the phylloxera struck. But the centre of gravity had shifted. Almost all the hillside vineyards had disappeared and production was now concentrated in the low-lying plains. These were now organized on a massive scale, with a high yield of wine of poor quality within a social structure balanced between large wine producers and small-scale vine-growers who only survived thanks to to the cooperative movement, which made it possible for them to develop highly technical methods that ensured a satisfactory yield. Changes in the hierarchy of these wine-producing departments reflected these geographical changes. The Hérault department, with 190,000 hectares, had not regained its earlier maximum area of vineyards, while the Aude had increased the area devoted to vineyards by 60 per cent and was now in second position as a producer of wine.

Repercussions on the vineyards of northern France

The development of the southern vineyards catering for a mass-market resulted in the irremediable decline of most of the vineyards of northern France. Only a switch to the production of quality wines could have saved them. But in the seventeenth and eighteenth centuries, the increasing consumption of wine among the popular classes of large towns, particularly the great city of Paris (see above, pp. 261–3) had, on the contrary, led to declining standards in the quality of their wines. Consequently, all the vineyards of the Ile-de-France and the Orléans region, which used to supply Paris, disappeared early on in the nineteenth century when faced with competition from the wines of southern France.

The pattern of events was similar, if slightly delayed, in the case of the vineyards of the Côtes de Meuse and the Côtes de Moselle, in Lorraine. These had been important producers up until the eighteenth century, while the State of Lorraine was still independent. In fact, as the wine produced here was of a quality superior to that produced around Paris, the vineyard continued to be profitable until late in the nineteenth century and even

The economic differentiation of space 351

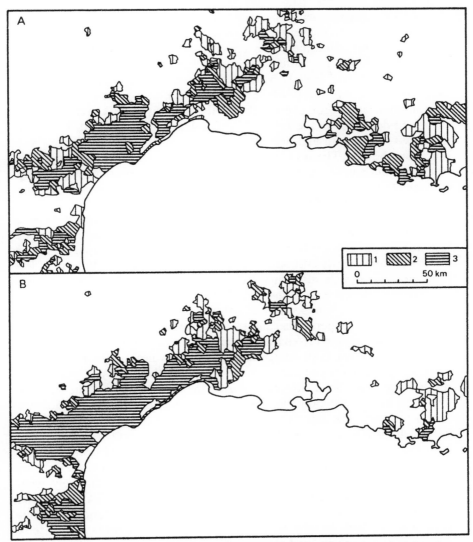

Figure 70. The development of the Languedoc and Roussillon vineyards (after the maps by Perpillou, 1977a and b).
The proportion of vines in relation to the total surface area: 1. 13–22%.
2. 22–31%. 3. Over 31%. A. First half of the nineteenth century. B. First half of the twentieth century.

enjoyed a fleeting renewal of prosperity in about 1880, when the phylloxera crisis had already hit the south, stimulating demand, but had not yet reached Lorraine. At this point, these vineyards even spread down to the plain at the foot of the hills. But thereafter a general decline set in. In the Moselle valley, where vineyards had already been on the retreat, faced with pressure from the expanding towns, even before the phylloxera disaster, they disappeared as industrialization took over. In the northern sector of the Côtes de Meuse, they were destroyed in the 1914–18 War and were never revived. Vineyards did survive on a thousand hectares in the Toulois region, more or less as a relic from the past. Thanks to an improvement in quality and the recognition of an *appellation controlée*, they are still managing to produce wine today. But essentially, that is due to the absence of any alternative solution in this, as yet, unindustrialized region.

The vineyards that survived in the north were first and foremost those of the great wines, which had turned to high-quality production at an early date (see above, pp. 231–2): Burgundy,[9] for example, and Champagne, which have been consistently successful down to the present day.[10] But other, less prestigious vineyards also survived, in Alsace for example, despite the particularly adverse conditions here.[11] Advantaged by favourable climatic conditions (for the Rhine valley constitutes a most positive anomaly, with its high summer temperatures) and by their exposed situation, the vineyards of Alsace also benefited from the German annexation from 1871 to 1918, for this protected them from competition from the wines of southern France, integrating them instead into a state in which they enjoyed one of the southernmost positions. However, during the German period, the Alsatian vineyards had considerably increased their production, precisely in order to satisfy the high demand, and had moved down to the plain, beyond their natural domain in the Vosges foothills. After the phylloxera attack, they had been reconstituted with hybrids and production had evolved towards quantity rather than quality. Logically, they might well have gone under. But they saved themselves and made a comeback thanks to an extremely well-organized sales promotion programme, within the framework of a social system which encouraged cooperation between, on the one hand, ambitious pilot schemes launched by large-scale wholesale vine-growers, who took charge of the publicity and image-making side, and on the other a large number of small-scale vine-growers (65,000 in the mid-twentieth century), producing 800,000 litres from a half-hectare, on average, who were responsible for the stability and quality of the harvest. Nevertheless, the vine-growing area had shrunk. Vineyards have completely disappeared where they constituted but one of a number of elements in a polyculture system. Only those which were already sufficiently specialized managed to survive in the evolution towards quality.

Complex social conditions of a similar nature explain the persistence of pockets of quality vine cultivation in the south-eastern Paris Basin (Chablis, Irancy, Saint-Bris in lower Burgundy; Pouilly-sur-Loire in Nivernais; Sancerre, Menetou-Salon, Quincy and Reilly in Berry), whilst the large vineyards of lower Burgundy (Auxerre, Tonnerre, Joigny) have disappeared.[12] In the nineteenth century, quality vine cultivation survived, thanks to meticulous methods which ensured high profits, wherever it constituted the fundamental livelihood of a deeply rooted society of vine-growers which remained faithful to this virtually exclusive activity. But it faltered and then disappeared in places where it was gradually swamped amid farms organized according to a polyculture system. Here, the owners left the vine cultivation to negligent employees, thereby paving the way for procedures which made for large quantities of cheap wine. In the last analysis, the vineyards that survived did so thanks to the efforts and merits of the individuals involved.

Dairy farming and livestock rearing

By increasing the demand for dairy produce and meat from the countryside, the spread of urbanization and the rising standard of living altered the map of pastoral farming just as much as or even more than that of agricultural production. Along with the appearance, in southern France, of vast areas of market gardens and orchards and of vineyards catering for mass production, the development of regions specializing in products from cattle and other animals has been a fundamental feature in the contemporary differentiation of the territory.

The selection of pastoral areas
In this process, factors of climate and soil were, of course, operative, but these were modified by demographic circumstances.

The first type of natural environment obviously favourable to the raising of animals was constituted by the wet mountains, where the presence of cereal cultivation was a nonsense dictated, in the traditional economy, by the absence of easy communication routes in these areas. Cereal cultivation has by now lost considerable ground in all the northern Alps, the southern Préalpes, the Jura, the Vosges and the western, central and southern parts of the Massif Central, as the revolution in fodder production in the valleys has assured the bases of a pastoral economy, sometimes complemented by a removal to Alpine pastures in the summer. Where rainfall is sufficiently heavy, limestone plateaux have also been pressed into service for this purpose. The cheese-making economy, based on the ancient institution of village *fruitières* which, by pooling milk in common make it possible to manufacture huge cooked cheeses (such as 'Gruyère'), took over virtually all the Jura plateaux in the nineteenth century. The Grands Causses of the southern Massif Central, formerly wheatlands, also switched to pastoral

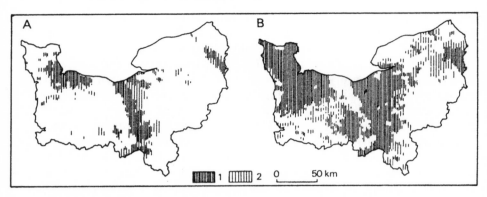

Figure 71. The development of grassland in Normandy (after Perpillou, 1971, pp. 344 and 347).
1. Grassland covering over 45% of the surface area. 2. Grassland covering from 25 to 45% of the surface area. A. First third of the nineteenth century. B. Last third of the twentieth century.

farming at about this period, concentrating on sheep and goats, within the orbit of the cellars of Roquefort, which had traditionally specialized in the veining and maturing of cheeses.

In the sedimentary plains and basins, where constraints due to high ground disappeared, the essential principle of differentiation was the nature of the soil. Clay inner lowlands and outcrops of impermeable soil generally evolved into pastureland. Thus, the hollows surrounding the Massif Central (Boischaut and Val de Germigny) and Morvan (Bazois, Auxois), still cereal-producing areas under the Second Empire, became either natural or artificial meadows and, from the last third of the century onwards, were devoted to the fattening (by the *embouche* system) of Charolais cattle destined for the meat market.[13] Lorraine, mostly devoted to dairy products, essentially evolved according to similar principles. The Vosges foreland, the Saulnois and the Pays des Etangs became meadowland, whilst the sandstone and limestone regions continued to produce cereals as their main crop. While lower Normandy turned predominantly to dairy produce and meadowlands, the Pays de Caux, on the opposite bank of the Seine, which is just as well watered but possesses silty soils, remained essentially devoted to cereal cultivation, as did the Campagne de Caen, with its Jurassic limestone soil (fig. 71).

But other factors must also be taken into account – human ones: the structure and capacities of the population and the general level of economic development. Until the first third of the twentieth century, the limestone plateaux of Châtillonais, in the south-eastern Paris Basin, were devoted to a system of cereal production, mainly wheat and oats, together with exploitation of the forests and the extensive raising of sheep for wool. But between the two world wars, this area went over to artificial meadowland for dairy farming and the production of butter and cheese (of the 'Gruyère' type), for the most part in cooperative dairies. There seems to be a psycho-social explanation for this conversion. The collapse of the metallurgical industry in the plateaux, between 1850 and 1870, had led to a massive exodus of rural worker-peasants. The land had been reorganized into large estates, most of which were owned by former master-blacksmiths. The moving force behind the conversion was the industrial tradition of these landowners, with all that this implied in openness of mind and technical expertise, the effect of which was to make them eminently adaptable to change.[14] Conversely, but on a far greater scale, in Brittany, the persistent rural overpopulation, general cultural backwardness and remoteness account for the continuing cultivation of cereals here until a much later date. For a long time, Brittany remained economically retarded compared to Normandy, which was far more open to the Parisian market. In the early twentieth century, Brittany was still an area of cereal cultivation which stood in sharp contrast to the wooded grassland of inland western France and Normandy; and this situation was to remain unchanged until the development, in a completely different economic context, of factory farms, after the Second World War.

Systems of pastoral production
The effects of these geographical limitations were never totally determining. There thus appeared in natural and human environments with particularly strong human characteristics systems of pastoral production, which rapidly freed themselves from dependence upon geographical factors and expanded considerably. Typical of such expansions is that of the production of Roquefort cheese. At the end of the eighteenth century, supplies for the cellars came from no further afield than the Causse du Larzac and the avant-Causse de Saint-Affrique. By the Second Empire, the dairy industry had established itself throughout the Causse Noir, the Causse Rouge and the Causse de Séverac and was infiltrating the Causse de Sauveterre. By the early twentieth century, the Causse Méjean and the western Causse de Sauveterre, followed by the 'vallons' of the Lot had also been absorbed into the Roquefort zone. The expansion was even to take in areas beyond the Causses. As early as 1875, Camarès had been taken over, followed by Lévezou; and in the 1880s, the rocky Ségala was reached. During the same period, the Société des caves et producteurs réunis (Company of associated cellars and producers), founded in 1842, which by 1881 was responsible for two-thirds of production, from the 1880s onwards set about rationalizing the manufacture of the cheese, creating a network of dairies to collect milk from the peasants. Thus organized, after the First World War the system spread widely throughout southern France, incorporating the Monts de Lacaune and

Albigeois, the Landes and the Basses-Pyrénées, the central Pyrenees and even Corsica. In similar fashion, the cheese-making economy that developed on the plateaux of the Jura took over the higher ridges and then spread extensively over the plateaux of the upper Saône. Yet another example is provided by the stock-raising of animals for meat in the lower rocky plateaux of the northern Massif Central. From Limousin to lower Morvan and Charolais, taking in Basse-Marche and Combraille in Bourbonnais, a system was developed which almost completely dominated the area: herds were raised in the open air throughout the fine weather, in enclosed meadows which were left as grassland for four to six years at a stretch and then used for the cultivation of cereals or cover crops for one year. The herds consisted of cattle which in the winter were fed in cowsheds, on hay, grain and root vegetables. The animals were sold exclusively for meat, usually at the age of two or three, or as calves which were then sent off to the meadows for fattening. By the eighteenth century, this system was solidly established on some estates in Limousin and it spread from west to east along the north of the Massif Central throughout the nineteenth century. By the end of the century, it had taken over the peripheral clay valleys where the calves born on the low, rocky plateaux were taken for fattening. But at the same time, it was also gradually taking over the upper areas of the Massif Central, even those higher than 500 to 600 metres: the Monts d'Ambazac, the Montagne Limousine, the hills of Marche and Combraille in Auvergne, the Montagne Bourbonnaise and Haute Marche. The system included a number of other forms of animal-raising too, such as the production of veal for butchers, pig-farming and even dairy-farming. The reason for these developments was that the greater the altitude, the shorter the growing season became and the longer the animals had to be kept in their sheds; as fodder supplies dwindled, the difficulties of feeding them increased. Some areas accordingly began to specialize in simply producing young animals, others in the system of herd-raising typical of the volcanic mountains of western Auvergne, where the animals were seasonally taken to graze in the high mountain pastures. But the boundary lines between the different variations remained vague and the transition between the two systems was very gradual because of the way that some of these farms were organized: those which used large quantities of fertilizer were able to adopt the methods of the lower regions even at high altitudes.[15]

The establishment of different breeds of animals
The development of these pastoral specializations was reflected in and based upon the establishment of different breeds of animals in different areas, a process which in France began to be methodically developed in the 1860s. Actually though, the constitution of groups of particular types of animals in particular areas, as a result of their adapting to natural homogeneous

environments and particular local farming practices for particular purposes, dates from much earlier. By the end of the eighteenth century, almost all the breeds of cattle known in France today were, to that extent, already to be found in the territory,[16] but their individual characteristics were not deliberately chosen and bred, for since all were used for multiple purposes, all were generally expected to perform in a number of different ways: as producers of milk, meat and also new stock.

The tentative specialization of breeds began with horses, the noble beasts *par excellence*. Because of their military role and their importance in ensuring rapid communications which facilitated the maintenance of the authorities' control over the country, those authorities took a special interest in horse-breeding and this led to the early development, in the eighteenth century, of an interregional horse trade. It has been shown that the Percheron breed of horses was already established by the late eighteenth century. A division of labour had already spontaneously been introduced between the wet Perche hills, which were reserved exclusively for mares in foal, and neighbouring Beauce, where the foals would be sold at an early age and, having been fed on its hay and grain, grow into powerful adult draught or carriage horses, before returning to their native Perche to work there as stallions. Against the official doctrine of the period, which recommended mixing in 'fresh stallions' from outside, de Fontenay, the inspector of stud farms, who held the post from 1766 to 1780, preserved the original breed of the local horses by applying a method of consanguineous reproduction.[17] However, this case was probably exceptional, possibly unique. At the end of the eighteenth century attempts were made to modernize sheep farming. At the instigation of Trudaine and with Daubenton in charge of the project, Spanish Merino sheep with exceptionally fine wool were introduced into France, the most spectacular episode in this operation being the establishment of the royal flock of Merino sheep at Rambouillet in 1786.[18] In the nineteenth century, between 1820 and 1860, attempts to improve the French stock of sheep involved the introduction of breeds from Britain, most of which had been established there since the eighteenth century, chief amongst them Durham sheep, which were particularly rapid meat producers.[19] Attempts at improvement such as these were systematically promoted by competitions for animals bred for meat. (The first regional competition took place in Lyons in 1842; and from 1844 on, national competitions were organized in Poissy, close to Paris.)[20]

Various deficiencies and difficulties of adaptation encountered with foreign animals were gradually recognized, leading to the suspension of these experiments. But the stimulus for improvement remained and eventually it led to new methods. When the revolution in methods of transportation massively increased the commercialization of animal products and specific regional specialities became distinguishable, it was realized that the only way to obtain stock perfectly adapted to those specialities was to improve and select local breeds. Thereafter, during the second half of the nineteenth century and the first two decades of the twentieth, the definition of particular breeds proceeded apace and genealogical records of the various breeds began to appear.[21] Some works on bovine breeds were already being produced by 1856 (black-muzzled Gascon cattle) and 1864 (Nivernais-Charolais cattle); but most of them appeared between 1880 and 1901. (Out of the twenty-five genealogical studies of native French breeds of cattle produced between 1856 and 1922, fifteen were published between those two dates.) The foreign breeds, particularly those with a high yield of dairy products, which acclimatized to French conditions (the Dutch or Friesian and the Flemish cows) were included in this work of definition and delimitation. Where horses were concerned, similar studies were made, particularly between 1880 and 1923. The majority of those concerned with the most important breeds appeared in the early years of that period (Percheron in 1883, Boulonnais in 1885). The geographical areas that were thus determined fell into two groups. First, there were those of the 'great breeds', whose specialized nature was clearly defined: in the case of cattle, the Normandy breed, the Friesian, the Red Piebald of eastern France, and the cattle of Abondance in the northern Préalpes were all milk-producing; Charolais and Limousin cattle and those known as the Blondes d'Aquitaine were meat-producing animals. These were the top breeds and they were associated with the great commercialized systems of farming without which they could never have emerged. The definition of the pure Charolais (1882) and the Limousin (1886) breeds played a decisive role in the development of the fattening system on the lower rocky plateaux of the northern Massif Central. Where breeds of sheep were concerned, the development of the manufacture of Roquefort cheese led to the expansion of the milk-producing Lacaune breed, (on which the first genealogical study appeared in 1905), which gradually supplanted the older Larzac breed, which was much less specialized.[22] Then there were also more localized breeds with less specific characteristics, less prestigious breeds with mixed qualities, many of which had evolved in the harsher mountain environments or in relatively isolated spots (for example, among the cattle, the Aubrac and Salers breeds in the volcanic mountains of the Massif Central; and the Tarine breed from Villard-de-Lans in the northern Alps, and so on). Breeds such as these were associated with areas less strongly integrated into the general economy.

Over the forty-year period between 1880 and 1920, pastoral France thus underwent a major process of regional differentiation in a number of particular sectors, where boundaries were in many cases clearly drawn and varied very little, particularly in mountainous regions where economic change was a slow process.[23] The situation remained more or less unaltered until after the Second World War, at which point super-productive breeds invaded virtually the entire French territory and many of the regional and local breeds were forced into retreat, in some cases to the point of being in danger of being wiped out altogether.

Reafforestation

Agricultural and pastoral developments could not be extended to the entire territory. Regions with poor soils or bad drainage that had no particular qualities to recommend them or were difficult to reach were by-passed. Their destiny was to be reafforestation. The landscape of today is strongly marked by this process. The forests of France, as we know them today, are to a large extent the result of the past century and a half of change. At the beginning of the nineteenth century the forests, in spite of royal rather than secular intervention, such as the great '*ordonnance*' of 1669, were still in a very run-down state. Their surface area had furthermore diminished greatly, as a result of the demographic pressure which was at its peak in rural areas at this time. By the end of the eighteenth century, only a few attempts at reafforestation had as yet been made, in a handful of locations: in the royal forest of Saint-Germain, close to Paris; and in a number of feudal forests; also in the Landes of Gascony, where stabilization of the active dunes was being tackled (see below). The process of degeneration had not been checked by the creation of the Forestry School of Nancy in 1824, nor by the reorganization of the forestry administration and legislation on forestry through the adoption of the Forestry Code of 1827, which brought communal forests and those belonging to public bodies under state authority. Much concern at the situation was expressed during this period and it was registered in two works published within a year of each other: *Les forêts de la France*, by Rougier de Bergerie (1817) and *La regénération de la nature végétale*, by F. Rauch (1818). But still little action was taken; in fact the surface area covered by forests seems to have reached an all-time low in about 1840: only about 8.8 million hectares of woodland remained, much of it in a very run-down condition.[24] It was at this point that, faced in particular with increasing erosion in the Alps, a powerful movement backed by scientific opinion began to press for reafforestation: the most influential publication that this produced was Surell's *Etude sur les torrents des Alpes* (1841). Under the Second Empire, the work of reafforestation began all over France. By the eve of the Second World War, the surface area covered by forests had risen to about 10.4 million hectares, that is to say 19 per cent of the land, as a result of the reafforestation efforts made in the preceding century. Their general state of health was incomparably better than it had been in the mid-nineteenth century.

The mountains of southern France

Here reafforestation had been particularly extensive, for the forests, on a fragile footing anyway, by reason of the climate, had been particularly run down by land clearance and the grazing of herds.[25] As a result of rural depopulation, large areas were available for reafforestation. In the Massif Central, various methods were followed. Through the initiative of enlightened large-scale landowners, activity in Auvergne began, even earlier than the mid-nineteenth century, with Montlosier's reafforestation of the *cheyre* of the Lac d'Aydat and a number of other projects launched by the general council of Puy-de-Dôme (1842). In the south, the administration took responsibility and the forestry engineer Georges Fabre restored 13,000 hectares of forest in Aigoual, planting a combination of beeches and pines. In Margeride, in Limousin, the task of reafforestation was essentially carried out by the peasants; in Lévezou and the Monts de la Madeleine, it was undertaken by reafforestation companies. Most of the trees planted by the peasants were Scots pines, which were native to the area, but these were complemented by a more artificial landscape, in particular of spruces, which were planted by the forestry workers. The Pyrenees and the Corbières were also affected by the reafforestation drive as were, above all, the southern Alps, where the impact made by forestry and hydraulic workers employed by the public services was particularly important. In the second half of the nineteenth century, 120,000 hectares were replanted here by the state, about 50,000 in the Basses-Alpes, 20,000 in the Hautes-Alpes, 30,000 in Drôme and Vaucluse and 8,000 in the Alpes Maritimes and the Var department. In the twentieth century, the task was largely taken over by private individuals (close on 100,000 hectares were replanted in the first half of the century, of which the state was responsible for only one third).

The plains

Here replanting was much more selective, concentrated mainly on regions of poor soil on which, in the old traditional economy, people used to scratch a living as best they could, but had given up the unequal task when better transportation methods made grain available everywhere. However, this pattern was not invariably followed. Replanting only took place where favourable socio-economic conditions existed – usually where the public authorities took action or where large landowners expanding their possessions did so.

The Landes of Gascony

These constitute a perfect example of the former type of case.[26] Up until the late eighteenth century there was only one natural forest of significant size here, the one which grew on the old established dunes lying a little inland from the shore. This was *la vieille forêt* (the old forest) and it consisted of maritime pines. The first attempts at reafforestation took place on the active coastal dunes, where Brémontier, an engineer employed by the Ponts et Chaussées in Bordeaux from 1766 to 1802, began in 1790 develop the technique of binding the active dunes by planting pine seeds among the deeply rooted marram grass (*Ammophila arenaria*). In 1801 he obtained a decree from the consular government, ordering the afforestation of the entire coastline. This was to be *la nouvelle forêt* (the new forest). But the interior of the Landes consisted of a great, sandy plain dating from the early Quaternary. This had been produced by the coastal ridge being blown inland by the strong westerly winds. The sand had fossilized and had sealed up the network of waterways which, meagre at the best of times for tectonic reasons, had consequently not been able to dispel it. This area was more or less dominated by leaching, resting upon an impermeable layer of iron-rich deposits a few dozen centimetres below the surface, which were a product of the podzolization of sand. In the eighteenth century and the first half of the nineteenth, the only pockets of cultivated land were those situated around the villages that were built on the valley slopes, for these were the only areas that were more or less satisfactorily drained. Here, a few patches of land were cultivated all the year round. In the winter, they produced rye, cultivated on mounds above the reach of the water, in the summer millet, cultivated in between the mounds. In the winter, the Landes were submerged in water and were virtually inaccessible. In the summer, when the water level fell, the natural vegetation was burned so as to obtain organic compost to fertilize the cultivated plots. Then the Landes were invaded by flocks of animals belonging to the local worthies. They were herded by specially skilled shepherds who moved around on stilts.[27]

If this region was to be reorganized, the first priority was to open it up, for movement within it was extremely difficult.[28] The navigable waterways were of little use; and, in the absence of supplies of stone in this sandy region, roads soon became little more than narrow tracks which were, anyway, flooded in winter. Pilgrims on their way to Santiago de Compostela were bound to make their way across this area, but it was not easy to do so:

> *Quand nous fûmes dedans les Landes*
> > *Bien étonnés*
> *Nous avions de l'eau jusqu'à mi-jambes*
> > *De tous côtés.*[29]
>
> (When we got into the Landes
> > What a surprise
> For we were knee-deep in water
> > On every side.)

Movement in this region was so laboriously slow that all economic development was blocked. Wood bought for 2.50 francs on the spot would be resold in Bordeaux for 9 francs: basically, the only way to move from one point to

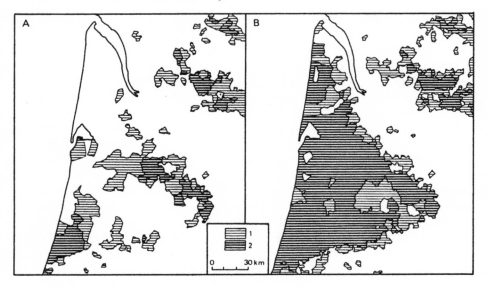

Figure 72. The development of the Landes forests (after the maps by Perpillou, 1977a and b).
The percentage of forests in relation to the total surface area. 1. 30–50%. 2. Over 50%. A. First half of the nineteenth century. B. First half of the twentieth century.

another was to take what was known as the 'Petites Landes route', which linked the ring of small towns on the outskirts of the Grande Lande: Langon, Bazas, Roquefort, Mont-de-Marsan, Tarbes and Dax. In the first half of the nineteenth century, all sorts of strenuous (but vain) efforts were made to resolve the problem: there were plans for canals to link Bordeaux and the Midouze, and the Garonne and the Douze, by way of the Baïse and the Gélise; and even a serious plan to acclimatize camels to the Grande Lande. On 26 June 1833, the 'Conseil d'Arrondissement' of Bordeaux pressed for the establishment of a stud farm for 'camels of the best breed' and a small, five-strong, camel train indeed arrived at Mont-de-Marsan in 1836 and was promptly entrusted to the tender care of the Justice of the Peace from Pissos! As to making the most of the land itself, vain attempts had been made to cultivate groundnuts and cotton, and water buffalo had been introduced in 1806.

In the end, two solutions to the Landes problem were found. The first was a railway system. A line linking Bordeaux and Dax was completed in 1854, another linking Bordeaux and Bayonne in 1855. These were followed by a branch line from Morcenx to Mont-de-Marsan and Tarbes in 1859, complemented between 1875 and 1880 by a number of local lines. It was now possible to sell produce outside the immediate area. The second solution was afforestation. A successful technique was cleverly worked out by Chambrelent, an engineer from the Ponts et Chaussées. In 1849, he had purchased 5,000 hectares of marshy land in the Gironde area and in under a year he had drained it using a system of *crastes*, ditches and canals, demonstrating that reafforestation was possible. The personal influence of Napoleon III gave the project the decisive boost. He bought 8,000 hectares of land to the north of Morcenx (an estate which was thereupon dubbed 'Solférino') and in 1857 and 1860 he had laws passed which obliged communes to cleanse their *landes* and plant trees on them, reserving for sale any plots that could be cultivated. The cleaning-up operation was soon completed and the sale of quinine fell by nine-tenths in fifteen years. In the course of the following half-century, a huge forest of pine trees became established. This covered a total surface area of 85,000 hectares which, together with the dune forests and the 30,000 or so hectares of oak forests which already existed, made up a block of 190,000 hectares, one tenth of the total area of forestland in France. This development was closely linked with that of the coal mines of northern Europe, for which the Landes forest supplied most of the pit props at the end of the nineteenth century. That was the first use to which these forests were put; later, they were also exploited for the distillation of resin, paper-mills and saw-mills.

The Sologne area[30]
This was a very different case, where attempts at reafforestation were no more than partially successful. This was an area of more or less clay-based Miocene granitic sand comprising a mixture of woods, ponds and pasture-land. In the Middle Ages, a few attempts at agricultural development had been made here, thanks to the extension of the neighbouring vineyards of Orléans. But a definite decline set in with the Wars of Religion. By the early seventeenth century, Sologne had become what it still was in 1850, a vast poverty-stricken area of sharecroppers' farms, half of which lay fallow. The basic difference from the Landes of Gascony was the size of the area covered by large landowners' properties, as opposed to the communal properties that were widespread in the Landes. Some plantations of pines were established by the beginning of the nineteenth century, but the real work of regeneration was not undertaken until the Second Empire, once again with the encouragement of Napoleon III, who himself took the lead by acquiring property in Sologne. The year 1859 saw the creation of the Comité Central Agricole de la Sologne, set up to promote land-drainage. A canal 45 kilometres long was dug from Blancafort (situated to the north of Bourges) to La Motte-Beuvron. It was known as the Canal de la Sauldre and its purpose was to make it possible to bring in the marly limestone soils from neighbouring Berry, for these were necessary to reconstitute Sologne. Here, agricultural development was a major element in the rehabilitation of the area. Reafforestation was another, but it suffered a catastrophe: in a single night,

during the harsh winter of 1879, 80,000 hectares of maritime pines, imprudently planted in this northern region, were killed by frost. However, the region did find agricultural stability in its meadows, where cattle were raised, potatoes were cultivated and poultry were farmed. By 1905, Sologne was twice as highly populated as in 1820, with a birth-rate much above average. The development of agriculture was halted after the First World War and afforestation once again became a priority. This time the particular aim in view was to create large hunting forests, since Sologne was close enough to Paris to become a leisure and recreation area for its inhabitants.

The Orléans forest and the Champagne plain

To the north of the Loire, the forest of Orléans, a vast crescent of sand similar to Sologne, of which it is a prolongation on the other side of the river, underwent far more complete afforestation, as it had in the past been a feudal forest. More or less cleared by the monks and exhausted by grazing herds and the exercise of traditional rights by the surrounding communities ever since the Middle Ages, by the beginning of the nineteenth century this area was little more than a stretch of neglected *landes*, brushwood and marshes. Through the efforts of the state forestry engineers, it was cleared, replanted with trees and put to use in a rational fashion. Today it is a vast area of woodland.[31]

In other low-lying regions with poor soil, where no such lead was given from above, no developments of this kind took place. Neither the regions of Dombes, Bresse, Puisaye nor the various Gâtinais or Gâtines of the Loire, nor the sandy Bourbonnais plains were replanted with trees to any significant extent. They remained cultivated areas with meagre yields and they developed hardly at all until very recently (although progress came a little sooner to Bourbonnais, which became more or less integrated into the economy based on raising cattle for meat, which had been adopted on the neighbouring low-lying rocky plateaux). On the other hand, reafforestation did take place in limestone and chalky areas. In the context of the old traditional economy, these had been prized for the regularity of their cereal harvests, but their soils had proved ill-suited to the fodder-crop revolution in agriculture. This had particularly affected, in the Paris Basin, the plateaux of the Montagne Bourguignonne and Haute-Marne, and the effects were most spectacular in the Champagne plain. In the traditional economy, this had been a typical 'good' region (see above, pp. 169–70), but in the eighteenth century it had become what the Encylopédistes (in 1753) called 'Champagne pouilleuse'. At this point, its chalky soils, which lacked fertilizers, prevented it from joining in the fodder-crop revolution in agriculture. That revolution did away with the system of leaving land fallow and replaced it with a triennial rotation scheme, in which fodder crops followed on after the cereals of winter and spring.[32] In these circumstances, the small-scale landowners were forced to emigrate, as pockets of agricultural land became increasingly concentrated and the most infertile areas of all, in particular the hill saddles, which had been used for sheep grazing in the traditional economy, now found an alternative solution in reafforestation. Here, the first attempts at reafforestation were made as early as 1755–60. Around the middle of the nineteenth century, they became general and the landscape that is still to some extent characteristic of today's Champagne plain began to take shape: copses of pine trees mark out the edges of elongated strips of land replanted individually by the peasants. These woods, composed of 90 per cent Scots pines and 9 per cent black pines, are submitted to an intensive 'tree-cultivation' process. They are cut down after forty years, at which point the stumps are burned, providing a few years' worth of fertilizer from the ashes for the crops that are grown there before the next batch of trees are planted. Thus, the Marne department, which had only 3,000 hectares of woodland in 1836, could boast 67,000 by 1912. Between 1830 and 1835, at the time of the first cadastral survey, the Petites-Loges commune,[33] which is a typical case, was composed of small-scale peasant landowners, unencumbered by indirect mortgages. However, no single property was larger than 25 hectares. The commune included no meadows and barely 3 hectares of woodland. In 1914, it comprised three properties of between 25 and 50 hectares, sixteen of between 5 and 25, and nine of between 1 and 5. Pine forests covered an area of 102 hectares. The last sheep disappeared from the area in 1921. We can learn a great deal about the socio-economic conditions of the old traditional system from the process of reafforestation in the limestone and chalky plateaux. In these areas, the local peasantry possessed the dynamism and the necessary means to tackle the conversion of infertile land on its own, and reafforestation certainly provided the solution. It was not until the Second Agricultural Revolution, essentially after the Second World War, when chemical fertilizers were introduced, that the process was reversed and the pine forests began to be cleared away (see below, chapter 11).

From dispersed industry to industrial regions

The geography of traditional industry, characterized by geographical dispersion and random specialization, had by the twentieth century undergone an even deeper transformation than the geography of agriculture. The above-mentioned effects of the revolution in transportation were compounded by the ubiquitous presence of the steam-engine. It was first used in 1779 to pump water from the Seine; then, in 1811, it was applied to the textile industry. From then on it found an ever-increasing number of uses. In 1815 only 880,000 tonnes of coal were produced in France and between 150 and 200 steam-engines were in use. By 1839 that number had risen to 2,450, by 1848 to 6,000. By 1870 there were 28,000 of these machines at work; coal production had exceeded 10 million tonnes by 1862 and 13 million tonnes by 1869.[34] These changes led to the abandonment of the scattered sites of industry based on water power and new concentrations of industry, dependent chiefly upon coal, were set up. The decline of dispersed industry coincided with the development of new industrial regions.

The decline of dispersed industry

In the course of the nineteenth century, the industries dispersed over the countryside gradually declined, having mostly peaked in the second third of the century: the maximum number of furnaces producing cast iron (433) was reached in 1837.[35] In the textile industry, the effects of the mechanization of spinning were felt at about the same time. Until this point, spinning had been widespread in the Vosges valleys, but it became concentrated in Mulhouse once coal started arriving there from Le Creusot in 1832, by way of the canal linking the Rhône and the Rhine, and from the Sarre region in 1854, by way of the canal linking the Marne and the Rhine.[36] However, at this time weaving, which was not yet mechanized, was still an activity that was spreading in the countryside. The considerable increase in cotton imports, which rose from 19,000 tonnes in about 1820 to 86,000 tonnes in about 1870,[37] was leading to an expansion of weaving, which was mostly carried out in rural areas. Between 1830 and 1850, 30,000 to 50,000 looms were scattered throughout the countryside of upper Normandy.[38] It was only from the 1870s onwards that the decline of rural industry really set in. Where wool and linen were involved, it was delayed until even later. It was by and large in the last quarter of the century that the textile industry, by now totally mechanized, became concentrated in the towns (fig. 73). Certain high-quality fabrics, closely associated with the skills of a specialized workforce, continued to be produced in the countryside for even longer. Resistance to change also continued in sectors that fell within the sphere of influence of a particularly dynamic town. Such was the case of the silk industry of Lyons, which was widely served by the rural areas of the Massif Central from 1830 on and by those of the Préalpes during the second half of the century and into the twentieth.

A similar pattern is detectable where other traces of the traditional dispersed industry survived. Those that did so were first and foremost particularly specialized industries which had early on evolved towards products of quality and precision and depended on a traditional workforce. These survived by setting up concentrations of small workshops and relying to a certain extent upon piece-work carried out at home. A typical example is provided by the industry in the Jura. In the mid-twentieth century, it operated essentially from workshops but these were widely scattered and piece-work at home still accounted for between 5 and 10 per cent of the

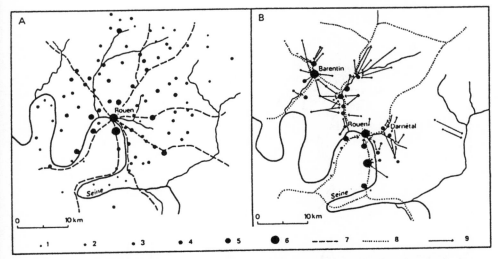

Figure 73. The concentration of the cotton industry in Rouen in the nineteenth century (after Levainville, 1913, pp. 229 and 231).
Number of cotton spinners: 1. 0–50. 2. 50–100. 3. 100–250. 4. 250–500.
5. 500–1,000 6. Over 1,000. 7. Roads. 8. Railways. 9. Workers commuting on a daily basis. A. Year III (1794–95). B. 1910.

industry.[39] Similar examples are to be found in the Alps, with screw-cutting, a direct legacy from the earlier clock-making industry, predominating in Haute Savoie, around Cluses. Many factories producing a variety of metal articles emerged out of the traditionally dispersed industries, one example being those of the valleys of the Saulx and the Ornain, where the old metallurgical industry of the Barrois plateaus was reconstituted (fig. 74). In the twentieth century, the Lyons silk industry became concentrated in small factories under the management of urban manufacturers. Their distribution around the countryside corresponded closely to that of the family looms of the old traditional system. In this instance initiative and capital from the town made the regeneration of small-scale industry possible.

Another type of rural industry was the 'substitute industry', where the disappearance of the old traditional industry coincided with the development of other activities which could call on the skills of the existing trained workforce. One case which has been carefully studied is the textile industry of the Pyrenees and Languedoc, which has now virtually vanished. In the absence of the motivating force provided by a centre such as Lyons, it was impossible to regenerate the textile industry here, except in Lavelanet,[40] situated in the Pyrenean foothills of the Ariège region, which benefited from the migration of the textile industry from Carcassonne at a point when this town was switching its activity to viticulture. But that instance of the

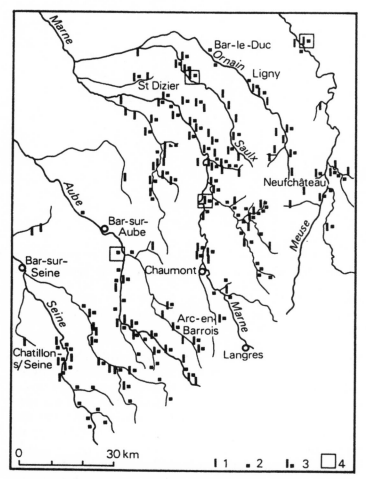

Figure 74. The concentration of the Barrois iron and steel industry (after Prêcheur, 1963, p. 107).
1–3. State in 1864: 1. Blast furnace. 2. Forge. 3. Blast furnace and forge. 4. Rolling mills in 1961.

Languedoc textile industry surviving in the Pyrenees is an isolated one. In most places, a substitute industry took over: the shoe-making industry at Mauléon, which became the espadrille capital,[41] and at Hasparren in the Basque country; the beret industry at Oloron; the felt-hat industry in the former centres of the textile industry in the Aude valley, where it continued to flourish up until the Second World War.

The new industrial centres

However, most contemporary industry developed on new and for the most part concentrated sites, which came to constitute veritable industrial regions. There were two distinct phases. The first was that of the coalfields. It might well have been expected that most of the new activities would be based on coal, which was now the dominant source of energy. But, as it turned out, the attraction of coal was far from universal. The development of the iron and steel industry provides a good example of this complex evolution. During the first two-thirds of the nineteenth century, the iron ore of the Lorraine seam, which is very phosphorous, was hardly used at all. The iron and steel industry developed, somewhat hesitantly, around the coalfields, where the puddling technique using coal was adopted (1824 in Raismes in the Nord department). But above all, the need for extremely pure ores for the manufacture of steel led to the industry being developed in coastal sites, for those pure ores had to be imported. A whole generation of factories thus sprang up in coastal areas: Marquise (1838 and 1839) and Outreau (1854–57) in Boulonnais; Isbergues (1883) in Pas-de-Calais; Rouen (1913); Caen-Mondeville (1910); Hennebont, close to Lorient (1860); Trignac, on the lower Loire (1879); Pauillac, on the Gironde estuary (1898); Le Boucau, at the mouth of the Adour (1880–83); and in the Mediterranean region at Balaruc, on the Etang de Thau (1880); Beaucaire, on the lower Rhône (1873). Many of these coastal factories were sooner or later forced to close, in the face of competition from Lorraine, once the discovery of the Thomas process (1878) made it possible to make direct use of the Lorraine ore. Within thirty years a new iron and steel region developed, supplying two-thirds of French production. But from 1866 on, the development of the Martin process (furnace with a basic sole-plate), which was well adapted to the manufacture of special steels, if necessary in small quantities, unlike the mass production of the Thomas converter, ensured the survival of quite substantial production based on coal in the north and in small pockets around the edges of the Massif Central. In the nineteenth century, a whole series of large industrial centres thus developed in this area: Le Creusot–Monceau-les-Mines and Saint-Etienne in the east; Montluçon in the north, where the coal from the Commentry seam could be brought to join the iron delivered by barge from Berry; Decazeville and Carmaux-Albi in the south-west. All these centres round the edges of the Massif Central, handicapped due to their distance from the ore, were obliged to move towards the manufacture of complex pieces, using special steels and a variety of metallurgical techniques (military equipment and boilers at Le Creusot; weapons and bicycles at Saint-Etienne and so on), whilst for a long time the north did not evolve beyond the stage of pure iron and steel production. The coalfields of the north, essentially around Lille and Roubaix-Tourcoing, similarly

Centralization and diversification

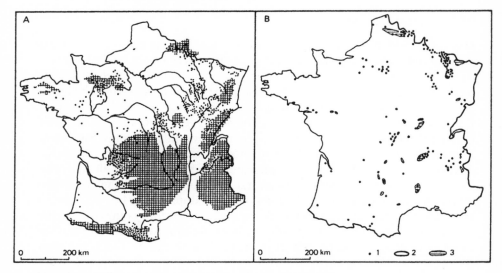

Figure 75. The concentration of the French iron and steel industry in the nineteenth century.
A. The situation in 1789 (after Bourgin, 1920, and Prêcheur, 1963, p. 9). Each dot represents a factory. Hatching represents mountainous areas. B. The situation in 1929 (in part from Prêcheur, 1963, pp. 72–3). 1. Iron and steel works. 2. Ore-bearing basins. 3. Coalfields.

retained a considerable proportion of the textile industry. But under the influence of Mulhouse, the Vosges mountains became the major regional centre for the cotton industry, while another very large centre survived at Rouen, where this industry had originally developed thanks to the port's relations with the outside world and its reception of raw materials. All in all, the coal-based phase made a relatively limited impact upon the industrial geography of France.

The effects of the second phase, that of hydro-electricity, were even more localized. Only in the northern Alps did this lead to complex industrial combinations. From the early twentieth century onwards, a number of industries that consumed large amounts of energy naturally enough established themselves here: first paper-mills, then electro-chemical and electro-metallurgical plants, which proceeded to evolve more and more in the direction of the chemical industry pure and simple. But the attraction of water power also operated indirectly. Around these industrial plants, a whole series of mechanical industries sprang up, producing the equipment for their waterfalls and factories. They did so largely as a result of the stimulus provided by the existence of the already dynamic urban environment produced by the large Alpine towns of both Savoy and Dauphiné

(Annecy, Chambéry, Grenoble) which were centres of economic and intellectual initiative. In the Pyrenees, where there was no such urban base, this stage was never reached. Development was limited to the energy-based industries. As for the Massif Central, here hydro-electricity produced no industrial developments at all. The area's electrical energy was soon being exported to other regions. The rubber industry of Clermond-Ferrand resulted from pure historical chance. The absence of any industrial tradition in the Massif was the determining factor in this lack of industrial activity.

The large urban industrial complexes

The major manifestation of the industrial concentration was constituted by, not so much the industrial regions that sprang up around the coalfields or near to other sources of energy, but rather the industrial complexes that developed in the large cities. The major industries of France were born not from the country's resources of natural energy but from trade. It was a type of industrialization very different from that of England or Germany, for it was connected far more with the currents of commercial exchange that stemmed from the existence of the isthmus of Gaul and the Parisian centralization.

The great industrial complex of Marseilles was linked with activity in the port and the raw materials that arrived there from overseas. These were the basis of a complex combination of food-processing industries of various kinds: flour-mills, originally dependent upon the temporary admission into France of foreign wheat, re-exported in the form of flour; semolina-mills; rice-processing plants and sugar refineries; industries based on oils and fats, which originally processed Provençal olive oil and then diversified into processing other, exotic, raw materials: producing oil, soap, glycerine, margarine, candles and stearin; chemical industries using pyrites, copper, sulphur and imported phosphates; and finally the pottery industry, whose bulky products constituted precious freight for the returning carrier ships and around which the usual collection of assorted metallurgical and other consumer industries developed.[42]

The great industrial complex of Lyons resulted from this city's position at a major junction of roads leading to Italy and also from its role as a fabrics market, which had equally been the cause of the silk-weaving industry that had developed here in the sixteenth century. Its industrial development was encouraged by the state, which granted it a number of privileges, including a monopoly over the marketing of raw materials. The great expansion here began in the eighteenth century, while the looms were still concentrated in the town, before being dispersed over the surrounding countryside in the nineteenth century. The presence of the considerable capital generated by the silk trade was a major factor in this industrial development, which was further encouraged by the many technical inventions and improvements that were devised in a number of fields (Danjon's wide loom in 1605; the Jacquard loom in 1804; piece-dying from 1880). In the last third of the nineteenth century, other textile industries took their place alongside the silk industry. Meanwhile, between 1835 and 1840, initially in connection with the needs of the textile industry, a powerful chemical industry was established, centred chiefly on pharmaceutical products and photography. Finally, at this geographical crossing of the ways and during this same period, a metallurgical industry also sprang up around the workshops constructing river-craft and railway equipment and this eventually diversified into a wide range of mechanical engineering plants and car factories.[43]

But the growth of Parisian industry is the major example of the contemporary development of large urban industrial complexes.[44] In the eighteenth century, Paris was not as yet an industrial town. According to the Orry Enquiry of 1745, only 22,000 of its population of 400,000 were workers or craftsmen. The first industries to be set up here were for the most part devoted to the production of luxury goods catering for the needs of the nearby court. These were already known as *articles de Paris*: clothing, ribbons, braid, soft furnishing materials, furniture, luxury leather goods, gold ornaments and jewelry, porcelain and carriages. Some of these were royal workshops producing luxury goods: Gobelin tapestries, mirrors from the Faubourg-Saint-Antoine, porcelain from Sèvres, fine fabrics from the Faubourg-Saint-Marcel, silk stockings from the Château de Madrid – all strictly in line with Colbert's policies. However, the monarchs did not encourage the development of industry in general, particularly since it would have engendered a working class that would inevitably have been a source of discontent and trouble.

The first spontaneous expansion took the form of chemical industries and was connected with two factors: in the first place, all scientific research took place here, for as early as the seventeenth century Paris was a major intellectual and scientific centre; secondly, the presence of abundant, cheap, raw materials produced by the urban environment – materials that could not be transported far and that had to be used on the spot: organic waste and fatty tissues from the slaughterhouses, and water and other products from the drainage system. Thus, at an early date, soap factories and perfume-producing workshops were set up. Next came the textile industry, above all cotton, which was in full swing by the beginning of the nineteenth century and operated in conjunction with the large local consumer market. Under the First Empire, Seine was the second largest cotton-spinning department (Nord being the largest) and the fourth largest for cotton-weaving. Next to develop were industries connected with the building trade. In the first half of the nineteenth century, the food-processing industry made its appearance, in the form of sugar refineries using the beet that was grown in the surrounding fields. The metallurgical industry also emerged at this point. At first and for quite a long time, it took quite limited forms, but these eventually developed into workshops manufacturing weapons, vehicle parts, and tools and machines for the textile industry. In 1843, Seine was the department with the sixth largest steel production. All in all, however, Paris had not yet become the major industrial centre in the country. In 1836, the Seine department was in third place, a long way behind Nord and Loire, in its use of horse-power.

It was the centralization of the railway system that provided the decisive stimulus for the metallurgical industry, creating a need for the manufacture of rails, engines and metal frames of various kinds, ~~ the industry expanded greatly under the Second Empire. By 1870, there were 800,000 workers in the Paris region. Between 1890 and 1910 the motor car took over. The initial demand was for de luxe sports models for the leisured classes of western Paris, but this was enough to determine the establishment of the first car factories. Then aviation and the cinema produced new industrial openings and meanwhile the use of electricity was becoming general, first in Paris, then elsewhere. Finally, all industries producing marketable goods were to benefit from the growing population and the consequent concentration of a very strong purchasing power in the capital: food-processing industries first and foremost, but also photographic, pharmaceutical and a number of other chemical industries. Furthermore, the international exhibitions regularly held in Paris attracted many buyers, creating a powerful extra stimulus. The final determining factor was the 1914–18 War, when Paris benefited from many new industrial opportunities at a time when the industrial regions of northern and eastern France were under enemy occupation. Lying some way back from the front yet sufficiently close to it, Paris played a crucial role in supplying the French armies. The supremacy of Parisian industry was henceforth assured. From this time on, over one quarter of the industry of France was located in the Paris region and produced correspondingly large industrial profits.

The gigantic development of Parisian industry illustrates the consequences of the processes of centralization studied in chapter 7. After the Second World War, one book which made a forceful impression[45] emphasized the negative aspects of this phenomenon. This industrial preponderance of the capital was unique among the major states of Europe. It came about much earlier and much more completely than in England, where the movement of industry to the capital was less overwhelming and came considerably later. It constituted a serious obstacle to the harmonious diversification of the territory as a whole, which has been only partially rectified by the policies of industrial decentralization introduced since the Second World War.

A great market: the geographical development and establishment of commercial brands

The nineteenth-century combination of industrial development and regional specialization in the agricultural regions of France, within the context of the great national market that now, thanks to the transport revolution, rapidly became accessible to all, was to produce a radical change in the conditions of commercialization and the distribution of products for immediate consumption – particularly foodstuffs. These were henceforth to be produced on a massive scale, achieving standardized levels of quality yet at the same time benefiting, in the eyes of their large nationwide clientèle, from the more or less flattering images attached to their places of origin.

The development of the fame of local products

The first consequence of the existence of this vast market was the development of the production of local and regional speciality foods. The range was extremely wide. Even in the days of the Ancien Régime, travel guides

---

Figure 76. Aspects of the establishment of local trade-names (*aires d'appellation d'origine*) in the nineteenth and twentieth centuries.
A. 'Crottin de Chavignol' (after Gilbank, 1987, p. 153). 1. Boundary of the *appellation d'origine contrôlée* (fixed in 1978). 2. Boundary of the Cher department. 3. The area of the *appellation contrôlée* of Sancerre wine. 4. Production and refinement of the 'Crottin' cheese. B. Establishments producing Camembert cheeses in 1958 (after Dionnet, 1987, p. 110). 1. Producers. 2. Boundary of the Pays d'Auge. 3. Boundary of Normandy. C. Spirits and liqueurs in France in about 1825 (after A. Huetz de Lemps, 1985: 1. From wine. 2. From *marc*. 3. From cider and pear-juice. 4. From grain and potatoes. 5. From cherries and plums. 6. Liqueur factories. D. The Lorraine *mirabelle* (after Peltre, 1985, p. 207). 1. Boundary of the *mirabelle appellation*, as defined in 1953. 2. Boundary of Lorraine. 3. Authorized distilleries of spirits listed in the *scel secret* of the duchy of Lorraine in 1715 and 1735. 4. Prefectorial authorization to establish distilleries in the Meurthe department (before 1871) and the Meurthe-et-Moselle department (after 1871). E. Calvados (after Brunet, 1985, p. 191). 1. Boundary of the 'calvados' *appellation*. 2. Boundary of the Calvados *department*. 3. Boundary of local appellations. 4. 'Calvados du pays d'Auge' *appellation d'origine contrôlée*. 5. Regulated *appellation d'origine* 'calvados du …'. 6. 'Eaux de vie de cidre' *appellations*.

The economic differentiation of space 375

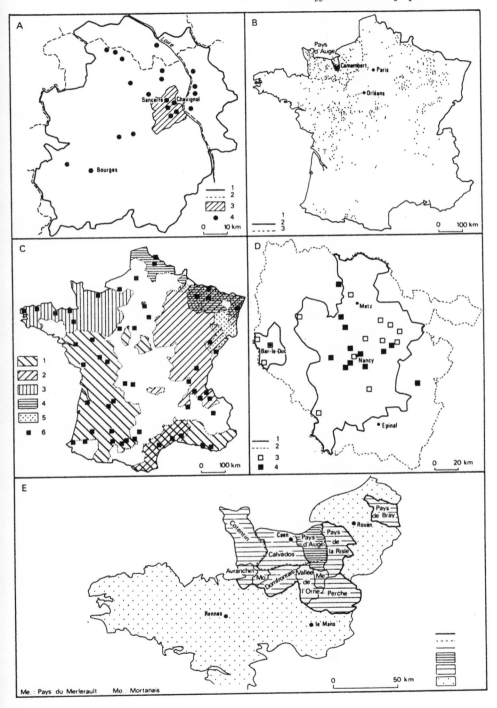

were at pains to mention them, urging travellers to sample them. But except for wines, these products left their regions of origin only very exceptionally and were hardly known outside them. The transport revolution and the emergence of a national market gave them a chance to win wider appreciation. Just as, through a subtle dialectical process, the cultural unification had provoked the appearance and establishment of local peculiarities, the constitution of a single national market encouraged the fashion for individualistic products, diversified by their own particular characteristics and origins.

The phenomenon was at its height in the second half of the nineteenth century. Two cases have recently been the subject of careful analyses, namely spirits and cheeses.[46] At the beginning of the nineteenth century, France was divided into a number of major geographical areas[47] in which the products from the dominant regional crops were submitted to distillation: grain and potatoes in the north and the Paris region; wine in the south-west (including the lower Loire) and the Mediterranean south; cider and pear juice in Brittany and Normandy, in the west; *marc* (from wine-making) in Burgundy and Lorraine, mixed with fruits (especially cherries and plums) in Alsace and Franche-Comté and with grain in northern Lorraine (fig. 76C). Only the spirits of the south and Aquitaine were exported on a massive scale and only Cognac and, by association, Armagnac had already established a solid reputation (see above, pp. 230–2). But a large number of liqueur makers (several hundred at least) were producing a wide range of other products. Only in the second half of the century did precise, recognized brand-names make their appearance. Some of these products were the result of strictly individual initiatives and bore the names of the particular individuals involved. One such was a particular Cointreau created in Angers in 1853,[48] another the Marie-Brizard liqueur,[49] which kept the name of the eighteenth-century spinster from Bordeaux and the commercial firm named after her (established in 1755) whose recipe for the anis-flavoured drink had made it a success at that date, before its sales plunged at the time of the Revolution and under the Empire, when it lost its colonial outlets. But the new factor now was the appearance and success of a whole range of regional names. The term 'calvados', the name of a department, was gradually adopted to denote the spirits made from cider and pear-juice in Normandy.[50] After 1870, the reputation of the *mirabelle* of Lorraine became established, reflecting the proliferation of orchards which had replaced the vineyards destroyed by the phylloxera.[51] In the last years of the century, a wholesaler from Bordeaux created the 'Izarra' liqueur in Bayonne, from a local recipe. *Izarra* means 'star' in the Basque language. The basis of the drink was spirits of the Armagnac type, flavoured by aromatic herbs and the image of this liqueur rapidly came to be associated with that of the Basque country.[52]

At about the same period, the types of cheese on offer in the Parisian market diversified and acquired individual reputations.[53] In the old days, apart from the distant cheeses of Roquefort and Holland, the only cheeses known by their local names of origin that were appreciated were those produced in the immediate neighbourhood: Brie, Montlhéry, Neufchâtel (from the Pays de Bray). Up until 1878, Brie retained its pride of place among the cheeses that found their way to the central markets. But as early as 1859, a whole range of new categories were added to the familiar commercial names: Livarot, a cheese with a soft rind, which had long been produced in Normandy; Mont-Dore (from Auvergne), a huge cooked cheese; and a number of other varieties. By 1871, the list of places of origin was so varied as to defy enumeration. This was when Camembert made its reputation. It had been produced for many years in the village of that name in the Pays d'Auge and had probably been made in its present-day form since the end of the eighteenth century. It is supposed to have been brought on to the Parisian market as a consequence of a partiality that Napoleon III displayed for it, in 1863.[54] It was at this point, at the end of the nineteenth century, that the definitive elaboration of the map of French cheeses began. The proliferation of names of all the varieties was then to win the admiration of all cheese connoisseurs and also to provoke the irony of politicians, who expressed scepticism as to the chances of efficiently governing a country of such complexity.

The expansion of manufacturing areas

These commercial labels indicating geographical origin were, for the most part, destined to be widely known, well beyond the original localities to which they referred, as their fame spread throughout the national market. Products which became established successes were soon imitated in the surrounding countryside and the manufacturing areas became much wider. Most of them were not defined precisely until relatively late and most legalized *appellations d'origine* date from after the Second World War. Thus the 'crottin de Chavignol',[55] a small round cheese, initially produced only in that village in the Sancerre region, from the milk of goats, the only animals able to graze on the steep slopes of the chalky ledges of these hills, is nowadays produced throughout more or less the entire department and even beyond it (fig. 76A). The Maroilles cheese which bears the name of an Abbey close to Avesnes, was in 1804 produced in thirty neighbouring communes. Today, the area defined in 1976 as qualified to use the name incorporates the whole of Avesnois and the part of Thiérache in the Aisne department.[56] Munster cheese,[57] from the Hautes Chaumes in the Vosges, which bears the name of a valley on the Alsatian side of the mountain range, is today mostly (90 per cent) produced on the Lorraine side, where this label has supplanted that of Géromé (from the town Gérardmer), which was a cheese well-known here as early as the sixteenth century. The boundaries of the area qualified to use the name, which were fixed in 1979, incorporate the whole of the Vosges and the Vosges foothills, the Vôge and the whole southern part of the plain of Lorraine. The fact that it extends so far testifies partly to the introduction of migratory grazing for the Alsatian flocks and herds on the Hautes Chaumes but also, and above all, to the commercial dynamism of the Alsatian people. Having conquered the Alsatian market in about 1890, at the expense of other local cheeses, when annexation came to an end, after the First World War, the Munster cheese went on to benefit from the sentimental attraction possessed by products from the newly recovered province. In the domain of spirits, while the area qualified to use the label 'mirabelle de Lorraine' may not yet incorporate quite all the departments of Lorraine (fig. 76D), that of 'Calvados' spreads well beyond the department of Calvados and even beyond the boundaries of Normandy. In some cases, the production of a particular item became so widely dispersed, reaching the farthest corners of the territory, that this made it impossible to define an area qualified to use the name of origin. So it was with the Camembert cheese which, even before the First World War, was being widely produced in the Loire region and also in the wet eastern part of the Paris Basin (above all in the Meuse department) and eventually spread as far afield as Aquitaine in the south (fig. 76B). In 1926, the Orléans Court of Appeal rejected a request to limit the use of the name of origin put forward by producers from Normandy, who were anxious to cut out the producers of the Loire region. This made it possible for the name to be used throughout France. The producers of the original region were obliged to fall back on the label 'Camembert de Normandie', which they acquired only in 1983 and which is accompanied by such strict conditions that in 1985 it was as yet only being used by 6 per cent of the Camemberts that truly were made in Normandy. One remarkable repercussion from this pattern of wide dispersal is, in some cases, the decline of the original centre of production, which finds it impossible to withstand the competition from its imitators. At the end of the nineteenth century, this was the fate suffered by the Brie de Meaux, which used to be particularly favoured by the Paris market but was subsequently submerged by the products of a vast region stretching all the way to the Meuse. It could no longer differentiate itself and eventually went under.[58] In other cases, the struggle is still of uncertain outcome. The producers of alcohols using the gentian,[59] traditionally made from the plant's roots dug up in the high mountains of Auvergne and associated with the image of this province, have quite blatantly, by using the trade-name 'Suze', or more discreetly that of 'Avèze', sought not to publicize the reference to Auvergne. Meanwhile, the 'Salers' version of this drink, which deliberately remained faithful to the regional association, was aimed at and sold to a quite different clientèle. The great national market is an area of competition in which the success of local products which became established at the end of the nineteenth century is, in the twentieth century, giving way more and more before the 'large organizations era' (see chapter 11), in which products are increasingly dissociated from their origins.

10
The rural exodus and urbanization

The exceptional development of the Parisian region, the shaping of the national culture under the influence of the central metropolis and the early emergence of a permanent dialectic between the centre and the periphery are all factors that are specifically French. They are complemented by the general trend towards urbanization that has affected the whole of Europe in contemporary times, resulting in a restructuring of space upon a new set of bases. France has not escaped that process which, although it has not done away with the imbalance between the capital and the provinces, certainly has modified its effects, by creating other sources of influence.

The completion of the urban network

The elaboration of a network of small towns in the late Middle Ages had made it possible to reach virtually every point in rural France without too much difficulty, even if that accessibility had not always been exploited to the full, for instance by the creation of urban markets. Subsequent additions to that network simply added extra touches of a secondary nature, answering new needs or reflecting new activities. But their importance has been by no means negligible.

The modern period: providing a frame for the national territory

The modern period has not created many new towns, but it has produced a few. Most, however, have been towns situated round the edges of the territory (fig. 77), making for a firmly delimited national area, rather than contributing to its internal structure, which was already in place.[1]

The most remarkable of the new foundations were the ports, designed to cater for the increasing tonnage of ships at a time when the French nation was beginning to expand overseas, and also to complete the coastal defences. The progressive silting up of the Seine estuary had led to a steady migration

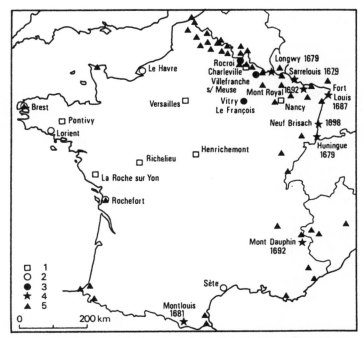

Figure 77. The completion of the urban network in the modern period (in part after Clout, 1977d, p. 485, with additions).
1. Towns created by princes, kings or emperors between the seventeenth and the nineteenth centuries. 2. Ports developed in the sixteenth and seventeenth centuries. 3. New fortified towns dating from the sixteenth century. 4. New towns created by Vauban (with the date of foundation). 5. Towns fortified by Vauban.

to the outer harbour downstream. Lillebonne (*Juliobona*), the capital of the Gallic city of the *Caleti*, was once a port on the Bolbec river, to the north of the estuary, but today it is 4 kilometres distant from the sea. It was succeeded by Harfleur, also to the north, on the Lézarde river, but in the fifteenth century this too became silted up and was in its turn succeeded by Honfleur, situated to the south, on the river Claire. In the Middle Ages, this was an important base from which Norman exploratory expeditions would set out and, thanks to its downstream position, it has more or less retained its importance to the present day. In 1517, Le Havre was created, as an outer deep-water military harbour. Though founded for the use of the navy, it was not long before it developed into a commercial port and a town. The seventeenth century saw the creation of a whole series of similar ports. Lorient, on the Scorff (Blavet ria), was created in 1666 on the southern coast of Brittany by the Compagnie des Indes Orientales, from which the town took its name. In the same year, in an attempt to salvage Languedoc trade, Colbert founded the port of Sète on the sandy coast of the plain. It was built around the canal that linked the Etang de Thau with the sea, on the site of a Roman port, at the foot of Mont Saint-Clair, on a spot where there was then nothing but a little fishing village. Another of Colbert's creations was Rochefort, designed to defend the inadequately protected coast between Lorient and the Pyrenees. It was built on a totally flat, marshy plain, on the estuary of the Charente. Finally, when Brittany was joined to France in 1532, Brest became a French military port, although when it was officially declared a town, at the end of the sixteenth century, it still consisted of no more than 1,500 inhabitants clustered in the precinct at the foot of the castle. It grew into a large city under the régimes of Richelieu and Colbert and eventually acquired its present-day aspect when the new quarter designed by Vauban was added in 1683.

A second series of frontier towns was constituted by the inland frontier-fortresses. In the sixteenth century, already, a number of these had been built: Rocroi, Villefranche-sur-Meuse (1544), close to Stenay, and Vitry-le-François (1545), built according to a geometric design, to replace Vitry-en-Perthois, which had been burned down in the war against the Empire. But the most important were the eight new, carefully planned fortress towns built by Vauban: Longwy (1679); Sarrelouis (1679); Mont-Royal (1692) on the Moselle, destroyed in 1697; Fort-Louis (1687), Neuf-Brisach (1698) and Huningue (1679) on the Rhine frontier; Mont-Dauphin (1692) on the Alpine frontier; Montlouis (1681) in the high Cerdagne basin on the Pyrenean frontier. Furthermore, essentially along the north and north-east frontiers, but also in the Alps and the Pyrenees, Vauban also remodelled fifty or so pre-existing towns, to adapt them to his new tactical ideas and set them up as fortresses.

In the interior, the few new urban creations of the modern period were essentially towns connected with the sovereigns – princes, kings or emperors. They constitute a reflection of the process of monarchical centralization, or at least of the centralization of authority, at a time when the function of the feudal castles was being eroded. These towns were geometrically designed and testified to a deliberate effort at town-planning on the part of the founders. At the end of the sixteenth century, Duke Charles III of Lorraine built the new town of Nancy, according to a chequerboard design, alongside the medieval town. In 1606, Charleville was founded by a minor princeling, Charles I of Gonzague, next to Mézières. In 1609, Sully, the minister of Henri IV, created Henrichemont (*Henrici mons*) in his principality of Bois-belle, in northern Berry, naming the town in honour of his master.[2] Thirty years later, Cardinal Richelieu, the Prime Minister, founded alongside his family castle in Poitou a town bearing his own name and attracted inhabitants to it by offers of many fiscal privileges. But the most prestigious of all these new towns was the royal town of Versailles, which Louis XIV began to build in 1671, establishing his residence there, 20 kilometres from Paris, in what had begun as a hunting lodge used by Louis XIII. Finally, in the early nineteenth century, Napoleon III created from practically nothing two new urban centres in western France. They were administrative and military towns designed to control these rural regions, which still posed something of a threat, and to ensure their pacification following the Vendée war and the *chouannerie* troubles: Pontivy in Brittany, where the new town was built alongside and at the foot of a tiny feudal country town and which was to undergo much new development in the nineteenth century; and La Roche-sur-Yon, created from scratch to be the administrative centre for the Vendée department, for which it became the *chef-lieu*.

The nineteenth and twentieth centuries: industrial towns and leisure towns

In all this period, virtually no industrial towns had been developed. Colbert had made a few attempts: the La Glacière quarter, close to Cherbourg, for the workers of a glass factory; and Villeneuvette, close to Clermont-l'Hérault, for the Cévennes weavers, neither of which was particularly successful. But industrial towns were to proliferate in the nineteenth century, when many mining towns were created in the iron and coal basins. Most were quite small and even today boast no more than 10,000 or so inhabitants, but many nevertheless acquired tertiary functions and all the attributes of proper towns. In the coalfields of the Nord and Pas-de-Calais departments, where mining towns were essentially juxtaposed alongside already existing villages, a whole constellation of small towns thus came into existence. A few centres which reached or exceeded 20,000 to 30,000 inhabitants stand out amongst them: Lens, Hénin-Liétard and Liévin. A similar pattern developed in the Lorraine iron basins, although here the main centres were rather smaller: Joeuf, Hayange, and Moyeuvre, none of which exceeded 10,000 to 15,000 inhabitants. Industrial quarters also sprang up in regions where industry was widely scattered: in the region of Montbéliard, and the Saint-Etienne basin between the Rhône and the Loire. In the coalfields around the edges of the Massif Central, urban centres, for the most part isolated but nevertheless powerful, developed: Le Creusot which, in conjunction with Montceau-les-Mines, has grown into a town with 50,000 inhabitants, having been no more than a village at the end of the eighteenth century; and Decazeville, created from scratch under the Restoration and named after a minister of Louis XVIII's. We may include in this category of industrial towns the railway towns which developed at important junctions just outside certain large towns: Saincaize, close to Nevers; Saint-Pierre-des-Corps, close to Tours; and Les Aubrais, close to Orléans; and towns born from the need to change engines at some point on long journeys: for example, Laroche-Migennes, half-way between Paris and Dijon. The presence of large repair workshops in such places lends them the air of small industrial towns.

The last series of urban creations consists of seasonal tourist resorts. There were two successive generations of these. The first was that of seaside towns which made their appearance before the mid-nineteenth century. In the Seine bay, Trouville, a simple fishing village, was recognized as a town under the July Monarchy. On the other side of the Touques, the twin town of Deauville began to be built from scratch in 1859. The population of the two towns taken together is today 12,000 permanent inhabitants. Paris-Plage – Le Touquet, on the left bank of the Canche, in Picardy, was also developed from scratch on a site which had formerly consisted of old water-meadows. On the Atlantic coast, La Baule and Les Sables-d'Olonne grew out of virtually nothing. In the south-west, Arcachon grew up from 1823 on, starting as a small hotel patronized by holiday-makers from Bordeaux. On the other hand, Biarritz, a little fishing port of 2,400 inhabitants in 1856, was not born from the tourism which provided its great boom under the Second Empire; no more was Royen (which had 1,500 inhabitants in 1840). The towns along the Côte d'Azur also all developed from ancient nuclei, even if these may be hard to find amid the amazing developments of recent years. It was only after the Second World War that new seaside towns were created from nothing on the sandy Languedoc coast and around the coastal lagoons of the Landes.

The last generation of new towns consists of mountain resorts. Winter sports centres really only began to be developed after the First World War and most remained at the level of seasonal resorts and never grew into proper towns. Some however, the oldest, which cater for a summer season as well as a winter one, did become real towns with all the central functions that belong to them. Though not common, this type of town is now tending to multiply, but it was extremely rare up until the Second World War. Chamonix, at the foot of Mont Blanc, is a quite different case, for it began to grow as early as the late eighteenth century and the early nineteenth, as a result of tourism in the high mountains and the popularity of mountain-climbing, so that by the end of the nineteenth century it already had the air of an established town.

All in all, these contemporary developments have made a substantial contribution to the urban network. At the beginning of the nineteenth century, natural conditions (for example, easy access) constituted a major factor in the location of towns. The map of towns with 5,000 inhabitants or more was still mostly based on that of the network of rivers (Seine, Loire, Garonne, Rhône). Away from these main valleys, only a few urban zones stand out: in the Nord department, the Languedoc plain and Normandy. The map of small towns (of between 3,000 and 5,000 inhabitants) was more complex. Some areas – mountain regions, for example – are avoided, above all in the Massif Central. But small towns were also sparsely scattered throughout the west, from Picardy to the Landes and including Brittany and Vendée. Conversely, sectors of more intense urbanization were to be found in the Lille region, Alsace and Provence, all with a history of ancient traditions and strong municipal activity. Throughout the century, small towns were to multiply and many of these relatively empty areas were to fill up. In the first half of the nineteenth century, between 1811 and 1851, the number of centres of population which could rank as towns (taking the bottom limit to be 3,000 inhabitants) had already increased by about one third. Between 1851 and 1911 it increased from 567 to 794,[3] that is to say by about 40 per cent.

From the countryside to the towns

Whether ancient or of recent foundation, the towns attracted vast numbers of people from the countryside. This was not an altogether new phenomenon. In the pre-industrial period, migrations had taken place towards Paris, the capital (see above, pp. 256–8) and, above all in southern France, towards the larger provincial villages (see above, pp. 211–13). The process was to speed up with the Industrial Revolution, leading to a major redistribution of the population, to the advantage of the towns. This took place in a number of stages.

Temporary migrations[4]

Even before the beginning of the definitive exodus and during its early stages, the countryside was the starting point for a number of major currents of temporary migration, which increased considerably in the course of the nineteenth century. An enquiry carried out under the First Empire calculated the number of these temporary migrants to be 120,000 (fig. 78A and B). The figure is certainly an underestimation: 200,000 would appear to be more plausible.[5] But an enquiry carried out in 1852 put the figure at 878,000 emigrants (526,000 men and 352,000 women) seeking agricultural work (some of whom no doubt counted twice or several times over in the course of a year) and a total of 901,000 temporary migrants. At a rough calculation, the number of individuals affected must have been at least 500,000: about 200,000 from the Massif Central, 70,000 from the Alps or the Pyrenees, and the rest mainly from the edges of the Paris Basin.[6] Temporary migrations were less of a feature in the Jura, where a wide range of dispersed industrial activities helped to keep the population occupied at home in the winter time.

In the first half of the century, most of these migrations were directed towards other rural areas. In the late eighteenth century and the early nineteenth, rural France, still densely populated and expanding both

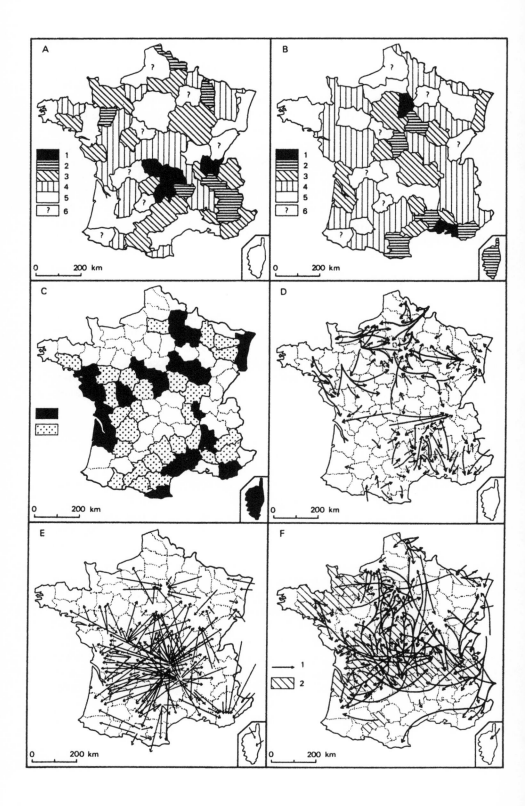

demographically and economically, offered a wide variety of jobs, first and foremost agricultural work, mainly in harvesting: grape-picking (fig. 78C), especially in Gironde and Charente, in western France, which was the main pole of attraction for work of this kind up until the mid-century (35,000 migrants worked as grape-pickers in Gironde and Charente-Inférieure in 1852);[7] harvesting (fig. 78D) in the plains where cereals were widely cultivated, in the heart of the Paris Basin (migrants from the belt surrounding the Basin, the north and the east), in Alsace, in the plains of the lower Rhône valley (particularly Camargue, still producing cereals at this period) and the Mediterranean south in general (migrants from the surrounding mountains, the Alps and the southern Massif Central); olive-collecting in the Mediterranean south; the harvesting of hemp in the mid-Garonne valley (migrants from the neighbouring hills); the cultivation of madder in the plains of the Comtat Venaissin (migrants from the Préalpes and the Cévennes).[8] For six or seven months in the year forest maintenance (fig. 78E) attracted 'pit sawyers', mainly from the eastern Massif Central (Forez and Livradois, Morvan and its margins), but also from the Ariège region of the Pyrenees and from Limousin.[9] Finally migrant workers, particularly those from mountain areas, practised countless crafts, trades and services, many of which were the particular specialities of their own native villages.[10] Some of these activities were closely connected with agriculture, such as the work of hemp-combers, large numbers of whom operated within a band running from the Gironde to the high Alps, extending down to lower Auvergne, up to the Paris Basin and as far afield as Lorraine and Alsace; wool-carders; silkworm producers from the Cévennes, the Causses and the southern Préalpes. Vivarais and the Ariège mountains produced muleteers; the countryside of the south-eastern Paris Basin and Franche-Comté produced waggoners (such as those from Saint-Claude in the Jura); the Nièvre and the Yonne produced 'floaters' to guide consignments of logs down-river to Paris. Tinkers and silverers came from Auvergne and Limousin, the Alps, the central Pyrenees and lower Normandy (the Bocage Normand); pork butchers came from Embrunais; mole-catchers from Orne. Pedlars came from the Oisans district, the Barcelonnette valley in the southern Alps and the eastern Cantal in Auvergne; rag-collectors and nomadic rag-and-bone men

---

Figure 78. Temporary migrations under the First Empire (1808–13) (after the maps and data provided by Châtelain, 1976).
A–B. Departures (A) and arrivals (B), by departments. 1. Over 5,000.
2. 3,000–5,000. 3. 1,000–3,000. 4. Under 1,000. 5. None. 6. No statistics.
C. Temporary migrations to the vineyards. 1. Departments receiving migrants.
2. Departments from which the migrants came. D. Migrations of harvesters.
E. Migrations of woodcutters. F. Migrations of construction workers.
1. Directions of these movements. 2. Departments from which they started.

came from the regions of Ambert, Mont-Dore and Cézallier in the Massif Central; cloth-sellers came from Auvergne (the Mauriac and Murat regions) but also from the countryside of western inland France; seedsmen from Faucigny in the northern Alps. Some of these traders also pursued their trade beyond the limits of the national territory. Natives of Cantal would travel to Spain; the people of Barcelonnette would ply their trade in Mexico and those from Queyras, in particular the umbrella-sellers of Aiguilles, would do so throughout South America. Amongst the migrations of service workers, it is worth mentioning the chimney sweeps from Savoy and the water-carriers from Auvergne and other curiosities too: teachers from Briançon and the Alps in general (Maurienne and Oisans); showmen with performing bears from the Ariège Pyrenees (the first is recorded in 1829), who, by the end of the century, ended up by buying their animals in central Europe and Russia, as they had more or less disappeared from the local mountains; singers from Haute-Marne. Some villages in the Massif Central and the Pyrenees had developed particular skills in begging: the people of Saint-Dier in Puy-de-Dôme and those of Saint-Jean-des-Ollières, who had been practising the art of sponging since the eighteenth century, claiming to be the victims of imaginary calamities attested by forged official documents; and the inhabitants of Chanterelle (in the canton of Condat in the Cantal department) who were still practising in Nice as 'deformed beggars' in 1926; and the people of the Massat valley in the Ariège, also still professional beggars at the end of the nineteenth century.

At the beginning of the nineteenth century, only a small proportion of these temporary migrants were making their way to the towns. At this period, the only towns that truly exercised a significant attraction were Paris – whose floating population under Louis XVI was estimated at 40,000, not counting 50,000 domestic servants – Lyons and Marseilles. Paris' sphere of attraction was already wide (see above pp. 256–8), extending as far as Limousin, the southern Massif Central, Savoy and Switzerland. Those of Lyons (which above all attracted people from Auvergne and the Cévennes, but also a number of masons from Limousin) and Marseilles (to which *gavots* – people from Gap and its region in the southern Alps – were drawn) were much more limited. However, the port-based activities of Marseilles and its international reputation also attracted migrants from outside France (for example, beggars from Piedmont and Neapolitans, on whom the police kept a strict eye). Apart from masons (fig. 78F) and beggars, most of these temporary migrants to urban centres were workers in the service sector: water-carriers from Auvergne, boot-blacks and chimney sweeps from Savoy, porters also from Savoy and the mountains around Lyons, ambulant hawkers and all kinds of craftsmen carrying out repairs; and domestic servants, migrants for life (in particular the Swiss, as porters in Paris).

In the second half of the century, the phenomenon of temporary migration underwent a number of important changes. The traditional destinations of agricultural migrations shifted.[11] Workers aiming for the vineyards now travelled to the Mediterranean, chiefly to the Languedoc Plain (see above, pp. 344–50), which was attracting 100,000 migrant workers by the beginning of the twentieth century; meanwhile, the number of migrant workers making for Bordeaux fell by half between 1850 and 1930. At the same time, the quality of the grape-picking workforce declined, for the work increasingly attracted a lower class of worker; in contemporary times most come from abroad, chiefly Spain. Over the same period, the area from which harvesters were recruited for the cereal-producing plains of the central Paris Basin expanded to incorporate Picardy, the Nord department (Camberlots), Belgium and – in the early twentieth century – Brittany too. At the same time, new streams of migrants were appearing, in search of various kinds of agricultural work (rather than harvesting jobs, as in the past): cultivating industrially produced crops such as beetroot in the central Paris Basin, for instance. These were drawn from the same geographical area as the cereal harvesters. Migrant workers also travelled to Alsace for the hop-picking and to the Côte d'Azur for the flowers grown for perfume; and the market gardens of the Paris suburbs attracted workers from lower Normandy, Brittany, Nivernais and lower Burgundy. The new horticultural regions of Vaucluse and the middle reaches of the Garonne (see above, pp. 343–4) also attracted migrant workers. But above all, temporary migrations to the towns increased and began to take somewhat different forms in the second half of the century.[12] Workers were drawn from ever-expanding areas. The attraction of Paris affected even the most distant sectors of France, in particular Brittany and Corsica. Professional specializations changed. Most migrants now went to work on building sites and on public constructions for rapidly expanding towns in need of new facilities. They came from three major regional sectors: a wide band made up of fifteen or so departments strung out between Lyons and the Gironde, centred on Limousin but including Auvergne and the margins of the Massif Central towards both the Paris and the Aquitaine Basins (in 1848, 34,000 migrants left the Creuse region as did 83,000, with families, in 1891);[13] in the west, Brittany, lower Normandy and Maine; and the northern Alps. Lower Normandy and the Massif du Giffre in the Préalpes du Nord produced stone-masons in particular. Migrating craftsmen and workers in trade, industry and services now tended to settle permanently, with upwardly mobile social aspirations. The water-carrier from Auvergne of the first half of the century tended, in the second half, to settle in Paris as a *bougnat* (a wine, wood or coal vendor) with a shop. Increasingly, workers would migrate for the duration of their working lives,[14] particularly those from Corsica, Gascony (postal workers) and Burgundy (lower grade civil servants and policemen). Two major areas of these working-life migrations may be defined on the map: one extends from the Sarthe region across to Franche-Comté and Savoy, incorporating the Loire region; the other, in Aquitaine, extends from the Basses Pyrénées to the edge of the Massif Central and to Poitou. Brittany, Normandy, Artois and the south-east were areas of secondary importance. Auvergne and Velay appear not to have gone in for working-life migrations but eventually switched to permanent ones. Until that happened, however, these regions remained faithful to temporary migrations for longer than either the Alps or the Pyrenees.

In this extremely complex picture, many aspects remain perplexing – in the first place the circumstances surrounding the emergence of specializations, both in individual villages and in particular regions. Perhaps it had something to do with individual or collective psychology: the chance successes of natives who had emigrated probably encouraged others to emulate their examples. Then there are some strange blanks in the map of migrations. The mountainous Grande Chartreuse and Vercors massifs in the Préalpes du Nord, poor regions from which there has been a considerable exodus in contemporary times, never, for reasons that are hard to determine, went in for temporary migrations. The same goes for the Préalpes of Nice and several sectors of the Préalpes of Digne and Grasse. No doubt the spur of initiative was lacking, but the underlying reasons escape us. However, over and above these imponderables, a typology does emerge, together with a number of clues, particularly in mountain areas where it is possible to distinguish certain clear correlations between various forms of migration and particular aspects of the geographical environment. As regards the timing, winter migrations appear to have constituted a way of complementing means of subsistence in mountain districts which practised a normal agricultural economy. Summer migrations, on the other hand, are in principle more indicative of overpopulation and a general regional poverty. The nature of the activities undertaken by migrants may also be explained by the characteristics of their native environment. The migrations of harvesters fitted conveniently into the cycle of agricultural tasks, taking into account the time-lag between periods of intensive activity in the mountains and in the plains. The migrations of craftsmen and traders, which were generally of longer duration, could for the most part only be fitted in when agricultural work was at a standstill. These factors help to explain, at least partially, how it was that migrants originated from particular localities in the Massif Central.[15] Temporary migrations tended to be much more a feature of the western massif which was wet and unsuitable for cereal cultivation. The eastern massif, which produced more cereals, only gave rise to winter migrations of 'pit-sawyers'. Migrations of traders and artisans were particularly characteristic of regions of meadows and pastureland, for instance the volcanic Auvergne. Yet much remains unclear. Why did the people from the *'planèze'* of Saint-Flour, a cereal-producing area, join the great emigrations to Spain of the traders of Cantal? The summer emigration of the Limousin masons is puzzling too. There was no question of overpopulation and no shortage of land here, for these mountains had always been amongst the most underpopulated parts of the Massif Central and when the men went off on their migrations, their skills were sorely missed. This emigration, of a 'contradictory' nature so far as agricultural tasks went, must have been prompted simply by a desire for higher profits and independence. One fundamental explanation for many of these local patterns of behaviour appears to have been the position and nature of the relevant poles of attraction. The eastern side of the Massif Central fell well within the orbit of a number of large industrial towns (Lyons, Saint-Etienne, Roanne and so on) which from an early date had played a stabilizing role in distributing work throughout the neighbouring countryside. There were no such industrial towns on the western side, and their absence encouraged migrations to distant places, to find alternative employment. In other cases, the geographical relationship with a particular urban centre produced direct effects. In the nineteenth century, Morvan, which had long looked towards Paris, as a result of the transportation of its wood to the capital by way of the river Yonne and its tributaries, also sent it a steady stream of wet-nurses. All in all, over and above quite straightforward correlations with the geographical environment from which these migrations originated and the type of life led there, the map of temporary migrations also reflects the infinite complexity of the historical movements of human beings and the aspirations that prompted them.

The part that these temporary migrations played in maintaining the economic and human balance of the national territory was of capital importance. It has sometimes been thought that they prepared the way for and encouraged the definitive migrations that followed.[16] But in truth, despite the fact that the latter did follow and in many cases adopted the very same routes, the initial effect of the temporary migrations was to slow down the rural exodus.[17] For a long time they made it possible to maintain communities of a relatively high density in mountain areas and to contain the demographic pressure of the first half of the nineteenth century. The Creuse department, the heartland of the emigrations of the Limousin masons, was supporting 51.5 inhabitants per square kilometre in 1851, as against 39 in 1801. The neighbouring Indre department, to the south of the Paris Basin, which produced far fewer migrants, had only 39.9 inhabitants per square kilometre in 1851, as against 30 in 1801.[18] The regular return of migrants furthermore slowed down the falling birthrate and helped to maintain balanced communities, which would have collapsed if all the young people had left for good. In nineteenth-century France, undergoing the process of industrialization, they slowed down the urban explosion, checking the overcrowding in the slums of the poorer parts of big cities. They thus played a crucial role in setting the scene for the gradual, relatively harmonious and crisis-free changes that characterized the urbanization of modern France, which stand in such striking contrast to the experience of Victorian England.

In the fullness of time, towards the end of the century, temporary migrations declined considerably. Essentially, they were a feature of a stage of economic evolution in which the areas to which the migrant workers made their way could not yet offer a wide enough range of activities to justify their settling there. When that situation changed, as industrialization proceeded, migrant labour for the most part no longer fitted the bill, although in some exceptional cases factories did set up hostels for their temporary workers and there continued to be a certain amount of seasonal work in the mines located round the edges of the Massif Central. Migrations of workers on alternate days was another solution tried out in certain areas in the years leading up to 1914. But eventually the most common upshot was a definitive exodus from the countryside. Another contributory element in the decline was the fact that many of the more lucrative occupations of the migrants were no longer profitable. Peddling, for instance, fell off in the face of improved transportation and the increasing number of traders fixed in their own shops. The decline of temporary migrations thus marks a major economic turning point.

Another factor in the decline was the general evolution of population numbers. The definitive exodus from rural areas rapidly affected such temporary migrations as were still taking place. Once the population in those areas had dropped and pressures on the communities involved diminished, seasonal migrations became less necessary. As soon as rural populations fell as a result of the exodus, the temporary migrations began to decline, although there was a certain time-lag between the two events. Rural overpopulation had peaked during the last few years leading up to the maximum demographic explosion and it took some time for the effects of less crowded conditions to be felt. In general, the decline of the temporary migrations thus seems to have begun a few years after the peak in the population curve, as many detailed analyses have shown, particularly in the Alps. In Faucigny, in northern Savoy, where the population curve reached its highest point in about 1860, the annexation to France, which also came at this date, sounded the knell for the major seasonal migrations.[19] In Maurienne,[20] where the population remained extraordinarily stable up until the end of the nineteenth century, 5.5 per cent of the local workers were still migrating temporarily in the early years of the twentieth century and definitive migration did not set in until this time, gathering enormous force during the 1914–18 War. The reason for that delay in this particular spot probably has to do with the fact that this high valley in the inner Alpine range enjoys a relatively dry climate which made it possible to cultivate cereals quite successfully and thus retain a higher proportion of its inhabitants, so long as temporary winter migrations continued. Almost everywhere else the decline started much earlier, during the second half of the nineteenth century. In the southern Préalpes, winter emigrations ceased between 1860 and 1880, but summer emigrations continued until the inter-war years.

In fact temporary migration never ceased altogether but has survived in various forms, not so much out of economic necessity but more for reasons of psychological tradition, out of habit, or even as a distraction. It is again in the Alps that the phenomenon has been carefully studied. In general, highly specialized activities demanding considerable technical skills have virtually totally disappeared and the continuing emigrations have involved a less highly qualified workforce, above all female domestic servants or relatively unskilled shop assistants. However, in a few cases quite the opposite situation has developed, as a result of particularly determined efforts to adapt to new commercial conditions (such as external trade). In the mid-twentieth century, the village of Mieussy, in the Massif du Giffre (Préalpes du Nord) were each winter still providing thirty to forty central heating technicians to work in Paris, in particular for the Elysée Palace, the various ministries and the big department stores. Prompted by the introduction of central heating in large institutions in 1855, this is an example of an ancient form of winter emigration, adapted to modern life. In 1952, in the village of La Côte-d'Arbroz, in Chablais, the seasonal emigration of seedsmen, which went back at least as far as the early eighteenth century, still involved forty individuals (that is to say one-fifth of the village population, or over one half if their families are included). Essentially, these men were wholesalers or semi-wholesalers who went off to sell their wares from their cars or vans in the Paris Basin and in eastern and central France, spending only three months of the year in their native villages.[21] Numerically, however, these curiosities are of little account; the only large temporary migrations to have survived down to the present day or recent times are those of agricultural workers employed in activities where mechanization has yet to be introduced: grape-picking, an occupation which, since the mid-twentieth century, has chiefly appealed to foreign labour; or working in the vineyards in the springtime. Around 1950, about 1,100 girls were still making the journey from Chablais down to the vineyards of Vaud, in June, to work as 'leaf-removers' for three weeks or a month.

The rural exodus[22]

On the whole, definitive departures from the countryside constitute a more recent phenomenon than temporary migration. But to a large extent the two movements have overlapped. It is not easy to date the phases of the definitive rural exodus precisely, for any analysis of this phenomenon is seriously complicated by specifically demographic factors, particularly the evolution of the birthrate.

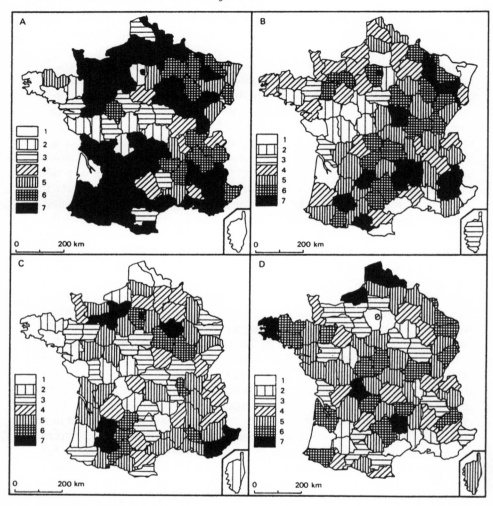

Figure 79. The rural exodus.
A. The date of the maximum rural population (after Merlin, 1971, p. 15). 1. 1911, 1906 or 1901. 2. 1896 or 1891. 3. 1886 or 1881. 4. 1876 or 1872. 5. 1866. 6. 1861. 7. 1856 and before. B. The percentage drop from the maximum population (Merlin, p. 15). 1. Up to 25%. 2. 25–30%. 3. 30–35%. 4. 35–40%. 5. 40–45%. 6. 45–50%. 7. Over 50%. C. Changes in the size of the population due to purely natural factors, 1872–1931 (Merlin, p. 28). 1. Over + 30%. 2. From + 20 to + 30%. 3. From + 10 to + 20%. 4. From 0 to + 10%. 5. From 0 to − 10%. 6. From − 10 to − 20%. 7. Under − 20%. D. Changes in the size of the rural population due solely to migrations, 1872–1931 (Merlin, p. 28). 1. Under − 50%. 2. From − 40 to − 50%. 3. From − 30 to − 40%. 4. From − 20 to − 30%. 5. From − 10 to − 20%. 6. From 0 to − 10%. 7. Increase.

Rural depopulation

Regular records kept in the nineteenth century clearly reveal a progressive rural depopulation. Its progress can be followed from 1856 onwards, for in that year the census for the first time distinguished between urban and rural populations (the latter being defined as villages of less than 2,000 inhabitants).[23] It is thus possible to determine both the relevant dates and the varying intensity of this rural depopulation (fig. 79A and B). As regards dates, nationwide the rural population was at its highest in 1861. But of course that is simply a national average. At departmental level, figures show that the depopulation of the countryside had started before that date in the whole of the south-west, the southern Alps, Normandy, Picardy and much of north-eastern France. However, it began much later in central France, Brittany and Vendée, where the rural populations were still growing between 1856 and 1886 (fig. 80A) and likewise in a few isolated departments such as Gironde, Seine-et-Oise and Corsica. The map showing the intensity of the phenomenon presents quite a different picture, however. On the whole, the south-eastern regions suffered the heaviest losses, with a number of exceptions (the Mediterranean south, Rhône and the Alpes du Nord). The hardest hit regions were the Alpes du Sud, the Massif Central, the Pyrenees, much of the Aquitaine Basin, and in particular the Lorraine departments. A comparison between dates and rates of intensity makes it possible to establish a roughly approximate typology of a number of different kinds of regions: (1) regions where populations peaked at an early date and the population loss was massive: almost all the south-west and the Massif Central, the Alpes du Sud and eastern central France; (2) regions where the population peaked early but the loss was slight: Normandy in particular; (3) regions where the peak came late and the loss was small: Brittany, the Loire region and isolated, relatively urbanized departments such as Pas-de-Calais, Gironde, Rhône, Vaucluse, Haute-Savoie and Corsica; and (4) regions where the population peaked late and the loss was heavy: certain departments of the Massif Central such as Lozère and Aveyron, and of eastern central France (Nièvre, Allier, Saône-et-Loire).

These results gain from being examined in detail outside the somewhat rigid framework of departments, in natural regions rather, taking account of how they correlate with the speed of agricultural progress. Such an analysis has been carried out for the Massif Central,[24] producing a series of most enlightening maps. Depopulation occurred early, between 1821 and 1846, in a number of fertile, overpopulated regions such as the vineyards of the Limagne and the Basses Cévennes, as well as in a number of regions at high altitudes (Cantal, Causses, Haute-Marne), where temporary emigration had been heavy ever since the eighteenth century and had soon prepared the way for definitive departures. Elsewhere, particularly throughout most of the regions of 'mediocre cultivation', situated in high, rocky areas, the rural population continued to increase, particularly where the cultivation of potatoes had been successful. Between 1846 and 1886, this movement continued in the industrialized east, in the Ségalas, Limousin, the upper Lozère and the Monts-Dore. But circumstances varied from region to region. In isolated spots such as the Montagne Limousine, the upper Lozère and the Monts-Dore, where the population density was low, the population continued to increase because the isolated situations made for retarded development. Here, small-scale agriculture continued to progress, encroaching on the *landes*. In contrast, in the wooded areas of lower Limousin and in the Ségalas, the growing populations were a sign of the early stages of an agricultural revolution that was starting to affect sectors not yet touched by depopulation, helping to keep the men where they were. From 1886 on, populations were shrinking everywhere but these same regions of the Ségalas, the margins of Limousin and the industrialized east were much less affected than most places. The greater than average resistance to depopulation in these areas, which was mainly a consequence of their natural features, was confirmed over the years that followed, up until the mid-twentieth century. Whereas the high, rocky regions of the south-east and the Limousin mountains became massively depopulated, the volcanic regions and the plateaux at middling altitudes, where agriculture was more progressive, were less affected. So once again, a typology of the various forms of depopulation can be established, taking account of both rhythms and intensity. There are early types of depopulation: slight (as in Cantal) or heavy (as in the Cévennes); middling types, as in most areas characterized by small-scale food-producing agriculture (Livradois); and late types, where depopulation started late as a result of the isolated location (Montagne Limousine, Margeride) or because economic progress helped to keep the population stable.

Emigration and natural population change

In the evolution of the rural population it is much trickier to distinguish between the effects of on the one hand emigration, on the other natural population change, for from the early nineteenth century onwards, the birthrate was falling in many parts of rural France. Direct calculations are possible only for the period 1872–1931, when census records showed births and deaths in rural communes separately from those in urban communes, making it possible to construct maps showing rural population changes due solely to natural causes in contrast to those occasioned by migratory movements (fig. 79C and D).[25] Figure 79D shows that the major regions where large migrations originated were the Massif Central in its entirety, Brittany and Vendée, the Nord department, the southern and eastern edges of the Paris Basin, and Franche-Comté; the number of people migrating from the Alps and the Pyrenees was slightly smaller. On the other hand, the major factor in the population decline throughout the heart of the Paris Basin, most of Normandy (except for the Pays de Caux), the Aquitaine Basin and the Mediterranean south was the falling birth-rate. These were by and large areas comprising the great plains and the principal communication routes, along which the falling birth-rate clearly spread, although this map certainly does not tally exactly with those showing other phenomena influenced by Paris, the main centre of attraction – religious practice, for instance (fig. 65B). But at any rate, there is clearly an inverted correlation between the two factors of depopulation, for the regions with a persistently high birth-rate and where the rural population was late in peaking clearly constituted the origins of the strongest waves of emigration. Figures 79C and D are to a large extent complementary.

To what extent are these results also applicable to earlier periods and may they be generally related to the phenomenon as a whole? The only way of answering that question would be by consulting the civil registers at commune level. That is quite a daunting task and it has not yet been undertaken in a systematic and coordinated fashion. However, a number of soundings have been taken, some of them relating to quite extensive regions, and these have produced interesting results. Thus, in the Massif Central, the respective parts played by an excess of deaths over births and by emigration in the depopulation of the southern Limagnes between 1836 and 1931 has been evaluated at 50 per cent for each. In the Grande Limagne, the excess of deaths accounts for 85 per cent of the population decrease.[26] But emigration is a much greater factor in mountainous regions. At Brousse-Montboisier, a Livradois commune, the natural population change accounts for only 10 per cent of the decline.[27] In the Ségala and the Châtaigneraie it accounts for no more than 5 per cent.[28] It was the uplands, particularly the rocky ones, where natural population increase continued the longest, that contributed greatly to the large migratory movements. Similar results were obtained for the Alps. Here, it has been possible to establish full demographic records for the various natural regions.[29] The total number of migrants in the nineteenth century has thus been calculated at 130,000 individuals for the whole of the Alpes du Sud and 160,000 for the Alpes du Nord. But an excess of deaths over births is detectable early on, in the mid-nineteenth century, in internal valleys such as the Grésivaudan and the lower Durance valleys. In the second half of the century this spread to the whole of the Grande Vallée and also to the transverse valleys of the Alpes du Nord. The demography of regions at high altitudes was consistently healthier than that of regions at middling and low altitudes. It was, in truth, not so much a matter of a high birthrate for, though high, this was not exceptional, but rather of a considerably lower deathrate in the higher valleys, where the mountain inhabitants, most of whom owned their own land, were relatively more affluent. Before 1850, Dévoluy thus had a mortality rate of 21.5 per thousand,

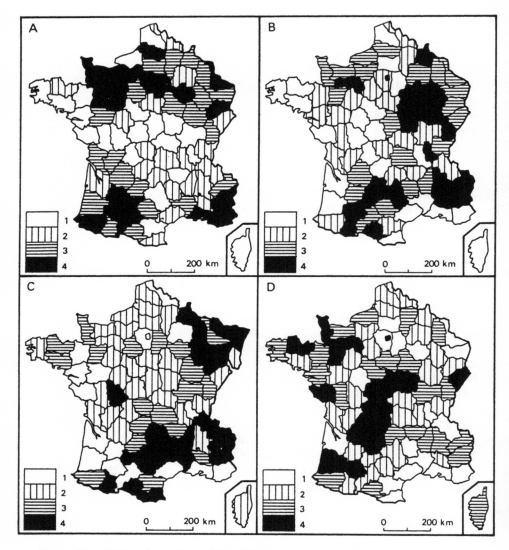

Figure 80. The rural exodus and mobility in the nineteenth century.
A. Population growth expressed as a percentage, 1856–86 (from the data provided by Merlin, 1971, and Le Bras, 1986, map on p. 195a). 1. > 0%. 2. From 0 to − 5%. 3. From − 5 to − 12%. 4. < − 12%. B. The same, 1886–1911 (Merlin and Le Bras, p. 195b). 1. > − 6%. 2. From − 6 to − 11%. 3. From − 11 to − 15%. 4. < − 15%. C. Departmental classification according to the size of emigrations between 1806 and 1856 (after E. Van de Walle, *The Female Population of France in the Nineteenth Century: a Reconstruction of 82 Departments*, Princeton, 1974, used in Le Bras, 1986, p. 198a). 1. From 84th to 64th rank. 2. From 63rd to 43rd. 3. From 42nd to 22nd. 4. From 21st to 1st. D. The relation between the emigrations of 1856–1906 and those of 1806–56, measuring the speed of the increase (Van de Walle, and Le Bras, p. 198b). 1. < 0.95. 2. 0.95–1.05. 3. 1.05–1.10. 4. > 1.10.

compared to 33.9 per thousand in the neighbouring mid-Durance area. Between 1823 and 1852, the Préalpes du Nord had a deathrate of 23.6 per thousand, compared to 25.6 per thousand in the transverse valleys, situated in between them. Between 1843 and 1902, the deathrate in the upper valleys of Maurienne and Tarentaise was 23.4 per thousand, compared to 27 per thousand in the lower valleys. However, these calculations based on altitude need to be complemented by other comparisons. The birthrate was always higher in Savoy than in Dauphiné. The falling birthrate spread to the Préalpes du Sud in the second half of the nineteenth century, while the deaths rose by 39,000 between 1852 and 1912, putting the deathrate at 27.3 per thousand compared with a birthrate of 24.2 per thousand: this accounted for over half the decrease in the population, calculated at 75,000 individuals.

The rhythms of migration

These detailed analyses make it possible to provide at least some answers to a particularly complex and controversial question, that of the role played by the railways: did they encourage the rural exodus or did they act as a brake on depopulation? The problem seems the more pressing given that the development of the railways was taking place more or less at the same time as depopulation was speeding up. The results of the enquiries that have been carried out on this subject are complex.[30] Positive correlations may certainly be found. The traditional temporary migrants used almost invariably to travel on foot. The railway simplified the business of moving from one spot to another. By making it possible for families, wives and children to join the migrant workers, it lent encouragement to definitive departures. Furthermore, by unifying markets of agricultural produce, it certainly helped to destroy the fragile economic balance of overpopulated rural regions and, at the same time, made it easier for migrant workers to return temporarily to their native villages and parade their increased affluence before those who had stayed at home. All this certainly encouraged more people to move away. On the face of it, then, the railway would appear to have encouraged rural depopulation, for in many places the increase in depopulation and large-scale definitive migrations certainly follows closely upon the arrival of the railway. But in truth, that is a somewhat simplistic interpretation of the situation. In plenty of cases, rural depopulation began as early as 1836–46 (Orne and Manche, Basses-Alpes and Hautes-Alpes), well before the arrival of the railways. In fact, by creating the possibility of finding industrialized or urban types of employment, the railways may well, on the contrary, have stabilized the population. It has been shown that, in the Masevaux valley (the Vosges of Alsace), the development of this means of transport, which put an end to isolation, slowed down the exodus. In the Massif Central, detailed regional analyses have shown a beneficial effect upon the population. In the southern Limagnes, there were four times as many emigrants during the decades before the arrival of the railways as there were in those that followed it;[31] and identical results have been obtained for the rocky plateaux of south-western parts of the Massif.[32] In the plateaux of upper Provence, continuous five-yearly records establish no direct link between the phases of maximum emigration and the establishment of railway lines and stations.[33] A number of particularly detailed studies have thus qualified earlier conclusions and produced some selective answers. A study of the influence of the railway line from Paris to Menton between 1866 and 1936 has shown that the populations of communes provided with railway stations increased greatly overall, although not in every individual case: in half of these places the populations actually decreased. The belt of dwellings less than 5 kilometres distant lost fewer inhabitants than the surrounding countryside did; and the greatest exodus occurred in the band situated at 5 to 10 kilometres distance from the railway or, in some cases, in the band 10 to 15 kilometres distant. At over 15 kilometres' distance, the presence of the railway made no difference at all.[34] A reasonable general conclusion would seem to be that the railway was mainly instrumental in redistributing populations, especially at a local level. But at a regional level, it helped to maintain the populations of the areas into which it was introduced. Like the temporary migrations, but at a later date, it had a moderating effect upon the process of French urbanization.

In sum, the overall maps of the phenomenon of workers' migrations show that it was a complex process. In the first half of the nineteenth century, migrations were particularly active in the southern part of the Massif Central, the Alps and the Pyrenees, but they also took place throughout Lorraine and Alsace, in the east. In the second half of the century, the process speeded up, particularly in the countryside of western France: lower Normandy, Brittany, Vendée and Maine; and in the north-west of the Massif Central and the southern part of the Paris Basin. This map makes two important points: it shows that there were some regions (such as Brittany and Vendée) that were hit by the rural exodus at a relatively late date (fig. 79A) without having been much affected by earlier temporary migrations; and that there were others (the north-western Massif Central) where temporary migrations to a large extent preceded and delayed the definitive exodus.

The geographical aspects of urban growth

The process of urbanization: rhythms and categories

The migratory movements fed the towns; and their growth may be precisely gauged in the nineteenth century, since regular records were kept from 1801 on and particularly after 1811.[35] If we take the lower limit for towns to be 2,000 inhabitants (which was the legal definition in France), their total population almost trebled in the nineteenth century, passing from 5.7 million in 1801 to 15.9 million in 1901: an increase of 180 per cent, compared with an overall population increase of no more than 48 per cent. This brought the urban population of France to 40 per cent of the population of the territory as a whole, at the beginning of the twentieth century. During this period, large towns grew much faster than small ones. The nineteenth century confirmed the status of the major cities which emerged out of the relative stagnation which affected the rest of the urban network. In 1801, there were 1.69 million people living in the thirty largest towns in the country, that is to say 6.4 per cent of the total population; by 1901, that figure had risen to 9.26 million, or 23.8 per cent. The share of those thirty towns in the total number of urban dwellers in France had meanwhile risen from 29.8 to 58.22 per cent. However, the rhythms of growth in the second half of the century were very different from those in the first half.

Between 1801 and 1851, the urban population of France increased by only 60 per cent, not so very much more than the total population of the country, which increased by 36 per cent. Taking into consideration only towns of 3,000 or more inhabitants, it rose from 4.2 million (14 per cent of the total) to 6.4 million (18 per cent) between 1811 and 1851. At this period, the rate of urban growth was possibly even lower than at the end of the Ancien Régime. What growth there was took place almost exclusively at the top end of the scale. The population of Paris doubled during the first half of the century, increasing to 1,247,000 by 1851. The population of towns of over 50,000 inhabitants rather more than doubled, increasing from 498,000 to 1,022,000 between 1811 and 1851. But over that same period, the population of towns of between 10,000 and 50,000 inhabitants increased by only 38 per cent, little more than the growth in the overall population, while the population of towns of between 3,000 and 10,000 inhabitants increased by only 31 per cent, which was less than the rate of increase for the overall population. Quite a few small and middle-sized towns shrank, in some cases by over 10 per cent, particularly in the Massif Central, which continued to be the very opposite of a pole of attraction (Riom, Billon, Clermont-Ferrand, Brioude), in the Mediterranean and Alpine south and also in Normandy (Valognes, Vire, Bayeux, Falaise), where a crisis of de-industrialization was already developing as the dispersed rural industries declined.

The urban revolution dates essentially from the second half of the century: between 1851 and 1901 the population of towns increased by 75 per cent, while that of the country as a whole grew by only 9 per cent. Between 1851 and 1911, the population of towns of 3,000 or more inhabitants increased from 6.4 million to 14 million, or from 18 to 35 per cent of the total population. Again, it was chiefly the major towns that were responsible for the rise. The population of the thirty largest towns rose from 2.9 million in 1851 to 9.26 million in 1901. The rates of growth in all these towns were

402 *Centralization and diversification*

higher in the second half of the century than in the first, except in the case of Toulon, which had expanded in the first half thanks to its growing military importance as a naval base. They were also higher than the average rates of urban growth. But in general the same was also true of other towns of over 10,000 inhabitants, although there were numerous exceptions. The process of urbanization was widely distributed. Nevertheless, at the bottom end of the scale, expansion was much slower. Between 1851 and 1911, the total number of towns of over 3,000 inhabitants rose from 567 to 794 and this was a steeper increase than that of towns of between 5,000 and 10,000 inhabitants and towns of between 3,000 and 5,000 which increased from 155 to 222 and from 304 to 365 respectively. Despite the absolute increase in their number, the proportion of towns of between 3,000 and 5,000 inhabitants, out of the total urban population, continued to fall. In 1811, there were 183 of them, that is 16.7 per cent of the overall urban population; in 1851, 304, that is 17.5 per cent. In 1911, their number had increased to 365 but they now accounted for no more than 10 per cent of the total. In the second half of the century, urbanization speeded up and extended its bases, but the smaller towns were still excluded from the expansion.

The factors of urbanization

What were the basic causes of this process? They become clear enough once we draw up a list of the towns in which growth was particularly rapid. In the first half of the century, they record was held by Roubaix, which increased its population fivefold, while Le Havre quadrupled its population, Mulhouse, Tourcoing and Saint-Etienne tripled theirs and thirteen other towns doubled in size. Out of the thirty largest towns in France, not counting military ports (Brest, Toulon), it was the industrial towns (Saint-Etienne, Limoges, Rheims) which developed the most rapidly – even faster than more complex major cities (Marseilles, Toulouse, Metz). Paris came in nineteenth place. Six out of the eight largest provincial towns expanded at a considerably lower rate than the average growth of the overall urban expansion: Nantes, Rouen, Bordeaux, Strasbourg, even Lyons. In two of them (Rouen and Nantes) and likewise in Aix, Troyes, Nancy and Amiens, the rate of growth was even lower than that for the total population of the country. In a number of other towns, it was barely equal. All this indicates clearly that in this period tertiary administrative functions and the status of regional capital were, on their own, insufficient to bring about urban expansion, which remained an extremely selective process. In this first half of the century, the principal motivating force for growth was industrialization. And, essentially, it was still the textile industry that was involved. The towns that headed the list of expanding centres were, apart from a few ports and metallurgical centres such as Saint-Etienne and Rive de Gier, mainly textile towns: Roubaix,

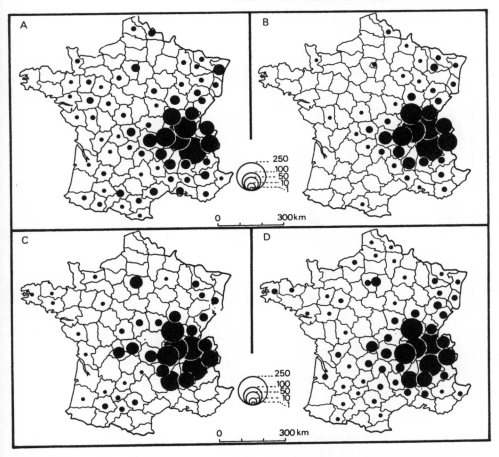

Figure 81. The evolution of the urban picture in the second half of the nineteenth century (after Lequin, 1977).
The origin of migrant workers who married in Lyons in 1851 (A and B) and in 1911 (C and D). On the left (A and C), the men. On the right (B and D), the women. Between the two dates, the mobility of women increases noticeably, while that of men shows little change.

Le Cateau, Mulhouse, Tourcoing, Elbeuf, Saint-Quentin, Bolbec and Mazamet. At this point, the effects of the presence of the railways were as yet imperceptible except in a few cases, such as Vierzon and Périgueux.

Between 1851 and 1911, industry was still a predominant factor. In the Nord–Pas-de-Calais basin, urban growth was remarkable. Roubaix came in sixth place, Hénin-Liétard in seventh, Tourcoing in ninth. Anzin, Halluin, Armentières and Denain were all included in the list of towns expanding at a

Centralization and diversification

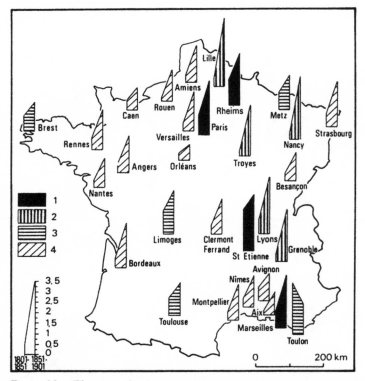

Figure 82. The rate of urban growth in the nineteenth century (after Clout, 1977d, pp. 529 and 533).
1. Growth rate greater than the average for the 30 largest towns during the 1801–1901 period. 2. The same, 1851–1901. 3. The same, 1801–51. 4. Growth rate lower than the average for the 30 largest towns during the 1801–1901 period. Scale of the diagram: 1. = 100% growth (doubling). 2. = 200% growth (tripling), etc. (Nice, which became attached to France in 1860, has not been taken into consideration).

rate of over 300 per cent, as also were the towns in the coalfields and the metal-producing areas on the edges of the Massif Central: Firminy, Montluçon, Decazeville, Le Creusot. But new factors were emerging. The records of relative growth were held by Nice (844 per cent) and Cannes (519 per cent), a fact that underlines the emergence of tourism as a factor of urbanization. The rankings of Wattrelos (in third place), close to Lille, Caudéran (tenth), close to Bordeaux, and Sotteville-lès-Rouen (eleventh) testify to the new role of suburbs. Meanwhile, the growth of the major cities was particularly rapid. The major phase of expansion in Paris occurred between 1851 and 1881. By 1911, with 4 million inhabitants, the capital accounted for one quarter of the urban population of the country. That was the peak point: from the 1930s on, that proportion began to slip. In the second half of the twentieth century, the rates of growth of Paris, Lyons, Lille and Nancy have all been higher than that of Saint-Etienne. Next come Marseilles, Rheims, Grenoble, Troyes and Strasbourg. The catchment area of the large cities expanded, particularly for women, many of whom found employment in domestic or other services (fig. 81). From this point on the urban phenomenon of expansion fuelled itself. Tertiary and administrative functions now complemented the effects of industrialization.

Many of the towns where tertiary functions developed at the end of the nineteenth century owed their initial spurts of expansion, in the first half of the century, to industry or maritime trade. Over the hundred years between 1811 and 1911, Le Havre (whose population increased by 700 per cent) and Lyons (440 per cent) grew faster than Paris, and the most extraordinary rates of growth were those of towns in the north and the coalfields in general, with the textile industry still unquestionably the predominant factor over the century as a whole (Roubaix 2,600 per cent; Tourcoing 1,400 per cent; Saint-Quentin, Roanne, Armentières, Cholet, Saint-Dié and Mazamet 300 to 500 per cent). Metallurgical industries and coal mining, together with maritime trade, were the next most important factors (Le Havre, Calais, Marseilles, Boulogne and Sète).

Over all, however, the importance of industrialization should not be exaggerated. A peculiar characteristic of urban growth in France, as compared with that in other European countries, is precisely that the development of towns was relatively gradual. The general law according to which the level of growth is proportional to the hierarchy of town sizes during phases of rapid urbanization (2 per cent or more per year) applied only briefly in France.[36] The variability of rates of urban growth in relation to the sizes of the towns involved has diminished over the years. Overall, the geographical face of France has been less marked by dramatic urbanization in the Industrial Revolution of the contemporary period than other countries in western and northern Europe. All the same, such as it was, that revolution certainly did produce a number of important repercussions on the inter-regional balance of the country.

Geographical consequences: new regional dynamism

If we map the respective rates of growth of the largest towns over both halves of the nineteenth century (1801–51 and 1851–1901), we notice that only a handful of cities show a greater than average rate of growth over both periods: Paris, Marseilles, Saint-Etienne and Rheims. All four are situated in the eastern half of the country. Equally, the towns with a higher than average growth rate between 1851 and 1901, the period when urbanization was most intense (Lille, Nancy, Troyes and Grenoble), are also all situated on the eastern side. Meanwhile, the cities with a lower than average rate of growth throughout the century are essentially situated in the west, the south-west and the southern interior. The same applies to middle-sized and small towns. And if we draw up a list of the towns whose populations dropped in the course of the century (headed by Arbois, Tarascon, Falaise and Riom), we notice that the greatest losses were suffered in Normandy, the inland Mediterranean south (both regions which were industrialized at an early date but whose importance in the French economy diminished in the nineteenth century) and in the west and the south-west generally.

A major geographical contrast thus appeared in the course of nineteenth-century urbanization. It set up an opposition between a dynamic France, by and large in the north and the east, and a more sleepy France in the west and the south, with the line of division running roughly from the Seine to the east of the Massif Central and onto the Rhône estuary. That is no longer quite the same as the line running from Saint-Malo to Geneva which marked out the border between progressive France and backward France at the beginning of the century. The new factor is the dynamism of the Lyons–Saint-Etienne–Grenoble region and Marseilles. In the twentieth century in particular, that dynamism spread to the Côte d'Azur. In the second half of the nineteenth century essentially, the railways brought new energy to the isthmus of France, with its Rhône–Saône corridor. The growth of Paris was accompanied by that of Lyons and Marseilles, both of whose growth rates over the century as a whole are greater than the national average. This pattern was to persist up until the point when, following the Second World War, new factors made their appearance: in particular, industrial decentralization and the growth of towns in the Loire valley. The geographical contrast that we have noted reflects above all a new economic organization of the country, based upon a distribution of industry infinitely more varied than that of traditional France (see chapter 6) and also upon the major arteries of communication linked with that Industrial Revolution.

The formation of the urban networks

As towns grew and increased in number, they were organized into hierarchical networks characterized by the establishment of regular links of subordination between them.

The historical development of the process

The establishment of an ordered hierarchy was an innovation of the second half of the century. As we know, nothing of the kind existed in pre-Revolutionary France (see above, p. 215). And even up until the mid-nineteenth century, the urbanization of France went ahead with no change in the relations between the various towns. Around 1850, there were still no urban networks organized as administrative structures. In Alsace, for example, the number of towns increased considerably between 1750 and 1850.[37] By 1850, in the lower Rhine plain, there were five towns and thirteen large villages with a variety of urban functions where, a century earlier, there had been but one town and four villages. Sundgau, in the upper Rhine valley, contained seven large villages in 1850, where in 1750 there had not been a single one. Elementary urban settlements were now scattered all over the plain of Alsace. With their creation, regions which in the past had lived according to a régime of almost total self-sufficiency awoke to a life of more active communications. Erstwhile central villages had been promoted to the status of proper little towns and their functions had multiplied. But the dissemination of town life had not been accompanied by any hierarchical organization. The scattering of towns had become more regular, covering more of the territory but, generally speaking, this did not mean that any links of subordination had yet been established. Strasbourg certainly dominated part of the region's commercial activity, but only a part, and the range of its commercial relations extended far beyond the limits of Alsace. Only around Mulhouse do we find an ordered hierarchy developing at this point. The early industrial development of this town, the establishment of a Stock Exchange there for cotton-dealing and its attraction for a large external workforce had already combined to confer upon it a manifest predominance over the surrounding area. But that was an isolated case. In the vast region comprising the southern Alps as a whole, the two most highly populated towns, Gap and Manosque, were only two or three times as large as the rest.[38] Only Gap possessed a relatively wide range of contacts, thanks to the fairs that were held there. Wholesalers were already numerous in the region, but one town was no more important than any other. The area consisted of fifteen small towns and many large villages that acted as local centres, but no organic links existed between them. There were numerous banks which, between them, already catered for virtually the entire rural area, but they had established no branches and each one served a strictly limited area. Other detailed analyses, carried out in a wide variety of regions, paint a similar picture.

In truth though, in the urban structure of pre-industrial France, it *is* already possible to make out several levels of importance, as regards a whole category of functions. Below Paris and half a dozen other major centres, three basic levels have been noted,[39] but the distinctions between them depend almost exclusively upon the nature of the administrative functions that they had acquired. By and large they divide up into prefectures, sub-prefectures and the *chefs-lieux* of certain cantons. These were assigned administrative functions at various levels within the organizational framework of the law, the church, the military and the universities. Parallel distinctions of a cultural order are also manifest: scholarly societies and academies, which played a role of prime importance in the intellectual life of the period, only existed in the highest category of towns. Even at this point, regular hierarchical links ensured the efficient grip of the centralized state. Despite the slowness of communications (in the mid-nineteenth century, a journey by coach from the north to the south, or from the east to the west, would take a minimum of 150 hours), this already functioned satisfactorily throughout the territory, thanks to the regular hierarchy of administrative centres, at the same time spreading the influence of the dominant culture through every literate layer of the population (see above, pp. 279–80). The rationalization of the spatial division of the land that the departmental organization represented and the centrally controlled systematization of the administrative organization, through reforms introduced by the Revolution and by Napoleon, had once and for all created a framework capable of coping with an harmonious hierarchy of functions. But throughout the first half of the nineteenth century, that framework remained remarkably underused. The effects of the hierarchical system that had been set up were felt hardly at all in daily life. Very seldom did the mass of the population in this pre-industrial society have recourse to the services provided by the state at higher than local levels. Economic life, that is to say the manufacture and commercialization of products, took place within the framework of a purely local system of unorganized exchange, apart from a tiny quantity of luxury goods which were exported to distant parts. Not until 1836 did provincial branches of the Banque de France, which had been founded at the beginning of the century, begin timidly to make an appearance in a number of important centres.

The urban networks, in the strict sense of the expression, came to life in the aftermath of the transport revolutions and the constitution of a single national market. These led to the emergence of a limited number of major towns and, concurrently, to the decline of smaller centres, which gradually lost their function of economic and social leadership and turned into mere intermediaries between those larger towns. In the second half of the nineteenth century, many small country towns even lost part of their population and became no more than points of departure for an emigrating workforce. The local bourgeoisie either disappeared or was forced to adapt. Manufacturers and autonomous workshops catering solely for the immediate neighbourhood were replaced by traders who distributed products brought from far afield, from the specialized agricultural and industrial regions which were gaining a firmer footing.[40] Meanwhile, as communications improved, the villages which had traditionally acted as local centres were short-circuited, supplanted by larger towns, and lost all their functions. In 1850, there were 285 *notaires* (notaries) in the Alpes du Sud, 230 all operating from different localities (one in four of its communes). By the mid-twentieth century, there were no more than 61 distributed among fifty localities.[41]

It is possible to pick out a number of decisive factors in this evolution. The establishment of banking networks has been carefully analysed in a study which, though centred on the field of influence of Lyons, is definitely of great general relevance.[42] In the mid-nineteenth century, many cantonal *chefs-lieux* had a bank of their own. The more remote and inaccessible a district was, the more autonomous financial agents it spawned. For example, the Dombes, an area studded with countless small ponds, held the record for the density of banking establishments. Local savings banks, which had proliferated since the legislation introduced under the Restoration, from 1818 on, and hosts of *notaires* and local merchants who acted as money-lenders, also played their part in the circulation of money. Local branches of the Banque de France were not set up until 1840 in Grenoble, 1848 in Lyons. And it was only in 1848 that a true embryonic financial unity was created throughout the national territory when – basically as a consequence of the revolutionary crisis – notes issued by the Banque de France were made the compulsory currency. The real breakthrough dates from the Second Empire and the creation of large deposit banks such as the Crédit Lyonnais (1863) and the Société Générale (1864). It was not long before a Lyonnais banking region became defined, stretching from Chalon-sur-Saône to Valence and from the Loire to Savoy, with agencies at Roanne on the one hand and Aix-les-Bains, Chambéry and Geneva on the other.[43] Meanwhile, a network of regional banks was becoming established. Trade was also becoming organized at this period. In upper Auvergne,[44] under the Second Empire, virtually every commune possessed a general store, which acquired its stocks in a totally disorganized fashion, drawing upon a wide variety of sources. However, the second third of the century saw the arrival of wholesale firms which gradually extended their sphere of influence, distributing products from all over the national territory. They first established themselves at Saint-Flour, which commanded the inclined plateau ('*planèze*') of Cantal and used to extend its influence over the Lozère region and the upper Loire; then at Mende and Le Puy, which both went on to become centres in their own right. In Aurillac, where the railway arrived between 1866 and 1868, they established themselves from 1875 onwards, acquiring a wide field of relations not only in the Massif but throughout western Aquitaine. As a result, this intermediary town, which was favoured by its fortunate position at the foot of the mountain, acquired a commercial importance out of all proportion to the size of its population. Aurillac's field of influence in respect of groceries was relatively limited (with a radius of about 100 kilometres), for a number of other towns competed with it in this domain. However, the predominance of its wholesale firms in haberdashery, confectionery, ironmongery and stationery was unrivalled as far afield as the limits of the zones of influence of

Bordeaux and Toulouse, and in some respects even up to the very gates of those cities.[45]

During the second half of the nineteenth century and the early years of the twentieth, hierarchical relations in every area of economic life developed between the various urban centres. Towns which had hitherto been largely independent now developed interrelations within organized networks. Urban functions multiplied and diversified and the larger centres increased the variety of the services that they offered and also the geographical range within which they offered them. Meanwhile, intermediate and smaller centres declined or even disappeared, as towns. However, this was a gradual and uneven process. In mountainous regions, where the altitude made for isolated communities, the pace was very slow. Some urban networks did come into being here, but they were affected by a series of particular circumstances. In the first place, a time-lag: in the Alpes du Nord banking networks were only set up in the period between the two World Wars.[46] In upper Auvergne, it was not until that same period that, as a result of the automobile revolution and the development of bus services, wholesalers managed to establish a firm hold, for the arrival of the railways, somewhat sparse in this region, had not been totally adequate for their purposes. Now they moved into manufacture and became the bankers of the retail merchants.[47] Secondly, in these areas, the system developed was incomplete: the classic schema of subordination at several levels assumed by central place theory was only partially realized in the Alpes du Nord. Even in the 1960s, many middling and small towns still manifested a measure of autonomy vis-à-vis the Savoyard *chefs-lieux*. A number of intermediary centres had held on to their tertiary functions. Finally, regional polarization remained incomplete, for no undisputed major centre had emerged here.[48] In the plains, in contrast, where access was easy, the process had already reached completion by the eve of the 1914 war, by which time the banking networks were more or less completely established.[49]

The principles of hierarchical development

What were the decisive factors in the selection of major urban centres? And to what extent was this process of increasingly clear hierarchical differentiation reflected in the varying rates of growth of the towns involved? Given that the phenomenon was a consequence of the new means of circulation, the most obvious idea is that the major selective factor was constituted by the expansion of relations and in particular by the development of the railway links, for the railway was the principal new means of circulation in this period. Initially, that idea was accepted without question. It was believed and stated that, in the nineteenth century, the railway acted as a factor of urban discrimination[50] and that 'rail' had been a 'very powerful element in the modification of the urban network'.[51] A study devoted to north-eastern France provided statistical confirmation of this view for the 1846–1936 period,[52] emphasizing the clear correlation between urban growth and railway accessibility measured according to the number and importance of the relevant lines. But what exactly did this positive correlation show? That is just what other research workers were wondering. To them it seemed that 'the railway above all confirmed previous tendencies', as is shown by the relatively modest development of towns born from railway junctions. These researchers were of the opinion that although the railway may well have promoted changes in the urban hierarchies, it was doubtful whether it was principally responsible for those changes, since the need for the railway companies to make a profit naturally led them to consolidate already viable economic positions rather than create new situations.[53] An in-depth quantitative analysis justified these doubts.[54] The populations of towns not served by the railway certainly fell, but even before the spread of the railway, the populations of those towns had already been falling to lower levels than those of the towns that were reached by the railway. The towns that were not included in the railway network were left out of it precisely because they were small and not expanding and they had simply retained those characteristics. The railway plan tended to be plotted in relation to the existing urban hierarchy. The larger the towns were to start with, the sooner they were reached by the railway; and the greater the number of lines that met at their station, the faster the towns grew. However, if the factor of the difference in size between the towns in question is discounted, the advantage of being served by the railway no longer appears absolute. If one compares the rates of growth of towns incorporated into the railway network at the same time, both before and after the arrival of the railway, the average level of growth turns out not to have been affected by the presence of a railway line. Overall, the presence or absence of the railway was not a factor in demographic evolution. Nor does the early arrival of the railway appear to have been a factor of greater growth, for that arrival seldom in fact coincided with the speeding up of a town's growth.

All in all, the railway did not make much difference to the varying growth rates of towns. However, the author of this authoritative analysis also studied the reorganization of the urban hierarchy. Towns operating within the same network could now be separated by longer distances and intermediary towns declined. The new infrastructure favoured an increase in both the sizes of the towns that fulfilled administrative functions at various levels and the spacing between them. But at the same time, towns in general were all growing and the range of their urban functions was undergoing considerable changes. The most satisfactory hypothesis is thus that it was the pre-existing pattern of urban functional distribution and the relative position of individual towns within that pattern that determined their

412 *Centralization and diversification*

subsequent evolution, and that the railway essentially simply adapted to that situation. If one considers the hierarchy of the 141 towns with over 10,000 inhabitants as it was in 1831, one realizes that on the whole little has changed since then and that the hierarchy has remained remarkably stable. The most important changes in urban status took place between 1841 and 1856 and such changes were very rare after 1876. As for the 1831 to 1911 period, the most significant change during those years was the development of second- to fourth-rank towns (Marseilles, Lyons, the Lille area) and those in the ninth to seventeenth ranks, for the second- to fourth-rank towns seem to have squeezed out those that ranked immediately below them. This stability is a striking characteristic of the French urban system and is connected with the manner in which the country became industrialized, for in France the industrialization process in the main made use of the already existing urban distribution, which the Industrial Revolution did little to change.[55]

It would thus seem that, although the hierarchization of economic functions actually took place in the second half of the nineteenth century, it did so essentially upon bases created in the first half of the century through the systematization of state control and the establishment of administrative functions, themselves based on a pattern of urban distribution which was to a large extent the legacy of two millennia of evolution. Of all the countries of western and northern Europe, France is probably the one in which the urban complexes of contemporary times owe the most to their heritage from earlier periods. The 'real' France of organized relations, the national territory as it is now lived in, which took shape during the second half of the nineteenth century, emerged directly from the 'willed' France organized through the deliberate action of the central authorities of the Revolution and the Napoleonic period and the constraints that they imposed. The Idea, embodied in the state, at last found expression in Life.

Conclusion: The France of Ardouin-Dumazet

Late nineteenth-century and early twentieth-century France was at once more united and more diverse than it had ever been. It was more united because the national Idea, already powerful in the past, but remote, had finally found expression in a common language and a common culture and in a constantly present image; and also because a web of interrelations had developed specifically within the national territory, which now constituted a single economic entity, dominated by a network of urban centres organized as an ordered hierarchy. At the same time, it was also more diverse, because the peasant masses, still numerous and numerically predominant, had reacted to pressure from the Parisian cultural model by declaring their need for an identity of their own and expressing it in countless local idiosyncrasies, in an atmosphere of vibrant social competition; also because major currents of ideas were now widely diffused and interpreted in the light of political and religious patterns of behaviour that varied from region to region; and because the establishment of the great national market had revealed the agricultural and industrial possibilities offered by the variations in climate and types of soil, the particular localities of raw materials and sources of energy and varying commercial situations. United though it was by common aspirations and the desire for revenge that, since 1871, had been fostered by the country's misfortunes, France was tearing itself apart, constantly confronting itself as it reinterpreted the great dramas of its national history, or even lesser incidents. The France of the Third Republic was an ordered and delimited territory, but a prodigiously complex one which was, moreover, affected by sometimes tragic tensions. Yet it became a concrete, living, functional reality. The Idea became flesh. But even as it did so, it presented a bewildering range of different aspects: some dull, others brilliant; an abiding rusticity alongside sophisticated elegance and urbanity; primitive conditions but at the same time the promise of a civilization of advanced technology. To provide an exhaustive description of such a disconcerting world might well have seemed an impossible task.

Yet one man dared to attempt it. In the sixty-six volumes of his *Voyage en France* (1890–1921), Ardouin-Dumazet became the rigorous and tireless analyst of this ebullient world, bearing witness to it in a most exemplary fashion. Michelet's *Tableau de la France* (1833) had hymned the glories of the land of France, celebrating its unity over and above the disparity of its various provinces, in unforgettable style ('A great light shone forth, and I perceived France'; 'France created France. It is the daughter of its liberty'). Elisée Reclus' *La France* (1879) belongs to the same tradition. Quite apart from providing a wealth of picturesque detail and statistical information, it stands as a work of great eloquence and feeling. Ardouin-Dumazet's monumental work, not finished until after the First World War but conceived at the beginning of the last decade of the nineteenth century, was produced over a period that corresponded exactly with the phase of maximum economic diversification, the years which saw the hierarchical ordering of the urban networks and the intensification of commercial relations which heralded the advent of the great national market. The very form of the work and the way that it evolved are significant. Before ever being collected into volumes, it was undertaken as a series of articles for the newspaper, *Le Temps*, intended to present a picture of French economical activity in 1890, by concentrating on a handful of typical regions. But it soon took on the air of a systematic composition designed to provide a description of the landscape and the people of the whole of France, together with an evaluation of their work. The organization of the sixty-six volumes is revealing in itself. No overall view is attempted; at the most the odd passage produced by way of a conclusion hazards a few general remarks on the subject of particularly unusual and individual fragments of the national territory, as for example in the case of the volume devoted to Corsica (14th series, 1898). The method adopted is that of an itinerary, but a very complex one. The work is divided into chapters each of which bears either the name of a small, in some cases microscopic, region, which is then further sub-divided, or else that of a particularly remarkable type of agricultural or industrial activity. The hive is thus divided into countless cells but ones which, unlike those made by the bees, are by no means uniform but on the contrary all different and all worth a special examination. The task was truly a labour of Sisyphus, for within the space of a single human generation, the situation changed, becoming ever more complex. In 1899, when the work had been in progress for a decade, it was planned to take the form of thirty-three volumes. In 1907, it was brought to an apparent conclusion (omitting Paris and its suburbs) with the completion of fifty volumes, the last three of which were devoted to the provinces that had been lost. Twenty years later, however, the author felt compelled to return to work and it was not simply to produce complementary sections and to bring the survey up to date. Instead, he undertook a total recasting of the earliest volumes. A comparison between the successive versions reveals how much they now gained both in detail and in depth, reflecting the extent to which the economy and the human landscape was constantly diversifying. Volumes 51 to 63 (preceding the last three, which were devoted to the Parisian suburbs) thus presented considerably more detailed pictures of Brittany and Normandy in the west, the Loire region, the Boulogne and Artois area, the Rhône valley, the Côte d'Azur, the Vosges and the provinces recently recovered.

But already new dynamic forces were appearing in the mosaic. Even as Ardouin-Dumazet's work was in progress, Paul Vidal de La Blache's *Tableau de la géographie de la France* (1903), a work of intelligence and thoughtful passion, was heralding a new phase of scientific thought. It involved description of a more organized and considered kind and a new overall view. This work sought to determine those features in the face of France that had resulted from influences both from overseas and from elsewhere on the continent. Ultimately, the dialectic between unity and diversity was, inevitably, to be resolved in favour of the former. The creation of the great national market, over and above the skin-deep reactions that had found expression in the development of local peculiarities and specialities, was logically bound to lead to a higher level of organization.

11
The France of large organizations

At the end of the nineteenth century France had, curiously, come to a halt mid-way through the process of modernization which characterized European societies generally. In many areas of technology, science and social organization, France had led the way with her innovations, but these had not upset the pattern of social life or the organization of the national territory as profoundly as they had in other countries. Late nineteenth-century England was already an urbanized nation, in which many traditional cultural variations had been erased. France, on the other hand, gives the impression of a country in which two different forms of diversification were at work: in the first place, traditional diversifications which resulted from small units closing in on themselves, by reacting against the surrounding society; secondly, modern diversifications, stemming from the specialization of work in a world where products could at last circulate easily and where population movements were speeding up.

Up until the Second World War, France was thus divided between the past and the future: at the time, it was thought that she was to be congratulated upon the moderation which warded off some of the effects of modernity: while Germany, with its Nazism, presented a terrifying image of what a distorted civilization of the masses may turn into, France remained protected by a thousand and one inertias. In a single generation, the old world was swept away. And then the 'glorious thirties', as Jean Fourastié called them, witnessed the disappearance of the peasantry and the depopulation of most of the countryside. The number of city-dwellers increased as much during the first three decades after the Second World War as it did between 1801, when the first census was taken, and the end of that war.

The France of today has lost some of its social and economic peculiarities: it is a less unusual country than it was in yesterday's Europe. That is an inevitable consequence of the institution of the Common Market, the development of a world economy and the speeding up of trade and travel. Yet, although it has clearly come to resemble other European countries in many ways, analysis shows that many deeply rooted individual characteristics have survived. To understand the France of large organizations it is not enough to turn to the countries from which it has imported some of its new models of organization – the United States, northern Europe, Germany and, latterly, Japan. We must also understand the forces that have been at work within French society itself ever since the early days of the Industrial Revolution: from that period on, large organizations began to stamp their mark upon French life.[1]

Modernization: the specific features of the French experience

The big business companies that have been multiplying over the last generation fit into the dynamic pattern of modernization that has been detectable in western society ever since the Renaissance and has been speeding up ever since the Industrial Revolution.

Conditions of unequal favourability

Modernization implies economic change: that is what first springs to mind – sometimes to the point of obscuring other aspects of the process. At the end of the eighteenth century, French geography was transformed by the creation of a system of royal highways and a proliferation of canals and rivers adapted for navigation. A national market had begun to develop even before the appearance of the railway. Within the space of a single generation, between 1840 and 1870, the construction of the railway network completed its evolution (fig. 57).[2]

Economic modernization made it necessary to mobilize new forms of energy. France had at her disposal waterfalls and coal. She was not as rich in coal as her neighbours to the north, Great Britain, Belgium or Germany, but her resources were sufficient to allow most of her industries to develop rapidly.[3] They were not distributed as conveniently as in some countries, since vast regions had no coalfields at all; but hydro-electric power to some extent corrected the imbalance, offering the mountainous regions of the south-east a chance.[4]

Secondly, economic modernization necessitates the establishment of an efficient system for the circulation of money and the collection of savings, with a view to encouraging productive investments.[5] The hazardous monetary experiences of the eighteenth century – Law's System and the promissory notes issued by the Revolutionary government – made for prudence. The establishment of a modern banking system is even more dependent upon the evolution of notions of business law than upon the development of transportation infrastructures. That is probably partly why progress in this area was so slow and hesitant, but it was also hampered by the heavy controls exercised by the Ministry of Finance and the Banque de France. In 1837, it was decided that all the funds accumulated in savings banks would be transferred to the Caisse des Dépôts et Consignations: it was money that was reserved for the use of the state instead of for boosting local investments. The Banque de France was very slow to set up branches in middle-sized towns, as a result of which it was not possible to reinvest commercial profits there until well into the nineteenth century. The banking system only really began to expand under the Second Empire, but the preponderance of state funds soon caused it to be centred on Paris, in contrast to what was happening in other European countries.

The use of mortgage loans facilitates the modernization of any economy by making it possible to mobilize wealth in the form of urban property or real estate, to help industry and commerce. France was not unfamiliar with that type of credit, but the conditions surrounding it made it very unattractive. It was used only as a last resort: here again, the state penalized French society, depriving it of such facilities, imposing conditions that were hard to tolerate and thwarting whole regions in their efforts to develop, by preventing them from diversifying from their rural bases into new sectors of activity.

Finally, economic modernization also implies the creation of firms capable of making the most of the new economies of scale that mechanization makes possible and of finding in large markets outlets for the growing volume of commodities that they are now in a position to produce. The problem is a legal one more than a matter of technology: the creation of limited companies makes it possible to multiply sources of capital. Without them, big business would not have been possible. The prudent nature of French legislation put a brake upon progress in this area: the commercial code of 1807 stipulated that no limited company could be created without authorization from the government. More modern legislation was not introduced until 1863.

It was only under the Second Empire that conditions at last became favourable for large companies. Before that, big business had not been totally impossible, but the most advantageous legal forms in which to develop were lacking; nor had it always been possible to raise from the banks the cash-flow that it needed, or from the general public, through the Stock Exchange, the capital vital for the launching of major projects. Once a French firm had been created, it was still difficult to market mass products owing to inadequate means of transport and the absence of large commercial companies. Within the space of about fifteen years, between 1850 and 1865, all this was to change.

Under the Second Empire, advantages began to be exploited

The creation of a world of big business and large organizations began to speed up under the Second Empire and many large-scale concerns were soon in operation: railways, shipping, coal-mining and the steel industry.[6] Commerce also began to be modernized: this was the period which saw the creation of the large department stores: Aristide Boucicaut's Bon Marché (1852), Chauchard and Hériot's Magasin du Louvre (1855), Jaluzot's Printemps (1865) and Mme Cognaq's Samaritaine (1870). There was no shortage of managerial staff and other personnel: the proliferation of large firms implied the existence of an élite interested in technology and rational management and workers and staff who were competent and disciplined. For such changes to take place, it was essential that society should be organized in such a way as to allot great influence to the best technicians and the best minds; and it was important that the masses should have access to the new culture and to a more varied range of products. If those two conditions were not met, tensions would accumulate and, sooner or later, growth would be compromised.

The natural and technical conditions in France were not unfavourable for rapid modernization. The legal framework, which had for a long time remained constricting, at last became more favourable under the Second Empire, when most of the obstacles disappeared. The only remaining handicap resulted from the large proportion of savings that the state continued to reserve for its own use.

Nor was there any shortage of the skills necessary to launch ambitious projects. The American historian Rondo Cameron has studied the role played by French businessmen and technicians in the modernization of Europe in the last century: the *grandes écoles* provided a technological training unparalleled elsewhere and the huge concentration of Paris and its surrounding districts favoured the development of a business centre capable of setting up great projects and managing them efficiently. Yet modernization in France continued to progress more slowly than in other countries: the trouble was that it was difficult for one group to overcome the resistance of the rest; furthermore, the tensions that existed absorbed energies which were thereby lost to the furtherance of growth.

Enduring social divisions and modernization

The movement of modernization that shook up the social body and multiplied the big business firms took place in a highly stratified society. There was a world of complex structures to be transformed. It is worth investigating those structures in order to understand the slow pace of change and the regrets and long periods of stagnation that punctuated the spurts of rapid progress.[7] The Revolution had made equality one of the aspirations of the French people, but it had not destroyed the hierarchies. To understand the forms of inertia that dogged French society, we must examine both its élite groups and its popular masses.

Complex and relatively divided élite groups

The élite groups of nineteenth-century France were more complex and more divided than those of most neighbouring countries.[8] The Ancien Régime had established a hierarchy consisting of the old nobility (*noblesse d'épée*), the legal nobility (*noblesse de robe*) and the bourgeoisie, all with different statuses, but whose patterns of behaviour were sufficiently homogeneous for them to be partially won over by the new ideologies. Not only did the Revolution and the Empire not wipe out those divisions but they furthermore introduced new ones stemming from people's support for the revolutionary ideology and their participation in the Montagne episode and/or the Napoleonic venture.

The élites owed their influence to their organizational abilities: they knew how to pool resources in order to exploit them efficiently, how to mobilize men and money in the service of the state and how to win sympathy and elicit loyalty. To understand any society, it is essential to remember that its hierarchies are born from the thousand and one ways of combining energies, exploiting workers and organizing large numbers of people.

The French élites were thus divided against themselves: they did not subscribe to the same principles and came into violent political opposition. Their fortunes had a variety of origins.

The land was still an essential source of wealth and prestige

There was a whole art in estate management, handing land over to sharecroppers or to tenant farmers and making the most of whatever marketing opportunities were available. The large-scale landowner had to be adaptable and the changing outlets of the second half of the nineteenth century tested his skills to the utmost. In the west at least, the old aristocracy had reconquered some of the ground that it had lost for a while under the Revolution, but there were now many new fortunes around: dynasties had arisen from the purchase of national possessions or from largesse handed out by Napoleon.[9] Gradually the weight carried by landowners in the French élite waned: new forms of wealth developed and from 1873 on the agricultural crisis combined with competition from new sectors brought about a rapid decline in the value of real estate.[10]

The nineteenth-century élites included some civil servants

A career as a prefect or a member of one of the state bodies brought considerable rewards. Generation after generation, a sense of public service impelled the members of certain families to enter the administrative classes. The old landed aristocracy and the new landowners provided a high proportion of the officers, magistrates and senior administrators, or married their daughters to the most prestigious representatives of those bodies. Alongside the bureaucracy, the representative political system also offered sought-after means of promotion: for many this was another way of serving the state.

For a long time, trade constituted a world apart
It was from trade that many families in the legal élite had sprung before the nineteenth century, but to scale the hierarchy of honours, it was necessary to pass into the service of the state. In the nineteenth century, wealth was enough to ensure success, but trade seldom proved a means of amassing a huge fortune: local markets were limited and the modernization of the economy really only opened up new opportunities in the ports, in a limited number of large towns which established themselves as regional markets and in Paris.

The manufacturing world
This was still very restricted in the eighteenth century and did not draw many of its members from the French élites, as the high proportion of foreigners and, in some provinces, Protestants in this walk of life testifies. From the time of the July Monarchy onwards, the industrial class began to grow wealthy. Some newcomers had moved on from trade. Most came from the category of craftsmen and skilled workers employed in such industrial enterprises as already existed. The various components of the French élite could only have united rapidly if it had been easy to mobilize the traditional forms of wealth – land in particular. There are quite a few examples of large-scale landowners becoming manufacturers – Buffon himself did in the eighteenth century – but their numbers did not increase much. It was more through marriages, dowries and legacies that contacts were made. Up until the Second World War, these were never enough to unite the bases of wealth and influence.

The patrimonies that the members of the various élites inherited were based upon a wide variety of components; and the aspirations and ideologies that they embraced frequently clashed. Yet they had more in common than they realized or than the representations of the press suggested. The type of education that their children received was similar. The boys went to public *lycées* or private schools, all of which dispensed the same classical education, with the same emphasis placed upon language proficiency and the kind of logical gymnastics promoted by the study of Latin and Geometry. Future leaders in the world of commerce would acquire their further training on the job; others would move on to the institutes of further education oriented towards professional specialization: either the faculties of Law or Medicine or the *grandes écoles*. For the training that these institutions provided, a taste for science and intellectual speculation was not essential – a fact that may account for the mediocrity of applied research in the development of French industry.

The members of the various French élites subscribed to a variety of ideologies, but they all agreed on one point: the exaltation of the French nation. The image of France nurtured from one extreme of the spectrum to the other may have varied, but nationalism was fervent on every side. For the Right, France stood for the monarchical tradition or, perhaps even more, was the eldest daughter of the Church. Catholicism was a powerful force in cementing together the national identity, as the cult of national saints – first Saint Marguerite-Marie Alacoque, then Joan of Arc – testifies. Even when they felt that they were unjustly excluded from the political sphere, the conservative élites were still willing to collaborate in the national endeavour. They operated within France as they did outside it, as did the Church, whose missions and schools promoted the use of the French language and made the French presence felt in every continent of the world.

For republican groups, obviously the image of France was different. It was identified with the memory of the Revolution and the ideas of liberty and equality. It was inseparable from the institutions that were set up from 1789 onwards. However, here too there was a messianic element – connected, in this case, with not religious faith but an enthusiastic commitment to the Enlightenment. In fact, the two ideals may have been somewhat similar, as is suggested by a certain Romantic openness of mind in which the Catholic mission and the mission of the Enlightenment were quite closely associated.

The common, or at least similar, values to which the élites subscribed helped to lend cohesion to the French social fabric. This must be the explanation for the striking rejection of socialist ideals in the nineteenth century. One section of the ruling classes, from the Right as well as the Left, was ready to make institutions more socialistic. The idea of solidarity, forged by Christian Socialists in about 1840, met with definite success among radicals as much as among Christians during the 1890s. How, then, is the hostility to socialism to be explained? The answer lies in the refusal of socialism to identify with a specifically national framework: the ideal of nationalism was what lay at the heart of that terrible drama of misunderstanding, the Dreyfus Affair.

The French élites were thus divided at the same time as they were united. They did not constitute an 'establishment' in the Anglo-Saxon sense of the term, because in most domains there simply was no general consensus: the national ideal constituted the sole area of agreement. In many ways, the French élites were relatively open: in many families, authority passed from a father to a son-in-law. When daughters were of an age to marry, balls were organized using the kind of lists that would be circulating in Parisian high society and these would include the names of the best students of the *grandes écoles*. In the provinces, that openness was less systematic, but most of these families had a foothold in Paris as well as in the provinces. The renewal of the élites was a process that truly did take place, although it generally took two to three generations to be completed. However, it never compromised the continuity of traditions since in many cases social success depended not only upon intellectual success but also upon an alliance with an already socially accepted family. The élites managed to turn to their advantage the system of promotion through education which had been increasingly developed ever since the Revolution. France profited greatly, since social advancement *was* possible, and this defused many tensions. On the other hand, education found itself somewhat distracted from one of its essential aims: the promotion of ideas and a respect for rigorous investigation. The whole of French society ended up by being contaminated by a kind of social pushiness, since the cream of those who could have become intellectuals preferred to climb into positions of power.

Strategies of continuity and survival

The most remarkable characteristic of the French élites was their determination to endure. Strategies of upward social mobility depended on families far more than upon individuals; those who merely succeeded as individuals acquired no more than a minimum of stability. Of all the advantages that were pursued, security was the most highly prized. The important thing was not so much to create the most thriving business or to have the most brilliant public career, but rather to make sure of a patrimony. The French élites were so anxious about permanent security, so concerned to set themselves upon solid bases that would guarantee their positions, that they sometimes forgot what it was that had justified the acquisition of those positions in the first place: namely an ability to organize groups and space. A solid fortune based on real estate was preferable to a high income constantly threatened by chance reversals. The economist Leroy-Beaulieu recommended managing one's fortune by spreading the risks and concentrating on the safest sources of income: one should invest partly in land and property, partly in state securities; only one quarter of one's fortune should be invested in shares.

The factors upon which the power of the élites rested were all of a kind to satisfy these aspirations. For those whose fortune was in real estate, permanence was easy to guarantee. But what of the others? For many, institutional life was the answer. For example, public service had inherited from the Ancien Régime the idea of office: it is true that positions could no longer simply be bought, but there was still an idea that state service was inconceivable without security. For many families, the major advantage of public service was that one enjoyed tenure: far more than abroad, the state services were organized around permanent posts – a fact that was making them increasingly cumbersome by the first half of the nineteenth century.

In France, the representative system was based more or less invariably upon a majority poll. This gave the advantage to candidates who were solidly established in their constituencies: once a clientèle of supporters was set up, it became much easier to win an election. This was how dynasties of local leaders were created. These could last for two or three generations or even longer, in the French provinces.

More mobility was required in business, as soon as it passed beyond the level of the family firm. That is why many businessmen hesitated to expand, for it would have meant seeking outside capital. They preferred the family to retain control: one could never be sure of the motivations of men whom external control promoted to the head of the business.

However, industrial circles soon learned how to minimize such risks by structuring limited companies in such a way as to whittle away the powers of small share-holders. Techniques leading to auto-control were developed very early on in French businesses. This explains the distrust of the Bourse that was clearly felt in many circles; even when the investments of small shareholders were profitable, the fact that they were allowed no influence gave them an unpleasant feeling of being manipulated. This slowed down the development of capitalism. However, to the captains of business, retaining control seemed more important than higher profits and more rapid expansion.

The concern for permanence shared by all the French élites thus resulted in the creation of a series of parallel and relatively independent pyramids at the head of the country's affairs. France was not a country that offered newly successful businessmen the keys to all the doors to advancement. They were obliged to take other influential groups into account and work with them in order to set up the institutions that they needed. By limiting the freedom of action of each group, the fragmentation of the élites, despite their common preoccupations, slowed down progress.

Popular classes passionately devoted to independence yet with little ambition

Modernization called for new projects designed to restructure the world and remodel space, strong national cohesion and social and economic progress for the popular classes.

National unity was already a reality: the attitudes towards the nation evinced by the élites predominated at every level, even the lowest (see chapters 4 and 8). This is something that occasionally escapes foreign historians, Theodore Zeldin and Eugen Weber for instance;[11] they are quite right to stress the extraordinary diversity of nineteenth-century France and to show that popular attitudes towards the government ranged from indifference to hostility; but wrong to conclude that the idea of France as a nation had not yet penetrated the popular strata. On the contrary, it had long been present everywhere and compulsory schooling resulted in its universal acceptance.

No modernization can take place unless every level of society is affected by economic change. And this can only happen with a basic level of education. As early as the beginning of the nineteenth century, rapid methods had been developed for learning to read and write and to acquire proficiency in the use of language and mathematics. The *écoles normales* saw to it that those methods were communicated to new generations of teachers. By the time that Jules Ferry made education compulsory, everything was set up for a rapid integration of the masses. Some of the disciplines fundamental to modernization were thus inculcated in a short space of time.

French society was deeply marked by the Revolution, but of all the Republican ideals it was the idea of liberty more than that of equality that took hold. What French people wanted was to lead their lives as they pleased; they were attuned to everything that gave them real independence, such as civil law and the possibility of acquiring property; and they appreciated being able to run their own local communal administration. But did they want anything more? Did they aspire to general upward social mobility? Apparently not.

Throughout the nineteenth century, French society was engaged in rethinking itself within a framework of diverse and hierarchical conditions. Liberty meant first and foremost the end of certain forms of dependence considered to be humiliating, not so much the possibility of acceding to the same status as those higher up the social ladder. Industrialization, the rural exodus and urbanization certainly introduced some changes in the course of the century, but not as many as is often claimed.

The peasant world was a world apart.[12] It was conscious of its own particular characteristics and its individuality but fortunes within it varied so widely that there was little sense of solidarity. Peasants knew that not all their children would find work on the land: they would have to try their chances elsewhere. Geographical mobility was the only alternative to poverty. This prevented the peasant world from turning in on itself.[13]

The workers' world was quite another matter. Even before the Industrial Revolution, a certain sense of solidarity had already existed here. The historians of *compagnonnage* and the Tour of France testify to that. Distrust of the authorities in power could only increase the cohesion of a group which tended to be marginalized in society. The workers' handbook, introduced in 1803 and officially valid up until 1890, helped to set the workers apart from other groups.

Revolutionary ideals might have popularized the idea of upward social mobility. But that did not happen: workers began life with the expectation of remaining workers and they brought up their children to believe the same: pointless to prolong schooling; when school was over, it was out to work. Once military service was out of the way, you started a family, and so it went on from one generation to the next. It was in the north that this model was the most widely accepted: this was partly as a consequence of early industrialization, but it also reflected ancient social traditions, the importance of the small cellular family and the custom of giving the young their independence at a relatively early age.

What most French people wanted was a livelihood, that is to say enough to eat (well) and enough to dress oneself (decently). Apart from that, they were not too fussy. Their aspirations were modest, a fact that explains the absence of permanent social tensions but also the lack of any internal urge to demand a better deal: the common reflex was to save when incomes increased, rather than to consume (except where food and drink were concerned). However, the moderate expectations of the working class did not mean that people were satisfied with the society in which they lived. They did not expect great material satisfactions from it, but they did resent its profound injustice which allowed a minority of rich people to live well at the expense of the masses who were poverty-stricken or lived in straitened circumstances. Hence the sudden bursts of anger and the abrupt explosions of a population which most of the time appeared united and content with its lot.

Urbanization and industrialization were accompanied by increasing prosperity for the middle classes. By the end of the nineteenth century their numbers had already increased greatly, but these were still people of modest pretensions: they were humble and identified with the people. Those sympathies were reflected in the radical or socialist orientation of the voting patterns of some of these strata. Socialist ideals had been violently rejected by the élites, on the grounds that they were both socially and nationally subversive. But when the middle classes adopted them, they were toned down: anti-militaristic references were dropped and the ideology really no longer threatened the national consensus.

At the end of the nineteenth century and the beginning of the twentieth, the working class was often combative – as the hard strikes of Fourmies and Carmaux testify – but its demands were really quite moderate. Its objectives were more political than economic and some were easily satisfied at little cost. Did this speed the process of modernization? Probably not: a society that accepts deeply unequal conditions is less powerfully motivated towards continuous development.

The creation of a society organized over an extensive territory presupposes the existence of some groups whose loyalties cease to be local. The peasant world was opened up by the rural exodus, but for a long time its only framework was provided by the local leaders of society and the Church. However, from the mid-nineteenth century onwards, new forms of sociability, such as associations and trade unions, began to make their appearance; and these were even more active in the world of workers. It was frequently the action of political and trade-union activists that succeeded in extending the bases of social life: French society was adapting to a wider framework, but as it did so its internal divisions were confirmed.

Checks upon the homogenization of the territory

The modernization of France depended upon the creation of large markets and upon specialization: so a measure of geographical homogenization was essential. The conflicting images that the various components of French society nurtured stood in the way of that process of homogenization.

The national territory was conceived as a pattern of a series of opposites. But the status of Paris was unique – and not simply because it was the largest town in France. Its uniqueness was also the consequence of political factors: France was composed of a series of diverse ancient elements some of which continued to clash up until the nineteenth century. The unity of France could not be justified on the basis of the existing situation. It depended upon reference to the past and the construction of a common future.

France was the heir to Gaul: in that sense it could look back to a common past with a homogeneous population, blotting out the troubles and divisions of the Middle Ages. But France had always been inconceivable without a plan for the future.

Was the Utopian ideal which was France's inspiration now beginning to take shape? The answer is 'yes', but only in the one spot which evolved beyond parochial interests and was now capable of identifying with a collective national destiny: namely Paris (see above pp. 273–5). Michelet put that point with his usual eloquence:

> But its own native and individual character was of secondary importance: a general character predominated. The name Paris conjures up the entire monarchy. How was it that a single town became the great and complete symbol of the country as a whole?... The spirit of Paris constitutes the most complex and the highest expression of France. One might have thought that something which resulted from the annihilation of all sense of local or provincial identity was bound to be negative. But that was not the case. Out of the negation of all these material, local and particular ideas there emerged a vibrant generality, something positive, a living force (*Tableau de la France*, vol. II of *L'Histoire de France*, 1833).

Was Paris really the symbolic place where France found full expression? In the French imaginary representation of reality, that is, in truth, but one aspect of the opposition between Paris and the provinces. Paris was also the place where the bridles that checked energies in deep rural France were broken and those with ambition could give free rein to their spirit of enterprise. '*A nous deux, Paris!* (It's between the two of us, Paris!)', exclaims Rastignac (in *Le Père Goriot*, 1835). That is certainly a fundamental aspect of the Parisian phenomenon. At the time when Balzac was writing, the slackening of all traditional constraints coincided with an explosion of immorality and criminality: Louis-Philippe's Paris was a twilight underworld. But it was also the place where new projects could be launched and one could meet people prepared to take a chance on even the maddest of schemes – people who would risk their money and launch new business ventures. The *grandes écoles* attracted all the best brains of the country and all those most in the know as to the latest possibilities offered by the new technology. Like all large cities, Paris was a place whose liberated energies generated innovativeness that flourished. During the first half of the nineteenth century, those two images combined to guarantee for the capital a predominant place in the minds of the people of France. But the situation was complicated by the break represented by the uprising of the Commune. Of course, Paris was already considered to be the centre of every kind of depravity, but in the first half of the nineteenth century that theme did not appear to be uppermost in conservative thinking: to be sure, Paris was the city of the Revolution and had launched the movement of dechristianization; on the other hand, it was also the city of the Empire and, subsequently, the Restoration of the monarchy. However, everything changed after the Paris insurrection. To many French people, it seemed something monstrous: it had taken place under the very nose of the enemy. Now the provinces lost their negative image: provincial France was where the deeper energies of the nation had been preserved intact. At the time of the Franco-Prussian conflict, France had just been coming to terms with its own diversity, as Taine's travel note-books testify and so, equally, does the emergence of one of the first regional movements, revealed by the announcement of the Nancy programme (see above, pp. 324–6). But the Commune episode infused provincialism with a new emotive charge. In 1904, Barrès wrote: 'My aim is not to prove any superiority but to get all differences recognized. In the last analysis, my ideas boil down to a reminder that there is a power in the land itself; a soul has accumulated in the native population; and wise administrations, instead of despising that power and opposing it, would do well to use it' (*Les Lézardes dans la maison*, Paris, 1904, p. 50).

At the end of the nineteenth century and in the first half of the twentieth, all three attitudes existed, engendering a measure of friction. The social conservation of the French provinces became even more pronounced following the Commune episode; and meanwhile the attraction of Paris as the centre of liberation in every form increased.

The provinces were, themselves, by no means homogeneous. Some were more, some less receptive to modernity. Most of the country had accepted the Revolutionary reforms: the civil Constitution of the clergy had not provoked a general reaction of rejection. But at that point some regions did refuse to follow the Parisian example. A distinction thus emerged between rural areas that were open to the new forms of social life and those that opposed them. It was in the west that the contrast was the most clear-cut: memories of the Vendée and *chouannerie* fostered a permanent tension between white areas and blue ones. In parts of the east, the Alps, Provence, the Massif Central and the south-west, there were similar reactions. The division that thus established itself reflected the unequal progress of the influences of Paris and other towns during the eighteenth century. The rejection of modernity took different forms in different places. In the west and some parts of the Massif Central, all forms of modernity, both economic and social, were rejected *en bloc*.[14] In the course of the nineteenth century, rural Brittany underwent many profound changes, but it still clung to its traditional self-sufficiency. Elsewhere, the rejection of dechristianization and Republican ideas did nothing to arrest technological progress: the farms of the upper Doubs, the most religious and conservative region in the Jura, were in the vanguard of technological innovation from the beginning of the nineteenth century onwards. In the Choletais region, the élite went in for rural industrialization in order to dissuade young people from abandoning the region for the towns, where their religious faith would come under threat (see above, pp. 342–3). The social cleavage provoked by reactions to the revolutionary period produced different effects in different places, but it certainly blocked France's evolution towards complete homogeneity.

On top of that first cleavage – which is the one that historians tend to emphasize the most – came that of 1871, the effects of which were, in truth, more general. Now it was a matter not of one region or a collection of regions rejecting certain forms of evolution, but of a mass reaction within French society as a whole. Faced with possible subversion, the countryside and small towns clung to social frameworks which in other countries were giving way to change. Whether one was a practising Catholic, a free thinker, a Jew or a Protestant, the essential thing seemed to be to salvage morality, for without it, society was in danger of collapsing altogether. Secular schools sang the praises of the family, the motherland, and work just as loudly as Church schools did. The works of many thinkers of the period – Renan and Durkheim, for instance – follow a similar line.

For almost a century, the modernization of France thus took place within what appeared to be an arrested institutional framework.[15] As a result, France acquired the air of a special case amongst the other countries of western Europe. At the same time, change was often impeded or limited to particular areas: the slow progress made in setting up France as a country of big business was to a large extent a consequence of this blocked situation.

The stages of social decongestion

It would clearly be mistaken to suppose that no change at all took place over such a long period, but obstacles to progress gave way only gradually.[16]

The first to go were the divisions that separated the French provinces as a consequence of the legacy of the Revolution. The Catholics' resistance to modernity, or certain aspects of it, began to crumble at the end of the nineteenth century. The Republic was no longer seen as an insurmountable obstacle to the spread of religion, even when it took up as aggressive a position as it did during the period of the separation of the Church from the state. But it was above all in the social and economic domains that attitudes were changing.

The Church was very much aware of the distress suffered in many popular circles: dechristianization had deeply affected towns and industrial regions. The Church now considered it its duty to make its presence felt in working-class circles and to fight against the poverty of all those excluded from the process of modernization. Its old method of providing a territorial ecclesiastical framework had worked perfectly in rural villages; but in a world where deeply felt solidarities had changed, new structures were needed. The territorial principle was replaced by a professional one and, in order to save the Church, efforts were concentrated upon the young. A number of movements sprang up and flourished: the Jeunesse Agricole Catholique (J. A. C.), the Jeunesse Ouvrière Catholique (J. O. C.) and the Jeunesse Etudiante Catholique (J. E. C.). Within these frameworks, the Church strove to maintain and strengthen the Christian presence. It also fought for the rights of every individual to decent working and living conditions. It thus played a major role in gaining acceptance for the objectives of development and progress from the part of France that had hitherto rejected them.

Therein lies the explanation for the remarkable social consensus without which the rapid progress made after the Second World War would not have been possible. From the Right to the Left, across the entire political spectrum, everybody now aspired to growth.

The regions that had remained traditional throughout the nineteenth century now came spectacularly into their own: they were more densely populated and families there were younger; in the agricultural domain they benefited more than many others did from the favourable conditions for growth.

The contrast between conservative rural areas and those that had long been more open to change began to disappear from 1930 on. But from 1955 on the types of social framework which had made possible the rapid changes that had taken place in some of the most Christian zones collapsed within the space of a few years: a different kind of transformation was taking place.

The people of France were no longer obsessed by stability.[17] The egalitarian ideal was beginning to take hold; and meanwhile the consumer goods available as a result of new forms of production were increasingly inducing the young to seek high wages. The peasant and working-class bastions upon which French society had long been based were collapsing rapidly: fewer and fewer adolescents starting their first jobs as factory workers expected their future to remain tied up with the employment that they had just obtained. They were unwilling to define themselves in relation to a rigid framework of working-class life in which they would be trapped for ever. Social mobility began to be valued in every walk of life: the sociological constraints that had affected the popular strata began to fall away.

Over the first half of this century such constraints have affected the national territory in a very uneven fashion. From 1900 on, with more administrative jobs available and the development of commerce and other tertiary activities, schooling became the means to accede to the middle classes. In the northern half of France working-class prejudices kept schooling to the minimum so that now a remarkable reversal took place. Ever since the beginning of the nineteenth century, the north and the east of the country had been characterized by a higher level of education. But from 1900 on, the south, the Massif Central and certain parts of Brittany – Finistère in particular – took over in this respect. The crumbling of the working-class bastion is of too recent date for the north to have had time to make up ground lost over the past three quarters of a century.

The growing strength of egalitarian aspirations is partly a reflection of the mass consumer society that has been created over the past generation, but it also prepared the way for that society and speeded up its creation.

The authoritarian system which used to legitimize the French social hierarchies took a long time to begin to change but when at last it did, in the early sixties, the upheaval was sudden: the Church, the army, the state and the schools were no longer accepted as a matter of course. The crisis of May 1968 was most significant in these domains: it marked the end of the rigid structures which had hitherto provided the framework of modernization.

A new geography of political attitudes

Nothing is more indicative of the slow pace of change in the French system and its recent acceleration than the evolvement of electoral behaviour. In the largely rural France of the late nineteenth century, the rural regions tended to elect local social leaders ('*notables*'): large-scale landowners, senior civil servants with local origins, doctors, wholesalers or industrialists. There were as yet no structured political parties, but those elected tended to come from the Right in some regions and from the Left in others. Within the relatively uniform social structures that existed then, this pattern reflected the existence of political attitudes whose stability has continued to be emphasized ever since André Siegfried's *Tableau politique de la France de l'Ouest* (1913).

The rise of parties for the masses (the Socialist Party, created in 1890, and the Communist Party, created in 1920) had already resulted in a certain

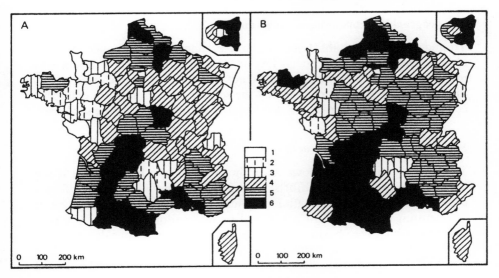

Figure 83. The presidential elections of 1974 and 1981.
Percentage of votes for François Mitterrand in A. 1974. B. 1981. 1. 30–35%;
2. 35–40%; 3. 40–45%; 4. 45–50%; 5. 50–55%; 6. Over 55%. Between these two elections, the traditional bastions of the Right lost ground in the east, the west and the southern Massif Central.

decline in the fortunes of local social leaders, but did not create uniformity in voting patterns. The old political cleavages were overlaid by others that stemmed from the extremely irregular distribution of the working-class population.

Over the past twenty years, the elections have displayed a trend towards standardization. The parties of the masses have been losing ground (particularly the Communist Party) or have been becoming platform parties, as have the parties of the centre and the Right. The political tendencies of the various regions of France are less sharply contrasted: their social composition is more uniform, social class is a matter of less importance and local political attitudes are less clear-cut. The voting pattern indicates that the public is now better informed, thanks to television, and that society is more mixed, thanks to new forms of mobility.

The process of standardization has speeded up, as can be seen from a comparison between the elections of 1974 and 1981 (fig. 83). But new factors of differentiation are emerging: for example, the rise of the National Front in southern France and the larger cities. They are a reflection of the specific problems of certain localities, for instance new suburban or urban zones that are faced with the problem of immigration.

The geographical conditions of modernization

To understand the establishment of large organizations in France, we must always bear in mind the social factors described above. The reason why industrialization and urbanization did not create such a ground-swell as they did in England and Germany is that the sphere in which modern enterprise could flourish opened up more slowly and less completely than elsewhere.

Specialization and social structures

The specialization that follows upon the creation of a large market is not usually the result of the efforts of isolated individuals or small family businesses. It implies the creation of commercial or industrial firms capable of operating at the level of a national or international market. The conditions in which such businesses can be created may vary considerably from one spot to another. In many places in France there was a serious lack of capital.[18] The traditional élites had no experience of business on the scale of modern industry and competent technicians were rare: of course, former pupils of the *grandes écoles* could be brought in, but to attract them to remote regions it was necessary to offer them large salaries: this was something that many firms were not in a position even to contemplate doing.

Places well-suited to the development of modern forms of business were thus relatively rare: Paris; the few towns that managed, at least for a time, to mobilize regional savings to their advantage (Lyons[19] and also Lille, Nancy, Grenoble[20] and Mulhouse, before the annexation); and regions where the natural conditions ensured better profits – the coalfields, where energy was cheap, for example.[21] As banking facilities were progressively concentrated in Paris, the possibilities of setting up firms in the provinces dwindled: by the beginning of the present century, it was already becoming difficult to make a breakthrough there.

What prospects were there for areas which had either not known how, or not been able, to create modern forms of organization in the fields where these first became essential, such as mining, industry and transport? The élites of such localities were not totally excluded from a share of power in the land: politics offered possibilities of promotion to the local leaders of every region in the country. Those in regions bypassed by commercial modernization strove to develop better primary schools and complementary courses. In this way, they opened up the national market of bureaucratic employment to the people of their localities, and they tried to place the humbler of their supporters in the most modest of such posts.[22]

There was one other possible course to take, up until the Second World War, namely, for a locality to turn in on itself and develop the kind of partial self-sufficiency practised in some parts of the west, the south-west and the

Massif Central. By so doing, small communities could maintain themselves on the fringes of modern society, without having to give up their values. In such cases, the local élites would do their utmost to preserve the isolation of their community, while at the same time trying to introduce a few modern improvements.

The role of the local environment

The creation of a large national market with outlets abroad brought unequally advantaged regions into competition with one another.[23] Up until the beginning of the nineteenth century, local self-sufficiency was common even in provinces where external relations were easy to maintain, in the north and the east in particular.[24] The ingenuity of farmers had led to the creation of extremely diverse combinations of crops: the cereal systems of the plains of northern and eastern France, in which wheat, barley and oats predominated; those of the south-west, where wheat and maize predominated; the poorer polyculture of the west, which concentrated on rye and buckwheat; the Mediterranean types of agriculture in both plains and mountains; mountain stock-raising; the exploitation of chestnut groves, and so on and so forth (see above, pp. 216–23). Opening up the market favoured regions able to provide the produce most in demand. Eating habits soon became standardized in towns: everyone ate white bread, potatoes and meat. As affluence increased, sugar was more widely used: jams and pastries were to be found on more and more tables. Zones which, to feed their dense populations, depended on products considered to be inferior, such as rye, buckwheat and chestnuts, found themselves at a disadvantage. Sometimes they complemented their crops by raising animals. Some were able to put off the moment of confrontation by turning in on themselves: that was how the contrast developed between Brittany and Normandy.

Up until the early nineteenth century living conditions seemed better in areas capable of cultivating a wide range of products within a relatively small area: here, it was easier to supply all the needs of a community or smallish region from close range. Mid-France, that is to say the band of territory that runs from the Pyrenees and the Loire across to the northern Alps and Alsace, thus used to offer plenty of opportunities for prosperity within frameworks of modest dimensions. But the opening up of the market soon led to disasters here: the small vineyards disappeared in the face of competition from those in the south, whilst the falling price of cereals compromised the survival of many farms.

Zones with less varied products, both in the north and the north-west and also in the Mediterranean south, were less affected by the opening up of the market, for their particular climates made it possible for them to specialize on a massive scale: animal raising in the wet western region and in the mountains, large-scale cultivation on the silty plateaux of the Paris Basin, vine or fruit tree cultivation in the south (see above, pp. 346–50).

The sectors that suffered most when the effects of the open market were fully felt were those that produced cereals and were unable to switch to other crops: the plains of the southern Paris Basin, the Aquitaine Basin and the plateaux of eastern France. Here the climate was too dry to switch to pasturage. Conversion to the raising of animals was successful only where the summers were relatively wet, on part of the plateaux of Lorraine, in Haute-Saône and the Montagne Bourguignonne, for example. It was also possible in western central France, thanks to the success of leguminous fodder: the Charente region accordingly began to specialize in dairy farming. But elsewhere, decline set in: the depressed prices fetched by wheat and wool undermined the prosperity of much of the country.

Before the Industrial Revolution the development of crafts and industry was a consequence of the abundant sources of energy – hydro-electric power or wood fires – the presence of mineral ore and a plentiful available workforce in rural zones. The regions where processing plants were important were widely dispersed over much of the territory: in the plains of the north, the ancient plateaux of the west and also in the mountains and hills of the east and the south-east. The Massif Central and the south-west were less involved in such activities, but they were not altogether absent from these regions either.

The eighteenth century had seen a new type of processing develop in new localities: urban factories had gradually proliferated. These were particularly prosperous in certain sectors of the textile industry, such as the emerging cotton and printed cotton mills. New factories had been set up inland from the large ports – Bordeaux, Nantes and Rouen – but there were also some in the Mediterranean south and in the east, where the influence of the neighbouring Swiss industries was strong (up until the Revolution, Mulhouse had been part of the Swiss Confederation, and Montbéliard also maintained many links with the Swiss cantons). Around Lyons, the manufacture of silk had given rise to similar factories.

Industrialization radically altered the map of processing zones in France, as in Britain, but in contrast to Germany, where there were fewer changes in the locations of industrial centres.

Before the Industrial Revolution the decisive factors in industrial success had always been easy access to sources of energy and cheap raw materials, and the availability of an efficient, relatively cheap workforce. Sites chosen by reason of their proximity to waterfalls now began to lose some of their attraction – although in a number of regions they did make it possible for dynamic firms to survive into the twentieth century. But it was coal that now became the economical source of energy. France was not without coalfields but they were very scattered and many were difficult to exploit. The coal that

they produced was more expensive than that of France's major competitors
– England, Belgium and Germany in the nineteenth century, Poland and the
United States in the twentieth. The French coalfields were unevenly distributed. The western part of the country had no coal resources at all. The
coalfields of the Massif Central supplied the industries of central, centraleastern and south-eastern France, but were unable to cope with any growth
in industries which were heavy consumers of energy. Eastern France was not
well supplied: the coal resources of Lorraine did not begin to be exploited
until after the First World War. The northern coal basin, the first to be
developed, initially appeared to be hardly more productive than the
coalfields of central France; in the mid-nineteenth century, however, it was
discovered to extend into the Pas de Calais.

It would have been possible to industrialize the Atlantic coast, using
imported coal: transport by sea was sufficiently cheap to do so. But Rouen
was the only city to exploit that possibility seriously. The use of hydroelectric power allowed the industrialization of a few thitherto disadvantaged
zones, particularly in the northern Alps.[25]

France's inadequate coal resources and their geographical distribution
undoubtedly hampered the modernization of much of western, southwestern and central France. But the persistence of small pockets of industry
in all those areas shows that this unfavourable factor did not, on its own,
explain the backwardness of those regions.

The effects of historical legacies and accidents

Late eighteenth-century and early nineteenth-century France was a by no
means homogeneous territory (see above, pp. 151–7). All the demographic
and social evidence underlines the contrasts between the areas on either side
of a line drawn between Saint-Malo and Geneva. The France of the openfields system of the north and the east was also the France with better health
conditions, more general education and greater professional diversification:
it was undoubtedly the easier part of the country to industrialize.

However, the earliest modern forms of industry, the marks of which are
still stamped upon the countryside,[26] were quite unaffected by that dividing
line. The large enterprises launched in the 1770s and 1780s, when France
lagged hardly at all behind England, were set up where coal was readily
available (Le Creusot), close to delivery routes (Imphy) or in the zone served
by the large ports of the west and the south. In the early 1780s, it looked
as though these regions might well be industrialized. The crisis of the last
years of the Ancien Régime and competition from England hit many of
these still fragile centres hard. But it was the period of the Revolution and
the Empire that ruined them. The slow-down or even total halt in longdistance commercial relations created a long-lasting crisis which reached its

climax during the Continental Blockade. This was sorely felt on the Atlantic and Mediterranean coasts, for it proved very difficult to smuggle in goods. Eastern and northern France were less affected: the North Sea allowed more British trade to get through and, by various routes, cotton continued to reach Rouen, the north and also the east. The regions of northern and eastern France were thus favoured by two sets of advantages: those connected with the heritage of a long tradition of rural progress[27] and those connected with the circumstances of the period of the Revolution and the Empire. Lyons, oriented towards Italy, suffered less from the period of the Revolution and the Empire than did more southern industrial centres.

In many cases large businesses were created by men who had been trained on the job, working in small workshops, but it was difficult for them to expand unless they found local backers. To find outlets for their products and to cope with the funding difficulties that this usually entailed, they needed the help of wholesalers willing to take a chance and skilful bankers. A combination of the technical and commercial skills essential to the making of outstanding fortunes was hard to find in early nineteenth-century France. Such commercial expertise as existed was mainly limited to the ports – which, as we have seen, were badly hit by the circumstances of the revolutionary period – and to the larger internal markets. Industry was dispersed throughout the territory and the spirit of innovation was unevenly distributed: in the iron and steel industry[28] and in glass manufacture, thinking was still virtually feudal: the main preoccupation of entrepreneurs in these fields was to retain control of the supplies of wood that they needed and of the small markets that they used as outlets. Master glass-producers and master smiths were mostly too close to the landed aristocracy from which many of them originated to be truly fired by a passion for technology. Buffon was a quite exceptional case.

In the textile industry, the impulse often came from the wholesalers, who kept the country looms working: that was certainly so around Lyons, in the east, the north and Normandy, as well as in the west and the south. Here too, at the beginning of the nineteenth century, not everyone involved felt the need to make the most of new technology: by the time they became aware of the powerful competition that they faced, it tended to be too late. Hence the importance, at the beginning of the nineteenth century, of areas that were more open towards technology and modern commercial methods. Certain rural regions and industrial cities constituted veritable forcing grounds for entrepreneurs in the early nineteenth century. Protestant centres were prominent in these zones of dynamism, although they did not monopolize them. The north and the Rouen region also enjoyed remarkable success, and here Calvinism was certainly not a factor.

As industry evolved over the nineteenth century, conditions changed.[29] Alongside the textile industry[30] and traditional metallurgy, which mostly produced cast-iron, steel works appeared,[31] as did increasingly diversified workshops and chemical industries. Two categories of people were now responsible for most of the innovations introduced. First engineers, whose scientific expertise was indispensable for all processes involving complex chemical reactions and for the organization of mechanical apparatus which had to be designed to be both economical and efficient; and secondly, technicians, who would think up new articles to produce and how to produce them. The technicians tended to operate in regions inclined to sub-contract work, where a proliferation of businesses set up by imaginative workers or foremen developed, clustered around a few pioneering workshops. The innovations of the engineers, on the other hand, had more chance of being developed in the larger urban centres. Hence, the growing role, in the course of the nineteenth century, played by large towns which had been able to set up technical and engineering schools and colleges. Paris, of course, made the most of the *grandes écoles*; but Lille's undeniable successes were due to its Catholic Institute and the Institut du Nord, and Nancy, Lyons, Saint-Etienne and Grenoble all founded similar institutions.

Economic growth had taken place against a background of peace for close on a century. But with the First World War, all that changed.[32] As the conflict dragged on, all resources had to be put to economical use and many innovations were introduced. Up until 1914, northern and eastern France had benefited enormously from the expansion of new forms of activity which had progressively brought greater prosperity to businesses set up under the July Monarchy or the Second Empire. As early as September 1914, the German advance had resulted in the occupation of quite a few industrial centres (in the north) and had brought others so close to the front line (around Nancy, Belfort and Montbéliard) as to rule out any idea of setting up new factories there. In most of western, central and southern France, the lack of industrialization was such that it was difficult to create factories rapidly. Most of the new manufacturing plants were accordingly concentrated in the large towns where engineers, capable of designing armaments, and skilled workers could be found without difficulty (see chapter 9). From this point on, essential industries were in the main set up in only two general areas: the Parisian district and the lower Seine, on the one hand, and the Lyons[33] district and the Alpes du Nord, on the other. The prosperity of the twenties was too short-lived to alter that pattern: the production of cars and aircraft, the cinematographical and pharmaceutical industries and the manufacture of artificial textiles were thus limited to those two major urban centres of attraction.

However, in the course of the thirties, the growing threat of war made it necessary to revise thinking on the location of strategic industries. German bombers were now capable of striking at anything within 500 kilometres from the frontier, and measures were accordingly taken to decentralize the aeronautical industry. In 1936, it was decided to set up new factories, shifting the centre of gravity for this industry to the west, the south-east and, in particular, the south-west.

Over a long period the modernization of the territory of France was unbalanced and incomplete. Some sectors preferred to turn in on themselves and aim for self-sufficiency, rather than open up and face competition. Up until the end of the Second World War that remained a viable option in the agricultural domain, although for it to be so, it was necessary that the price of wheat and other foodstuffs should not fall too low. If they did, the proportion of produce that these areas reserved for the commercial market, so as to fund the few outside purchases that they needed to make, to pay taxes and to accumulate such profits as would one day make it possible to expand production, would have been insufficient.

The new centres of business did not control economic life totally, and in the political domain the influence of the more traditional areas prevailed for a long time. Hence the hesitations of French economic policy, its sometimes luke-warm character and the refuge soon sought in a protectionism whose essential effect was to prolong the life of social forms already on the way out in other countries.

The stages of progressive modernization and the establishment of big business

The hesitancies of the early nineteenth century

Modernization started in France, in the industrial sector, in the 1780s. At this point, France lagged only slightly behind England. But the upheaval of the Revolution and the Empire arrested progress, which really only resumed with the Restoration, or even later in western and southern France.

During the first decades of the nineteenth century, the prevailing circumstances made for difficulties and governments were as prudent as possible. The tools indispensable for modernizing the economy simply did not exist:[34] the Banque de France was loath to move outside Paris, credit was scarce in the provinces, the creation of shareholders' companies was still subject to strict controls, and the absence of any procedures of expropriation in the public interest was delaying the creation of a railway network. The first Industrial Revolution began as a very modest affair. It was generated by family firms, particularly dynamic in the east, Alsace, the Lyons area, the north and Normandy. It was these that were responsible for the development of the textile industry. Commercialization at this point still operated through the traditional channels of wholesale or semi-wholesale trade in the large cities or in smaller centres. Most of the efforts which led to the creation of large organizations originated in the mines[35] – where the Compagnie d'Anzin, with a century of activity behind it, set an example – or in the iron and steel industry,[36] the modern expansion of which began with the development of Le Creusot, when it was taken over by two Alsatian industrialists, the Schneider brothers, in 1836.

The installation of the railway network speeded up industrialization and the emergence of large business companies (fig. 57).[37] This was a domain in which, for technical reasons, massive investment was essential. Family businesses had no place here. The only way to amass the necessary capital was to sell company shares to the public. During the last years of the July Monarchy, against a background of crisis, recession and social unrest, the indispensable structures for the furtherance of big business were set up. Only Paris could offer all the conditions necessary for success: a Stock Exchange, business banks prepared to venture into risky enterprises and a supply of competent engineers.

The turning point of the beginning of the Second Empire
Following a long period of gestation, the first wave of modernization began in France between 1850 and 1865. The railway network had definitively prepared the way for a national market. Banks, credit and shareholders' companies made it possible to embark upon big business ventures. When the Franco-British trade treaty of 1860 was signed, the threat of outside competition forced further concentrations of capital and more large joint-stock companies came into being. As a result, the iron and steel industry was seriously shaken up. However, apart from the railways,[38] the most remarkable successes were in the commercial and banking sector and in that of shipping.[39] The capitalism that thus came into being was more geared to trade than to production.

Paris' advantages were considerable, but not yet exclusive: regional banks were active; and the transfer of company headquarters to Paris had not yet begun. That is no doubt how it was that centres such as Lyons, Lille, Nancy and Mulhouse played such an important part in this phase of French expansion.

All these changes turned France into a country where travel was no problem and commercial relations manifested all the characteristics of modernity, even though the country was not thoroughly industrialized. A large proportion of the resources made available from savings, which the banks drained off and put at the service of business, was used to promote new forms of consumption. The English fashion of holidaying at the seaside, at spas or in the mountains was spreading. The major projects that were carried out during this period were as likely to be at Vichy, Deauville,[40] Arcachon,[41] or Biarritz, or in towns remodelled after the style that Haussmann had popularized in Paris and other cities, as they were to be in the industrial centres of the north, the east, Normandy or the Lyons region. In Paris, workshops still outnumbered modern factories (see chapter 9).

1865–85: twenty difficult years

Studies of economic history emphasize the surprising stagnation of French growth between 1865 and 1885.[42] This was the second time that a gap developed between France and its competitors. What is the explanation for this halt in development? Economists point to falling profits, possibly occasioned by over-rapid increases in workers' pay over the first decade of the Second Empire. But other factors were also involved: rising living standards stimulated a demand for food products which French agriculture was incapable of fully meeting; the 1870 war, the defeat and the indemnities that France was forced to pay to Germany also played their part. From 1873 on, a recession set in: after twenty-five years of rapid and continuous growth, the economy entered a period of difficulties. In the case of France, this was marked by agricultural stagnation: the wheat from southern Russia and the American mid-west that arrived in the French ports was so cheap that whole regions of the country were plunged into disarray. The phylloxera catastrophe ravaged the vineyards from 1868 on in both the Gard and the Bordeaux areas, but its impact was particularly grave between 1872 and 1888: in the space of fifteen years, most of the vineyards were destroyed and as yet there were no means of repairing the damage.

There were no doubt other causes too for the slackening in growth. The boom of the 1850s had been produced by the creation of a modern transport system and the emergence of new forms of consumption far more than by an upsurge in production. That is probably the best explanation for the slackening rate of growth. In the course of two decades, the rates of increase *per capita* income fell from about 1.5 per cent to about 0.6 per cent.

The flaccid economic situation was reflected in a marked slowing-down in the expansion of large organizations. Investments for the promotion of distribution and consumption stagnated and the process of industrialization marked time. Only one important change is worth noting during this period. French industry shifted definitively towards the north and the east. The coal deposits of the Massif Central were too small to support an expansion of heavy industries which were heavy consumers of energy. In 1863, the production of the northern coalfields for the first time exceeded that of the Loire basin. From 1880 on, the north supplied over half of all French coal.[43]

In Lorraine, deposits of iron ore had long been known to exist, but appeared difficult to exploit on account of the ore's high phosphorous content. At the time of the annexation, the Germans had judged it unprofitable to mine very far down the slopes of the Moselle hills as the yield seemed so small. But the whole situation was transformed once Gilchrist and Thomas' process had been perfected: France now possessed sufficient resources to warrant developing a heavy iron and steel industry. Capital from Lorraine itself was invested in this development, notably capital from the Wendel family on both sides of the frontier. However, profits could only be speeded up by appealing to the entire national market. Large companies were accordingly set up which made Lorraine, even more than the Nord and Pas de Calais departments, an area of industrial big business.[44]

The major wave of innovations: 1885–1914
During the thirty years between 1885 and 1914, the strength of industrial France was confirmed: in the north and the east, mineral deposits were exploited intensively and the heavy iron and steel industry made great strides. And in the north, which was more diversified, the engineering and textile industries also developed.

There were other industrial centres that were not linked with mining – old sectors of light industry modernized at the beginning of the nineteenth century, for instance, and certain urban centres. Lyons and Paris began to be known for their many firms oriented towards engineering and other innovative sectors of industry – the organic chemical industry, the car and aeronautical industries and the cinema.

But the industrial momentum was not strong enough to extend to the whole of France. The west, the south-west, the centre and the south were unaffected, except for the ports, where development was rapid, and a few isolated centres, some of them old industrial ones, such as Cholet, Vierzon and Limoges, others more recent, such as Clermont-Ferrand.

Modernization would probably have been more sudden and more complete had it not been for the return to protectionism: Méline introduced it in 1892. The great upheaval which had just shaken the rural world was subsiding. The exodus continued[45] but, for one generation at least, it involved craftsmen, agricultural labourers, extremely small-scale landowners and the children of over-large families, rather than the real peasants.

During this phase, when industrialization was progressing swiftly, the creation of large companies was more rapid in industry than in the sectors of transport and distribution. In these domains, the Second Empire's wave of innovation seemed to have petered out. France remained a land of small-scale shopkeepers. Only the establishment of a number of branches testified to the growing force of big business in this sector.

The First World War and the inter-war period
The First World War and the period between the wars hit an economy which was certainly dynamic and which had been growing at a pace pretty well equal to that of the other countries of western Europe. At this point, the situation of France's modern engineering, car, chemical and pharmaceutical industries augured well.

The first effects of the war were the destruction that it brought and the redistribution of major industrial sectors that it occasioned. The hostilities mainly affected the coalfields of the Pas-de-Calais. With reconstruction, industry soon got going again, now using more modern equipment. But the shift of the centres of innovation towards Lyons, the lower Seine and Paris, already detectable at the beginning of the century, speeded up. The return of the annexed provinces of Alsace and Lorraine was not enough to diversify the economy of eastern France. Coal mining got going seriously around Forbach and the iron and steel industry expanded, but without the slightest diversification. Montbéliard was the only eastern centre of industry to come out of the situation well: Peugeot had profited from the war effort, thanks to its factories; then, in 1927, it concentrated all its car factories on the Sochaux site.

The 1920s period of prosperity was too short-lived for the economy to improve much upon the results obtained in 1913, and the crisis of the thirties was swiftly to reduce levels of production. The record production figures of 1929 were not to be matched again until 1948 or 1949!

The industrial growth of France should not be allowed to mask the situation in other sectors of the economy. Agriculture suffered from the slump in many products and was still unable to supply all the country's food. Agricultural exports were limited to a few quality products, such as wines and spirits. The construction business, active ever since the Second Empire, now faltered as a consequence of the rent freeze.[46] The Loucheur law, which made it easier for civil servants to acquire property, was not enough to reactivate the market, even if it did change the face of the suburbs by encouraging them to spread. One million houses built over a period of twenty years in a country engaged in the process of urbanization was a derisory figure.

Were there no changes, then? Well, there were a few, but only in particular areas. The urbanization of the coasts and the mountains, as a result of tourism, advanced in leaps and bounds. The institutionalization of paid holidays speeded up the democratization of holidays, which already involved most of the middle classes, or at least those in a position to acquire a holiday residence or rent a villa for the holiday season.

The final phase of change developed after the Second World War: for a whole century two Frances had coexisted, the one traditional, the other modern. Only now did the constrictions that had made that possible at last give way.

The 'glorious thirty years' or France at last won over to modernity

Eminently favourable circumstances

The conditions that stood in the way of the total modernization of France gave way before, during or immediately after the war. The invisible barriers that had made many French people reluctant to opt for progress disappeared. Right-wing France was no longer committed to economic conservatism.[47] The action of the Catholic movements had proved decisive. Young people everywhere claimed the right to take part in progress. The American model was now what inspired them. The level of industrial aspirations rose. Everyone now wanted to eat the right food and dress decently, to have somewhere comfortable to live, a car, medical care when they were ill and a good education for their children. The barrier separating the working class from the rest of society lasted up until the 1960s,[48] but the industrial workers were then the first to claim higher living standards. The consensus on growth that was eventually reached made it possible to mobilize energies unused up until then. For the space of a generation, the state's major concern was to boost the country's economy.[49]

The traditional components of the élite groups were still in place, but the balance had shifted. Fortunes based on real estate now counted for little. Post-war legislation struck a heavy blow at landowners who rented to tenant farmers and quite ruined those who used the sharecropper system.[50] The rent-freeze, partially maintained for lodgings in old houses by the 1948 law, greatly reduced the incomes of housing landlords.

The outlook for industrial growth was excellent:[51] the country needed reconstruction, demand was high and continued to rise as needs multiplied. Businesses had only to be bold in order to succeed – but they needed funds and these were no longer freely available. State ownership was increasing. Electricity, major savings banks, insurance companies, the aeronautical and car industries were all nationalized, partially at least. Through its grip on credit, the state effectively controlled business as a whole. Flexible advance planning, introduced from 1948 on and promoted by the Monnet Plan, fixed development objectives. During the early post-war years, many industrialists still tended to underestimate the possibilities for growth, for Malthusian reflexes were still common enough. The introduction of a system of coordinated planning led them to revise and raise their sights in many sectors.

The division between élites of the public sector and those in the service of the state virtually disappeared. Private companies were extremely sensitive to the attitude of the public authorities. The already long-standing practice of employing senior civil servants became more general, and the creation of the Ecole Nationale d'Administration speeded up the process. Traditional oppositions within the élite were thus disappearing: the best civil servants might finish up in either the private sector or the nationalized sector or the para-public sector. Some allowed themselves to be tempted by a political career. Alongside the consensus on growth, a greater unity now prevailed amongst the ruling classes and this considerably strengthened the hands of senior state administrators and the captains of industry by bringing technocrats closer to the authorities and – for a time, at least – sheltering them from criticism.

In 1945, the state of France's natural resources was as it had been in 1930: limited reserves of both coal and iron ore, the latter of mediocre quality.[52] But the outlook had changed: petroleum now worked out cheaper than other sources of energy and its attraction grew rapidly from the early 1950s onwards. Supplies seemed assured and another of the obstacles in the way of change was swept aside. The growing use of petroleum and natural gas made it easier to opt for new industrial sites. The strong advantage of the eastern and northern regions disappeared.

The setting up of the European Coal and Steel Community followed by the Common Market forced industrialists to change their pattern of behaviour. In an extended market, cautious Malthusianism was no longer possible. So long as only heavy industry was affected the impact of the new open market remained limited, for only sectors dominated by very large organizations were affected; here the creation of cartels had been endemic ever since the beginning of the century. The ECSC did not break down national solidarities and left the member states with considerable direct control over these sectors. But the Treaty of Rome was quite another matter: it brought competition into every domain.[53] Mental attitudes do not change overnight: French farmers and businessmen almost all continued to think in terms of the national market, but they knew that 'outsiders' might threaten them while, on the other hand, the community was bound to absorb their surpluses.

All sectors were affected by the transformations of the 'glorious thirty years': inward-looking tactics were useless in an enlarged market.

The new face of agriculture

The mechanization of French agriculture only really started at the beginning of the 1950s.[54] Nobody could have guessed what an impact it was to make. The tractors which arrived with the Marshall Plan and those that emerged from European factories, many of which were inspired by trans-Atlantic models, were reliable, sturdy and relatively cheap. They simplified agricultural tasks to such a degree that within fifteen years all resistance had been swept away. In many cases, older farmers allowed themselves to be won over in the hope of keeping their sons with them.

Once motorization had begun, there was a chain reaction: there could be no question of holding back now. The farm was in debt and the hire-purchase instalments had to be paid off. Things would go better with a bit more land. Farmers accordingly set about renting land or buying it if the chance arose, and that often meant calling upon more credit. After the tractor, there was all the rest of the mechanical equipment to buy. In the old days, threshing had been managed by neighbours all lending each other a hand. Now, as soon as one farmer used a combine-harvester, all the rest followed suit.

The fragile balance that had been preserved in parts of southern, south-eastern, central and western France was now upset. The smallest farming units soon went to the wall:[55] they were too small for mechanization and in their cases the Crédit Agricole, which provided loans for modernization, was reluctant to commit itself. Battery farming offered one solution, but it was not popular, except in Brittany.[56]

By the end of the 1950s, the need for fundamental change was clear. The agricultural trade-unions persuaded the government to adopt their plans: these favoured a peasant-based agriculture and discouraged urban investment into real estate, by prohibiting the accumulation of land. How could the young be offered viable prospects in such conditions? Public aid was essential and the SAFER (Sociétés d'Aménagement Foncier et d'Etablissement Rural) came into being. Their role was to buy up the land of those who were leaving or abandoning farming and then to organize it into viable farming units. The guiding principles followed varied from region to region and depending on the types of agriculture involved. Analyses show that in most rural zones the rural exodus – in many places over 2 per cent a year – was so rapid that even without the intervention of the SAFER, these objectives would have been realized within ten, fifteen or twenty years.

Modernization led to a deep upheaval in agrarian structures and habitat. If machinery was to be used, fields had to be as large as possible, with regular contours and devoted to a single crop. Thicket hedgerows got in the way. Tractors needed a space for turning at the end of each row and the presence of a growing hedge meant that extra land was wasted. Everywhere, modernization encouraged the spontaneous reorganization and regrouping of fields and the suppression of many hedges (fig. 84).[57]

The fragmentation of holdings in rural France was extremely varied. In openfield regions, in eastern France, particularly in Alsace, the fragmentation was intense,[58] whereas large-scale farmers had managed to create huge fields in the central Paris Basin.[59] Elsewhere, fields were tiny. The land consolidation had begun well before 1939, but progress had been slow. Little had been achieved except in places where it was not necessary, that is to say where large-scale agriculture already existed. After the war, the combination of conditions necessary for success changed, and the process speeded up. The areas involved were first and foremost northern France, but also the wooded west and the old open fields. The south-west, the Massif Central and the plains of the Saône and Rhône valleys were also affected.

The rural exodus had left the buildings of both the farms and the villages of France in the state in which they had been in 1860 or 1880. Brittany and the vine-growing south were virtually the only areas where more recent changes had taken place. The type of reconstruction completed in zones that had been devastated in the two wars reflected a strong desire to stick to tradition.[60]

Machinery needed to be housed in large barns. In animal-raising areas,

446 *Centralization and diversification*

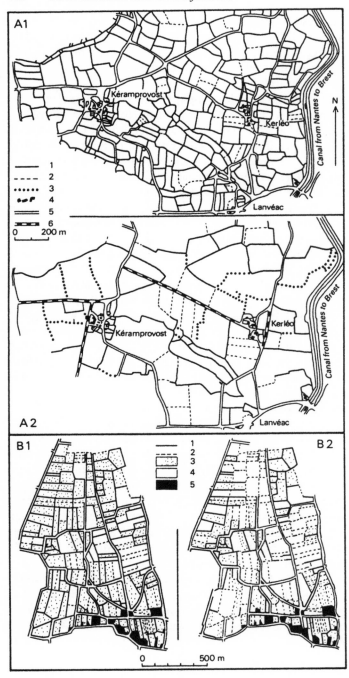

flocks and herds no longer comprised five or ten beasts per farm, but thirty, fifty, or even seventy or a hundred. With a small workforce, they could only be looked after if everything was arranged in such a way as to make the feeding of the animals and the cleaning of the sheds as easy as possible. The optimum number of animals raised in battery sheds increased rapidly and buildings 50 or more metres long soon came into use.

Up until about 1965, the modernization of agriculture affected the built environment hardly at all, but from that time on everything began to change. Equipment for the building of modern cowsheds speeded up the transformation of regions devoted to the raising of animals in western, central and eastern France, whilst in zones devoted to the production of cereals, where animal-raising was losing ground, barns and cowsheds were allowed to fall into ruins, while efforts were concentrated on the building of silos.

The landscape, which since the end of the nineteenth century had changed so little that it seemed immutable, was now undergoing a rapid transformation.[61] In the west, '*bocage*', or the closed field system was shrinking and more and more hedgerows were being destroyed, so the land was becoming increasingly bare. Everywhere fields were increasing in size. The switch to specialization was particularly rapid in areas where the self-sufficiency system had lasted longer than elsewhere.

The establishment of the Common Market made a decisive impact upon the evolution of the countryside of France: agriculture was now geared to Europe, not just France, so the specialization of crops was now planned according to different criteria. At the same time, technological progress had overcome many differential factors which had been important to cultivators in the past. Farm machinery could cope with land which previously, after rain, had proved too much for even the strongest teams of horses or oxen. The intrinsic richness of soils was now less important than their structure: provided the latter was good, fertilizers could make up for any deficiencies. Selection and cross-fertilization extended the boundaries of a number of crops: maize was no longer restricted to the south-west, the plains of the Saône and Rhône areas and Alsace. Modern techniques reduced the effects of frost upon a whole range of delicate crops. Artificial insemination speeded up the selection of animal breeds and soon improved average productivity:

Figure 84. The consolidation of fields and the suppression of hedgerows in western France.
A. The commune of Paule (Côtes-du-Nord) (after A. Guellec, 1977 and Pitte, 1983, II, p. 136). The situation before (A1) and after (A2) the 1970 regrouping.
1. The limits of plots with banks. 2. The same, without banks. 3. Banks preserved within plots. 4. Habitation. 5. Old roads. 6. New roads. B. The commune of Valcanville (Manche) (after P. Brunet, 1968). The situations in 1955 (B1) and 1963 (B2). 1. Banks. 2. The limits of plots without banks. 3. Pastures. 4. Ploughed land. 5. Houses and gardens.

448 *Centralization and diversification*

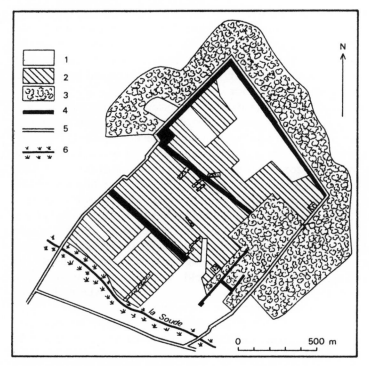

Figure 85. Contemporary land clearance in 'Champagne pouilleuse' (after Leroux, 1962, p. 31).
1. Land under cultivation before the clearance. 2. Tree clearance. 3. Remaining pine forests. 4. Screens of trees kept in and around the cultivated area. 5. Roads. 6. River. It is noticeable that a number of small strips of land, untouched by the clearance operations, were unaffected by the new arrangement.

richer and better balanced feed promoted growth, giving more meat, milk or eggs per calorie consumed.

The Common Agricultural Policy created a system of guarantees: the community undertook to buy at an agreed price all that was put on the market. The protection that this afforded was uneven: total for cereals and sugar beet and strong for meat and dairy produce; weaker for other products and virtually non-existent for fruits and vegetables. The choice of farmers, in France at least, fell upon crops which guaranteed them the most regular incomes, even if these were less profitable than others. This put northern France in an advantageous position.[62] The northern Paris Basin produced more and more wheat and beet and Champagne was also taken over by large-scale cultivation. The misjudged reafforestation using Austrian black pines that had taken place there was cleared within ten years (fig. 85). Many of those who cleared the area were the sons of large-scale farmers in the Ile-de-France, where there was no room left for them to set up on their own. Elsewhere in the plains of France, maize and barley were now introduced to complement the wheat harvests. In many places, animal-raising disappeared as large-scale cultivation took over virtually everywhere.

Regions where animals were raised also profited from the expanding market. Thanks to the production of French cheeses, dairy products were exported throughout the community and beyond. The Italian taste for veal lent encouragement to the traditional tendencies of the Massif Central. In Brittany, where farms were too small for straightforward specialization, battery farming made it possible to adapt to new circumstances. The farms of Finistère, the Côtes-du-Nord and Morbihan enjoyed a long period of prosperity, raising chickens and pigs.[63]

The south benefited less from these changes because it was oriented more towards less protected markets. The growing role of imported feed components in poultry and pig farming reduced the traditional advantage of the south-west: the centres of production shifted towards Brittany in France and the Low Countries and Denmark further afield. The opening-up of the market did not benefit Mediterranean countries as much as the Low Countries, where the proportion of delicate crops increased and efficient organization and a more central position combined to produce high profits.

In the space of thirty years, agricultural France had been totally transformed. The map of local specialization was considerably simplified. One area increasingly resembled the next, each with the same vast fields and the same prefabricated farm buildings. Everywhere the workforce indispensable for cultivating the land shrank, precipitating the rural exodus. Even in the regions that produce the most spectacular results, ten inhabitants per square kilometre is quite sufficient these days.[64]

In the end, mechanization proved too much for the peasant bastion.[65] Agriculture no longer constitutes a relatively isolated sector, a world apart. It is increasingly integrated into the general economy. Hence the suburbanization of many regions and the ever closer relations between farms and all the urban activities without which the productivity of the farms would be low and their products hard to sell.

An industrial France at last

At the end of the Second World War, France was still only partially industrialized.[66] Fewer workers were employed in this sector than in West Germany or England. In France, industry was concentrated in the eastern part of the country. A line stretching from Le Havre to Marseilles more or less traced the boundary between the France that had been transformed by large industrial organizations and the France that remained virtually

450 Centralization and diversification

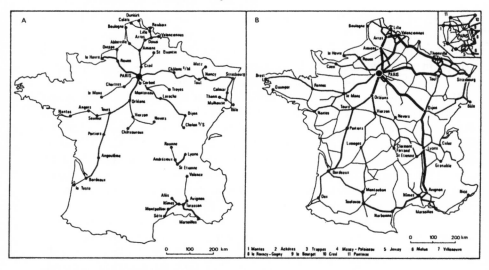

Figure 86. Rail traffic (freight).
A. In 1854. B. In 1963 (from the *Atlas historique de la France contemporaine 1800–1965*, pp. 76 and 78). In 1854, the railway network did not yet cover the entire national territory. Paris dominated the transport system. The north and the east were not yet linked and heavy industry only affected the flow between Saint-Etienne and Lyons. The network shown on map B was constructed at the end of the nineteenth century and is still in use. It shows that the heaviest traffic involves the regions of heavy industry and their essential outlet, Paris (the north-east–Paris triangle), and indicates the role of the Paris–Lyons–Marseilles line, which links a number of essential centres of diversified industries.

untouched by industrial development. Even the activity of ports such as Bordeaux, Caen and Nantes and the burgeoning aeronautical industry still hardly modified that overall image.

In the ten years following the war, gigantic efforts were made to provide France with the sources of energy and the heavy industry essential to its prosperity. Investments in these domains above all favoured the north, Lorraine,[67] the Alps and the Rhône Valley (as can be seen from figure 86B). This was where the essential coal, iron and hydro-electric power was to be found, and also where the major refineries were located: 85 to 90 per cent of the tonnage of imported petroleum passed through the lower Seine, the Etang de Berre and Dunkirk.

At the time of reconstruction and up until 1955, the heavy industries were given priority,[68] but the industrial upsurge was mainly due to the diversification of the engineering industries, electrical construction, the car industry and aeronautical manufacture, armaments and the chemical and pharmaceutical industries. In these fields, large organizations provided the motivat-

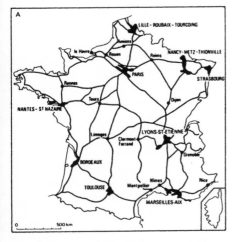 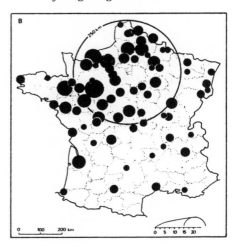

Figure 87. Decentralization in France.
A. shows the plan for establishing a new balance by setting up new *métropoles d'équilibre* (centres of equilibrium): the weight of Paris to be balanced by the development of 7 (or 8, including Strasbourg) *métropoles d'équilibre*. A number of transverse motorways are included in the plan. B. shows developments during the first phase of decentralization. Decongestion took place inside a 250 km radius around the capital, as can be seen from the new installations created in various localities between 1955 and 1970. (B. is taken from Jean Bastié, *Annales de géographie*, 1972.)

ing force and also work for many smaller firms. Big business, whether private or nationalized, thus played an increasingly important role in the industrialization of France. And large organizations never even considered establishing their headquarters anywhere but in Paris. It was there that they instinctively installed most of their laboratories, technical services and manufacturing plants. In the early 1950s, the industrial supremacy of Paris looked all set to be strengthened even further.

However, the plan to reorganize the territory aimed to distribute industrial activities more evenly and to limit their congestion in the Paris area (fig. 87A). The scaled system of aid for new industries reflected that aim: the highest subsidies went to the sectors that had been the most neglected in the past: Brittany, the Massif Central and the south-west. Conversely, the establishment or expansion of industry in the Paris region was strictly controlled. Businesses were thus forced to move into thitherto neglected areas.

New centres of industry certainly were set up, but not all according to plan. Many small factories were operating in cramped conditions in Paris (fig. 87B) and were obliged to move out of the centre in order to expand.

Centralization and diversification

These were only too glad to sell their land at a huge profit. Even without the official strategy, decongestion would have taken place, but businesses would probably have moved into the suburbs or the closest towns around. As it was, the decentralization regulations forced them to move slightly further afield. However, it was still essential for them to remain in close touch with the capital: in many cases the headquarters of the business would be there; that was where clients were to be found and also where the orders came from. In the 1950s, without means of rapid transport, there was no possible solution except to remain within reach of Paris. Most of the new industries were thus set up within a radius of 250 kilometres from the capital.

Brittany, the south-west, the Massif Central and Languedoc benefited hardly at all during the first ten years of decentralization. But in 1965, the scene began to change. D.A.T.A.R. (Délégation à l'Aménagement du Territoire et à l'Action Régionale), which had taken over planning policy for decentralization, made great efforts to persuade large firms to move into thitherto neglected zones. It did obtain results, but mostly in the larger or middle-sized towns or in their vicinity rather than in smaller towns or in the countryside. The movement was helped by the establishment of internal flights: Brest, Bordeaux, Pau, Biarritz, Toulouse, Perpignan and Montpellier could now be reached more quickly than Rheims or Rouen, as figure 88, showing the evolution in means of contact, demonstrates clearly.

The dwindling of external markets, increased productivity levels in France and the appearance of increasingly severe foreign competition led to decline in some of the longest established industries: the textile industry,[69] for one, was affected from the mid-1950s on. The process of deindustrialization had actually begun even before the completion of industrial decentralization, but up until the early 1970s the balance remained on the whole positive. The stagnation beginning to affect the little Pays Noirs of the Massif Central, the Pas de Calais, the Nord and the Vosges area was still limited. The concern to make reconversion succeed here was as yet not hindering the promotion of industrialization elsewhere.

Given the overall structure of the European industrial world, many sectors of French industry were dependent upon Germany, for it produced a huge proportion of the equipment, components and products of value that France needed. French manufacturing successes were remarkable above all in the field of durable consumer products, cars in particular, aeronautics and also armaments, the latter being an area in which Germany, a defeated nation, was for a time handicapped.

Between 1950 and 1960, the industrialization of France completed a decisive phase, but this did not stamp its mark upon the country as strongly as it did upon France's neighbours to the north and the east. Productivity increased so rapidly that the increase in products hardly provoked any

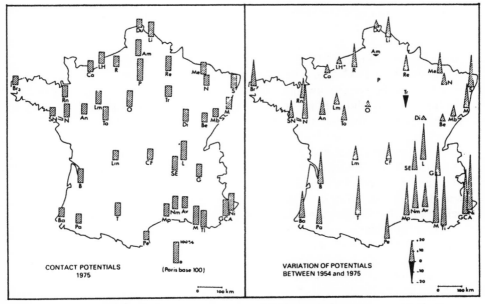

Figure 88. Contact potentials in France in 1954 and 1975.
Contact potentials vary according to the number of individuals employed in high-level tertiary activities and in inverse proportion to the distance-time separating two towns. In 1975, following the establishment of French internal airlines, there was not much difference between one town and another. Before the establishment of these contacts by air, the industrial centres closest to Paris were the most advantageously placed. But between 1954 and 1975, it was they that failed to develop or even lost ground.

increase in employment. The development of the working population was not keeping up with industrial production.

From 1970 on, the new means of circulation became determining factors in the siting of new industries, and meanwhile the traditional factories were struck by crisis. Deindustrialization now became a major problem for the older industrial regions and most of the urban centres.

A new universe of movement and communications

During the 'glorious thirty years', the modernization of France definitely owed more to the improved means of travel and communications than to industrialization itself. The car, radio and, later, television rapidly changed the way that life was lived. Even before the war, there were over two million cars on the roads of France and to possess a car was the ambition of most young people. The network of roads, though dense enough and of high quality, was not of a kind to make for particularly rapid journeys or a high-speed flow of traffic, but everywhere it improved mobility over short or medium distances. Towns became easier to reach and, in many peasant families, the young people who had to find work in the towns made the most of the situation and continued to live at home in the country.[70] The car also made it possible for city dwellers to get away in holiday periods and explains the fashion for holiday homes that characterized the 1960s and 1970s. But in the absence of investment into a system of motorways – these only really made their appearance from 1960 on – the impact of the car upon the national territory as a whole was relatively limited. The choice of the routes to be developed tended to be dictated by economic considerations. As the most heavily used of them all converged on Paris, new roads followed suit, repeating the star-like configuration of the railways built over a century before and that of the royal highways constructed even earlier (fig. 87A).

Radio and television now relayed the news universally, bursting through the closed horizons of small villages. Peasants were no longer people who knew nothing of what was happening in the wider world. They were now as well-informed as anyone else.

From the point of view of the organization of the national territory, the improvement in long-distance communication routes completely revolutionized the choice of industrial sites, making it possible for industries to operate with installations in a number of different localities.

The post-war years saw the creation of special energy distribution networks – high tension cables, oil and gas pipelines – which brought cheap supplies to all the centres that were heavy consumers of energy. The electrification of the railways speeded up the delivery of merchandise and meanwhile the navigable waterway network was developing routes for vessels of large dimensions: the Seine as far as Montereau, the Rhine and the Moselle up to Toul, the Rhône and the Saône downstream from Chalon-sur-Saône. As for passenger travel, the progress of the railways did little to upset the geography of France until the beginning of the eighties. But the same cannot be said of the airlines. In order to take part in nationwide business, it was essential to be within reach of the major hub, namely Paris: all the large towns situated over 500 km from the capital provided sufficient traffic to justify numerous, regular flights. To operate within the European and international market, direct contact with the major European centres, North America and the Far East was indispensable. In a country of average size, it is difficult to set up more than a handful of international airports. France is served by those of Paris and Nice and can also use Geneva. Finally, Lyons is the centre of a region sufficiently dynamic and densely populated to maintain services of flights to and from other European centres.

The present geography of France thus reflects a balance between two tendencies: cars, the telephone, television and radio all make for increasing decongestion and decentralization; air communications, on the other hand, favour the larger towns, international airports and urban centres capable of maintaining profitable relations with Paris. The phase of decongestion which was limited to the heart of the Paris Basin was thus followed by a new distribution of industry. This favoured the larger cities on the borders of France, particularly in the south-east, where the only alternative international airports are situated. Only there can firms contemplate setting up their study centres or headquarters (fig. 88). Otherwise, Paris remains the only possible location. Consequently, more peripheral sites are home only to factory maintenance firms and a few laboratories and research centres.

Telecommunications and information services[71] have altered the conditions determining the siting of many secondary and tertiary activities, but they have not led to the total dispersion that some people expected, because of the continuing reliance upon face-to-face contacts and meetings.

In this new state of affairs, northern and eastern France have lost most of the advantages that their centrality in relation to the rest of Europe might have afforded them: Rheims, Amiens, Nancy and Lille are further from London and Frankfurt than are Nice or Lyons or even, passing by way of Paris, Toulouse or Bordeaux. To the extent that industries now have a freer choice as to their location than they used to, since transport costs for energy and manufactured goods are less of a consideration, the heart of old Europe has lost many of its trump cards. Business firms are aware of that fact. To attract dynamic managers and good workers and staff, it makes more sense to set up in regions where everyone dreams of living, where you can enjoy an outdoor life in the sun, go swimming and practise all kinds of sports. The seaside and the mountains are beginning to attract business, as can be seen from the continuing growth of Provence, the Côte d'Azur and Languedoc-Rousillon and, even more, the northern Alps: the zone most favoured by the young these days is the area between Grenoble and Annecy or Thonon. In the past twenty years, the *chef-lieu* of Haute-Savoie has become a centre of modern industry that is mostly unplagued by crises and recession.

The introduction of the T. G. V., *trains à grande vitesse* (high-speed trains), evens out the advantages of many situations. Airlines miss out medium-sized towns. It is true that high-speed trains do not make many stops, but they have restored to towns situated less than 500 km from the capital – or even further afield, as can be seen in the Rhône valley and the Alps – advantages that they were on the way to losing. The introduction of the T. G. V. Ouest should restore the balance in Brittany, the Loire valley, central western France and even Bordeaux. To the north, prospects are even better, since high-speed rail links should restore opportunities to Picardy and the northern regions, well-placed as they are at the junction of routes between Paris, London, Brussels, Amsterdam and Cologne. But everything

Centralization and diversification

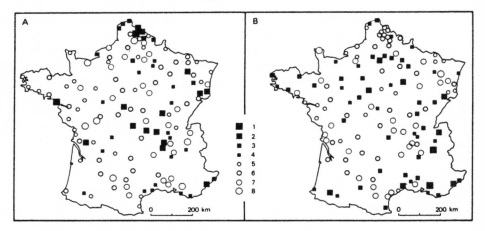

Figure 89. Changes in the geography of urban growth between the nineteenth and the twentieth centuries (after Pumain, 1982, p. 183).
A. The 1831–1911 period. B. The 1954–75 period. The coefficients of differential growth represent the gap in relation to average urban growth. 1: 2.4; 2: 1.8; 3: 1.3; 4: 1.1; 5: 0.9; 6: 0.7; 7: 0.4; 8: 0.1. In the nineteenth century, urban growth was essentially a feature of the north, the east, the centre-east and the south-east. But in the second half of the twentieth century it becomes much more widespread all around Paris, in the west (particularly Brittany) and in the south-west; and it remains particularly strong in the south-east.

depends upon the Channel tunnel, which is essential for the financial success of the overall operation.

The new geography of transport and communications routes has revolutionized the urban network (fig. 89).[72] Between 1950 and 1960, growth outside Paris occurred particularly in medium-sized towns with populations ranging between 50,000 and 200,000 inhabitants. In these it proved easier to resolve housing and traffic problems than in the larger cities. Many public facilities were introduced here and decentralized industries were attracted by them. In the Paris Basin, smaller towns also benefited from increasing industrialization.

In the 1970s, new tendencies appeared. The movement of decongestion continued, with suburbs spreading further and further into the countryside and zones with urban and industrial populations clearly enjoying a new lease of life. In some regions, the proliferation of second homes was followed by owners moving into them definitively upon retirement or earlier, and this restored vitality to zones that had been sinking into lethargy. That is certainly true of Provence, but it also applies to the entire Atlantic seaboard and southern France as a whole. At the same time, the larger towns found circumstances becoming more favourable to them. Their efforts over the past twenty years to set up better public facilities were paying off. Large cities now provided fast roads, underground transport (Lyons, Lille and Marseilles) or efficient tramways (Nantes and Grenoble). The number of women in work was steadily increasing, so that it paid couples to settle in towns large enough for both partners to find employment easily. The improved air and high-speed rail links have made it possible for major companies to set up large factories which can remain constantly in contact with other company sites.

The new geography of mobility that thus took shape in France upset the traditional frameworks of life. Up until the early 1970s, interregional migrations were everywhere smaller in scale than migrations towards Paris. The capital represented the only real melting pot for settlers from other regions. Now though, more and more people were moving into the south-east and the south.

Mobility was highest amongst the managerial classes. In the past, the traditional élites had often been deeply rooted in a particular region, even when they maintained links with family in Paris. Now however, careers unfolded within the framework of particular companies or bureaucracies which insisted on frequent changes of address. In his study on Normandy, Armand Frémont (1978) has shown the extent to which the spatial horizons of the new bourgeoisie differ from those of more traditional strata of society.

This increased mobility is not the only sign of the profound changes that have taken place in social structures – the progressive crumbling of all that used to make for both the moderation and the rigidity of France. But it is by analysing the ways in which people set up home, behave as consumers and live generally that we can assess the changes that have come about in all these domains – and the consequences of those changes for the landscape.

The art of living and the landscape

In the rush towards modernization, France has certainly not played as important a part as England, Germany or the United States: its industrial power has never been great enough for it to assume a dominant role. However, this relative weakness should not obscure the brilliant successes that the country has achieved.

The first wave of machinery and industrial technology came from England, and France's contributions were slight: Vaucanson was born too early to see the methods that he had imagined brought into use. Only Jacquard and Marc Seguin made a noteworthy impact: the former mechanized the weaving of the most complicated fabrics; the latter invented the suspension bridge and perfected the steam engine boiler. Both men were natives of Lyons, a fact that indicates where the centre of French industry lay at the time. The elaboration of photography was a lengthy process, but here too Niepce and Daguerre, both Frenchmen, played an essential part.

In the second half of the nineteenth century, France appeared as a singularly innovative country. Ideas were springing up on every side: the telephone, colour photography, the gramophone and cinema were all new concepts although, unlike the Lumière brothers, not all inventors had the ability to perfect their creations and put them into industrial production. The history of the car and that of aviation are also studded with French names.

The *grandes écoles* were producing engineers whose skill was universally admired. Eiffel's career is a splendid example. His great viaducts were wonders of technology, but it was the construction of the tower that bears his name that brought him unquestioned fame. France's brilliance lay more in the style in which it made use of modern technology than in the development of great industrial empires. The discoveries of Pasteur testify to the excellence of French science, yet the Institut Pasteur remained a relatively modest affair because it refused to patent its vaccines. In the second half of the nineteenth century the French everywhere, both in the provinces and in Paris, were manifesting a prodigious inventiveness, but only a proportion of those inventions were put to use. Rather than material success, what mattered above all was a certain elegance, a sophisticated art of living, an ease of tone and style. The prudence of the French ruling classes and their determination to endure was offset by this sense of quality which marked every aspect of life.

Encouraged by Napoleon III and thanks to the genius of Haussmann, France developed a skill in the art of town-planning which at the time won it unrivalled acclaim.[73] During the first half of the nineteenth century, Paris had grown so quickly that it had been impossible to adapt road-systems and services to the needs of its inhabitants and all its new activities. The team of planners that Haussmann mustered, in which Alphand played a role of the first importance, approached a town as a single entity. It transformed the built-up environment, road layout, traffic flows, domestic and industrial water supply, sewers, drains and the supply of food for the capital; it produced more open green spaces by creating new squares and parks and opening up the Bois de Vincennes and the Bois de Boulogne. Leisure and entertainment were also catered for, as Garnier's Opera House testifies.

The town plan for Paris did not emerge from a long tradition of reflection on the subject. It was produced by engineers who had learned to analyse the problems and find adequate solutions to each of their aspects. The existing built-up environment owes much to architects whose Beaux-Arts training had already accustomed them to adapt classical forms and devise plans based upon functionalist principles. This was the period in which the role of colour in ancient Greek architecture was discovered, as is reflected in some of the most remarkable of Paris' monumental buildings – the Opera House first and foremost. This was something quite new; and it made little impression upon ordinary buildings before 1880, or indeed upon any buildings outside Paris itself.

Haussmann's Parisian town-planning answered the needs and aspirations of the age precisely: it was sufficiently technical to resolve the difficult problems posed by the traffic-flows through the city; and sufficiently faithful to the classical style not to offend the taste of the period. At the same time, by avoiding the excesses of neo-Gothic Romanticism, it laid the emphasis on rationalism, a concern for economy and rigour, and modernity. This was exactly the right setting for the triumphant bourgeoisie which truly was modelling a new civilization.

Haussmann was imitated in all the large cities of France – Marseilles, Lyons, Bordeaux, Lille, Nantes. Local styles of architecture for the first time gave way before a relatively uniform national style. Except for their flatter roofs, plenty of houses in Nîmes, Montpellier and Marseilles could well have passed for houses in Paris or Lille. In northern France, the systematic uses of slates and zinc accentuated the air of uniformity.

Gradually the impetus that Haussmann had given to urbanization in the mid-nineteenth century died away.[74] There were many explanations for this. The costs of expropriation for the benefit of the public had risen steeply since the 1858 ruling of the Court of Appeal in Paris, which had based calculations on the value of the land *after* construction work rather than before it. The municipal law of 1878 made mayors more timorous, since now they were elected rather than nominated. The possibilities opened up, in Paris at least, by the underground railway lines made unnecessary the transformation of the quarters that Haussmann had left untouched. And the essential spirit of Haussmann's town-planning was soon forgotten once the construction teams which had given it realization broke up in the absence of further employment. As frequently happened in nineteenth-century France, the initiatives had been launched by engineers well-trained but out of touch with university circles, so that their new ideas were not integrated into any teaching traditions. The Ecole des Beaux-Arts remained absorbed in the compositional logic dreamed up by Durand, taking no interest in the problems that town-planning à la Haussmann had learned how to tackle – the organization of traffic-flows and green open spaces, for example.

The Haussmann tradition nevertheless lived on in the major cities up until the First World War and also came into its own in a number of post-war reconstruction operations, for example in Rheims.[75] The forms of buildings were changing but more gradually than in other countries. The impact of the 'modern style' was weaker than in Brussels. A few innovative buildings did appear in the provinces. At the end of the century, Nancy was a centre of

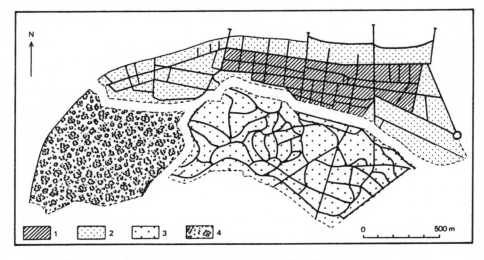

Figure 90. Plan of Arcachon in 1926 (after *La Ville d'hiver d'Arcachon*, p. 47).
1. Densely built-up area with a geometrical layout. 2. Less densely built-up area with a regular layout. 3. Sparsely built-up area with country paths. 4. Parc Pereire. Most seaside resorts were built to a geometrical layout with more, and less, densely built-up areas, like zones 1 and 2 in the summer town here. Arcachon was complemented by areas considered as parks: the large grounds of the Pereire villa (4) and the winter town (3).

remarkable artistic and architectural innovation. In Lille and other northern towns, reflections of British and Flemish styles were discernible in the use of brick.

Working-class housing remained mediocre. The example set by the Société Philanthropique de Mulhouse was more influential abroad than in France as a whole. It was again in the north that the employers took innovative action. By the late nineteenth century the quality of workers' housing was improving rapidly.

The suburban landscape and the holiday homes scene is worth a mention. The idea of settling down in a natural environment surrounded by gardens did not originate in England. It inspired Palluel's plans for the development of building-land around Le Vésinet in 1860; and it was also behind the design of the 'winter town' of Arcachon (fig. 90).[76] Nevertheless, the success of the Haussmann style of planning slowed down the process of suburban spread as did the railway companies' reluctance to introduce reduced fares and convenient timetables for commuters. By the end of the century, however, suburban spread was becoming general all over France.[77] At this period the suburbs were a pleasant place to live. Rectangular mansions with slate Mansard roofs were gradually being replaced by yellow millstone grit cottages with red tiles: the 'coloured' style was reaching a more democratic clientèle.

What was the best model to adopt at the seaside? Houses in the British style were very popular on the Channel coast and gave rise to a 'Norman' style which was frequently very charming.[78] There was no shortage of other models, however, to introduce variety. The finest examples of late nineteenth-century architectural flights of fancy are to be found in fashionable holiday resorts. In the course of the reconstruction of many of the earliest resorts and the destruction of the war years in Normandy, many examples of this heritage were, sadly, lost. But Biarritz and the Côte d'Azur also boast many villas, parks and gardens. Urban spread has eaten into many of these unusual properties but not all have vanished yet.

Modernization was a feature of the towns. In the late nineteenth century, traditional forms of architecture still prevailed in the countryside.

At the beginning of the twentieth century, systematic attempts were made to come to grips with town-planning. The Institut d'Urbanisme was set up in Paris and a cautious welcome was given to the ideas that Benoît-Lévy put forward, favouring the garden-city. The First World War caused the vote on the first major law on town-planning to be delayed for a few years, but in 1919 the Cornudet law made it compulsory to draw up plans to coordinate developments in all communes of over 10,000 inhabitants. The law came into force at the point when all construction work was paralysed by the freeze on rents.

French enthusiasm for architecture and town-planning obviously waned during the period between the wars. Construction work did not seem a high priority in a country whose population growth was stagnating. The behaviour of the middle classes still conformed too closely with popular tastes for the quality of housing to be regarded as a priority. Scientifically trained town planners came into conflict with young iconoclasts whose intellectual terrorism had fascinated the public. Their doctrine was based on a few simple postulates: as to form, cubism inspired a preference for pure geometric shapes, while functionalism ruled out all ornamentation. Absolute priority went to the promotion of hygiene which, in its turn, justified the idea of abandoning streets with unbroken lines of buildings and encouraged scorn for all forms of truly urban sociability. The solution to problems of traffic flow was sought in isolating roads – this was the most successful part of the programme – but that was not enough to ensure harmonious planning.

The very facelessness of the buildings that went up was such that the theories behind them did not seem threatening. Le Corbusier, the most famous of the Young Turks, was himself responsible for very few constructions between the wars, and the villas that he did build preserved a human scale. Not until the creation of the Popular Front (1936) was the new style used in the public building projects which proliferated as an antidote to

462 Centralization and diversification

unemployment. Their strict geometry and lack of ornamentation were not found shocking, but meanwhile the colourist tradition was still influential, as the taste for brick testifies.

The art of creating parks and gardens goes hand in hand with the art of town-planning. The nineteenth century had transformed the formula of the park à l'Anglaise to adapt it to the need for gardens open to the public. In the second half of the nineteenth century and the first half of the twentieth, there was no lack of imaginative artists, but they could only find employment on private estates, in southern France in particular or abroad, as in the case of Forestier. Meanwhile, urban parks were managed by gardeners who were competent enough, but the Versailles Ecole d'Horticulture, where they were trained, for years did nothing to develop the artistic capacities of its students. Perhaps that accounts for the staggering indifference that greeted Monet's experiments at Giverny. The tonal garden, of which he was one of the inventors, only made its mark in France after the Second World War. In the meantime, all that the disciples of the new modernism could come up with were the meanest of open spaces consisting of monotonous lawns dotted with a few spindly poplar trees: truly nothing to spark the imagination.

After the Second World War, France thus found itself faced with daunting problems in the field of town-planning:[79] the destroyed towns of Normandy, the north and the east had to be rebuilt; the increasingly numerous working class had to be decently housed; extra supplies of water had to be laid on for the population as a whole, both urban and rural; means of communication had to be improved and inner-city decay had to be overcome.

The solution to the problem was provided by the massive construction of social dwellings. Low-rent housing (*habitations à bas loyers*: H.B.L.) had been introduced in the early 1890s by the Siegfried law, but very few dwellings were built before the First World War and not many more during the period between the wars. Two new measures were introduced to speed up construction: (1) massive funding, thanks to employers subscribing 1 per cent of all the sum set aside for wages and the commitment of the Caisse des Dépôts et Consignations; and (2) new compulsory purchase procedures which soon made available indispensable land at a relatively low cost.

France was faced with the problem of the industrialization of housing. The engineers and technicians responsible discovered in the Charter of

---

Figure 91. The evolution of 'grands ensembles' (huge housing estates) (from the Centre de Recherche d'Urbanisme, *L'Urbanisation française*, Paris, 1964, central fascicule of unnumbered pages).
A. Sarcelles. B. Poissy-Beauregard. C. Chevilly-Larue. *Grands ensembles* such as these, which were built in increasing numbers from 1953 on, within a few years evolved towards gigantism, both in the scope of the projects and the volumes involved, as can be seen from these examples taken from the Paris suburbs.

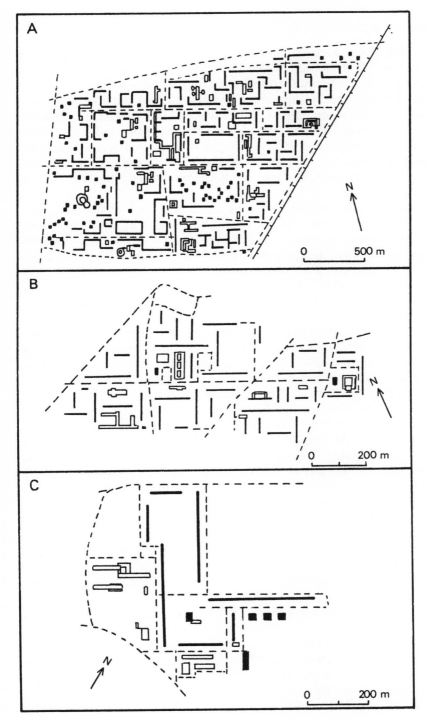

Athens principles which authorized them to mass-produce blocks and towers in which to house people. The writings of Le Corbusier and his disciples provided magnificent justification for the decisions taken. The huge powers then enjoyed by the French technocrats and the absence of any organized opposition on the part of the élites, all of whom accepted the same notion of progress, did the rest. France was now covered with *grands ensembles*. The wave of construction began virtually everywhere at roughly the same time, between 1953 and 1958, and led to a monotonous repetition of ugly forms, the presence of which emphasized social contrasts in built-up areas (fig. 91).[80]

The same conditions characterized the modernization of the rural landscape. Before and during the war, great care had been taken to respect regional traditions, but now the graceless blocks of the new C.E.S. (collèges d'enseignement secondaires) and H.L.M. (habitations à loyers modérés) appeared everywhere, even in the smallest country towns. Land consolidation conceived with a technocratic perspective led to similar simplifications in field patterns. The heavy forms of industrial constructions for agricultural purposes invaded even the most rural zones.

From 1965 on, a reaction set in. It expressed many French people's rejection of the environment that had been built for them and the preference of most new urban dwellers for a more individualistic habitat. It testified, albeit more discreetly than in Britain and Germany, to the influence of the new ecological watch-words. More generally, it was a reaction against environments so cold that they threatened to obliterate people's identities.

In the crises of the late 1960s, France, like the rest of the western world, began to question the bases of its social system. People no longer believed in movements that were supposed to lead humanity into a future of happiness, free from all conflict. But by rejecting such Utopian references, social life was in danger of losing its sense of direction. The quest for identity now led to a high value being set on everything that recalled the pre-industrial roots of the western world.[81] Suddenly everybody began to appreciate the countryside, nature, folklore, the past, everything authentic, even – paradoxically enough – the urban patterns of the past, which seemed more conducive to conviviality and communication. During the 1970s, these confused, often contradictory trends inspired a number of measures aimed at improving the quality of life.

In truth, the movement had begun some years before. One of the first signs had been the extraordinarily prevalent fashion for second homes in the country. A whole generation worked enthusiastically at restoring to their original glory old farmhouses tucked away in the remotest corners of the French countryside. During the same period, the institution of certain protected sectors, under the Malraux Law of 1962, granted recognition to the pedigrees of local urban styles of the past. The concern now was not simply

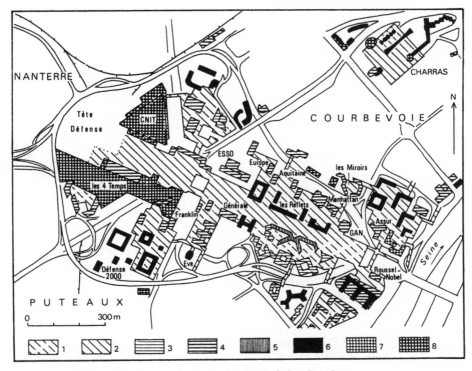

Figure 92. Paris - La Défense (from *Le Guide Michelin, Paris*).
1, 3, 5, 7: under construction or planned; 2, 4, 6, 8: completed. 1, 2: flagged; 3, 4: offices; 5, 6: housing; 7, 8: collective facilities. La Défense is the most characteristic example of the recent French business centres.

to save a few isolated monuments, but to bring back to life whole complexes, in which popular dwellings invariably held a place of importance.

Around 1970, the movement became wider. People became aware that their mountains[82] and coastlines were in danger of being concreted over by developers, building with the tourist trade in mind. For it was no longer just a matter of huge housing estates. Property developers were engaged in vast schemes to exploit the mountains and such stretches of the coastline as had so far been spared on account of their stark, unprepossessing air: the Languedoc shore and the Landes of Gascony, for example.

Planning for new towns, launched in the mid-sixties, closely resembled that for the huge housing estates. As demographic predictions dropped and new tastes in the matter of habitation developed, that planning was revised, making way for many new experiments ranging from individual-oriented housing to horizontal blocks of collective housing and quarters of high-density housing inspired by traditional models. Dense outlying business centres also made their appearance, the prototype being La Défense, in Paris (fig. 92).

The concern to protect nature led to the creation of nature reserves and various types of parks: national parks in the mountains (the western Pyrenees, Mercantour, Ecrins, Vanoise, Cévennes) and also along the coast (Port-Cros); regional parks at high altitudes in the mountains (Queyras, Corsica), at medium altitudes (Haut-Languedoc, Lubéron, Pilat, Vercors, Volcans d'Auvergne, Morvan, Vosges du Nord) and also in the plains (Lorraine, the Forêt de Brotonne, the Forêt de Saint-Amand et Raismes in the north, the Forêt d'Orient, Normandie-Maine, Brière, Armorique, Landes of Gascony) and on islands (Ouessant). The Conservatoire du Littoral has acquired a number of coastal sectors that were under threat and has thereby saved them from development.

All in all, modernization has had a very uneven impact on regional organization and on the rural landscape.[83] It has left whole towns,[84] rural areas[85] and regions[86] quite unspoilt. Generally speaking, modernization has increased the size of towns,[87] especially the larger cities.[88] Regional structures have been affected above all by the progressive industrialization which has overturned traditional urban hierarchies and created new solidarities.[89] The most noticeable changes in the landscape as a whole are particularly to be found where new developments and infrastructures are concentrated – especially in the Rhône valley.[90] Beyond a certain level, depopulation threatens the foundations of social life and compromises the entire organization of the territory, as can be seen in certain mountain regions.[91]

Over the last forty years, the geography of France has changed more than it did over the preceding century. Modernization has imposed a certain uniformity on the landscape, introduced a different scale in all types of buildings and apparently forgotten all that long French traditions in the art of living had discovered. Over the past fifteen years or so, a reaction has set in. People are rediscovering the charms of a more human scale and turning away from the gigantism that was fashionable in the 1950s and 1960s. But even as it dilutes the built-up scene with new styles, that reaction is itself speeding up change. Modernization has a different face but it is still weaning the land of France from its traditional forms.

General conclusion

France was a creation of the Mind. From a territorial entity suggested, to be sure, by nature, but defined only by an external political will, there gradually emerged an Idea which for a long time remained vague and remote but eventually, somewhat late in the day, fell into place in the daily reality of human relations.

France, as such, known by that name, came into being in about 1000 AD, within a geographical framework which was shaped by the major axes of circulation that were, in their turn, dictated by the lie of the main features of relief and rivers.[1] This was the territory that had been defined and united by the Roman civilization. The influx of barbarian peoples and the ensuing political structures had made no fundamental difference to that heritage, which was reinforced by the conversion to Christianity and the persistence of the imperial ideal. The new 'foundation myth' expressed in the legend of the Trojan origins of the Franks took on life in the territory which was a continuation of Gaul, a territory in which the humanists of the sixteenth century were also to promote that notion of the origins of the French people. But the traditional geography of this France was based solely upon a few major contrasts – north/south or north-east/south-west – inherited essentially from the cleavage introduced by the Germanic invasions and the socio-economic developments that followed upon the establishment of rural feudal settlements in this varied geographic environment. These vast, homogeneous tracts of land composed of interminably repeated agricultural patterns, with the murmur of scattered craftsmanlike activities on every side, were traversed here and there by a few axes of general circulation and bore the weak imprint of very moderate traffic flows oriented around the provincial and local centres of the feudal divisions of the territory. For centuries, the only thing that really held together this great amorphous territory, devoid of functional unity, was a sense of nationality long established in the populations' thoughts and dreams.

The Industrial Revolution, the development of first mass, then rapid, means of transport and Parisian centralization replaced this simple structure by an infinitely more complex geography, characterized by the emergence of regional agricultural specialization, the concentration of industry in certain specific areas and the appearance of a hierarchy of urban networks which gradually spread to include the entire territory. The effects of all these upheavals were fully manifest towards the end of the nineteenth century, after the completion of the railway network had produced a single national market and before the rural exodus had made a serious impact upon the country areas, which were still sufficiently densely populated to respond fully to outside economic pressures. This was the period – between 1875 and 1914, roughly – when the territory of France was at its most diversified. Ardouin-Dumazet's masterly scientific description of France at this period is an extraordinarily minutely researched study of the whole range of local peculiarities. The homogeneous upper level of culture was a product of the Ultramontane Counter-Reformation of the seventeenth and eighteenth centuries together with the establishment of the great royal highway system; but at a lower level, a prodigious fragmentation of popular culture now expressed a desire for individualism that was felt by even the smallest communities, both in the plains and in the most remote mountain valleys.

468 *General conclusion*

The political and intellectual divisions produced by the traumas of first the Revolution, then the Empire which followed it, were expressed through a variety of anthropological structures deeply rooted since the Germanic invasions at least and possibly since prehistory. These eventually produced a religious and political geography which was of a largely enigmatic nature but was to remain amazingly stable throughout the first half of the twentieth century. Through its infinite diversities, late nineteenth-century France, at once united and heterogeneous, expressed the individualism of small, naturally closed communities and a wide range of social responses to the political process of centralization and the movement towards the establishment of a single national market.

In the twentieth century, the picture reverted to greater simplicity. The second Agricultural Revolution,[2] brought about by chemical fertilizers, did away with the disparity between the natural qualities of the soils in different regions but introduced inequalities of a different kind, namely in levels of technological investment. Gradually, many regional specializations died out. Since the mid-twentieth century, descriptive and critical studies of French agriculture have adopted as their starting point not the geographical locations of farming enterprises but their economic and social typology.[3] The ever-increasing use of the car has made it possible to short-circuit many local centres and so has reduced the complexity of the urban hierarchies. Meanwhile, in the now largely deserted countryside, a whole series of busy small towns, activated by a more fluid stream of traffic, have come to resemble one another more and more. In the aftermath of the Second World War, determined policies of decentralization benefited many medium-sized towns which were gradually becoming more or less independent of the constraints of nature, in a wide variety of locations. The growth of such towns as these outstripped that of larger cities in the 1970s (a new move towards metropolization became evident in the 1980s) and, meanwhile, successful measures were being taken to diminish the excessive Parisian centralization. Finally, the success of contemporary audio-visual communication media has resulted in a deep unification of cultures and behaviour patterns. Regional contrasts of a political or religious nature are being replaced by stereotyped reactions which transcend geographical diversities and manifest themselves in more or less similar forms throughout the national territory. A new organizational model is emerging for the country, now considered as a single entity in a voluntarist policy of comprehensive planning which has very little to do either with the layout of the isthmus of France out of which the national territory grew or with the major cleavages bequeathed by the human upheavals of history. Nevertheless, history is still very much present, lying beneath the surface and masked by the triumphant march of uniformization. It will be a very long time before it will be possible to understand the face of France other than in the light of its past.

Notes

Preface

1 This point has been explained in detail in Planhol, 1972.
2 L. and A. Mirot, 1947.
3 Fierro-Domenech, 1986.
4 All the maps were drawn in the Department of Geography of the Université de Paris-Sorbonne, by Mlle Véronique Boquet.
5 In particular, Sinclair, 1985.
6 Most recently that by Braudel, 1986–87.
7 Clout, 1977.
8 This work was first published in French as *Géographie historique de la France*, by Fayard, 1988.
9 The first ten chapters, in an abridged form, were given as lectures to my geography students at the Université de Paris-Sorbonne.
10 See in particular Jean Peltre's definitive demonstration (1975) on the agrarian measures that played such a determining role in the shaping of land-plots. He defines one unique system, that he calls the *toise-jour* system, which functioned roughly from the thirteenth century until the progress made in surveying techniques in the modern period and the reforms introduced at the time of the Revolution, the diffusion of which was inseparable from that of the community arrangement of open fields.
11 On the work of Roger Dion, see Gulley, 1961; Planhol, 1972, pp. 40–4; and a number of necrological articles, in particular Broc, 1982.

1 The Isthmus of Gaul

1 Vidal de La Blache, 1908, p. 8.
2 Strabo, IV, 1, 14.
3 *Ibid.*, IV, 1, 2.
4 *Ibid.*, IV, 1, 14.
5 Guilaine, 1976; Bourdier, 1967.
6 Camps, 1982, p. 342.
7 Caesar, *The Gallic War*, V, 12, 2 (trans. H. J. Edwards, The Loeb Classical Library, London and Cambridge, Mass., 1966).

8 Hawkes and Dunning, 1930, pp. 176, 222; Duval, 1961, p. 64.
9 Hubert, 1932a, p. 149.
10 See, most recently, Chevallier, 1980, pp. 2–6.
11 Dion, 1959.
12 Jannoray, 1955, pp. 377–87; Clavel, 1970, pp. 55–7.
13 See Jullian, 1908–26, I, pp. 262–7; Hubert, 1932, pp. 363–7.
14 Polybius in Strabo, IV, 2, 1 (trans. Horace Leonard Jones, Loeb Classical Library, London and Cambridge, Mass., 1969).
15 On the location of *Corbilo*: Dion, 1963, pp. 407–10; Ramin, 1965, pp. 119–21; 1975. Iberian etymology: Schulten, 1914, pp. 82–3. Celtic etymology: Dion, in Ramin, 1965, p. 120.
16 Dion, 1963, pp. 396–7.
17 Clavel, 1970, pp. 130f.
18 It has been likened to the name of the Basque 'Euskualdun' people and also to that of the *Ausci* tribe, after which the town of Auch is named. See Jullian, 1908–26, I, pp. 276–7.
19 Lot, 1963.
20 On all that follows, Duval, 1974.
21 *Argonautica*, IV, 641.
22 In Strabo, I, 4, 3 and 5; III, 2, 11; see Jullian, 1908–26, I, pp. 3–4.
23 Polybius, VII, 9, 6 and 17.
24 *Ibid.*, III, 51, 7.
25 Dion, 1968, pp. 515–17.
26 However, the popularization of the 'Gallic cock' as national animal came much later, at the time of the Revolution.
27 Catullus, 29, 3.
28 Hubert, 1932a, pp. 147–9.
29 The fact that the territory of the *Sequani* extended up to the sources of the Seine, which is a point made by Schmitt, 1981, p. 11, is not an adequate explanation for their name.
30 Lambert and Riouffreyt, 1981, p. 127.
31 Josephus, *The War of the Jews*, II, 16, 4; Plutarch, *Caesar*, 16. See Hubert, 1932a, p. 146.
32 Jullian, 1908–26, II, p. 19.
33 *Ibid.*, pp. 22–3. We know the exact number of these groups only from the Roman period onwards. Out of 72 ethnic names that Caesar mentions in Gaul, 43 are certainly the names of nations (or cities), 15 are probably the names of tribes; in the case of 14 it is not possible to say to which of those two loose categories they belong. Meanwhile, 4 or 5 large groups are not mentioned at all by Caesar (Jullian, 1908–26, II, p. 21).
34 Duval, 1961, p. 222.
35 Provost, 1981, p. 188.
36 Dion, 1947, chapter 1, 'The frontier-deserts'.
37 Caesar, *The Gallic War*, IV, 3; VI, 23, 1.
38 Legros, 1981, p. 175.
39 Dion, 1947, pp. 8–9.
40 *The Gallic War*, VI, 10, 5.

41 Dion, 1947, pp. 14–15, gives many examples: the frontier-forest between the *Carnutes* (Chartres) and the *Senones* (Sens); the Auge forest between the *Baiocassi* (Bayeux) and the *Lexovii* (Lisieux); the forests of upper Perche between the *Carnutes* and the *Sagii* (Sées), and so on.
42 Lambert and Riouffreyt, 1981, p. 136.
43 Deyber, 1981, pp. 34–5.
44 Beaujard, 1981, p. 102.
45 Provost, 1981, pp. 187–8.
46 Audin, 1981, pp. 80–1.
47 Dion, 1947, pp. 23–4. On *equoranda, ibid.*, p. 34; the literature on this subject is quite extensive. Recent bibliographies appear in Lambert and Riouffreyt, 1981, p. 171; Audin, 1981, pp. 88–9.
48 J.-M. Desbordes, cited by Burin, 1981, in *FG*, p. 113–14. On ethnic and tribal patriotism and internal quarrels, see Jullian, 1908–26, II, pp. 16, 37–43.
49 Jullian, 1908–26, II, pp. 3–8; cf. Hubert, 1932a, pp. 167–8. Other evaluations are much less convincing. Most recently, Harmand, 1967, p. 265; Braudel, 1986–87, II, 1, p. 60.
50 Polybius, II, 35, 8; cf. Livy, V, 34: *abundans multitudo*.
51 Essentially, the next paragraph reproduces Planhol, 1978.
52 Caesar, *The Gallic War*, VI, 30.
53 *Ibid.*, I, 29.
54 *Ibid.*, I, 5.
55 One of these passages refers to the *Menapii* (Caesar, *The Gallic War*, IV, 38), in connection with whom Caesar elsewhere (VI, 60) mentions *aedificia vicosque*.
56 A synthesis may be found in Martin, 1971, chapter 3, 'Economie rurale et technique agricole en Gaule. Tentative de mise au point'.
57 Jullian, 1908–26, II, pp. 71–4, 277; Thévenot, 1949, pp. 64f.
58 Pliny, *Naturalis Historia*, XVIII, 296; Palladius, VII, 2; see Lebel, 1959; Kolendo, 1960.
59 Jullian, 1908–26, II, pp. 64–6.
60 Lot, 1948, pp. 22–3; Harmand, 1960, p. 360.
61 Harmand, 1960, p. 360.
62 Johnson, 1961, p. 168; Butlin, 1971, pp. 60–1.
63 Agache, 1971, p. 33 and figs. 362–5.
64 *Ibid.*, pp. 33–5.
65 *Ibid.*, fig. 342.
66 Meynier, 1945, 1966, 1972; Soyer, 1970.
67 Soyer, 1970; see in particular pp. 89–96.
68 Caesar, *The Gallic War*, II, 22.
69 *Ibid.*, I, 5, 2.
70 *Ibid.*, II, 4, 6–7.
71 Plutarch, *Caesar*, 16.
72 As we learn from Caesar, *The Gallic War*, VII, 4, 2; see Jullian, 1908–26, II, 62.
73 *The Gallic War*, VII, 15, 4.
74 Kruta, 1980, pp. 204–5.
75 Jullian, 1908–26, II, pp. 254–9, on all this.

76 There is a good synthesis in Pitte, 1983, I, pp. 56–8; see Jannoray, 1955; Benoît, 1957; Gallet de Santerre, 1978; Goudineau, 1980.
77 Gallet de Santerre, 1965, pp. 627–8.
78 *The Gallic War*, VII, 28.
79 Vendryès, 1955; see Duval, 1961, p. 76.
80 *The Gallic War*, VII, 13, 3; see Dion, 1952a, p. 553.
81 Caesar, *The Gallic War*, VII, 62, 9; see Dion, 1947, pp. 20–1.
82 Audin, 1981, pp. 80, 83.
83 On all that follows, see Jullian, 1908–26, II, pp. 437–48; 1930–31, I, pp. 172, 186; he tends to exaggerate the sense of unity.
84 Caesar, *The Gallic War*, VII, 77, 8.
85 *Ibid.*, VII, 77, 7.
86 Bayet, 1930, pp. 221–44; Harmand, 1967, pp. 345–6. *Cf.* Duval, in Jullian, 1963, pp. 35f.
87 Braudel, 1986–7, II, 1, p. 66.
88 Diodorus, IV, 19, 1–2.
89 Livy, V, 34.
90 As far as the Atlantic Ocean and the Rhine: 'καὶ μέχρι Ὠκεανοῦ καὶ Ῥήνου' for the first *Arverni* empire in the second century BC, according to Posidonius, in Strabo, IV, 2, 3; in the next generation, in about 80 BC, Celtill, the father of Vercingetorix, '*principatum totius Galliae obtinuerat*' (Caesar, *The Gallic War*, VII, 4, I).
91 Jullian, 1908–26, II, p. 548.
92 *Ibid.*, p. 236.
93 Caesar, *The Gallic War*, VII, 3, 3; see Jullian, 1908–26, II, p. 229.
94 Jullian, 1908–26, II, pp. 97–8, 445–7.
95 Caesar, *The Gallic War*, VI, 13, 10.
96 Jullian, 1908–26, II, p. 447.
97 *Ibid.*, p. 97.
98 Caesar, *The Gallic War*, VI, 13, 11–12; Jullian, 1908–26, II, pp. 113–15; 1922, p. 105.
99 Jullian, 1908–26, II, pp. 394–6; Hubert, 1932a, p. 301.
100 Desbordes, 1971 and Dehn, 1965 explain them essentially in terms of physical geography.
101 Vadé, 1972–74, 1976; Guyonvarc'h, 1961; Le Roux, 1961.
102 Jullian, 1908–26, II, pp. 392–6.
103 Clark, 1955, pp. 388–97, and maps pp. 388, 393; Navarro, 1925.
104 Jullian, 1908–26, I, pp. 195–213.
105 *Ibid.*, pp. 213–20, 396–402; Benoît, 1965, pp. 99–134.
106 If, as is held by Bérard, 1927, I, pp. 423–6, Μόνοικος is indeed a Greek pun on the Semitic *Menoha*; but experts of toponymy nowadays are more inclined to regard the name as pre-Semitic (Rostaing, 1945, p. 37). See Benoît, 1965, p. 96. As for Nice, its foundation is attributed to the people of Marseilles by Strabo, IV, 1, 9, and Ptolemy (III, 1, 2) calls it Νίκαια Μασσαλιωτῶν. But powerful arguments against this idea have been put forward, most recently by Ricolfis, 1980, I, pp. 776–82. They are based on evidence of a phonetic and, above all,

archaeological nature: no Greek remains have been discovered between the Ligurian and the Roman strata, both of which are well attested and appear to be continuous.
107 Déchelette, 1931–34, II, 3, p. 1577. Many contrary arguments are to be found in Busquet, 1954; Benoît, 1965.
108 Map in Ramin, 1965, p. 137.
109 Gaudron and Soutou, 1961, map p. 585; Soutou, 1963, map p. 383.
110 Diodorus, V, 22 and V, 38; Posidonius, around 100 BC, in Strabo, III, 2, 9.
111 Extensive literature exists on this subject: Broche, 1936; most recently, Dion, 1977, pp. 175–222.
112 Dion, 1952a; 1968, pp. 503–15 (with considerably modified conclusions); Carcopino, 1957, in particular pp. 43–69; Ramin, 1965, *passim* and in particular pp. 82–90 and 96–108.
113 Dion, 1952; Ramin, 1965.
114 Diodorus Siculus V, 22.
115 *Ibid.*, 38.
116 Possibly confused with the Isle of Wight by Diodorus; see Carcopino, 1957, pp. 37–40.
117 Carcopino, 1957, pp. 62–9.
118 Villard, 1960, pp. 143f., 157.
119 Clark, 1955, p. 403.
120 *Ibid.*, p. 405.
121 Carcopino, 1957, pp. 103–4.
122 Benoît, 1965, pp. 191–213; on wine, Dion, 1959a, pp. 65–98.
123 Diodorus, V, 26.
124 Polybius, II, 19, 4; XI, 3, 1; Livy, V, 44, 6.
125 Livy, V, 33; Plutarch, *Camillus*, XVI; Pliny the Elder, XII, 5.
126 Dion, 1959a, pp. 65–76; Drouot, 1952 and 1952a.
127 Cited by Dion, 1959a, p. 76.
128 André, 1954.
129 Dion, 1959a, pp. 95–6.
130 Benoît, 1953, p. 104; cited by Dion, 1959a, p. 96.
131 Diodorus, cited by Dion, 1959a, p. 102.
132 Jullian, 1908–26, II, p. 551 and V, pp. 551–4; to be complemented by the pertinent reflections of Duval, in Jullian, 1963, p. 41–3.

2 The impact of Rome

1 In 43 BC, in Cicero, *Ad familiares*, X, 9.
2 Duval, 1974, pp. 412–13.
3 Clavel, 1970, pp. 132–5.
4 Jullian, 1908–26, III, p. 23.
5 Schmittlein, 1954.
6 Livy, *Perioch.* XCVII for the year 71 BC.
7 In about 56, Cicero, *De Provinciis Consularibus, XIII*, included the Germans, as well as the Helvetii, among the Gauls. Strabo (IV, 4, 2) considers the Germans

and the Gauls to share the same origins: 'συγγενεῖς ἀλλήλοις οὗτοι'; and even at the end of the second century AD, Dio Cassius was of the same opinion: 'Some Celts, whom we call Germans'; see Schmittlein, 1954, p. 24.

8 On the complex question of the distribution of the Celtic and the German peoples at the time of the Roman conquest, and on the ethnic relationship between the Germans and the *Belgae*, see most recently Maréchal, 1981.
9 Caesar, *The Gallic War*, I, 28, 4.
10 Schmittlein, 1954, p. 59.
11 Fragment II, Book I, cited by Duval, 1974, p. 412.
12 Jullian, 1908–26, IV, p. 35.
13 Jullian, 1908–26, V, pp. 81–129; Cavaillès, 1946, pp. 16–20; Chevallier, 1972, pp. 181–96, with an important bibliography, pp. 264–8; Pitte, 1983, I, pp. 77–80; an exemplary regional monograph is to be found in Thévenot, 1969.
14 Jullian, 1908–26, IV, pp. 129–40.
15 Compare, for example, those given in figures 9 and 10 with Chevallier, 1972, p. 182, from F. Benoît.
16 Jullian, 1908–26, IV, pp. 67–9.
17 Strabo, IV, 2, 1.
18 Jullian, 1908–26, IV, pp. 70–2.
19 The number must have varied and nine was only reached by progressive stages; see Jullian, 1908–26, IV, pp. 71–2.
20 Strabo, IV, 3, 2.
21 Petry, 1981.
22 Picard, 1970–76; Agache, 1978, pp. 415–17; Audin, 1981, p. 81; Lambert and Riouffreyt, 1981, p. 133.
23 Agache, 1981.
24 They may be compared to the 'Sunday towns' of Brazil, which have been studied by Deffontaines, 1938, which are used for religious ceremonies and for Sunday meetings of the *fazendeiros*, in a region where the colonial type of dispersed habitat based on large ranches is quite reminiscent of the *villae*.
25 Jullian, 1908–26, IV, pp. 31–4, 76–8.
26 *Ibid.*, p. 262.
27 *Ibid.*, V, pp. 46–7, 98ff.; Mertens, 1983, 1985a.
28 Jullian, 1908–26, V, pp. 42–4.
29 *Ibid.*, IV, p. 73.
30 *Ibid.*, p. 74.
31 That is, if one accepts the classic identification of Gergovia with the Merdogne plateau. Another plausible solution identifies it with the Plateau des Côtes, to the north of Clermont, a height of 650 m. Cf. Busset, 1933; Eychart, 1961, 1969.
32 Strabo, IV, 2, 3, already refers to the capital of the *Arverni* as Νεμωσσός. Cf. Jullian, 1908–26, IV, p. 76.
33 Jullian, 1908–26, VI, pp. 377–82.
34 *Ibid.*, III, pp. 42–7.
35 *Ibid.*, pp. 250–3.
36 *Ibid.*, I, pp. 35–7; Vidal de La Blache, 1908, pp. 227–30.
37 IV, 6, 11.
38 See Dion, 1952a, pp. 555–6.

39 Table by Peutinger, s. 1, 4, cited by Jullian, 1908–26, IV, p. 449. See also *ibid.*, VI, pp. 515–27; Audin, 1979.
40 Jullian, 1908–26, IV, pp. 425–51.
41 IV, 4, 2.
42 Ammianus Marcellinus, XV, 9–12; *Expositio totius mundi*, 58; see Jullian, 1908–26, IV, p. 449.
43 Jullian, 1908–26, IV, pp. 449–51.
44 For all that follows, see the contributions collected in *La patrie gauloise*.
45 7, 216; see also '*mea Gallia*' (*ibid.*, 5, 356).
46 The most recent synthesis may be found in Pitte, 1983, I, pp. 59–67. Goudineau, 1980a, is definitely opposed to the idea of a general urban model.
47 Pitte, 1983, I, pp. 70–4.
48 Audin, 1979, p. 2.
49 Jullian, 1908–26, V, pp. 35–8; Lot, 1945–53.
50 Pitte, 1983, I, p. 65.
51 On the whole of the next paragraph, see Planhol, 1978; Pitte, 1983, I, pp. 80–8.
52 Chevallier, 1962; Chouquer *et al.*, 1982.
53 Piganiol, 1962.
54 Callot, 1980, pp. 223–53; with an exhaustive bibliography of the reconstructions that had been proposed at this date.
55 Soyer, 1973–74.
56 The fundamental synthesis on the Gallo-Roman *villa* is still that produced by Fustel de Coulanges, 1914, pp. 1–96; also for the Merovingian period, pp. 196–238, 438–61. See also Jullian, 1908–26, IV, pp. 373–8; VIII, pp. 130–49; Grenier, II, 2, 1934, pp. 727–35f.
57 Fustel de Coulanges, 1914, p. 41.
58 Grenier, II, 2, 1934, pp. 782–96.
59 Fossier, 1968, I, pp. 134–8.
60 Leday, 1976, p. 250, from Grenier, p. 695.
61 Fournier, 1962, pp. 127–9. Fustel de Coulanges had minimized the importance of the *vici* (1914, pp. 40–2, 198–220). Cf. Flach, 1886–1904, II, pp. 47–54.
62 *VGR*, 1976.
63 Statistics (the first to have been produced) relating to 159 *vici* (*VGR*, p. 5) suggest that only 3 were agricultural.
64 Audin, 1976.
65 Petry, 1976, p. 290.
66 Ternes, 1976.
67 Grenier, II, 2, 1934, p. 869.
68 Agache, 1971, 1978.
69 Agache, 1971, p. 186.
70 *Ibid.*
71 Grenier's chapter on the distribution of *villae*, written sixty years ago (1934, pp. 858–75) thus needs to be completely revised.
72 Agache, 1971, p. 200.
73 Palladius, VII, 2.
74 Clavel, 1970, pp. 296–324, and map p. 301.
75 Planhol, 1978, pp. 239–41; 1982.

76 Sidonius Apollinaris, *Epistolae*, II, 2; VIII, 4.
77 *Epistolae*, II, 2; 'Before reaching the place where a space opens up that is large enough to serve as the entrance to a dwelling'. We know the exact connotations of these large open spaces ('*patens et spatio suo porrectus*', as Isidore of Seville writes in *Etymologies*, XIV, viii, 23), which are increasingly connected with the term *campus* in the late Latin period.
78 Fournier, 1962, p. 202.
79 *Epistolae*, VIII, 4.
80 *Ibid.*, II, 9, for a *villa* close to Alès, on the banks of the Gardon.
81 See *Digeste*, XXXIII, 7, 18.
82 Brunet, 1960, cf. pp. 56–8 and the photographs on pp. 23 and 25.
83 Planhol, 1978, p. 240; 1982, pp. 404–5.
84 *Epistolae*, IV, 21.
85 *Panegyrics*, VIII, 6. For an identification of this place, see Landriot and Rochet, 1854, p. 312, n. 4.
86 Agache, 1971, p. 203; 1978, p. 343.
87 Fournier, 1962, p. 324.
88 Meyer, 1972, pp. 455–7 and 464.
89 Recent progress in this exploration is gradually making it possible to distinguish a whole typology (see *VR*, 1982). For a good recent synthesis, see Pitte, 1983, I, pp. 84–7.
90 Dion, 1946.
91 Ausonius, *Epistolae*, XIV, 2.
92 *Ibid.*, XXIII, 125.
93 Jullian, 1908–26, II, pp. 265–9; V, pp. 180–3.
94 The Emperor Julian, *Misopogon*, ed. Hertlein, 1875–76, II, p. 438, cited by Dion, 1959a, p. 165.
95 Cicero, *De re publica*, III, 9, 16; cf. Aymard, 1948; Dion, 1959a, pp. 98–9.
96 *Epigrams*, XIII, 107.
97 Suetonius, *Domitian*, VI.
98 *Historia Augusta*, 18, 8.
99 Strabo, IV, 1, 2.
100 *Ibid.*, II, 1, 16.
101 Columella, III, 9.
102 Dion, 1959a, pp. 125–6.
103 Pliny, *Naturalis Historia*, XIV, 26–7.
104 On all this, Dion, 1952b; 1952c; 1959a, pp. 139–47; 1960; Thévenot, 1952, 1961.
105 Dion, 1951.
106 Planhol and Lacroix, 1963; Planhol and Pérardel, 1969.
107 *Liber de Caesaribus*, XXXVII, 3, cited by Dion, 1959a, p. 149.
108 The Emperor Julian, *Misopogon*, ed. Hertlein, 1875–76, II, p. 438, cited by Dion, 1959a, p. 151.
109 Dion, 1959a, pp. 161–4.
110 *Ibid.*, pp. 155–8.

3 From Gaul to France

1. The major synthesis, using new methods, is still that of Salin, 1950–59; excellent critical views in L. Musset, 1965 and 1965a.
2. A good synthesis appears in L. Musset, 1965, pp. 171–81.
3. *Ibid.*, pp. 295–6.
4. *Ibid.*, pp. 174–6.
5. *Ibid.*, p. 176.
6. *Ibid.*, pp. 191–4; Specklin, 1979. Recently, on the questions raised for anthropology, rather than for the answers that it provides, *Le phénomène* ...
7. From the recent excavations carried out by Christian Pilet, reported by Yvonne Rebeyrol in *Le Monde* of 29–30 December 1985.
8. L. Musset, 1965, pp. 194–5.
9. There is quite an extensive bibliography. The principal syntheses on Germanic contributions are those by Longnon, 1920–29, pp. 175–300; Dauzat, 1926 (1963), pp. 133–41; Vincent, 1937, pp. 133–58; Rostaing, 1945, pp. 73–9; Nègre, 1963, pp. 95–109. Cf. Bonnaud, 1981, II, pp. 129–34. For a (possibly excessive) regional use of toponymy in the study of settlements, Broëns, 1943, 1956.
10. Bonnaud, 1981, I, pp. 135–7.
11. Lot, 1933; Johnson, 1946; Salin, I, 1950, pp. 271–3; Bonnaud, 1981, I, pp. 502–5.
12. Most recently, Bonnaud, 1981, I, pp. 76–8; useful earlier studies by Tulippe, 1934; Bloch, 1939; Dauzat, 1939; Soyer, 1933–52; summarized in Kroemer, 1976, pp. 28–40.
13. This was developed on the basis of toponymic arguments by Dauzat, 1939; Bloch, 1939, basing his comments on historical texts, concluded that clearance was only completed between the eleventh and the thirteenth centuries.
14. Blottière, 1973. Of course, this does not mean that the original vegetation was the kind that grows on the steppes, as Tulippe, 1934, and Bloch, 1939, still believed. It was certainly mainly forests, perhaps with a few natural clearings of plants that flourish in dry conditions.
15. Soyer, 1933–52.
16. Perrenot, 1942.
17. Bonnaud, 1981, I, pp. 131–5.
18. *Ibid.*, pp. 134 and 503.
19. Rostaing, 1945, p. 80.
20. All the monographs produced before this date are listed in Salin, I, 1950, pp. 319–409.
21. *Ibid.*, pp. 400–6.
22. *Ibid.*, pp. 385–7.
23. *Ibid.*, pp. 333–7.
24. Chaunu, 1982, p. 99.
25. *Ibid.*, pp. 97, 112.
26. Pitte, 1983, I, p. 93, photograph 16 and fig. 8.
27. Fournier, 1962.
28. Fossier, 1968, I, p. 159.
29. Salin, 1950, pp. 369–74.
30. *Ibid.*, pp. 390–3, from Broëns, 1943.

31 L. Musset, 1965, pp. 265–7.
32 Jullian, 1908–26, VI, pp. 458, 499.
33 Lestocquoy, 1953; Vercauteren, 1934. The ancient precinct walls of Autun, 6 kilometres long, which in the early twentieth century were once again to contain close on 11,000 inhabitants, were reduced to 1,300 metres in the third century. Rheims was reduced to the town centre, about 20 to 30 hectares in area, which could accommodate barely 5,000 to 6,000 inhabitants, a figure that was not exceeded in either Bordeaux or Sens. Amiens and Soissons appear to have been reduced to 2,000 inhabitants. Périgueux, with an area of 5.5 hectares (as against the 50 hectares that it included before the invasions), was reduced to about 1,500 inhabitants. The only cities to have escaped this process appear to have been those of the south-east, to the south of Grenoble and east of Carcassonne. See Février, 1964, p. 44.
34 Rouche, 1969; Jullian, 1908–26, IV, pp. 525–30.
35 Jullian, 1908–26, IV, pp. 323–6.
36 According to Lestocquoy's (1953) suggestive title.
37 Agache, 1971, pp. 203–4.
38 *Germania*, XVI: '*Ne pati quidem inter se junctas sedes*'.
39 Verhulst, 1966.
40 Analyses in Perrin, 1942, pp. 32–77; Duby, 1962, I, pp. 100–29.
41 Like the *butinae aut mutuli* in the Law of the *Ripuares*, XXXIV, 4; *butinae* is a corruption of *botontini* found amongst the *gromatici*, and indicated mounds of earth. *Mutuli* (corbel in architecture; here, standing stone) appears in the *Lex parieti faciundo* (*Corpus inscriptionum latinarum*, X, 1781). See Fustel de Coulanges, 1914, p. 111.
42 Planhol, 1959, pp. 417–18.
43 This fact is mentioned in Haudricourt, 1955, pp. 48 and 338, but no discussion of its cultural significance is attempted.
44 Higounet, 1980.
45 The earliest mention dates from 877; Fournier, 1962, pp. 298–9.
46 Meynier, 1943, p. 170; Guilcher, 1946; Flatrès, 1971.
47 In Lorraine, normally fifteen feet of enclosing wall for each half-*manse*, but occasionally as much as thirty (Perrin, 1935, p. 393; '*censier*' of the Abbaye de Chamousey, between 1109 and 1128; p. 456: '*censier*' of the Abbaye de Bouzanville, third quarter of the twelfth century). See Perrin, 1942, pp. 63–4, 71; Planhol, 1959, p. 418.
48 Rainbow: Merian, 1914; Milky Way: Rotzler, 1913; cited by Specklin, 1979, p. 27. On carriages: Deffontaines, 1936; Marcel Gautier, 1971, pp. 239–69. The texts on camels are collected in Salin, I, 1950, pp. 142–3 and 470.
49 Jullian, 1908–26, IV, pp. 595–7; VIII, pp. 19–25; Mirot, 1947, pp. 29–37.
50 Rouche, 1979. Maps of blood groups set alongside Basque linguistic elements can be found in Fierro-Domenech, pp. 143–4, from *Annales, Economies, Sociétés, Civilisations*, 1976.
51 Coville, 1928, pp. 79–236; Jarry, 1948, pp. 15–24; L. Musset, 1965, pp. 111–15; Guichard, 1965.
52 P. E. Martin, 1933; Duparc, 1958.
53 Frédégaire, *Chronique*, IV, 44.

54 Jarry, 1948, p. 38.
55 *Ibid.*, p. 42.
56 Chaume, 1922; cf. Lot, 1935, pp. 172–8.
57 Jarry, 1948, pp. 70–1; Poupardin, 1907.
58 That, at least, is the most plausible thesis, upheld in particular by Grand, 1965. It certainly accounts for the distinctive features of the Franks and their persistent differences from the other Germanic peoples. A synthesis of earlier studies appears in Musset, 1965, pp. 117–32.
59 Grand, 1965, pp. 150–1.
60 On all this: Servais, 1889; Bourquelot, 1865; cf. Pfister in Lavisse, II, part I, pp. 135, 156. *Vie de Saint-Didier* cited from Werner, 1984, p. 330.
61 Salin, 1950–59, I, pp. 47 and 285–300; Grand, 1965, pp. 75–81; Musset, 1965a, p. 47; Guinet, 1967.
62 For example, toponyms in *-fleur*, which are very common on the Normandy coast, are sometimes attributed to them. They are thought to stem from *fleot*, meaning 'creek, bay', whose relation to Scandinavian languages is considered doubtful; see Longnon, 1920–29, pp. 181–7; Rostaing, 1945, p. 71; Dauzat, 1926 (1963), pp. 144–6; L. Musset, 1965a, p. 258. It is mistaken to attribute to them the names in *-ic* to be found around the mouth of the Loire, for these are Breton; see Dauzat, 1926 (1963), p. 179.
63 On Boulonnais: Salin, 1950–59, I, pp. 290–2. Normandy: Alduc-Le Bagousse, 1983.
64 Reinaud, 1836; Rey, 1971; Poupardin, 1901, pp. 243–73; Manteyer, 1908, pp. 238f.; Lot, 1937, I, pp. 78–85; Lacam, 1965; L. Musset, 1965a, pp. 273–7; Sénac, 1980.
65 Sénac, 1980, pp. 83–5. On the other hand, the present name of the Massif des Maures is probably derived from the Provençal *mourre* (= muzzle), used to describe a squashed-looking mountain summit, so the assimilation to the Saracens is the result of a false popular etymology. The same name is also found in the neighbouring Tanneron massif, which was never occupied by the Saracens.
66 *Ibid.*, p. 104.
67 *Ibid.*, p. 101.
68 Haudricourt and Hédin, 1943, p. 123; Maurizio, 1932, pp. 300–3.
69 Rey, 1929; Latouche, 1931.
70 Livet, 1983.
71 This is the title of Fleuriot's synthesis, 1980, which supersedes all earlier literature; see, nevertheless, L. Musset, 1965, pp. 166–70. The pioneering study was that by Loth, 1883.
72 Dauzat, 1926 (1963), pp. 175–6.
73 Flatrès, 1977, p. 308.
74 L. Musset, 1965a, pp. 253–61.
75 Dauzat, 1926 (1963), pp. 146–8; Rostaing, 1945, pp. 71–2.
76 Cf. n. 62.
77 L. Musset, 1965a, pp. 260–1.
78 Maréchal, 1981, pp. 50–1.
79 L. Musset, 1965a, pp. 129, 134–5, 259.
80 *Ibid.*, p. 259.

81 *Ibid.*, p. 257.
82 In particular by Loth, 1883.
83 See Fleuriot's recent demonstration of survival, 1980, pp. 55–9.
84 Falc'hun, 1963.
85 Fleuriot, 1980, pp. 67–9.

4 The birth of France

1 Mirot, 1947, p. 84.
2 Dion, 1947, pp. 71–8; 1950, pp. 461–5.
3 Pirenne, 1909; cf. Dion, 1947, p. 73.
4 Ermold le Noir, *Poème sur Louis le Pieux et épîtres au roi Pépin*, edited and translated by Edmond Faral, Paris, 1932, 1. 1117, p. 86, cited by Dion, 1950, p. 463.
5 *Ibid.*, ll. 2488–2499 (pp. 188–90), cited by Dion, 1950, p. 463.
6 R. de Lasteyrie, *Cartulaire général de Paris*, I, 1887, p. 2, cited by Dion, 1950, p. 463.
7 In Bouquet, *Recueil des Histoires de la France*, vol. VII, p. 185, note *a*, cited by Dion, 1947, p. 75.
8 Nithard, IV, 4, ed. and trans. P. Lauer, Paris, 1926, cited by Dion, 1950, p. 462.
9 Nithard, *ibid.*; Dion, 1950, p. 463.
10 *Ibid.*
11 Nithard, IV, 1; Dion, 1950, p. 464. This is the division between Louis and Charles, March 842.
12 In his review of R. Dion's book published in 1947; Febvre, 1947, p. 207.
13 Ganshof, 1956.
14 In the *Historia Augusta*: Vopiscus, *Aurelius*, XXV, 7, probably for the years 256–57 only; see Grand, 1965, pp. 40–1.
15 The references are collected in L. Musset, 1965, p. 118; Grand, 1965, pp. 21, 29–30.
16 L. Musset, 1965, p. 118.
17 Isidore of Seville, *Etymologies*, XVIII, 6, 9.
18 L. Musset, 1965, pp. 298–9.
19 L. Musset, 1965, pp. 297–8; Grand, 1965, pp. 32–7.
20 The literature on this subject is extensive. Particularly useful are: Guérard, 1849; Bourquelot, 1865; Longnon, 1875; Kurth, 1895; Gallois, 1908, pp. 180–92; Lugge, 1960.
21 Cited by L. Musset, 1965, p. 119.
22 *Ibid.*, pp. 219–20.
23 *Ibid.*
24 Martin, 1949, p. 97.
25 Pfister, in Lavisse, 1911, II, 1, pp. 371–2.
26 *Ibid.*, pp. 363, 371–2.
27 Cited by Martin, 1949, p. 98.
28 Pfister, in Lavisse, 1911, II, 1, p. 372.
29 Beaune, 1985, p. 310.

30 *Ibid.*
31 Petit-Dutaillis, 1950, pp. 7–8.
32 Lot, 1928; Louis, 1960; Menéndez Pidal, 1960, pp. 331–2 and 335–6; Ewig, 1982.
33 *Les Narbonnais*, ed. H. Suchier, Paris, 1898, 2 vols., I, ll. 1856–57; cited by Dion, 1959, p. 220.
34 Dion, 1959, pp. 220–3.
35 Gallois, 1908, p. 192.
36 See above, n. 32.
37 Gautier, 1868; 1872, II, pp. 54–5; 1890, p. 58.
38 For example in Ibn Hauqal, *Configuration de la terre (Kitab Surat Al-Ard)*, French translation by J. H. Kramers and G. Wiet, Beirut–Paris, 1964, I, p. 199; the Djabal al-Fulal (= Djabal al-Qilâl, Massif des Maures; see above, p. 85 and n. 65) is in France. Cf. B. Lewis and J. F. P. Hopkins, article *Ifrandj* in *Encyclopédie de l'Islam*, 2nd edn., III, pp. 1070–3: Miquel, 1975, pp. 354–9.
39 Lugge, 1960.
40 Martin, 1949, pp. 98–9, 124–5.
41 Jullian, 1930–31, II, p. 140, from the *Historia Francorum* by Aimoin; cf. Benda, 1932, pp. 64–5.
42 Werner, 1984, pp. 30–2.
43 L. Musset, 1965, p. 207; Martin, 1949, p. 69.
44 Chaunu, 1982, p. 108.
45 *Ibid.*, pp. 126–7.
46 Beaune, 1985, p. 208.
47 Cited by Gautier, 1890, pp. 65–6.
48 Beaune, 1985, p. 209.
49 *Ibid.*, pp. 211–12.
50 *Ibid.*, pp. 212–29.
51 *Ibid.*, p. 226.
52 *Ibid.*, pp. 108–10.
53 *Ibid.*, pp. 161–2.
54 *Ibid.*, pp. 58–74.
55 *Chanson de Roland*, ll. 16, 109, 702, 1695, 2379, etc.
56 *Ibid.*, l. 1695.
57 *Ibid.*, l. 2311.
58 *Ibid.*, l. 1861.
59 *Couronnement Looys*, BN, fr. 774, fo. 184; cited by Gautier, 1890, p. 61.
60 *Renaus de Montauban*, ed. Michelant, Stuttgart, 1862, p. 404, l. 20, cited by Gautier, 1890, p. 60.
61 *Girard de Viane*, cited by Gautier, 1878–94, I, p. 409.
62 Philippe Mouskes, *Chronique*, ll. 8063–8067, cited by Gautier, 1890, p. 60.
63 Girart d'Amiens, *Charlemagne*, BN, fr. 778, fo. 30r, cited by Gautier, 1890, p. 60.
64 *Berte aux grans piés*, ed. Scheler, Brussels, 1874, p. 9, l. 219, cited by Gautier, 1890, p. 60.
65 *Chanson d'Antioche*, I, 141, cited by Gautier, 1890, p. 60.
66 For example, E. R. Curtius, cited by Menéndez Pidal, 1960, p. 506; Zumthor, 1954, p. 160.

482   Notes to pages 101–7

67 At the very beginning of the first eclogue, 1. 3: '*Nos patriae fines et dulcia linquimus arva*'.
68 *Chanson de Roland*, ll. 2377–2379.
69 X, 782.
70 *Chanson de Roland*, ll. 16, 1194, 1223, etc.
71 Menéndez Pidal, 1960, pp. 505–9.
72 *Cligès*, ll. 30–35 (trans. J. Gardiner, London and New York, 1912).
73 Beaune, 1985, pp. 309, 313.
74 *Continuation du discours des misères de ce temps* (1562).
75 Beaune, 1985, pp. 264–90.
76 *Les Regrets*, IX.
77 Beaune, 1985, pp. 264–90.
78 *Oeuvres complètes*, ed. Viallaneix, VI, 1978, p. 121, cited by Chaunu, 1982, p. 252.
79 Beaune, 1985, pp. 112–13.
80 *Ibid.*, pp. 237–63.
81 *Ibid.*, pp. 291–308.
82 Brunetto Latini, *Le Livre du trésor*, ed. P. Chabaille, Paris, 1863, p. 3, cited by Beaune, 1985, p. 295.
83 Nicole Oresme, cited by Beaune, 1985, p. 295.
84 Beaune, 1985, p. 296.
85 Gilles Le Bouvier, alias the Herald Berry, *Le Livre de la description des pays*, ed. E. T. Hamy, Paris, 1908, p. 52, cited by Beaune, 1985, p. 298.
86 Beaune, 1985, p. 299.
87 *Ibid.*, pp. 307–8.
88 Gervais du Bus, *Le Roman de Fauvel*, ed. A. Langfors, Paris, 1914–19, II, p. 115, cited by Beaune, 1985, p. 318.
89 Beaune, 1985, pp. 318–21.
90 See Chaunu, 1982, pp. 285–6.
91 Werner, 1970, p. 26; Beaune, 1985, pp. 8, 324.
92 Gautier, 1868; 1878–94, I, pp. 159–60; 1890, pp. 55–66. Cf. already Thibaudet, 1938, pp. 210–11, 226–9.
93 Suger, *Vita Ludovici sexti*, ed. H. Waquet, Paris, 1964, pp. 221–2, cited by Beaune, 1985, pp. 321, 334; cf. Benda, 1932, pp. 86–7; Beaune, 1985, p. 325.
94 Cited by Werner, 1970, p. 295.
95 As Chaunu puts it, 1982, p. 106.
96 Beaune, 1985, p. 78.
97 Chaunu, 1982, pp. 196–200.
98 Duby, 1973a; cf. Beaune, 1985, p. 348.
99 Joinville, *Histoire de Saint Louis*, ed. N. de Vailly, 1874, CXXXVII, pp. 374–5; cf. Benda, 1932, p. 73.
100 Cited by Coville, in Lavisse, IV, 1, p. 155.
101 *Les Oeuvres de Maistre Alain Chartier*, ed. A. Duchesne, Paris, 1642, p. 410, cited by A. France, 1909 (1969), p. 60. On the fifteenth century, see in particular Lestocquoy, 1968, pp. 27–34.
102 Jules Quicherat, *Procès de condamnation et de réhabilitation de Jeanne d'Arc*, Paris, 1841, II, p. 436, cited by A. France, 1909 (1969), p. 60.

103 *Ballade contre les mesdisans de la France.*
104 Beaune, 1985, pp. 331–4.
105 Chaunu, 1982, p. 33.
106 Dion, 1947, pp. 79–85.
107 On the frontier to the west of the Meuse, Marot, 1939, p. 105; on the Rhône frontier, Longnon, 1922, pp. 19–20 and 176. See the map p. 39 in Sinclair, 1985.
108 Benda, 1932, p. 67.
109 Petit-Dutaillis, 1950, pp. 9–11; Calmette, 1947, pp. 9–10; Dion, 1947, pp. 40–7. The literature on the development of the theory of natural frontiers is extensive, in particular, Zeller, 1932; 1933; Ancel, 1938, pp. 68–70, 147–50; Fierro-Domenech, pp. 24–6.
110 Dion, 1947, pp. 88–9.
111 Giraldus Cambrensis, *De principis instructione liber*, ed. G. F. Warner, in *Giraldi Cambrensis opera*, VIII, London, 1891, p. 294; cited by Petit-Dutaillis, 1950, p. 240; Benda, 1932, p. 115.
112 A. Dumas, *Charles VII chez ses grands vassaux*, Paris, 1831, p. 89; for the historical basis, which goes back to the intervention by the king and the dauphin Louis in Lorraine, Alsace and Switzerland in 1444, *Ordonnances des rois de France*, XIII, p. 408; cited by Babelon, II, 1917, p. 207 and Braudel, 1986–7, I, p. 289.
113 Calmette, 1947.
114 Daveau, 1959, pp. 84–5.
115 Zeller, 1926, I, p. 420, cited by Benda, 1932, p. 17.
116 Zeller, 1926, I, pp. 425f.; see Benda, 1932, pp. 113–14.
117 A recent assessment in Carmona, 1984, pp. 46–50.
118 Zeller, 1926, I, pp. 425f., cited by Benda, 1932, p. 114.
119 *Ibid.*, p. 115.
120 Cited by Sée, 1891, p. 25.
121 Du Bellay, ed. E. Courbet, Paris, 1918, I, p. 140; Régnier, ed. Prosper Poitevin, Paris, 1873, p. 216.
122 Cited by Carmona, 1984, p. 49.
123 Longnon, 1922, p. 325, cited by Dion, 1947, p. 92.
124 Dion, 1947, p. 92.
125 *Ibid.*, pp. 99–101.
126 *Ibid.*, pp. 92–3.
127 Lovie, 1979; after Lavallée, 1864, which is still useful.
128 Dion, 1947, p. 99, citing article 7 of the treatise of 14 July 1795.
129 Plandé, 1938; Higounet, 1946; see Calmette, 1947, p. 37.
130 Cavaillès, 1910, 1931; see Dion, 1947, pp. 30–1, 99.
131 Daveau, 1959, pp. 94–106.
132 Lentacker, 1974, pp. 13–24; Girard d'Albissin, 1970.
133 Letter from the chevalier de Bonneval to the marquis de Puiseulx, Minister for Foreign Affairs, dated 29 April 1747, cited by Lentacker, 1974, p. 14.
134 Lentacker, 1974, *passim*; Dion, 1947, pp. 101–4.

5 The major divisions

1. Wolff, 1970, p. 139.
2. *Ibid.*, p. 146.
3. Bonnaud, 1981, I, p. 238; II, pp. 194–5; Müller, 1971, 1974.
4. Brunot, 1905, pp. 124–9.
5. Bonnaud, 1981, I, pp. 227, 238f.
6. Dondaine, 1972.
7. Pougnard, 1952; Pignon, 1960; Chaussée (de la), 1966; cf. Bonnaud, 1981, I, pp. 570–1.
8. Boutruche, 1935.
9. Bonnaud, 1977, p. 55.
10. Bonnaud, 1977, p. 46, here reproduced in fig. 25.
11. Bonnaud, 1981, II, p. 192.
12. Cited by Bonnaud, 1981, I, pp. 244 and 583.
13. Rohlfs, 1970; cf. Bonnaud, 1981, I, pp. 249f.
14. That is certainly the opinion of Bec, 1963; cf. Bonnaud, 1981, I, pp. 244f.
15. Apart from the synthesis by Krüger, 1972, see the virtually exhaustive and extremely erudite inventory of the present state of research work in Specklin, 1979, 1982, which, however, is marred by the confusions to which the attention of the reader is now drawn. Cf. his parallel study, Specklin, 1979a.
16. Mapped by Brunhes, 1920, p. 441; a map revised and corrected following local research undertaken in connection with his first synthesis appears in Brunhes, 1947, p. 309.
17. To borrow the words of Dauzat, 1941, pp. 49–54.
18. Godfrin and Planhol, 1966.
19. On threshing: Parain, 1939; Jeanton, 1938. On hives: Armbruster, 1928; Schier, 1939; cited by Specklin, 1979, p. 41.
20. Rivals, 1984.
21. Klimrath, 1843.
22. It has been possible to study these thanks to the findings of Gilliéron, 1902–20.
23. Gernand, 1928; Kläui, 1930; cited by Specklin, 1979, p. 37.
24. Dauzat, 1939a, pp. 37, 57.
25. Wandruszka, 1939, pp. 15, 21.
26. As Léon Gautier has definitively shown, 1878–94, I, pp. 129–46.
27. Brutails, 1923.
28. A remark made by Jacques Chaumié in *Les Marges*, 15 April 1920 and cited by Dupouy, 1942, p. 172.
29. See the discussion in Dupouy, 1942, pp. 179f. Text from Mme de Staël: p. 153, published by Garnier, Paris, n.d.; Lanson, 12th edn., Paris, 1912, p. 884.
30. In chronological order: Bloch, 1931; Roupnel, 1932; Dion, 1934.
31. The idea comes from Roupnel, 1932.
32. A conclusion that is not formulated, but implicit in Dion, 1934.
33. See in particular the synthesis by Krenzlin, 1959.
34. Clearly set out in Duby, 1973.
35. Chapelot and Fossier, 1980, pp. 139–44.
36. Fournier, 1978.

37 This was probably invented north of the Alps (Rhetia?) in the first centuries AD; see Haudricourt and Jean-Brunhes-Delamarre, 1955, pp. 348–68.
38 Duby, 1973, p. 219.
39 Lefebvre des Noëttes, 1931.
40 Fossier, 1968, I, p. 253; Duby, 1973, p. 212.
41 According to the German expression *Vergetreidung*.
42 Fossier, 1968, I, p. 335.
43 Higounet, 1956.
44 Juillard, 1953, p. 44.
45 Reitel, 1966.
46 Planhol, 1959, p. 419.
47 Dion, 1946.
48 The line was plotted by Dion, 1934, p. 10, who claims, somewhat inaccurately, that it is based on Young's remarks. Cf. Specklin, 1979, pp. 19–20.
49 Gay, 1967, pp. 137–41.
50 Niemeier, 1961; Uhlig, 1961; see the review by Planhol, 1961.
51 On all this Souillet, 1943; Thomas-Lacroix, 1954; cf. Bonnaud, 1981, I, pp. 148–59.
52 Chiffre, 1985, in particular, pp. 18, 83–95.
53 Bonnaud, 1981, I, pp. 138–9 and 151–3.
54 *Ibid.*, pp. 95–9.
55 Meynier, 1945a.
56 The subject is now dominated by the studies of C. Higounet, in particular by his synthesis of 1975, with a basic map, p. 396 of the French text. Another good account, from a different point of view, appears in Clout, 1977a, pp. 92–103.
57 Higounet, 1949.
58 Dion, 1946a.
59 Livet, 1962, 1983, is particularly insistent upon these ancient examples.
60 See the studies collected in Flatrès and Planhol, 1983.
61 Rey, 1929.
62 Livet, 1962, pp. 200–5.
63 Fel, 1959; 1962, pp. 46–9, 127–33.
64 Faucher, 1950; Chevalier, 1956, pp. 272–8.
65 Derruau, 1949; see the map, constructed on the basis of his information, by Prince, 1977, p. 166.
66 Merle, 1958.
67 Perpillou, 1971.
68 Synthesis and map by A. Meynier, pp. 78–89, in Juillard, Meynier *et al.*, 1957.
69 Furet and Ozouf, 1977.
70 *Ibid.*, pp. 60–1, to be complemented by those of Le Roy Ladurie and Dumont, 1971, pp. 439, bottom, and 440, top, which also provide information on lower Alsace.
71 Furet and Ozouf, 1977, I, p. 326.
72 On this subject, see Flatrès, 1963.
73 Le Roy Ladurie and Dumont, 1971.
74 Chamla, 1964.
75 Dupin, 1827; d'Angeville, 1837; cf. Le Roy Ladurie and Dumont, 1971, p. 412.

6 The secondary divisions

1. G. Bloch, 1911, pp. 190–1.
2. In Mirot's list, 1947, pp. 291–7, based on Longnon's *Atlas*.
3. According to Moreau's list, 1972, pp. 386–401.
4. According to Jarry, I, 1942, pp. 131–45 and Fierro, 1986, pp. 308–11.
5. Bancal, 1945, pp. 34–57; cited by Fierro, 1986, p. 95.
6. Fierro, 1986, based on Brette, 1904, 1907.
7. 375 in *pays d'Elections* and 78 in *pays d'Etats*: Mirot, 1947, pp. 369–74.
8. Durand, 1984, p. 80, based on Dupâquier, 1979.
9. Esmein, 1901, pp. 587–90; Mirot, 1947, pp. 356–66; Fierro, 1986, pp. 97–8.
10. Esmein, 1901, pp. 590–7; Mirot, 1947, pp. 367–74.
11. Fierro, 1986, p. 98.
12. *Ibid.*
13. Mirot, 1947, pp. 309–24; Fierro, 1986, pp. 91–4.
14. Mirot, 1947, pp. 375–81; Fierro, 1986, p. 99.
15. List in Mirot, 1947, pp. 380–1.
16. Fierro, 1986, pp. 100–4.
17. Dupont-Ferrier, 1929, 1929a.
18. Fierro, 1986, pp. 100–1; the lists are compared pp. 312–16.
19. Jarry, III, 1948, p. 117.
20. Chaume, 1944.
21. C. Pfister in Lavisse, 1911, II, 1, pp. 178–9.
22. Jarry, III, 1948, p. 113.
23. Fiétier, 1979.
24. Cited by Durand, 1984, p. 18.
25. On towns that grew out of castles and on *châtellenies*, Fournier, 1978, *passim* and in particular pp. 129–36. Towns and monasteries: Latouche, 1956, pp. 271–309 and in particular 279f. The text from Estienne is cited by Mandrou, 1961, p. 7.
26. Jullian, 1908–26, VI, p. 499: from the name of the *Elosedienses*, 'the people established on the Ill', or from a mysterious *Pagus Alisacensis*.
27. The *Pagus Alionensis*, known in the Carolingian period, derived its name from the *castrum Alionis* > Châtelaillon, a town ruined in the thirteenth century, 12 kilometres to the south of La Rochelle. The growth of the latter explains the establishment, in the eighteenth century, of the *gouvernement* of Aunis, which was separate from that of Saintonge.
28. Richard, 1954; Devailly, 1973; Bur, 1977.
29. The only attempt at a systematic analysis of the geographical structure of the provinces is that by Brunhes, 1920, pp. 342–68. Overall, it is not very convincing, but it contains many new and rewarding perceptions of detail.
30. Dion, 1947, pp. 40–5, which draws upon: Chénon, 1892; Stein and Legrand, 1905; Dupont-Ferrier, 1942; Lemarignier, 1945.
31. Stein and Legrand, 1905, cited by Dion, 1947, p. 45; see *Histoire de la Lorraine*, 1939, pp. 104–5.
32. Dion, 1961.
33. *The Gallic War*, V, 24; VII, 90.
34. Bur, 1977.

35 Le Lannou, 1940.
36 Devailly, 1973.
37 Demangeon, 1946, I, pp. 81–2.
38 Dion, 1947, pp. 8–9.
39 Agache, 1975, 1978.
40 Crampon, 1985.
41 Schneider, 1939, p. 162.
42 On the historical geography of Lorraine, Gain, 1937 and the studies collected in *Histoire de la Lorraine*, 1939.
43 Cited by Vivien de Saint-Martin, *Nouveau Dictionnaire de géographie universelle*, I, Paris, 1879, p. 561.
44 Flatrès, 1977, p. 305, from the studies of J.-B. Colbert de Beaulieu.
45 Dion, 1938.
46 Specklin, 1979, pp. 52–7.
47 Cited by Marc Bloch, 1940, p. 175.
48 Claval, 1978, p. 16.
49 *Ibid.*, p. 69.
50 Sion, 1934, p. 176.
51 Kolodny, 1962.
52 Mirot, 1947, p. 415.
53 Cited by Durand, 1984, p. 202.
54 For the sixteenth century: the account given by the Venetian ambassador Marin Giustiniano in 1535, cited by Dupuy, pp. 38–9; for the Merovingian period: C. Pfister, in Lavisse, 1911, II, 1, p. 172; for the Carolingian period: A. Kleinclausz, *ibid.*, p. 306; for the period of the early Capetians: A. Luchaire, *ibid.*, II, 2, p. 177.
55 Pierre Goubert, cited by Pierre Chaunu, in Durand, 1984, p. 10.
56 Durand, 1984, p. 85.
57 From the 'Table of customs of France' established by Father Lelong on the basis of the *Nouveau Coutumier général* by Bourdot de Richebourg, cited by Durand, 1984, p. 81.
58 Esmein, 1901, pp. 741–2.
59 *Ibid.*, pp. 742–8; Bloch, 1939a, p. 188.
60 Cited by Durand, 1984, p. 81.
61 Esmein, 1901, p. 714.
62 Durand, 1984, p. 229.
63 Foisil, 1970; cf. Durand, pp. 232–6.
64 Fléchier (Esprit), *Mémoires sur les Grands Jours d'Auvergne en 1665*. This work was not published until 1844.
65 Cited by Durand, 1984, pp. 29–30.
66 Tenèze, 1955.
67 On *Girart*..., Bédier, 1926–29, II, pp. 4f.; more recently, Louis, 1947. On *Gaydon*, Subrenat, 1974, p. 51 and, for the portraits of the 'Angevins', pp. 198–268.
68 Ll. 39–41; cf. Rossi, 1975, p. 588.
69 Gautier, 1890, p. 753.
70 BN, fr. 25518, fo. 4r, cited by Bédier, 1926–29, I, p. 464.

71 Bloch, 1940, p. 197.
72 Cited by Durand, 1984, p. 29.
73 In the *De excidio Romae et magnificentia Catonis*, ed. A. Boutemy, Brussels, 1943, p. 53, cited by Lestocquoy, 1968, pp. 21–2.
74 Bonnaud, 1981, I, p. 191.
75 Brutails, 1923.
76 *Ibid.*, p. 16.
77 Brunhes and Deffontaines, 1926, maps pp. 640–52.
78 Brutails, 1923, p. 26.
79 Ménard, 1980, used by Durand, 1984, pp. 216–23.
80 Planhol, 1968, 1971; Planhol and Popelard, 1976. On the phenomenon as a whole, Itzin, 1983.
81 Gallois, 1908, pp. 35f., which gives many examples of modern scholarly resurrections of the names of *pagi*.
82 *Ibid.*, pp. 213–14. The map of names of *pays* used in the west, provided by Flatrès, 1977, shows clearly that much of the area was not named at all.
83 Gallois, 1908, p. 208; Dauzat, 1926–63, p. 207.
84 Rostaing, 1945, p. 28.
85 On the development of the name of the Jura, Poupardin, 1907, pp. 6–7. The term, used to denote the mountain that separated the *Sequani* from the *Helvetii*, already appears in Caesar, *The Gallic War*, I, 2, 6 and 8.
86 Rostaing, 1950, p. 129; 1945, p. 118; cf. Clavel, 1970, p. 143.
87 Caesar, *The Gallic War*, VII, 8: '*Mons Cebenna, qui Arvernos ab Helviis discludit*'.
88 *Ibid.*, V, 3; VI, 29; VI, 31.
89 Moreau, 1972, p. 20.
90 L. 2017, ed. Ferdinand Castets, *La Chanson des Quatre Fils Aymon d'après le manuscrit La Vallière*, Montpellier, 1909 (Publications de la Société pour l'étude des langues romanes, XXIII), p. 335.
91 *Ibid.*, l. 3195, p. 374.
92 *Ibid.*, ll. 1786–1787, p. 328.
93 Rostaing, 1945, p. 117.
94 *Ibid.*, p. 118.
95 Levainville, 1909, pp. 4–5; Jullian, 1908–26, II, p. 536.
96 Moreau, 1972, p. 188.
97 Gallois, 1908, pp. 112–16.
98 Derruau, 1949, p. 8.
99 Musset, 1919; Pélatan, 1985.
100 Rostaing, 1945, p. 27.
101 Marres, 1935.
102 Blottière, 1973.
103 Cited by Gallois, 1908, p. 82.
104 All these quotations and references are taken from Kroemer, 1976, p. 36.
105 All these quotations and references are taken from Plavinet, 1974, pp. 2–3.
106 Gallois, 1908, p. 78, already according to Charles Estienne, *La Guide des chemins de France*, Paris, 1552: 'which is fertile in rye (*seigle*), after which Sologne is named'.

107 L. 5980.
108 Bloch, 1952, p. 58.
109 Perhaps Laon.
110 Ed. Castets, line 2009.
111 Flatrès, 1977, p. 305.
112 As was certainly noted for the vine by Chabot, 1943.
113 Specklin, 1979, p. 30.
114 Gallois, 1908, pp. 83–100.
115 *Ibid.*, pp. 253–60.
116 Moreau, 1972, p. 70.
117 All this is according to Jullian, 1908–26, I, p. 67.
118 Feuchère, 1954.
119 Rézeau, 1976; Durand, 1984, p. 38.
120 *L'Aubrac*, 1970–81.
121 Mandrou, 1961, pp. 299–310, for numerous examples.
122 Durand, 1984, p. 201.
123 Marcel Gautier, 1971, pp. 25–6, with a map showing the catchment area of the Rostrenen fair, in the Côtes-du-Nord.
124 Allix, 1914.
125 Durand, 1984, p. 201. For the Alps, see the map in Arbos, 1922, p. 664, which indicates all the places where fairs were held at least four times a year.
126 Durand, 1984, pp. 115–16.
127 Schwab, 1980, pp. 141–64.
128 *Ibid.*, p. 153.
129 *Ibid.*, p. 164.
130 Durand, 1984, p. 79.
131 A clear idea of the movements of a country lord in the seventeenth century is conveyed by an extraordinary document, the *Journal de Paul de Vendée, capitaine huguenot, 1611–1623*, published by A. Benoni Drochon, Niort, 1880 (extract from the *Mémoires de la Société de statistique, sciences, lettres et arts des Deux-Sèvres*, XVII).
132 Schwab, 1980, pp. 224–30.
133 Elhaï, 1965.
134 Dugrand, 1956.
135 Armand, 1974.
136 Pouthas, 1956; see Clout, 1977c, p. 500.
137 Schwab, 1980, pp. 51–3.
138 Clout, 1977c, p. 488.
139 Gascon, 1980.
140 Bonnaud, 1981, I, p. 178, according to Crubellier, 1975.
141 Bonnaud, 1981, I, p. 178, according to Delafosse, 1955.
142 Perrot, 1974, pp. 219–75.
143 *Ibid.*, p. 684.
144 Huetz de Lemps, 1975.
145 Durand, 1984, p. 169.
146 *Ibid.*
147 Schwab, 1980, p. 207.

148 Faucher, 1934.
149 Planhol, 1965.
150 Devailly, 1980.
151 William the Breton, *Philippide*, V, 5: *ciseraeque tumentis Algia potatrix*; VI, 334: *Non tot in autumni rubet Algia tempore pomis/Unde liquore solet siceram sibi Neustria gratam*. The translation is by Guizot, in *Mémoires relatifs à l'Histoire de France*, 12, Paris, 1825, p. 164. The text on Azincourt is cited by Sion, 1909, p. 154, and Dion, 1934 (1981), p. 116.
152 J. Peltre, verbal communication.
153 Papy, 1948.
154 This idea, put forward by Deffontaines, 1932, has been considered by some to be overcategorical. D. Faucher speaks simply of the introduction of maize into the polycultivation system, not of a positive revolution in the agricultural system. Cf. R. Brunet, 1965, pp. 332–3; Lerat, 1963, pp. 147–8.
155 Cited from Fauchez, 1949, p. 108. The text appears in vol 2, pp. 352–3 in the second edition of the French translation, Paris 1794 (Year II).
156 Pitte, 1986, for all that follows.
157 Le Roy Ladurie, 1966, p. 760, cited by Pitte, 1986, p. 110.
158 Pitte, 1986, pp. 101–3.
159 *Ibid.*, pp. 106–10, and, in more detail, Pitte, 1978.
160 Pitte, 1986, p. 90.
161 Augé, 1955, pp. 16–18; Fel, 1962, pp. 171–2; Morineau, 1970; Clout, 1977b, pp. 410–11.
162 Chevalier, 1956, pp. 237–41 and map p. 242 for 1840.
163 Morineau, 1970; map in Clout, 1977b, p. 411.
164 East, 1939, pp. 292–9.
165 On this early over-population, Blanchard, 1906, pp. 341f.; Dion, 1938a.
166 Cited by Faucher, 1949, p. 83; the text appears in vol. 2, pp. 241–2 in the 1794 French translation.
167 Juillard, 1953, pp. 40–2.
168 Pédelaborde, 1957.
169 *Elégies*, III (1565).
170 Dion, 1934a.
171 Huetz de Lemps, 1975, pp. 247–54 and 260–6.
172 Caster, 1954, 1962; Brunet, 1965, pp. 326–7.
173 Rondeau, 1964, *passim*; Renucci, 1974, pp. 56–9; Fel, 1975.
174 Alcuin, *Epistolae*, in *Monumenta Germaniae historica, Ep. merov. et karol. aevi*, II, Berlin, 1895, p. 318, cited by Dion, 1959a, p. 172.
175 Dion, 1959a, pp. 184–5.
176 *Ibid.*, p. 200.
177 *Ibid.*, pp. 201–6.
178 *Ibid.*, pp. 207–43, on all this.
179 *Ibid.*, p. 245, citing the account, by Lambert d'Ardes, of a banquet held in 1178 in honour of the Archbishop of Rheims by the comte de Guines, in Calais.
180 *Ibid.*, pp. 244–308, on all this.
181 *Ibid.*, pp. 384–98.
182 *Ibid.*, pp. 336–98, on all this paragraph.

183 *Ibid.*, pp. 323–4.
184 *Ibid.*, pp. 308–35, on all this paragraph.
185 *Ibid.*, pp. 417–91, on all this.
186 Apart from Dion, 1959a, for all that follows, see in particular Enjalbert, 1975, pp. 81–129.
187 Enjalbert, 1974; Pijassou, 1980.
188 Enjalbert, 1975, pp. 120–3.
189 Clout, 1977b, for all this paragraph. This is the first overall treatment of the subject.
190 Demangeon, 1946, II, p. 651.
191 Bourgin, 1920; Gille, 1960.
192 There is a detailed analysis of the early years of the modern period in Rose-Villequey, 1970.
193 Markovitch, 1968; Woronoff, 1970.
194 Clout, 1977b, p. 460.
195 Not the population density, as in Clout, 1977b, p. 460.
196 Peltre, 1978.
197 Chevalier, 1956, pp. 541–603 and in particular pp. 543–4.
198 Veyret-Verner, 1948.
199 Chevalier, 1961.
200 Peltre, 1980.
201 Durand, 1984, p. 265.

Conclusion to Part II: The beginnings of a territory with a spirit of its own?

1 H. Lemonnier, in Lavisse, 1911, V, i, p. 277.
2 Mandrou, 1961, p. 93, for the range of the merchant from Amiens. On woad, Caster, 1954.
3 Bergeron; Earden; Cocula; in Léon, 1975, for all this.
4 Guichonnet, 1948.
5 Robert-Muller and Allix, 1923; Allix, 1929, 1932.
6 Fel, 1962, pp. 150–4; Delaspre-Parquet, 1954.
7 Cohen, 1948, pp. 73–4.
8 C. Petit-Dutaillis, in Lavisse, 1911, IV, 2, pp. 217–19.
9 Dontenville, 1966, 1973, 1973a.
10 On dragons in general, Dontenville, 1973a, pp. 149–72; on the Tarasque, Louis Dumont, 1951.
11 Lecouteux, 1982, pp. 35–9.
12 Dontenville, 1973a, pp. 173–98.
13 *Ibid.*, pp. 651–68; the limits of Gargantua's domain.
14 Tristan, 1979, pp. 37f.
15 Viollet, 1893, pp. 201–36.
16 Esmein, 1901, pp. 778–83.
17 Mandrou, 1961, p. 134.
18 Cited by Dupuy, p. 40.

7 Paris and the Parisian centralization

1. Dion, 1951b.
2. To borrow the words of A. Cholley, 1943, pp. 91–4 and map p. 89.
3. Duval, 1961, pp. 251–3.
4. Dion, 1951a, p. 17.
5. Most recently, Duval, 1961.
6. As Pitte points out, 1983, I, p. 73.
7. Dion, 1952a, p. 556.
8. Strabo, IV, 3, 5.
9. Dion, 1952a, p. 555.
10. Martin, 1949, p. 102.
11. Maçoudi, *Les Prairies d'or*, French translation by Barbier de Maynard and Pavet de Courteille, revised by Charles Pellat, II, Paris, 1965, p. 343; see Miquel, 1975, pp. 357–8.
12. Dion, 1951a, p. 24; Timbal, 1973; and above all, for the above, Bautier, 1978.
13. Olschki, 1913; Dion, 1949; 1953a and b.
14. Cited by Dion, 1949, p. 46.
15. *Roman de Berte aux grands pieds*, 1. 266, ed. A. Scheler, Brussels, 1874; cited by Olschki, 1913, p. 281.
16. Dion, 1952d, p. 25.
17. L. 756; cited by Dion, 1952d, p. 44.
18. Olschki, 1913; cited by Vossler, 1953, p. 41.
19. *Ibid.*
20. In Le Roux de Lincy and Tisserand, *Paris et ses historiens aux XIVe et XVe siècles*, Paris, 1867; cited by Olschki, 1913, p. 25.
21. *Orlando furioso*, XIV, 104.
22. In W. Wattenbach, 'Aus den Briefen des Guido von Bazoches', *Neues Archiv der Gesellschaft für Ältere Deutsche Geschichtskunde*, 16 (1891), 69–113; cited by Wright, 1965, pp. 331–2.
23. Ballade 'Des femmes de Paris', in the *Grand Testament* (trans. Lewis Wharton, London, 1935).
24. *Pantagruel*, I (1532), chapter 5; see Barrière, 1961, p. 49.
25. *Elegies*, I, xxxii; ed. Vaganay, Paris, 1923, V, p. 178.
26. *Les Regrets*, sonnet CXXX.
27. *The Essays of Michael, Lord of Montaigne*, trans. John Florio, London and Toronto, 1928, p. 216.
28. Ballade 'par laquelle Villon crye mercy à chascun', in the *Grand Testament* (trans. Lewis Wharton, London, 1935).
29. Favier, 1974.
30. Mercier (Louis-Sébastien), *Tableau de Paris*, chapter 57 (1781 text).
31. In *Le Pauvre diable* (1758).
32. Bergeron *et al.*, 1970; Bertillon, cited by L. Chevalier, 1950; see pp. 57–60 and 164–71.
33. *Ibid.*
34. *Ibid.*
35. Phlipponneau, 1956, pp. 24–119 for all this paragraph.

Notes to pages 259–77   493

36 *De la richesse territoriale du royaume de France*, ed. Schelle and Grimaut, Paris, 1825, pp. 149f., cited by Phlipponneau, 1956, p. 31.
37 Philipponneau, 1956, pp. 33, 42.
38 Dion, 1959a, pp. 492–607 for all that follows.
39 *Ibid.*, p. 387.
40 Demangeon, 1946, I, p. 226.
41 Pitte, 1985.
42 Brunet, 1960, who is followed in the whole of the next paragraph.
43 *Gargantua*, XVI (trans. Sir Thomas Urquart and Peter Motteaux, Chicago and London, 1952). On Beauce in the fifteenth century, references in Anatole France, *Vie de Jeanne d'Arc* (reprinted Geneva, 1969), I, p. 436.
44 (Charles Estienne), *La Guide des Chemins de France*, Paris, 1552.
45 *Eclogue I*, Margot (1565).
46 Cavaillès, 1946, for the whole paragraph.
47 At a date that remains uncertain; Cavaillès, 1946, pp. 35–7.
48 BN, Maps, Ge D D 960 (104), cited by Cavaillès, 1946, p. 39.
49 Cavaillès, 1946, p. 72.
50 Blanchard, 1942, pp. 36–62; Demangeon. 1946, I, pp. 408–13.
51 Blanchard, 1942, pp. 44–5.
52 *Ibid.*, pp. 163–7.
53 H. Lemonnier, in Lavisse, 1911, V, 2, pp. 264–6.
54 Barrière, 1961, p. 137.
55 *Ibid*, p. 152.
56 *Ibid*, pp. 140–3.
57 *Ibid*, pp. 300–4.
58 *Lettres familières*, IV, to Abbé Niccolini (6 March 1740).
59 *L'Ancien Régime et la Révolution*, Paris, 1866 (*Oeuvres complètes*, IV), p. 108.
60 Arthur Young, *Travels in France*, Cambridge, 1929, pp. 428–9.
61 *Les Déracinés*, chapter 3.
62 Léon Ferrero, cited by Halévy, 1933, p. 215.
63 Réau, 1951.
64 See the accounts and impressions analysed in Dupuy, c. 1960.
65 Barral, 1968; cf. Chevalier, 1947, and in particular pp. 115–33.

8 Cultural action and reaction: unity and diversity

1 To borrow the expression used by Zeldin, 1980–81.
2 Coirault, 1941, pp. 11–13.
3 Coirault, 1921–33, 1941, 1953–63.
4 Coirault, 1953–63, I, pp. 45, 60.
5 *Ibid.*, I, p. 60.
6 Coirault, 1941, pp. 345–6.
7 E. Scribe, Inaugural lecture at the Académie française, cited by Emile Littré, *Dictionnaire de la langue française*, Paris, 1873, I, p. 551, s.v. *chanson*.
8 Coirault, 1941, p. 131.
9 *Ibid.*, p. 304.
10 Delarue, 1957, p. 27.

11 *Ibid.*, p. 29.
12 *Ibid.*, pp. 34–45.
13 To borrow P. Coirault's expression, cited by Soriano, 1968, p. 94.
14 Mandrou, 1964; Bollème, 1971; Darmon, 1972, p. 15, following the pioneer work of Nisard, 1864.
15 Alphonse Dupront, in *Livre et société dans la France du XVIIIe siècle*, Paris–The Hague, 1965, p. 230; cited by Darmon, 1972, p. 15.
16 Darmon, 1972, p. 140.
17 Cited by Durand, 1984, p. 19.
18 Alexis de Tocqueville, *L'Ancien Régime et la Révolution*, 7th edn., Paris, 1866 (*Oeuvres complètes*, IV), p. 116.
19 Balibar, 1985, p. 195.
20 *Ibid.*, p. 196.
21 *Ibid.*, p. 276.
22 Darmon, 1972, p. 38, with map.
23 *Ibid.*, graph p. 51.
24 *Ibid.*, pp. 122–8.
25 According to Sansot, 1985.
26 *Ibid.*, p. 29.
27 *Ibid.*, p. 23.
28 Sagnac in Lavisse, 1920, I, pp. 170–4; Brunhes, 1920, pp. 399–411; Mage, 1924; Brun, 1939; Bancal, 1945; Mirot, 1947, pp. 396–440; Jarry, 1948, III, 2, pp. 404–11; Fierro, 1986, pp. 105–11. Of the many departmental monographs, Porée, 1905; Mettrier, 1911.
29 Bancal, 1945, pp. 144–5; Fierro, 1986, pp. 105–6, on all that follows.
30 Cited by Fierro, 1986, p. 107.
31 Brun, 1939, p. 60; Fierro, 1986, p. 108.
32 Sagnac, in Lavisse, 1920, I, p. 173.
33 Edmund Burke, *Reflections on the French Revolution*, Harmondsworth (Penguin), 1968, p. 297.
34 *Ibid.*, p. 133.
35 *Ibid.*, pp. 126–9.
36 Mirot, 1947, p. 399.
37 Compare the examples collected in: Association pour l'étude du fait départemental, *Actes du Colloque de Rennes (janvier 1982): le rôle décisionnel dans l'espace départemental*, Poitiers–Rennes, 1983; see Part V. On the department as an autonomous economic unit: Guellec, 1979; 1980.
38 A. Coville, p. 12, in Lavisse, 1911, IV, 1.
39 The royal itinerary was recorded in detail by a 'servant' of the king, Abel Jouan, *Recueil et discours du voyage du roy Charles IX de ce nom*, Paris, 1566.
40 Coirault, 1980.
41 According to Mongrédien, 1965, I, pp. 67–101.
42 Coornaert, 1966, pp. 356–81, see p. 380.
43 *Ibid.*, p. 36.
44 *Ibid.*, p. 142.
45 *Ibid.*, pp. 188ff.
46 *Ibid.*, p. 142–6.

Notes to pages 289–312   495

47 *Ibid.*, p. 142.
48 *Ibid.*
49 Durand, 1984. p. 36.
50 See the centenary edition, Paris, 1977, Librairie Eugène Belin, with an Afterword by Jean-Pierre Bardos, pp. 304–31, which skilfully sets the work in its historico-cultural context. For the publication dates, see p. 312.
51 *Ibid.*, pp. 297, 300.
52 Cited by Halévy, 1933, p. 230.
53 Bruno, *Le Tour de la France par deux enfants*, Paris, 1877, pp. 247–8.
54 Cuisenier, 1977, for all this paragraph. Of the regional syntheses, Dufournet, 1981.
55 Cuisenier, 1977, pp. 58–9.
56 *Ibid.*, pp. 34–5.
57 *Ibid.*, p. 211.
58 *Ibid.*, p. 187.
59 Tardieu, 1950, cited, 1977, p. 99.
60 Creston, 1974, pp. 26–7.
61 Creston, 1974. Another regional analysis, less full but very perceptive, for the Berry: Favière, 1957.
62 On the Bresse hat, Jeanton, 1937; see pp. 16, 24–7, from 'La rustique de Brèce', fig. 21. in F. Desprez, *Recueil de la diversité des habits*, Paris, 1567; see also Alcan, 1960, p. 6. On the West, the demonstration by Durand, 1984, pp. 40–1.
63 Creston, 1974, pp. 30, 33.
64 Favière, 1957, pp. 169–70.
65 Creston, 1974, pp. 32–3.
66 *Ibid.*, pp. 60, 71.
67 Dufournet, 1981, pp. 131–3.
68 Creston, 1974, p. 38.
69 *Ibid.*, p. 116.
70 *Ibid.*, pp. 93ff., 124ff.
71 *Ibid.*, pp. 259ff.
72 J.-M. Guilcher, 1963, 1967, 1969, for all the next paragraph.
73 J.-M. Guilcher, 1967, p. 5.
74 This now appears to be established (J.-M. Guilcher, 1969, pp. 18ff.), despite the hypothesis, long defended, of an autochtonous origin.
75 J.-M. Guilcher, 1969, p. 96.
76 J.-M. Guilcher, 1967, p. 11; 1969; pp. 194ff.
77 J.-M. Guilcher, 1965, pp. 3–4, 13–14, 28–32.
78 Parain, 1957.
79 *Ibid.*
80 *Ibid.*
81 Mauriange, 1955.
82 P. de Martin, 1979.
83 Riff, 1923; Denis and Groshens, 1978.
84 Alcan, 1960, p. 13.
85 J.-M. Guilcher, 1969a, 1984.
86 Cuisenier, 1977, pp. 105–9.

87 Carlos Garcia, *L'Opposition et la conjonction des deux grands luminaires du monde, oeuvre plaisante et curieuse, où l'on traite de l'heureuse alliance de la France et de l'Espagne et de l'antipathie des Espagnols et des Français*, Rouen, 1617, p. 338, cited by Giordan, 1980, p. 109; Arthur Young, *Voyages en France*, 2nd edn., Paris, 1794, I, pp. 81, 141; (*Travels during the years 1787, 1788 and 1789*, Dublin 1793). Michelet, *Histoire de France*, definitive edn., Paris, n.d., II, pp. 38–9.
88 Michel Etchevery, 'Paris met des gants d'osier', *Le Monde*, 20 September 1986.
89 J.-M. Guilcher, 1967, p. 11.
90 Reuss, 1920, pp. 188–92, 360–7.
91 Lafont, 1970; Lafont and Anatole, 1970.
92 Lafont, 1970, p. 13.
93 *Ibid.*, pp. 27–8.
94 *Ibid.*
95 Lafont and Anatole, 1970, II, p. 422.
96 Cited from Lafont, 1970, p. 86.
97 *Ibid.*, pp. 223, 236.
98 *Ibid.*, pp. 109–44.
99 *Ibid.*, pp. 118–20.
100 *Ibid.*, p. 143.
101 *Ibid.*, pp. 270–307.
102 *Ibid.*, pp. 292.
103 *Ibid.*, p. 299.
104 *Ibid.*, p. 204.
105 Mistral, *Mireille*, I, verse 2.
106 Le Braz, 1904, p. 232.
107 *Ibid.*, pp. 251, 268.
108 The list was drawn up by Hersart de la Villemarqué in his *Essai sur l'histoire de la langue bretonne*, 1847, reproduced by Tanguy, 1977, II, pp. 79ff.
109 A. Léon, 1909; see the introduction, pp. 11–102, *passim* and in particular pp. 13, 16, 29, 54ff., 65ff., 91.
110 There is a good synthesis on Corsican literature, both scholarly and popular, in Marcaggi, 1926, especially the introduction, pp. 5–83.
111 *Ibid.*, pp. 12–13.
112 *Ibid.*, pp. 34–6.
113 *Ibid.*, pp. 7–8.
114 *Ibid.*, p. 14.
115 *Ibid.*, p. 10.
116 *Ibid.*, pp. 13–14.
117 Lafont and Anatole, 1970, II, 498–502.
118 Jacoubet, 1923, 1929.
119 Lafont, 1954, p. 143.
120 *Ibid.*, p. 91.
121 Van Tieghem, 1917; on celtomania see I, pp. 192–201 and II, pp. 28–34.
122 Tanguy, 1977, I, pp. 44, 129ff., 341.
123 Cited by Lafont, 1970, p. 247.

124 *Mireille*, Book XI.
125 Lafont, 1954, pp. 180, 183.
126 Lasserre, 1918, p. 163.
127 Lafont, 1954, p. 144.
128 Lafont and Anatole, 1970, p. 598.
129 Lafont, 1971, pp. 134–5.
130 *Ibid.*, p. 150.
131 Lafont, 1954, pp. 144–5.
132 Cited by Richard Kleinschmager, in Lacoste, 1986, I, p. 961.
133 *L'Alsace est-elle allemande ou française? réponse à M. Mommsen*, Paris, 1870, reprinted in *Questions historiques*, Paris, 1893, pp. 505–12.
134 Cited by Kleinschmager, in Lacoste, 1986, I, p. 951.
135 Latest synthesis in Pomponi, 1979.
136 Reclus, 1879, p. 87.
137 Cited by Tanguy, 1977, I, p. 45.
138 *Ibid.*
139 *Ibid.*, pp. 129ff.
140 Cited by Tanguy, I, p. 171.
141 *Ibid.*, p. 46.
142 Cited by Tanguy, 1977, I, p. 385.
143 Cited by Lasserre, 1918, p. 161.
144 Lafont, 1971, p. 117. Soboul, p. 44, in Gras and Livet, 1977, shows clearly that the Girondins' insistence upon decentralization concentrated upon the departmental level.
145 1912 edn., IV, pp. 421ff.; cited by Flory, 1966, p. 35.
146 Flory, pp. 113–14.
147 *Ibid.*, p. 2.
148 Phlipponneau, in Lacoste, 1986, II, p. 437.
149 Flory, p. 36.
150 Phlipponneau, in Lacoste, 1986, II, p. 437.
151 Flory, pp. 113–14.
152 Boulard, 1947, 1954, for all that follows.
153 Boulard, 1954, pp. 35–9.
154 *Ibid.*, pp. 44–6.
155 The pioneering work on the subject, within a regional framework, is by Siegfried, 1913. A basic synthesis appears in Goguel, 1951, and a monumental but uneven analysis in Lacoste, 1986. Many studies of electoral geography, mostly within a departmental context, have been produced by historians over the past thirty years, but most are of little interest from our point of view. From a geographical standpoint, the most perceptive monograph is that by Masseport, 1960.
156 Boulard, 1954, pp. 36, 44–5.
157 *Ibid.*, pp. 40–1.
158 This is one of the major themes of Siegfried's book, 1913.
159 Muriel Le Roux, 'Les origines complexes d'un bastion de la gauche', in Lacoste, 1986, II, pp. 869–88.
160 Le Bras, 1986, pp. 79–81.

161 To judge by the map showing the cafés that remained open during church services; Guermond in Lacoste, 1986, II, p. 263; see Siegfried, 1913, pp. 232–53.
162 Phlipponneau, in Lacoste, 1986, II, pp. 319, 333; see Siegfried, 1913, pp. 173–80.
163 Siegfried, 1913, pp. 392–400.
164 Kleinschmager, in Lacoste, 1986, I, pp. 1024–7.
165 Masseport, 1960, pp. 148–54.
166 Le Bras and Todd, 1981.
167 *Ibid.*, pp. 40–4.
168 *Ibid.*, pp. 36–40.
169 *Ibid.*, pp. 44–53.
170 *Ibid.*, pp. 53–5.
171 *Ibid.*, p. 46.
172 *Ibid.*, pp. 109–10.
173 Michel Grosclaude, 'Deux tempéraments politiques', in Lafont, 1971a, pp. 181–204.
174 Arambourou, 1954, cited by Grosclaude, in Lafont, 1971a, p. 197.
175 Grosclaude, in Lafont, 1971a, p. 202.
176 *Ibid.*, p. 201.
177 A. Rébelliau, in Lavisse, 1911, VIII, 1, pp. 378–83.
178 See the maps published in Mandrou, 1961, pp. 108 and 177. On the conditions of the conversion, Le Roy Ladurie, 1966, I, pp. 348–56.
179 Secher, 1986, pp. 31, 253–92.
180 *Ibid.*, pp. 31–3.
181 G. Pariset, in Lavisse, 1920–22, II, p. 75, presents a 'geography of resistances'.
182 Durand, 1984, pp. 240–8.
183 Siegfried, 1913, p. 22.
184 Cited by Michelet, *Histoire de France*, definitive edn., Paris, n.d., II, p. 16 and Appendix, 6.
185 Jean de Viguerie, cited by Durand, 1984, pp. 214–15. Bibliography on the Montfortains: p. 339.
186 Secher, 1986, pp. 43–53, 96.
187 Durand, 1984, p. 214.
188 Jean Yole, *La Vendée*, Paris, 1936, p. 17, cited by Secher, 1986, p. 31.
189 Latest analysis: Renard and Chauvet, 1979.
190 *Revue des Deux Mondes*, 92 (4), 15 April 1889, pp. 884–912.
191 Siegfried, 1913, pp. 22–36; Secher, 1986, pp. 301–5.
192 Toujas-Pinède, 1960, where the earlier bibliography is to be found.
193 J.-C. Martin, 'Les réactionnaires progressistes de l'Ouest', *Le Monde*, 14 February 1982, pp. xi–xii.

9 The economic differentiation of space

1 Arbellot, 1973; cf. Clout, 1977c, p. 463.
2 *Ibid.*, p. 464.
3 The maps produced by A. Perpillou, 1977 and 1977a, have made it possible to study this evolution in detail.
4 Bénévent, 1938.

5 Castela, 1968, pp. 28–9.
6 According to Galtier, 1960, a work which is fundamental for the whole of this paragraph; see I, pp. 112–49.
7 Galtier, 1960, I, p. 114.
8 Enjalbert, 1975, p. 119.
9 Gadille, 1967, Laurent, 1957.
10 Colin, 1968.
11 Lucius, 1922; Tricart *et al.*, 1949; Juillard, 1949, 1950; Marthelot, 1950; Riedinger, 1950.
12 Gilbank, 1981; especially pp. 419–32.
13 Wolkowitsch, 1946.
14 Dubois, 1961; Debesse-Arviset, 1928.
15 Fel, 1954.
16 Théret, 1983, p. 162.
17 Mulliez, 1982.
18 There is abundant contemporary literature on the subject: Carlier, *Traité des bêtes à laine*, Paris, 1770; Daubenton, *Instruction pour les bergers et pour les propriétaires de troupeaux*, Paris, 1782 (frequently reprinted up until 1810); Tessier, *Instruction sur les bêtes à laine et particulièrement sur la race des mérinos*, 2nd edn., Paris, 1811; C. P. Lasteyrie, *Traité sur les bêtes à laine d'Espagne*, Paris, n.d. (1797), pp. 161–87, and 229ff.
19 Théret, 1983.
20 Théret, 1981.
21 Bougler, 1982.
22 Flament, 1981.
23 *L'Aubrac*, I, 1970, pp. 39–52, for a good example.
24 Devèze, 1965, pp. 73–7, 82–6; Demangeon, 1946, I, p. 128.
25 Sclafert, 1961.
26 On the paragraph as a whole, Arqué, 1939, pp. 93–5.
27 Papy, 1948.
28 Cavaillès, 1933.
29 Song cited by Cavaillès, 1933, p. 567.
30 Bidault, 1952.
31 Domet, 1892.
32 Dion, 1961.
33 Guilcher and Tricart, 1954.
34 Clout, 1977b, pp. 470–1 (from Fohlen, 1970 and Léon, 1970).
35 *Ibid.*, p. 475 (from Fohlen, 1970).
36 *Ibid.*, p. 477.
37 *Ibid.*
38 Sion, 1909; map for 1863, from Sion, in Clout, 1977b, p. 459.
39 Chevalier, 1961.
40 Fohlen, 1949; Chevalier, 1951; Tanière, 1954, cf. Chevalier, 1956, pp. 930–9.
41 Viers, 1958.
42 Demangeon, 1946, II, pp. 644–6.
43 *Ibid.*, pp. 742–50; Laferrère, 1960.
44 Poète, 1919–20; Pinkney, 1950; Bastié, 1970.

45 Gravier, 1947.
46 Huetz de Lemps and Roudié, 1985; Brunet, 1987.
47 Huetz de Lemps, 1987.
48 Jeanneau, 1985.
49 Roudié, 1985.
50 Brunet, 1985.
51 Peltre, 1985.
52 Laborde, 1985.
53 Garnier, 1987.
54 Dionnet, 1987.
55 Gilbank, 1987.
56 Dion and Verhaegue, 1987.
57 Dietrich, 1987; Wackermann, 1987.
58 Delfosse, 1987.
59 Fel, 1985.

10 The rural exodus and urbanization

1 There is a good analysis in Clout, 1977d, pp. 484–7.
2 Pitte, 1983a.
3 Dupeux, 1981.
4 On the whole of the next paragraph, Châtelain, 1976, with a bibliography that was exhaustive at that date.
5 *Ibid.*, p. 42.
6 *Ibid.*, p. 43.
7 *Ibid.*, p. 114.
8 *Ibid.*, pp. 155–246, 670.
9 *Ibid.*, pp. 249–335.
10 *Ibid.*, pp. 377–545.
11 *Ibid.*, pp. 669–773.
12 *Ibid.*, pp. 775–1005.
13 *Ibid.*, p. 821.
14 *Ibid.*, pp. 977ff.
15 Cf. Fel, 1962, pp. 143–58.
16 Raoul Blanchard, cited by Merlin, 1971, p. 34.
17 Philippe Arbos, cited by Merlin, 1971.
18 Châtelain, 1976, pp. 1010ff.
19 Guichonnet, 1945.
20 Onde, 1942.
21 Guichonnet, 1952.
22 For all this paragraph, Merlin, 1971.
23 *Ibid.*, pp. 7–16.
24 Fel, 1962, pp. 167–80 and 186–90.
25 Merlin, 1971, pp. 16–24.
26 Gachon, 1939.
27 Gachon, 1939a.
28 Meynier, 1931.

29 Blanchard, 1958, contains a synthesis of the results.
30 For a synthesis see Merlin, 1971, p. 40; Châtelain, 1976, pp. 663–7.
32 Gachon, 1939.
32 Meynier, 1931.
33 Merlin, 1969.
34 Châtelain, 1947.
35 All this paragraph is based on Carrière and Pinchemel, 1963; Clout, 1977d; Dupeux, 1981.
36 Pumain, 1982, p. 189.
37 Rochefort, 1960.
38 Barbier, 1969.
39 Juillard, 1970.
40 Dugrand, 1953, provides an excellent analysis of the contemporary decline of a small town.
41 Barbier, 1969, p. 179.
42 Labasse, 1955, p. 46.
43 *Ibid.*, p. 44.
44 Durand, 1946.
45 *Ibid.*; see the map p. 148.
46 Labasse, 1955, p. 46.
47 Durand, 1946, p. 126.
48 Armand, 1974, p. 914.
49 Labasse, 1955, p. 13.
50 George, 1968.
51 Auphan, 1975.
52 Juillard, 1971.
53 Wolkowitsch, 1980.
54 Pumain, 1982a.
55 Carrière and Pinchemel, 1963, p. 308. See also above, p. 405.

11 The France of large organizations

1 On the transformations brought about by the growing role of large organizations, it is useful to refer to the maps in the *Atlas historique de la France contemporaine*, 1965; cf. also Clout, 1977.
2 Caron, 1973.
3 Gillet, 1973.
4 Léon, 1954.
5 Caron, 1954.
6 *Ibid.*, and Lévy-Leboyer and Bourguignon, 1985.
7 For an overall view of the economic evolution and its social aspects and effects: Braudel and Labrousse, 1979–80, IV.
8 Dupeux, 1972.
9 Dugrand, 1963.
10 Clout, 1983.
11 Weber, 1983; Zeldin, 1978–79.
12 Faure, 1966; Frémont, 1981.

13 Merlin, 1971.
14 Weber, 1983.
15 Mendras, 1980; Hoffmann, 1963.
16 Braudel and Labrousse, 1979–80, IV.
17 Mendras, 1980.
18 Thuillier, 1966.
19 Laferrère, 1960.
20 Léon, 1954.
21 Gillet, 1973.
22 Chevalier, 1985.
23 Price, 1983.
24 Clout, 1980.
25 Léon, 1954.
26 Daumas, 1980.
27 Demonet, Dumont and Le Roy Ladurie, 1976.
28 Gille, 1968.
29 Braudel and Labrousse, 1979, IV, 1.
30 Fohlen, 1968.
31 Gille, 1968.
32 Braudel and Labrousse, 1980, IV, 2.
33 Laferrère, 1960.
34 Thuillier, 1966.
35 Gillet, 1973.
36 Gille, 1968.
37 Caron, 1973.
38 *Ibid.*
39 Caron, 1981.
40 Rouillard, n.d.
41 *La ville d'hiver d'Arcachon*, 1983.
42 Caron, 1981; Lévy-Leboyer and Bourguignon, 1985.
43 Gillet, 1973.
44 Prêcheur, 1959.
45 Merlin, 1971.
46 Bastié, 1964.
47 Mendras, 1980; Hoffmann, 1963.
48 Mendras, 1980.
49 Jeanneney, 1959.
50 Brunet, 1965.
51 Malinvaud, Carré, Dubois, 1972; Guibert, 1975.
52 Jeanneney, 1959; Chardonnet, 1954.
53 Prêcheur, 1969.
54 Duby and Wallon, 1976, III and IV.
55 Béteille, 1981.
56 Flatrès, 1963.
57 Flatrès, 1979.
58 Juillard, 1953.
59 Brunet, 1960.

60 Duby and Wallon, 1976, III and IV.
61 Pitte, 1983, II.
62 Flatrès, 1964; Foutrein, 1976.
63 Flatrès, 1963.
64 Béteille, 1964.
65 Chapuis and Brossard, 1986; Faure, 1966; Mendras, 1984.
66 Jeanneney, 1959.
67 Prêcheur, 1959.
68 Chardonnet, 1954.
69 Battiau, 1976.
70 Chapuis and Brossard, 1986.
71 Nora and Minc, 1978.
72 Duby, 1984–86, V and VI; Pumain, 1982; Saint-Julien, 1982.
73 Duby, 1979, IV.
74 Evenson, 1979; Sutcliffe, 1970.
75 Duby, 1985, V.
76 *La ville d'hiver d'Arcachon*, 1983.
77 Bastié, 1964.
78 Rouillard, n.d.
79 Duby, 1984, IV.
80 Noin, 1984.
81 Léger and Hervieu, 1979; Marie and Viard, 1977; Mendras, 1979.
82 Pitte, 1983, II.
83 *Ibid.*
84 Castex, Céleste and Panerai, 1980.
85 Lebeau, 1955.
86 This is very clear in the Loire Valley as soon as one moves beyond the suburbs of the main towns. The countryside described by Dion, 1934, is unchanged.
87 Dugrand, 1963.
88 R. Brunet, 1965.
89 Bruyelle, 1965; Juillard, 1953; Schwab, 1980.
90 Béthemont, 1972; Bravard, 1987.
91 Chevalier, 1985; Fel, 1962.

General conclusion

1 Despite Braudel's views, 1986–87, I, pp. 239–71, ('We should not exaggerate the role of the "French isthmus"'). Without really discrediting the concept of the French isthmus, essentially he takes issue with the over-use of it in the modern period and contemporary times. He is apparently unaware that the idea dates back to Strabo (p. 239; 'the French "isthmus", an expression coined, I believe, or at any rate used by French geographers themselves').
2 In the sense of Flatrès, 1963.
3 R. Dumont, 1951.

Guide to further reading

G. Bois, 1984, *The crisis of feudalism: economy and society in Eastern Normandy, c. 1330–1550*, Cambridge, Cambridge University Press.

F. Braudel, 1972–3, *The Mediterranean and the Mediterranean world in the age of Philip II*, 2 vols., London, Collins.

1988–90, *The identity of France. (I. Environment and history; II. People and production)*, 2 vols., London, Collins.

M. C. Cleary, 1989, *Peasants, politicians and producers: the organisation of agriculture in France since 1918*, Cambridge, Cambridge University Press.

H. D. Clout (ed.), 1977, *Themes in the historical geography of France*, London, Academic Press.

1980, *Agriculture in France on the eve of the railway age*, London, Croom Helm.

1983, *The land of France, 1815–1914*, London, Allen and Unwin.

G. Duby, 1991, *France in the Middle Ages 987–1460: from Hugh Capet to Joan of Arc*, Oxford, Basil Blackwell.

C. Heywood, 1992, *The development of the French economy 1750–1914*, London, Macmillan – a short but authoritative pamphlet in the Macmillan Series in Economic and Social History.

O. Hufton, 1974, *The poor in eighteenth-century France 1750–1789*, Oxford, Oxford University Press.

P. M. Jones, 1988, *The peasantry in the French Revolution*, Cambridge, Cambridge University Press.

E. Le Roy Ladurie, 1976, *The peasants of Languedoc*, Urbana, University of Illinois Press.

1984, *Love, death and money in the Pays d'Oc*, Harmondsworth, Penguin.

1987, *The French peasantry 1450–1660*, Aldershot, Scolar.

R. Magraw, 1983, *France 1815–1914: The bourgeois century*, Oxford, Oxford University Press.

P. McPhee, 1992, *A social history of France, 1780–1880*, London, Routledge.

J. Merriman, 1982, *French cities in the nineteenth century*, London, Hutchinson.

1991, *The margins of city life: explorations on the French urban frontier 1815–1851*, Oxford, Oxford University Press.

L. P. Moch, 1983, *Paths to the city: regional migration in nineteenth-century France*, London, Sage.

A. Moulin, 1991, *Peasantry and society in France since 1789*, Cambridge, Cambridge University Press.
P. E. Ogden and P. E. White (eds.), *Migrants in modern France: population mobility in the later nineteenth and twentieth centuries*, London, Unwin Hyman.
R. Price, 1981, *An economic history of modern France 1730–1914*, 3rd edn., London, Macmillan.
  1983, *The modernization of rural France: communications networks and agricultural market structures in nineteenth-century France*, London, Hutchinson.
  1987, *A social history of nineteenth-century France*, London, Hutchinson.
E. Weber, 1976, *Peasants into Frenchmen: the modernization of rural France, 1870–1914*, London, Chatto and Windus.
T. Zeldin, 1973, *France, 1848–1945*, 2 vols., Oxford, Oxford University Press.

Bibliography

N. B. The collection of works given below by no means claims to be a bibliography of the entire subject, for it would be impossible to define the limits of any such list. The present list simply provides the references of the works mentioned in the notes and, occasionally, in the text. It does not contain the sources, which are cited at the points where they are used.

Agache, Roger, 1971, *Détection áerienne de vestiges protohistoriques, gallo-romains et médiévaux dans le bassin de la Somme et ses abords*, 2nd edn., Amiens (= *Bulletin de la Société Préhistorique du Nord*, no. 7).
    1978, *La Somme pré-romaine et romaine* (*Mémoires de la Société des Antiquaires de Picardie*, Quarto series, no. 24), Amiens.
    1981, 'Les sanctuaires antiques et les limites de la cité des Ambiens', in *FG*, pp. 52–69.
Agache, Roger and Bréart, Bruno, 1975, *Atlas d'archéologie aérienne de Picardie*, Amiens.
Alcan, Louise, 1960, 'Coiffes des pays de France', *Arts et traditions populaires*, 3–21.
Alduc-Le Bagousse, A., 1983, 'La présence anglo-saxonne en Basse-Normandie au VIe siècle: colonisation ou commerce?', in *Le phénomène* . . . , pp. 51–61.
Allix, André, 1914, 'La foire de Goncelin', *Recueil de Travaux de l'Institut de Géographie Alpine*, 299–334.
    1929, *L'Oisans*, Grenoble.
    1932, 'Anciennes migrations dauphinoises', *Revue de Géographie Alpine*, 119–26.
Ancel, Jacques, 1938, *Géographie des frontières* (*Géographie humaine*, 12), Paris.
André, J. 1954, 'La vigne et le vin en Provence dans l'Antiquité', in *Mélanges géographiques offerts au doyen Ernest Bénévent*, pp. 361–8, Gap.
Angeville, A. D. d', 1837, *Essai statistique sur la population française*, Bourg-en-Bresse.
Arambourou, Robert, 1954, 'La Gavacherie de Monségur au temps de la IIIe République', in *Nouvelles Etudes de Sociologie Electorale*, pp. 81–131 (*Cahiers de la Fondation Nationale des Sciences Politiques*, no. 60).
Arbellot, G. 1973, 'La grande mutation des routes en France au milieu du XVIIIe siècle', *Annales, Economies, Sociétés, Civilisations*, 765–91.
Arbos, Philippe, 1992, *La vie pastorale dans les Alpes françaises*, Grenoble.

Armand, Gilbert, 1974, *Villes, centres et organisation urbaine des Alpes du Nord*, Grenoble.
Armbruster, Ludwig, 1928, *Die alte Bienkenzucht der Alpen*, Neumünster.
Arqué, Paul, 1939, *Géographie du Midi Aquitain*, Paris.
Atlas historique de la France contemporaine, 1966, Paris.
L'Aubrac, étude ethnologique, linguistique, agronomique et économique d'un établisement humain, 1970–81, Paris, 7 vols. in 9.
Audin, Amable, 1979, *Lyon, miroir de Rome*, Paris.
Audin, Pierre, 1976, 'Luynes, Langeais et Ingrandes: trois villages gallo-romains du Val de Loire tourangeau', in *VGR*, pp. 196–204.
1981, 'La Civitas Turonum et ses limites au Nord de la Loire', in *FG*, pp. 70–93.
Augé-Laribé, Michel, 1955, *La Révolution agricole* (*L'Evolution de l'Humanité*, 83), Paris.
Auphan, E., 1975, 'Les nœuds ferroviaires, phénomène résiduel ou points forts de l'espace régional', *L'Espace Géographique*, 127–40.
Aymard, André, 1948, 'L'interdiction des plantations de vignes en Gaule transalpine sous la République romaine', in *Mélanges géographiques offerts ... à M. Daniel Faucher*, I, pp. 27–47, Toulouse.
Babelon, Ernest, 1916–17, *Le Rhin dans l'Histoire*, 2 vols., Paris.
Balibar, Renée, 1985, *L'institution du français. Essai sur le colinguisme des Carolingiens à la République*, Paris.
Bancal, J. 1945, *Les circonscriptions administratives de la France*, Paris.
Barbier, Bernard, 1969, *Villes et centres des Alpes du Sud*, Gap.
Barral, Pierre, 1968, *Les agrariens français de Méline à Pisani* (*Cahiers de la Fondation Nationale des Sciences Politiques*, 164), Paris.
Barrière, Pierre, 1961, *La vie intellectuelle en France du XVIe siècle à l'époque contemporaine* (*L'Evolution de l'Humanité*, no. 96), Paris.
Barruol, G. 1969, *Les peuples pré-romains du Sud-Est de la Gaule, étude de géographie historique*, Paris.
Bartels, D. 1964, *Das Maures-Bergland*, Wiesbaden.
Bastié, Jean, 1964, *La croissance de la banlieue parisienne*, Paris.
1970, *Paris, ville industrielle* (*Notes et Etudes Documentaires*, no. 3690–91), Paris.
Battiau, M. 1976, *Les industries textiles de la région du Nord-Pas-de-Calais*, Paris.
Bautier, Robert-Henri, 1978, 'Quand et comment Paris devint capitale', *Bulletin de la Société d'Histoire de Paris et de l'Ile-de-France*, 105: 17–46.
Bayet, Albert, 1930, *La morale des Gaulois*, Paris.
Beaujard, B., 1981, 'Calètes et Véliocasses à l'époque gallo-romaine', in *FG*, pp. 94–107.
Beaune, Colette, 1985, *Naissance de la nation française*, Paris.
Bec, P. 1963, *La langue occitane*, Paris.
Bédier, Joseph, 1926–29, *Les légendes épiques*, 4 vols., Paris.
Benda, Julien, 1932, *Esquisse d'une histoire des Français dans leur volonté d'être une nation*, Paris.
Bénévent, Ernest, 1938, 'La vieille économie provençale', *Revue de Géographie Alpine*, 531–71.
Benoît, Fernand, 1953, 'Antiquités historiques, XIIe circonscription, Bouches-du-Rhône. Epave du IIe siècle avant J.-C', *Gallia*, 103–6.

1957, *Entremont*, Aix-en-Provence.

1965, *Recherches sur l'hellénisation du Midi de la Gaule*, Aix-en-Provence.

Bérard, Victor, 1927, *Les Phéniciens et l'Odyssée*, 2 vols., Paris.

Bergeron, L., 1975, 'Paris dans l'organisation des échanges intérieurs français à la fin du XVIIIe siècle', in P. Léon (ed.), *Aires et structures*..., pp. 237–64.

Bergeron, L. et al., 1970, *Contribution à l'histoire démographique de la Révolution française*, 3rd series (*Commission d'histoire économique et sociale de la Révolution française, Mémoires et documents*, XXV), Paris.

Béteille, Roger, 1981, *La France du vide*, Paris.

Béthemont, Jacques, 1972, *Le thème de l'eau dans la vallée du Rhône*, Saint-Étienne.

Bidault, René, 1952, 'Un terroir de chasse: la Sologne orientale', *Centre de Documentation Cartographique et Géographique, Mémoires et Documents*, II, pp. 213–21.

Blanchard, Marcel, 1942, *Géographie des chemins de fer* (*Geographie Humaine*, no. 17), Paris.

Blanchard, Raoul, 1958, *Les Alpes et leur destin*, Paris.

Bloch, G., 1911, *Les origines. La Gaule indépendante et la Gaule romaine*, vol. 1, Part 2 of E. Lavisse, *Histoire de France*.

Bloch, Marc, 1912, 'L'Ile-de-France (les pays autour de Paris)', *Revue de Synthèse Historique*, 209–310.

1931, *Les caractères originaux de l'histoire rurale française*, Oslo (2nd edn., 2 vols., Paris, 1952–56).

1939, 'Le problème du peuplement beauceron', *Revue de Synthèse Historique*, 17: 62–73.

1939a, *La société féodale. La formation des liens de dépendance*, Paris (*L'Evolution de l'Humanité*, no. 34).

1940, *La société féodale. Les classes et le gouvernement des hommes*, Paris (*L'Evolution de l'Humanité*, no. 34a).

Blottière, J. 1972–73, 'Contribution à l'étude du toponyme Beauce', *Revue Internationale d'Onomastique*, 177–81.

Bollème, Geneviève, 1971, *La Bibliothèque Bleue. La littérature populaire en France du XVIe au XIXe siècle*, Paris.

Bonnaud, Pierre, 1977, 'Peopling and the origins of settlement', in H. D. Clout (ed.), *Themes in the historical geography of France*, pp. 21–72.

1981, *Terres et langages. Peuples et régions*, 2 vols., Clermont-Ferrand.

Bougler, J. 1982, 'La race et les livres généalogiques', in *Le Concept de race en zootechnie*, (*Ethnozootechnie*, no. 29), pp. 69–72, Paris.

Boulard, Fernand, 1947, 'Ebauche d'une carte religieuse de la France rurale', *Cahiers du clergé rural*, 92 (November): 403–14.

1954, *Premiers itinéraires en sociologie religieuse*, Paris.

Bourdier, Franck, 1967, *Préhistoire de la France*, Paris.

Bourgin, H., 1920, *L'industrie sidérurgique en France au début de la Révolution*, Paris.

Bourquelot, Félix, 1865, 'Sens des mots France et Neustrie sous le régime mérovingien', *Bibliothèque de l'École des Chartes*, 566–74.

Boussard, J. 1957, *Atlas historique et culturel de la France*, Paris.

Boutruche, R., 1935, 'Les courants de peuplement dans l'Entre-Deux-Mers', *Annales d'Histoire Economique et Sociale*, 13–36 and 124–55.

Bowen, E. G., 1969, *Saints, Seaways and Settlements in the Celtic Lands*, Cardiff.

Braudel, Fernand, 1986–87, *L'identité de la France*, 2 vols. in 3, Paris.
Braudel, Fernand and Labrousse, Ernest (eds.), 1970, 1979–80, *Histoire économique et sociale de la France*, II, 1970; IV, 2 vols., years 1880–1914, 1914–50, Paris.
Bravard, Jean-Paul, 1987, *Le Rhône du Léman à Lyon*, Lyons.
Brette, A., 1904, *Atlas des bailliages ou juridictions assimilées ayant formé unité électorale en 1789*, Paris.
1907, *Les limites et les divisions territoriales de la France en 1789*, Paris.
Broc, Numa, 1982, 'Roger Dion', *Annales de géographie*, 205–17.
Broche, G. E., 1936, *Pythéas le Massaliote*, Paris.
Broëns, Maurice, 1943, *Le peuplement de l'Albigeois des temps préhistoriques à l'époque féodale*, Albi.
1956, 'Le peuplement germanique de la Gaule entre la Méditerranée et l'Océan', *Annals du Midi*, 17–38.
Brun, M., 1939, *Départements et régions*, Paris.
Brunet, Pierre, 1960, *Structure agraire et economie rurale des plateaux Tertiaires entre la Seine et l'Oise*, Caen.
1968, 'Evolution des bocages herbagers en Basse-Normandie', in H. Jäger *et al.*, *Beiträge zur Genese der Siedlungs- und Agrarlandschaft im Europa*, pp. 12–26.
1985, 'Le Calvados, une eau-de-vie à la recherche de la qualité', in Heutz de Lemps and Roudié, *Eaux-de-vie et spiritueux* ..., pp. 189–204.
1987, *Histoire et géographie des fromages. Actes du colloque de géographie historique, Caen, 1985*, published under the direction of ..., Caen.
Brunet, Roger, 1965. *Les campagnes toulousaines. Etude géographique*, Toulouse.
Brunhes, Jean, 1920, *Géographie humaine de la France, I*, Paris (= Gabriel Hanotaux, *Histoire de la nation française*, I).
1947, *La Géographie humaine*, abridged and edited by M. Jean-Brunhes-Delamarre and Pierre Deffontaines, Paris.
Brunhes, Jean and Deffontaines, Pierre, 1926, *Géographie humaine de la France*, II, *Géographie politique et géographie du travail*, Paris (= Gabriel Hanotaux, *Histoire de la nation française*, II).
Brunot, F., 1905, *Histoire de la langue française*, I, Paris.
Brutails, J. A., 1923, *La géographie monumentale de la France aux époques romane et gothique*, Paris, separately and in *Le Moyen Age*, 2nd series, 25: 1–46.
Bruyelle, Pierre, 1965, *L'organisation urbaine de la région du Nord – Pas-de-Calais*, 3 vols., Marcq-en-Bareul.
Bur, Michel, 1977, *La formation du Comté de Champagne (v. 950–v. 1150)*, Nancy.
Busquet, Raoul, 1954, 'Marseille a-t-elle ou n'a-t-elle pas civilisé la Gaule?', *Revue historique*, 211: 1–10.
Busset, Maurice, 1933, *Gergovia, capitale des Gaules et l'oppidum du plateau des Côtes*, Paris.
Butlin, R. A., 1971, 'Irish agrarian history: inventory and prospect', in Dussart, *L'Habitat* ..., pp. 59–68.
Callot, Henri-Jacques, 1980, *La Plaine d'Alsace. Modelé agraire et parcellaire*, Nancy.
Calmette, Joseph, 1947, *La question des Pyrénées et la Marche d'Espagne au Moyen Age*, Paris.
Camps, Gabriel, 1982, *La Préhistoire*, Paris.

Carcopino, Jérôme, 1957, *Promenades historiques au pays de la dame de Vix*, Paris.
Carmona, Michel, 1984, *La France de Richelieu*, Paris.
Caron, François, 1973, *Histoire de l'exploitation d'un grand réseau: la Compagnie du chemin de fer du Nord*, Paris.
　1981, *Histoire économique de la France. XIXe–XXe siècles*, Paris.
Carrière, F. and Pinchemel, Philippe, 1963, *Le fait urbain en France*, Paris.
Castela, Paul, 1968, *La fleur en Europe occidentale. Etude géographique de la production et du commerce des plantes ornementales*, Paris.
Caster, Gilles, 1954, 'Types économiques et sociaux du XVIe siècle. Le pastelier toulousain', *Annales, Economies, Sociétés, Civilisations*, 63–74.
　1962, *Le commerce du pastel et de l'épicerie à Toulouse (1450–1561)*, Toulouse.
Castex, Jean, Céleste, Patrick and Panerai, Philippe, 1980, *Versailles. Lecture d'une ville*, Paris.
Cavaillès, Henri, 1910, 'Une fédération pyrénéenne sous l'Ancien Régime. Les traités de lies et passeries', *Revue Historique*, 1–34 and 241–76.
　1931, *La vie pastorale et agricole dans les Pyrénées des Gaves, de l'Adour et des Nestes*, Paris.
　1933, 'Le problème de la circulation dans les Landes de Gascogne', *Annales de Géographie*, 561–82.
　1946, *La route française. Son histoire. Sa fonction. Etude de géographie humaine*, Paris.
Chabot, Georges, 1943, 'Toponymie et géographie: les toponymes dérivés de la vigne en Bourgogne', *Annales de Géographie*, 53–6.
Chamard, Henri, 1939–40, *Histoire de la Pléiade*, 4 vols., Paris.
Chamla, M. C., 1964, 'L'accroissement de la stature en France, 1880–1960', *Bulletin de la Société d'Anthropologie de Paris*, 261–78.
Chapelot, Jean and Fossier, Robert, 1980, *Le village et la maison au Moyen Age*, Paris.
Chapuis, Robert and Brossard, Thierry, 1986, *Les ruraux français*, Paris.
Chardonnet, Jean, 1954, *La sidérurgie française*, Paris.
Châtelain, Abel, 1947, 'Démographie du grand tronc ferré Sud-Est (Paris–Lyon–Méditerranée), 1866–1936', *Revue de Géographie de Lyon*, 35–82.
　1976, *Les migrants temporaires en France de 1800 à 1914*, 2 vols., Villeneuve d'Ascq.
Chaume, M., 1922, 'Le sentiment national bourguignon', *Mémoires de l'Académie de Dijon*, 195–308.
　1925–37, *Les origines du duché de Bourgogne*, 4 vols., Dijon.
　1944, 'Une question: la succession des circonscriptions sur un même coin de terre', *Annales de Bourgogne*, 162–74 and 229–40.
Chaunu, Pierre, 1982, *La France*, Paris.
Chaussée, F. de la, 1966, *Les parlers du Centre-Ouest de la Vendée*, Paris.
Chénon, E., 1892, 'Les marches séparantes d'Anjou, Bretagne et Poitou', *Nouvelle Revue Historique de Droit Français et Etranger*, 16: 18–62 and 165–211.
Chevalier, Louis, 1947, *Les paysans. Etude d'histoire et d'économie rurale*, Paris.
　1950, *La formation de la population parisienne au XIXe siècle*, Paris (Institut National d'Etudes Démographiques, Travaux et Documents, no. 10).
Chevalier, Michel, 1951, 'L'industrie textile pyrénéenne et le développement de Lavelanet', *Revue Géographique des Pyrénées et du Sud-Ouest*, 43–60.

1956, *La vie humaine dans les Pyrénées Ariégeoises*, Paris.
Tableau industriel de la Franche-Comté (1960–61), (*Cahiers de Géographie de Besançon*, no. 9), Paris.
1985, *L'Ariège*, Rennes.
Chevallier, Raymond, 1962, 'La centuriation et les problèmes de la colonisation romaine', *Etudes rurales*, 3: 54–80.
1972, *Les voies romaines*, Paris.
1980, *Les Ligures* (*Caesarodunum*, supplement no. 35), Tours.
Chiffre, Jean, 1985, *Les aspects géographiques des communautés familiales de France centrale* (*Publications de l'Université de Dijon*, 62), Dijon.
Cholley, André, 1943, 'Recherches sur les surfaces d'érosion et la morphologie de la région parisienne', *Annales de Géographie*, 1–19, 81–97, 161–89.
Chouquer, Gérard, Clavel-Lévêque, Monique and Favory, François, 1982, 'Cadastres, occupations du sol et paysages agraires antiques', *Annales, Economies, Sociétés, Civilisations*, 847–82.
Clark, J. G. D., 1955, *Prehistoric Europe: the economic basis*, London.
Claval, Paul, 1978, *Atlas et géographie de la Haute Bourgogne et de la Franche-Comté*, Paris.
Clavel, Monique, 1970, *Béziers et son territoire dans l'Antiquité*, Paris.
Clout, Hugh D. (ed.), 1977, *Themes in the historical geography of France*, London.
1977a, 'Early urban development', in H. D. Clout (ed.), *Themes ...*, pp. 73–106.
1977b, 'Agricultural change in the eighteenth and nineteenth centuries', in H. D. Clout (ed.), *Themes ...*, pp. 407–46.
1977c, 'Industrial development in the eighteenth and nineteenth centuries', in H. D. Clout (ed.), *Themes ...*, pp. 447–82.
1977d, 'Urban growth 1500–1900', in H. D. Clout (ed.), *Themes ...*, pp. 483–540.
1980, *Agriculture in France on the eve of the Railway Age*, London.
1983, *The Land of France 1815–1914*, London.
Clozier, René, 1963, *Géographie de la circulation. I. L'économie des transports terrestres*, Paris.
Cocula, Anne-Marie, 1975, 'Pour une définition de l'espace aquitain au XVIIIe siècle', in P. Léon, (ed.), *Aires et structures ...*, pp. 301–30.
Cohen, Gustave, 1948, *Chrétien de Troyes et son œuvre*, 2nd edn., Paris.
Coirault, Patrice, 1927–33, *Recherches sur notre ancienne chanson populaire traditionnelle*, Paris (reprinted Paris–Geneva, 1955).
1941, *Notre chanson folklorique*, Paris.
1953–63, *Formation de nos chansons folkloriques*, 4 vols., Paris.
Coirault, Yves, 1980, 'Le Tour de France des ducs de Bourgogne et de Berry d'après quelques lettres et relations du temps (décembre 1700–avril 1701)', in *La découverte ...*, pp. 15–32.
Colin, Georges, 1968, 'Le dynamisme du vignoble champenois', *Revue Géographique de l'Est*, 337–53.
Coornaert, Emile, 1966, *Les compagnonnages en France du Moyen Age à nos jours*, Paris.
Coville, Alfred, 1928, *Recherches sur l'histoire de Lyon du Ve au IXe siècle*, Paris.
Crampon, Michel, 1985, 'Quatre régions en une. Des frontières imprécises et toujours contestées', *Le Monde*, 7 February, p. 16.

Creston, R. Y., 1974, *Le costume breton*, Paris.
Crubellier, M. et al., 1975, *Histoire de la Champagne*, Toulouse.
Cuisenier, Jean, 1977, *L'Art populaire en France. Rayonnement, modèles et sources.* Fribourg/Paris.
Dainville, F. de, 1952, 'Un dénombrement inédit au XVIIIe siècle: l'enquête du contrôleur-géneral Orry, 1745', *Population*, 49–68.
Darmon, Jean-Jacques, 1972, *Le colportage de librairie en France sous le Second Empire*, Paris.
Daumas, Maurice, 1980, *L'archéologie industrielle de la France*, Paris.
Dauzat, Albert, 1926, *Les noms de lieux*, Paris (5th edn., 1963).
  1939, *La toponymie française*, Paris (New edn., 1946).
  1939a, *Les noms de personnes*, Paris.
  1941, *Le village et le paysan de France*, Paris.
  1945, *Traité d'anthroponymie française. Les noms de famille de France*, Paris.
Daveau, Suzanne, 1959, *Les régions frontalières de la montagne jurassienne, étude de géographie humaine* (*Institut des Etudes Rhodaniennes de l'Université de Lyon, Mémoires et Documents*, 14), Trévoux.
Debesse-Arviset, Mme, 1928, 'Le Châtillonnais', *Annales de Géographie*, 428–51.
Déchelette, Joseph, 1931–34, *Manuel d'archéologie préhistorique, celtique et gallo-romaine*, 2 vols. in 4, Paris.
La découverte de la France au XVIIe siècle, Neuvième colloque de Marseille ... 25–28 janvier 1979 (*Colloques internationaux du Centre National de la Recherche Scientifique*, no. 590), Paris, 1980.
Deffontaines, Pierre, 1932, *Les hommes et leurs travaux dans les pays de la Moyenne Garonne*, Lille.
  1936, 'Note sur la répartition des types de voitures', in *Mélanges de Géographie et d'Orientalisme offerts à Emile-Félix Gautier*, Tours, pp. 169–85.
  1938, 'The origin and growth of the Brasilian network of towns', *Geographical Review*, 379–99.
Dehn, Wolfgang, 1965, 'Mediolanum, Lagetypen spätkeltischer Oppida', *Studien aus Alteuropa*, II (Beihefte der *Bonner Jahrbücher*, 10–11), pp. 117–28, Cologne.
Delafosse, M., 1955, 'Noms de famille rochelais en 1224', in *Recueil de travaux offerts à M. Clovis Brunel*, Paris, 339–44.
Delarue, Paul, 1957, *Le conte français*, I, Paris.
Delaspre-Parquet, Suzanne, 1954, *L'émigration temporaire en Basse Auvergne au XVIIIe siècle* (*Publications de l'Institut de Géographie de Clermont-Ferrand*, 5), Clermont-Ferrand.
Delfosse, Claire, 1987, 'Une richesse locale menacée: la production de Brie dans la région de Meaux à la fin du XIXe siècle', in P. Brunet, *Histoire et géographie des fromages...*, pp. 43–52.
Demangeon, Albert, 1946, *Géographie économique et humaine de la France*, 2 vols., (= *Géographie Universelle* under the direction of P. Vidal de la Blache, VI), Paris.
Demonet, M., Dumont, P. and Le Roy Ladurie, E., 1976, 'Anthropologie du conscrit: une cartographie cantonale (1819–1830)', *Annales, Economies, Sociétés, Civilisations*, 700–60.
Denis, Marie-Noële and Groshens, Marie-Claude, 1978, *L'Architecture rurale française. Alsace*, Paris.

Derruau, Max, 1949, *La Grande Limagne auvergnate et bourbonnaise*, Paris.
Desbordes, J. M., 1971, 'Un problème de géographie historique: le Mediolanum chez les Celtes', *Revue Archéologique du Centre*, 187–201.
Desjardins, Ernest, 1876–93, *Géographie historique et administrative de la Gaule romaine*, 4 vols., Paris.
Devailly, Guy, 1973, *Le Berry du Xe siècle au milieu du XIIIe*, Paris–The Hague.
 1980, 'Le cidre en Bretagne au IXe siècle', in P. Flatrès and X. de Planhol, *Paysages arborés* ..., pp. 65–73.
Devèze, Michel, 1965, *Histoire des forêts*, Paris.
Deyber, Alain, 1981, 'Les frontières des peuples pré-romains dans l'Est de la Gaule à la fin de l'époque de La Tène', in *FG*, pp. 28–44.
Dietrich, Geneviève, 1987, 'Le Munster, paysages et systèmes d'élevage', in P. Brunet, *Histoire et géographie des fromages* ..., pp. 67–77.
Dion, Raymond and Verhaegue, Raymond, 1987, 'Le Maroilles, "le plus fin des fromages forts"', in P. Brunet, *Histoire et géographie des fromages* ..., pp. 79–107.
Dion, Roger, 1934, *Essai sur la formation du paysage rural français*, Tours (reprinted Paris, 1981).
 1934a, *Le Val de Loire*, Tours.
 1938, 'Orléans et l'ancienne navigation de la Loire', *Annales de Géographie*, 128–34.
 1946, 'La part de la géographie et celle de l'histoire dans l'explication de l'habitat rural du Bassin Parisien', *Publications de la Société de Géographie de Lille*, 6–80.
 1946a, 'Effets de l'insécurité sur le choix des sites d'habitat rural', *L'Information Géographique*, 143–6.
 1947, *Les frontières de la France*, Paris.
 1949, *Paris dans les récits historiques et légendaires du IXe au XIIe siècle*, Tours.
 1950, 'A propos du traité de Verdun', *Annales, Economies, Sociétés, Civilisations*, 461–5.
 1951, 'Vin de Chalon et vin d'Autun, ou rapports entre l'histoire des villes et les caractères du vignoble en Bourgogne', *Bulletin de l'Association de Géographes Français*, 218–19 (May–June): 125–33.
 1951a, 'Paris dans la géographie: le site et la croissance de la ville', *Revue des Deux Mondes*, 1 January, pp. 5–30.
 1952, 'Le problème des Cassitérides', *Latomus*, 306–14.
 1952a, 'Les routes de l'étain', *Hommes et Mondes*, 547–57.
 1952b, 'Métropole et vignobles en Gaule romaine: l'exemple bourguignon', *Annales, Economies, Sociétés, Civilisations*, 1–12.
 1952c, 'A propos des origines du vignoble bourguignon', *Annales de Bourgogne*, 47–52 (Reprinted in *Revue Historique*, 1960, 321–5).
 1952d, 'La leçon d'une chanson de geste: les Narbonnais', *Fédération des Sociétés Historiques et Archéologiques de Paris et de l'Ile-de-France, Mémoires, I (1949)*, pp. 23–45.
 1959, 'Géographie historique de la France', *Annuaire du Collège de France*, 485–508.
 1959a, *Histoire de la vigne et du vin en France des origines au XIXe siècle*, Paris.
 1961, 'Le "bon" et "beau" pays nommé Champagne pouilleuse', *L'Information géographique*, 209–14.

1963, 'Géographie historique de la France', *Annuaire du Collège de France*, 389–410.
1968, 'Géographie historique de la France', *Annuaire du Collège de France*, 503–18.
1977, *Aspects politiques de la géographie antique*, Paris.
Dionnet, Marie-Claude, 1987, 'Evolution de la production du Camembert en France', in P. Brunet, *Histoire et géographie des fromages...*, pp. 109–17.
Domet, P., 1892, *Histoire de la forêt d'Orléans*, Orléans.
Dondaine, C., 1972, *Les parlers comtois d'oïl*, Paris.
Dontenville, Henri, 1966, *La France mythologique*, Paris.
1973, *Histoire et géographie mythiques de la France*, Paris.
1973a, *Mythologie française*, Paris.
Drouot, H., 1952, 'Aucune vigne gauloise', *Annales de Bourgogne*, 112–13.
1952a, 'La vigne en Gaule avant les Romains', *Annales de Bourgogne*, 272ff.
Dubois, Maurice, 1961, 'L'économie rurale du Châtillonnais', *Centre de Documentation Cartographique et Géographique, Mémoires et Documents*, VII, pp. 7–118.
Duby, Georges, 1962, *L'économie rurale et la vie des campagnes dans l'Occident médiéval*, 2 vols., Paris.
1973, *Guerriers et paysans: VIIe–XIIe siècles. Premier essor de l'économie européenne*, Paris.
1973a, *Le dimanche de Bouvines*, Paris.
(ed.), 1980–85, *Histoire de la France urbaine*, 5 vols., Paris.
Duby, Georges and Wallon, Armand (eds.), 1975–76, *Histoire de la France rurale*, 4 vols., Paris.
Dufournet, Paul, 1981, *L'art populaire en Savoie*, Le Puy.
Dugrand, Raymond, 1953, 'Ganges, étude d'une petite ville sub-cévenole', *Bulletin de la Société Languedocienne de Géographie*, 24: 7–142.
1956, 'La propriété foncière des citadins dans le Bas-Languedoc', *Bulletin de l'Association de Géographes Français*, 259–60 (May–June): 133–44.
1963, *Villes et campagnes du Bas-Languedoc*, Paris.
Dumont, Louis, 1951, *La Tarasque*, Paris.
Dumont, René, 1951, *Voyages en France d'un agronome*, Paris.
Dupâquier, J., 1979, *La population rurale du Bassin Parisien à l'époque de Louis XIV*, Paris–Lille.
Duparc, Pierre, 1958, 'La Sapaudia', *Comptes rendus de l'Académie des Inscriptions et Belles-Lettres*, 371–83.
Dupeux, Georges, 1972, *La société française*, Paris.
1981, *Atlas historique de l'urbanisation de la France (1811–1975)*, Paris.
Dupin, Charles, 1827, *Forces productives et commerciales de la France*, Paris.
Dupont-Ferrier, G., 1929, 'Sur l'emploi du mot "Province", notamment dans le langage administratif de l'ancienne France', *Revue Historique*, 241–67.
1929a, 'De quelques synonymes du terme "Province"', *Revue Historique*, 278–303.
1942, 'Incertitudes des limites territoriales en France du XIIIe au XVIe siècle', *Comptes rendus de l'Académie des Inscriptions*, 73–6.
Dupouy, Auguste, 1942, *Géographie des lettres françaises*, Paris.
Dupuy, Aimé, c. 1960, *Voyageurs étrangers à la découverte de l'ancienne France 1500–1850, n.p.n.d.*, Paris.
Durand, Alfred, 1946, *Aurillac, géographie urbaine*, Aurillac.

Durand, Yves, 1984, *Vivre au pays au XVIIIe siècle*, Paris.
Dussart, Franz (ed.), 1971, *L'habitat et les paysages ruraux d'Europe*, Liège (*Congrès et Colloques de l'Université de Liège*, 58).
Duval, Paul-Marie, 1961, *Paris antique*, Paris.
 1974, 'Les noms de la Gaule', in *Littérature gréco-romaine et géographie historique. Mélanges offerts à Roger Dion publiés par Raymond Chevallier*, Paris, pp. 407–16.
East, Gordon, 1939, *Géographie historique de l'Europe*, Paris.
Elhaï, Henri, 1965, *Recherches sur la propriété foncière des citadins en Haute Normandie*, Paris (*Mémoires et Documents*, X, 3).
Emmanuelli, François-Xavier, 1977, 'De la conscience politique à la renaissance du "provençalisme" dans la Généralite d'Aix à la fin du XVIIIe siècle', in C. Gras and G. Livet, *Régions et régionalisme . . .*, pp. 116–38.
Enjalbert, Henri (ed.), 1974, *La seigneurie et le vignoble de Château-Latour*, 2 vols., Bordeaux.
 1975, *Histoire de la vigne et du vin. L'avènement de la qualité*, Paris.
Esmein, A. 1901, *Cours élémentaire d'histoire du droit français*, 4th edn., Paris.
Evenson, Norma, 1979, *Paris, a Century of Change, 1878–1978*, New Haven.
Ewig, Eugène, 1982, 'Xanten dans *la Chanson de Roland*', in *La Chanson de geste et le mythe carolingien*, pp. 481–90. *Mélanges René Louis*, I, Saint-Père-sous-Vézelay.
Eychart, Paul, 1961, *L'oppidum des Côtes, Augustonemetum, Gergovie*, Clermond-Ferrand.
 1969, *Gergovie, légende et réalité*, Clermond-Ferrand.
Falc'hun, F., 1963, *L'histoire de la langue bretonne d'après la géographie linguistique*, Paris.
Faucher, Daniel, 1934, 'Polyculture ancienne et assolement biennal dans la France méridionale', *Revue Géographique des Pyrénées et du Sud-Ouest*, 241–58.
 1949, *Géographie agraire. Types de cultures*, Paris.
 1950, 'Assolements collectifs et aménagement du paysage: un exemple pyrénéen', in *Livre jubilaire offert à Maurice Zimmermann*, pp. 103–10, Lyons.
Faure, Marcel, 1966, *Les paysans dans la société française*, Paris.
Favier, Jean, 1974, *Paris au XVe siècle*, Paris.
 1982, *François Villon*, Paris.
Favière, Jean, 1957, 'Documents sur les costumes ruraux du Berry', *Arts et Traditions Populaires*, 159–81.
FCC = *Frontières et contacts de civilisation* (*Colloque universitaire franco-suisse, Besançon-Neuchâtel, October 1977*), Neuchâtel, 1979.
Febvre, Lucien, 1947, 'Un livre: la diversité des frontières de la France', *Annales, Economies, Sociétés, Civilisations*, 205–7.
Fel, André, 1954, 'Problèmes de limites entre les systèmes d'élevage. Exemples tirés du Massif Central français', *Bulletin de l'Association de Géographes Français*, 243–44 (May–June): 97–103.
 1959, 'Réflexions sur les paysages agraires des hautes terres du Massif Central français', in *Géographie et histoire agraires* (Mémoire no. 21 in *Annales de l'Est*), pp. 155–67, Nancy.
 1962, *Les hautes terres du Massif Central*, Clermont-Ferrand.

1975, 'Paysages agraires et civilisation rurale de la vieille Corse', in *I Paesaggi Rurali Europei* (*Deputazione di Storia Patria per l'Umbria, Appendici al Bollettino*, no. 12), pp. 183–95, Perugia.

1977, 'Petite culture 1750–1850', in H. D. Clout (ed.), *Themes* ..., pp. 215–45.

1985, 'Les alcools de gentiane et l'Auvergne', in A. Heutz de Lemps and P. Roudié, *Eaux-de-vie et spiritueux* ..., pp. 321–7.

Feuchère, Pierre, 1954, 'Dans le Nord de la France. La permanence des cadres territoriaux', *Annales, Economies, Sociétés, Civilisations*, 94–100.

Février, P. A., 1964, *Le développement urbain en Provence de l'époque romaine à la fin du XIVe siècle*, Paris

FG = Actes du Colloque '*Frontières en Gaule*', Tours, 1981 (*Caesarodunum*, 16).

Fierro-Domenech, Alfred, 1986, *Le pré carré. Géographie historique de la France*, Paris.

Fiétier, Roland, Locatelli, René and Moyse, Gérard, 1979, 'La frontière au Nord-Est de la Franche-Comté durant le haut Moyen Age (aux origines du comté de Montbéliard)', in *FCC*, pp. 97–113.

Flach, J., 1886–1904, *Les origines de l'ancienne France aux Xe et XIe siècles*, 3 vols., Paris.

Flament, J. C., 1981, 'Les concours spéciaux des races ovines du rayon de Roquefort et l'évolution de l'amélioration génétique des brebis laitières', in *Les concours de bétail* (*Ethnozootechnie*, 28), pp. 45–54, Paris.

Flatrès, Pierre, 1963, 'La deuxième "Révolution agricole" en Finistère', *Etudes Rurales*, 8: 5–56.

1964, 'L'évolution de l'agriculture dans la région du Nord', *Hommes et Terres du Nord*, 7–22.

1971, 'Les anciennes structures rurales de Bretagne d'après le Cartulaire de Redon', *Etudes Rurales*, 41: 87–93.

1971a, 'Réflexions sur la débocagisation', in F. Dussart (ed.), *L'habitat* ..., pp. 129–44.

1977, 'Historical geography of Western France', in H. D. Clout (ed.), *Themes* ..., pp. 301–42.

1979, 'L'évolution des bocages: la région de Bretagne', *Norois*, 303–20.

Flatrès, Pierre and Planhol, Xavier de (eds.), 1980, *Paysages arborés et complantés*, Paris (*Publications du Département de Géographie de l'Université de Paris–Sorbonne*, no. 9).

(eds.), 1983, *Etudes sur l'habitat perché*, Paris (*Publications du Département de Géographie de l'Université de Paris–Sorbonne*, no. 11).

Fleuriot, Léon, 1980, *Les origines de la Bretagne*, Paris.

Flory, Thiébaut, 1966, *Le mouvement régionaliste français: sources et développement*, Paris. (*Travaux et Recherches de la Faculté de Droit et des Sciences Économiques de Paris*, series '*Sciences Politiques*', no. 6).

Fohlen, Claude, 1949, 'En Languedoc: vigne contre draperie', *Annales, Economies, Sociétés, Civilisations*, 290–7.

1956, *L'industrie textile au temps du Second Empire*, Paris.

1970, 'The industrial revolution in France', in R. E. Cameron (ed.), *Essays in French Economic History*, pp. 201–25, Homewood.

Foisil, M., 1970, *La révolte des Nu-Pieds et les révoltes normandes de 1639*, Paris.

Foncin, Myriam, 1919, 'Versailles, étude de géographie historique', *Annales de Géographie*, 321–41.
Foncin, P., 1916, *Les Maures et l'Esterel*, Paris.
Fossier, R., 1968, *La terre et les hommes en Picardie jusqu'à la fin du XIIIe siècle*, 2 vols., Paris–Louvain.
Fournier, Gabriel, 1962, *Le peuplement rural en Basse Auvergne durant le haut Moyen Age*, Paris.
1978, *Le château dans la France médiévale, essai de sociologie monumentale*, Paris.
Foutrein, Christiane, 1976, 'Modernisation de l'équipement agricole en Flandre intérieure française (conséquences, problèmes)', *Hommes et Terres du Nord*, 83–99.
France, Anatole, 1909, *Vie de Jeanne d'Arc*, 2 vols., Paris (cited from the edition by Jacques Suffel, Geneva, 1969).
Frémont, Armand, 1978, *Espace et cadre de vie: l'espace vécu des caennais*, under the direction of..., Caen.
1981, *Paysans de Normandie*, Paris.
Furet, François and Ozouf, Jacques, 1977, *Lire et écrire: l'alphabétisation des Français de Calvin à Jules Ferry*, 2 vols., Paris.
Fustel de Coulanges, 1914, *Histoire des Institutions politiques de l'Ancienne France. Vol. IV. L'alleu et le domaine rural pendant l'Époque Mérovingienne*, 2nd edn., Paris.
Gachon, Lucien, 1939, *Les Limagnes du Sud et leurs bordures montagneuses*, Tours.
1939a, *Une commune rurale d'Auvergne du XVIIIe au XXe siècle: Brousse-Montboisier*, Clermont-Ferrand.
Gadille, Rolande, 1967, *Le vignoble de la Côte Bourguignonne*, Paris.
Gain, André, 1937, 'Géographie historique de la Lorraine', in *Géographie Lorraine*, pp. 1–40, Nancy.
Gallet de Santerre, Hubert, 1965, 'Les civilisations classiques en Languedoc méditerranéen et Roussillon, principalement d'après les fouilles d'Ensérune', in *Le rayonnement des civilisations grecque et romaine sur les cultures périphériques (huitième congrès international d'archéologie classique)*, pp. 625–38, Paris.
1978, *Ensérune*, Paris.
Gallois, Lucien, 1908, *Régions naturelles et noms de pays*, Paris.
Galtier, Gaston, 1960, *Le vignoble du Languedoc méditerranéen et du Roussillon, Etude comparative d'un vignoble de masse*, 3 vols., Montpellier.
Ganshof, François, L., 1956, 'Zur Entstehungsgeschichte und Bedeutung des Vertrages von Verdun (843)', *Deutsches Archiv für Erforschung des Mittelalters*, 313–30.
Garden, Maurice, 1975, 'Aires du commerce lyonnais au XVIIIe siècle', in P. Léon, *Aires et structures*..., pp. 265–300.
Garnier, Bernard, 1987, 'Paris et les fromages frais au XIXe siècle. Un demisiècle de vente en gros aux halles', in P. Brunet, *Histoire et géographie des fromages*..., pp. 123–35.
Gascon, B., 1980, 'Immigration et croissance au XVIe siècle: l'exemple de Lyon', *Annales, Economies, Sociétés, Civilisations*, 988–1001.
Gaudron, G. and Soutou, A., 1961, 'Les racloirs triangulaires de la fin de l'âge du Bronze et la route de l'étain de Nantes à Narbonne', *Bulletin de la Société Préhistorique de France*, 583–93.

Gautier, Léon, 1868, 'L'idée politique dans les chansons de geste', *Revue des Questions Historiques*, 79–114.
   (ed.), 1872, *La Chanson de Roland*, 2 vols.
   1878–94, *Les épopées françaises*, 2nd edn., 4 vols.
   1890, *La chevalerie*, Paris, n.d.
Gautier, Marcel, 1971, *Chemins et véhicules de nos campagnes*, Saint-Brieuc.
Gay, François, 1967, *La Champagne du Berry*, Bourges.
George, Pierre, 1968, 'Chemin de fer et développement urbain', in *Mélanges offerts à Maurice Pardé*, pp. 229–37, Rennes.
Gernand, Karl, 1928, *Die Bezeichnungen des Sarges im Galloromanischen*, Giessen.
Gilbank, Gerald-Jack, 1981, *Les vignobles de qualité du Sud-Est du Bassin Parisien, évolution économique et sociale*, Paris.
   1987, 'D'une production de survie à une production commerciale: l'histoire du "Crottin de Chavignol"', in P. Brunet, *Histoire et géographie des fromages . . .*, pp. 151–6.
Gille, Bertrand, 1960, *Les forges françaises en 1772*, Paris.
   1968, *La sidérurgie française au XIXe siècle*, Geneva.
Gillet, M., 1973. *Les charbonnages du Nord de la France au XIXe siècle*, Paris.
Gilliéron Jules, 1902–20. *Atlas linguistique de la France*, 18 vols. Paris.
Giordan, Henri, 1980, 'Des images affectives des Gascons et des Provençaux au XVIIIe siècle', in *La découverte . . .*, pp. 103–17.
Girard d'Albissin, Nelly, 1970, *Genèse de la frontière franco-belge. Les variations des limites septentrionales de la France de 1659 à 1789*, Paris.
Godfrin, Jean and Planhol, Xavier de, 1966, 'Découverte d'un type primitif de toit de "tuiles romaines" en Lorraine', *Revue Géographique de l'Est*, 287–9.
Goguel, François, 1951, *Géographie des élections françaises de 1870 à 1951*, Paris, (*Cahiers de la Fondation Nationale des Sciences Politiques*, no. 27).
Goudineau, Christian, 1980, 'La Gaule méridionale', in G. Duby, *Histoire de la France urbaine*, I, pp. 141–93.
   1980a, 'Les villes de la paix romaine', in G. Duby, *Histoire de la France urbaine*, I, pp. 237–390.
Gourvil, Francis, 1960, *Théodore-Claude-Henri Hersart de la Villemarqué (1815–1895) et le 'Barzaz-Breiz' (1839–1845–1867). Origine-Édition-Sources-Critique-Influences*, Rennes.
Grand, Roger, 1965, *Recherches sur les origines des Francs*, Paris.
Gras, Christian and Livet, Georges (eds.), 1977, *Régions et régionalisme en France du XVIIIe siècle à nos jours*, Paris.
Gras, Solange and Christian, 1982, *La révolte des régions d'Europe occidentale de 1916 à nos jours*, Paris.
Gravier, J. F., 1947, *Paris et le désert français*, Paris.
Grenier, Albert, 1931–60, *Manuel d'Archéologie Gallo-Romaine*, Paris, 4 vols. in 7 (following J. Déchelette, *Manuel d'archéologie préhistorique, celtique et gallo-romaine*).
Guellec, Agnès, 1979, *Département et unité rurale: l'exemple des Côtes-du-Nord*, Poitiers (supplement to *Norois*, no. 26).
   1980, *Les Côtes-du-Nord, espace rural?*, 2 vols., Lille.

Guenée, Bernard, 1967, 'Etat et nation en France au bas Moyen Age', *Revue Historique*, 237: 17–30.
  1968, 'Espace et Etat dans la France du bas Moyen Age', *Annales, Economies, Sociétés, Civilisations*, 744–58.
Guérard, Benjamin, 1849, 'Du nom de France et des différents pays auxquels il fut appliqué', *Annuaire de la Société de l'Histoire de France*, 152–68.
Guibert, B. et al., 1975, *La mutation industrielle de la France*, Paris.
Guichard, René, 1965, *Essai sur l'histoire du peuple burgonde. De Bornholm (Burgundarholm) vers la Bourgogne et les Bourguignons*, Paris.
Guichonnet, Paul, 1945, 'L'émigration saisonnière en Faucigny pendant la première moitié du XIXe siècle', *Revue de Géographie Alpine*, 465–535.
  1948, 'L'émigration alpine vers les pays de langue allemande', *Revue de Géographie Alpine*, 533–76.
  1952, 'Deux types d'émigration préalpine: les marchands de graines de La Côte d'Arbroz et les chauffeurs de chauffage central de Mieussy', *Comité des Travaux Historiques et Scientifiques, Bulletin de la Section de Géographie*, 73–96.
Guilaine, Jean, 1976, *Premiers bergers et paysans de l'Occident méditerranéen*, Paris–The Hague.
Guilcher, André, 1946. 'Le finage des champs dans le Cartulaire de Redon'. *Annales de Bretagne*, 140–4.
Guilcher, André and Tricart, J., 1954, 'La XXXVIe Excursion Géographique Interuniversitaire, Champagne et Lorraine (1er–5 mai 1953)', *Annales de Géographie*, 1–21 and 88–98.
Guilcher, Jean-Michel, 1963, *La tradition populaire de danse en Basse-Bretagne*, Paris–The Hague.
  1965, 'Les formes anciennes de la danse en Berry', *Arts et Traditions Populaires*, 3–34.
  1967, 'Conservation et renouvellement dans la culture paysanne ancienne de Basse-Bretagne', *Arts et Traditions populaires*, 1–18.
  1969, *La contredanse et les renouvellements de la danse française (Etudes Européennes, 6)*, Paris–The Hague.
  1969a, 'Les formes basques de la danse en chaîne', *Arts et Traditions Populaires*, 1–54.
  1984, *La Tradition de danse en Béarn et Pays Basque français*, Paris.
Guinet, Louis, 1967, *Contribution à l'étude des établissements saxons en Normandie*, Caen.
Gulley, J. L. M., 1961, 'The practice of historical geography: a study of the writings of Professor R. Dion', *Tijdschrift voor Economische en Sociale Geografie*, 169–83.
Guyonvarc'h, Ch. J., 1961, '*Mediolanum Biturigum*, deux éléments de vocabulaire religieux et de géographie sacrée', *Ogam*, 13: 137–58 (completed in *Celticum*, I, 1962).
Halévy, Daniel, 1933, *Courrier d'Europe*, Paris.
Harmand, Jacques, 1967, *Une campagne césarienne: Alésia*, Paris.
Harmand, L., 1960, *L'Occident romain*, Paris.
Haudricourt, André G. and Hédin, Louis, 1943, *L'homme et les plantes cultivées*, Paris (*Géographie Humaine*, 19).

Haudricourt, André G. and Jean-Brunhes-Delamarre, Mariel, 1955, *L'homme et la charrue à travers le monde*, Paris (*Géographie Humaine*, 25).
Hawkes, Christopher and Dunning, G. C., 1930, 'The Belgae of Gaul and Britain', *The Archaeological Journal*, 87: 150–335, 531–41.
Higounet, Charles, 1946, *Le Comté de Comminges, de ses origines à son annexion à la couronne*, 2 vols., Toulouse.
   1949, 'La frange orientale des bastides', *Annales du Midi*, 359–67.
   1956, 'L'assolement triennal dans la plaine de France au XIIIe siècle', *Comptes rendus de l'Académie des Inscriptions et Belles-Lettres*, 507–10.
   1975, 'Pour l'histoire de l'occupation du sol et du peuplement de la France du Sud-Ouest, du XIe au XIVe siècle', in Higounet, *Paysages et villages neufs ...*, pp. 374–97.
   1975a, *Paysages et villages neufs du Moyen Age*, Bordeaux.
   1980, 'Les grandes haies forestières de l'Europe médiévale', in 'Le paysage rural: réalités et représentations. X^e colloque des Historiens médiévistes', *Revue du Nord*, 62, (244): 213–17.
Hoffmann, Stanley (ed.), 1963, *À la recherche de la France*, Paris.
Hofmann, B., 1976, 'Diffusion en Gaule romaine des marques sur sigillée italique depuis l'époque de César jusqu'au règne de Tibère', in *VGR*, pp. 39–46.
Hubert, Henri, 1932, *Les Celtes et l'expansion celtique jusqu'à l'époque de La Tène*, Paris (*L'Evolution de l'Humanité*, 21).
   1932a, *Les Celtes depuis l'époque de La Tène et la civilisation celtique*, Paris (*L'Evolution de l'Humanité*, 21a).
Huetz de Lemps, Alain, 1985, 'Les eaux-de-vie et liqueurs en France vers 1825', after Cavoleau, in Huetz de Lemps and Roudié, *Eaux-de-vie et spiritueux ...*, pp. 329–40.
Huetz de Lemps, Alain, and Roudié, Philippe, 1985, *Eaux-de-vie et spiritueux, Actes du colloque de géographie historique. Bordeaux-Cognac 1982*, published under the direction of ..., Paris.
Huetz de Lemps, Christian, 1975, *Géographie du commerce de Bordeaux à la fin du règne de Louis XIV* (*Civilisations et Sociétés*, 49), Paris–The Hague.
Itzin, Ulrich, 1983, *Das ländliche Anwesen im Lothringen* (*Tübinger Geographische Studien* 86), Tübingen.
Jacoubet, Henri, 1923, *Le comte de Tressan et les origines du genre troubadour*, Paris.
   1929, *Le genre troubadour et les origines françaises du romantisme*, Paris.
Jäger, H., Krenzlin, A. and Uhlig, H., (eds.), 1968, *Beiträge zur Genese der Siedlungs- und Agrarlandschaft im Europa* (*Geographische Zeitschrift*, Beiheft 18), Wiesbaden.
Jannoray, J., 1955, *Ensérune: contribution à l'étude des civilisations préromaines de la Gaule méridionale*, Paris.
Jarry, Eugène, 1942–48, *Provinces et pays de France (I. Formation de l'unité française. II. Monographies provinciales. Agenais à Béarn. III. Monographies provinciales. Bourgogne)*, 3 vols., Paris.
Jeanneau, Jacques, 1985, 'La société Cointreau, d'Angers au marché mondial', in A. Huetz de Lemps and P. Roudié, *Eaux-de-vie et spiritueux ...*, pp. 307–12.
Jeanneney, Jean-Marie, 1959, *Forces et faiblesses de l'économie française*, Paris.
Jeanton, Gabriel, 1937, *Costumes bressans et mâconnais*, Tournus.

1939, 'Les seuils. Contribution à l'étude de la limite septentrionale en Bourgogne de la zone du battage du blé en plein air', *Romanica Helvetica*, 14 (= *Mélanges A. Duraffour*): 10–15.
Johnson, J., 1946, *Étude sur les noms de lieux dans lesquels entrent les éléments court, ville et villiers*, Paris.
Johnson, J. H., 1961, 'The development of the rural settlement patterns of Ireland', *Geografiska annaler*, 165–73.
Juillard, Etienne, 1949, 'Ascension et déclin de la viticulture en Basse-Alsace depuis le XVIIIe siècle', *Revue d'Alsace*, 57–62.
1950, 'Un problème: le déclin de la vigne dans le Nord de l'Alsace', *Annales de la Société d'Ethnographie Française*, 95–101.
1953, *La vie rurale dans la plaine de Basse-Alsace*, Strasbourg–Paris.
1970, 'L'armature urbaine de la France préindustrielle. Pour une carte du réseau urbain et de l'organisation régionale à la veille de l'établissement du réseau ferré', *Bulletin de la Faculté des Lettres de Strasbourg*, 299–307.
1971, 'Croissance urbaine et accessibilité des villes', *Revue Géographique de l'Est*, 257–69.
Juillard, Etienne, Meynier, André, Planhol, Xavier de and Sautter, Gilles, 1957, *Structures agraires et paysages ruraux. Un quart de siècle de recherches françaises* (Mémoire no. 17 of *Annales de l'Est*), Nancy.
Jullian, Camille, 1908–26, *Histoire de la Gaule*, 8 vols., Paris.
1922, *De la Gaule à la France*, Paris.
1930–31, *Au seuil de notre histoire*, 3 vols., Paris.
1963, *Vercingétorix*, 2nd edn. by P. M. Duval, Paris.
Kläui, Hans, 1930, *Die Bezeichnungen für Nebel im Galloromanischen*, Aarau (Diss. Zürich, 1929).
Klimrath, Henri, 1843, *Travaux sur l'Histoire du Droit français*, 2 vols., Paris–Strasbourg.
Kolendo, J. 1960, 'La moissonneuse antique en Gaule romaine', *Annales, Economies, Sociétés, Civilisations*, 1009–14.
Kolodny, Yerahmiel, 1962, *La géographie urbaine de la Corse*, Paris.
Knafou, Rémy, 1978, *Les stations intégrées de sports d'hiver des Alpes françaises*, Paris.
Krenzlin, Anneliese, 1959, 'Blockflur, Langstreifenflur und Gewannflur als Ausdruck agrarischen Wirtschaftsformen in Deutschland', in *Géographie et histoire agraires* (Mémoire no. 21 of *Annales de l'Est*), pp. 343–52, Nancy.
Kroemer, Doris, 1976, *Die Agrarlandschaft der Beauce. Historische Entwicklung und heutiges Bild einer hochtechnisierten Getreidebaulandschaft*, Aachen (*Aachener Geographische Arbeiten*, 10).
Krüger, Fritz, 1950, *Géographie des traditions populaires en France*, Mendoza.
Kruta, Venceslas, 1980, 'La Gaule intérieure', in G. Duby, *Histoire de la France urbaine*, I, pp. 196–229.
Kühn, Herbert, 1940, *Die germanischen Bügelfibeln der Völkerwanderungs in der Rheinprovinz*, Bonn.
Kurth, Godefroy, 1895, 'La France et les Francs dans la langue politique du Moyen Age', *Revue des Questions Historiques*, 337–56.
Labasse, Jean, 1955, *Les capitaux et la région*, Paris.

Laborde, Pierre, 1985, 'La société Izarra de Bayonne', in A. Huetz de Lemps and P. Roudié, *Eaux-de-vie et spiritueux* ..., pp. 313–20.
Lacam, J., 1965, *Les Sarrasins dans le haut Moyen Age, français*, Paris.
Lacoste, Yves (ed.), 1986, *Géopolitiques des régions françaises*, 3 vols., Paris.
Laferrère, Michel, 1960, *Lyon, ville industrielle*, Paris.
Lafont, Robert, 1954, *Mistral ou l'illusion*, Paris.
  1970, *Renaissance du Sud. Essai sur la littérature occitane au temps de Henri IV*, Paris.
  1971, *Clefs pour l'Occitanie*, Paris.
  (ed.), 1971a, *Le Sud et le Nord. Dialectique de la France*, Toulouse.
Lafont, Robert and Anatole, Christian, 1970, *Nouvelle histoire de la littérature occitane*, 2 vols., Paris.
Lambert, Cl. and Riouffreyt, J., 1981, 'Jalons pour une frontière des Cénomans et des Diablintes', in *FG*, pp. 123–74.
Landriot, Abbé and Rochet, Abbé, 1854, *Traduction des discours d'Eumène*, Autun.
Lasserre, Pierre, 1918, *Frédéric Mistral, poète, moraliste, citoyen*, Paris.
Latouche, R., 1931, 'Les idées actuelles sur les Sarrasins dans les Alpes', *Revue de Géographie Alpine*, 199–206.
  1956, *Les origines de l'économie occidentale*, Paris (*L'Evolution de l'Humanité*, 43).
Lauer, P., 1915, 'De la signification du mot "France" aux époques mérovingienne et carolingienne', *Mémoires de la Société de l'Histoire de Paris et de l'Ile-de-France*, 42(8).
Laurent, R., 1957, *Les vignerons de la 'Côte d'Or' au XIXe siècle*, Paris.
Lavallée, Théophile, 1864, *Les frontières de la France*, Paris.
Lavisse, Ernest, 1911, *Histoire de France depuis les origines jusqu'à la Révolution*, 17 vols., Paris.
  1920–22, *Histoire de France contemporaine*, 10 vols., Paris.
Lebeau, René, 1955, *La vie rurale dans les campagnes du Jura méridional*, Lyons.
Le Bras, Hervé, 1986, *Les Trois France*, Paris.
Le Bras, Hervé and Todd, Emmanuel, 1981, *L'invention de la France*, Paris.
Le Braz, Anatole, 1904, *Le théâtre celtique*, Paris.
Lecouteux, Claude, 1982, *Mélusine et le chevalier au cygne*, Paris.
Leday, A., 1976, 'Trois *vici* du Cher', in *VGR*, pp. 237–55.
Lefebvre des Noëttes, Cdt, 1931, *L'attelage, le cheval de selle à travers les âges*, 2 vols., Paris.
Léger, D., and Hervieu, B., 1979, *Le retour à la nature*, Paris.
Legros, Roger, 1981, 'Les frontières des Rèmes', in *FG*, pp. 175–9.
Le Lannou, Maurice, 1940, Le Berry, *L'Information Géographique*, V, 1.
Lemarignier, J. F., 1945, *Recherches sur l'hommage en marche et les frontières féodales*, Paris.
Lentacker, Firmin, 1974, *La frontière franco-belge. Etude géographique des effets d'une frontière internationale sur la vie de relations*, Lille.
Léon, Albert, 1909, *Une pastorale basque: Hélène de Constantinople, étu. de historique et critique*, Paris.
Léon, Pierre, 1954, *La naissance de la grande industrie en Dauphiné*, Gap.
  1970, 'La réponse de l'industrie', in F. Braudel and E. Labrousse (eds.), *Histoire économique* ..., II, pp. 217–65.

(ed.), 1975, *Aires et structures du commerce français au XVIIIe siècle*, Paris.
Lequin, Yves, 1977, *Les ouvriers de la région lyonnaise (1848–1914): la formation de la classe ouvrière régionale*, Lyons.
Lerat, Serge, 1963, *Les pays de l'Adour (Structures agraires et économie agricole)*, Bordeaux.
Le Roux, F., 1961, 'Le Celticum d'Ambigatus et l'omphalos gaulois', *Ogam*, 159–84.
Leroux, R., 1962, 'Reboisement et défrichement de la Champagne crayeuse', *Revue Générale des Sciences Pures et Appliquées*, 19–32.
Le Roy Ladurie, Emmanuel, 1966, *Les paysans de Languedoc*, 2 vols., Paris–The Hague.
Le Roy Ladurie, Emmanuel and Dumont, Paul, 1971, 'Quantitative and Cartographical Exploitation of French Military Archives, 1819–1826', *Daedalus*, 397–441.
Lestocquoy, J., 1953, 'De l'unité à la pluralité: le paysage urbain en Gaule du Ve au IXe siècle', *Annales, Economies, Sociétés, Civilisations*, 159–72.
1968, *Histoire du patriotisme en France*, Paris.
Levainville, J., 1909, *Le Morvan, étude de géographie humaine*, Paris.
1913, *Rouen, étude d'une agglomération urbaine*, Paris.
Lévy-Leboyer, Maurice and Bourguignon, François, 1985, *L'économie française au XIXe siècle*, Paris.
Livet, Roger, 1962, *Habitat rural et structures agraires en Basse-Provence*, Gap.
1983, 'Recherches sur les villages perchés dans la France méridionale', in P. Flatrès and X. de Planhol (eds.), *Etudes . . .* , pp. 11–34.
Longnon, Auguste, 1875, 'L'Ile-de-France. Son origine, ses limites, ses gouverneurs', *Mémoires de la Société de l'Histoire de Paris et de l'Ile-de-France*, 1: 1–43.
1878, *Géographie de la Gaule au VIe siècle*, Paris.
1907, *Atlas historique de la France*, Paris.
1920–29, *Les noms de lieux de la France*, 4 vols., Paris.
1922, *La Formation de l'unité française*, Paris.
Lot, Ferdinand, 1928, 'Etudes sur les légendes épiques françaises, V, la Chanson de Roland', *Romania*, 357–80.
1933, 'De l'origine et de la signification des noms de lieu en-ville et en-court', *Romania*, 199–246.
1935, *Les invasions germaniques. La pénétration mutuelle du monde barbare et du monde romain*, Paris.
1937, *Les invasions barbares et le peuplement de l'Europe*, 2 vols., Paris.
1945–53, *Recherches sur la population et la superficie des cités remontant à la période gallo-romaine*, 4 vols., Paris.
1948, *La France des origines à la guerre de Cent Ans*, Paris.
1963, *La Gaule*, 2nd edn., Paris.
Loth, J., 1883, *L'émigration bretonne du Ve au VIIe siècle de notre ère*, Rennes.
Louis, René, 1942, *Girart, comte de Vienne, dans les chansons de geste: Girart de Vienne, Girart de Fraite, Girart de Roussillon*, 2 vols., Auxerre.
1960, 'La grande douleur pour la mort de Roland', *Cahiers de Civilisation Médiévale*, 3: 62–7.
Lovie, Jacques, 1979, 'Où en sont les frontières françaises?', in *FCC*, pp. 195–202.
Luchaire, A., 1911, 'Les premiers Capétiens (987–1137)', in E. Lavisse, *Histoire de France depuis . . .* , II, Part 2.

Lucius, A., 1922, 'Le vignoble alsacien', *Annales de Géographie*, 205–14.
Lugge, M., 1960, *Gallia und Francia im Mittelalter: Untersuchungen über dem Zusammenhang zwischen geographisch-historicher Terminologie und politischen Denken vom VIsten bis XVsten Jahrhundert*, Bonn.
Mage, G., 1924, *La division de la France en départements*, Toulouse.
Malinvaud, E., Carré, S. J. and Dubois, P., 1972, *La croissance française*, Paris.
Mandrou, Robert, 1961, *Introduction à la France moderne. Essai de psychologie historique, 1500–1640* (*L'Evolution de l'Humanité*, 52), Paris.
  1964, *De la culture populaire aux XVIIe et XVIIIe siècles. La Bibliothèque bleue de Troyes*, Paris.
Manteyer, G. de, 1908, *La Provence du Ier au XIIe siècle*, Paris.
Marcaggi, J. B., 1926, *Lamenti, voceri, chansons populaires de la Corse*, Ajaccio.
Maréchal, J. R., 1981, 'Les frontières des Aduatiques et des Germains cisrhénans. Situation de leurs *oppida*', in *FG*, pp. 45–51.
Marie, Michel and Viard, Jean, 1977, *La campagne inventée*, Le Paradou.
Markovitch, T. H., 1968, 'L'industrie française au XVIIIe siècle. L'industrie lainière à la fin du règne de Louis XIV et sous la Régence', *Cahiers de l'I.S.E.A., Economies et Sociétés*, II, pp. 1517–697.
Marot, P., 1939, 'Le Moyen Age de 959 à 1477', in *Histoire de Lorraine*, pp. 93–148, Nancy.
Marres, Paul, 1935, *Les Grands Causses*, 2 vols., Tours.
Marthelot, P., 1950, 'Le recul de la vigne dans les vallées vosgiennes', *Annales de la Société d'Ethnographie Française*, 103–10.
Martin, Marie-Madeleine, 1949, *La formation morale de la France*, Paris.
Martin, P. E., 1933, 'Le problème de la *Sapaudia*', *Revue Suisse d'Histoire*, 183–205.
Martin, R., 1971, *Recherches sur les agronomes latins et leurs conceptions économiques et sociales*, Paris.
Martin, Pierre de, 1979, *La maison rurale des hautes terres de l'Auvergne sud-orientale*. Unpublished thesis for a Doctorat ès Lettres, Université de Paris–Sorbonne.
Masseport, Jean, 1960, 'Le comportement politique du Massif du Diois: essai d'interprétation géographique', *Revue de Géographie Alpine*, 5–167.
Mauriange, Edith, 1955, 'Le mobilier bressan à deux tons de bois', *Arts et Traditions Populaires*, 97–115 and 227–37.
Maurizio, A., 1932, *Histoire de l'alimentation végétale*, French translation, Paris.
Ménard, Michèle, 1980, *Une histoire des mentalités religieuses aux XVIIe et XVIIIe siècles: mille retables de l'ancien diocèse du Mans*, Paris.
Mendras, Henri, 1979, *Voyages au pays de l'Utopie rustique*, Le Paradou.
  (ed.), 1980, *La sagesse et le désordre. France 1980*, Paris.
  1984, *La fin des paysans. Vingt ans après*, Le Paradou.
Menéndez-Pidal, Ramón, 1960, *La Chanson de Roland et la tradition épique des Francs*, 2nd edn., Paris.
Merian, Samuel, 1914, *Die französischen Namen des Regenbogens*, Halle (Diss. Phil. Basle, 1912).
Merle, L., 1958, *La métairie agraire de la Gâtine poitevine de la fin du Moyen Age à la Révolution*, Paris.
Merlin, Pierre, 1969, *La dépopulation des plateaux de Haute-Provence*, Paris.
  1971, *L'exode rural* (*Institut National d'Etudes Démographiques, Travaux et Documents*, no. 59), Paris.

Mertens, J., 1985, 'The military origins of some Roman Settlements of Belgium', in *Rome and her Northern Provinces, Papers presented to Sheppard Frere*, pp. 155–68, Gloucester.
   1985a, 'L'urbanisation dans le Nord de la Gaule Belgique à l'époque romaine', in *Les débuts de l'urbanisation en Gaule et dans les provinces voisines* (*Caesarodunum*, XX), pp. 261–80.
Mettrier, H., 1911, *La formation du département de la Haute-Marne en 1790, étude de géographie politique*, Chaumont.
Meyer, Jean, 1972, 'L'évolution des idées sur le bocage en Bretagne', in *La pensée géographique française contemporaine. Mélanges offerts au Professeur André Meynier*, pp. 453–67, Saint-Brieuc.
Meynier, André, 1931, *Ségala, Lévezou, Châtaigneraie*, Aurillac.
   1943, 'Champs et chemins de Bretagne', in *Conférences Universitaires de Bretagne 1942–1943*, pp. 161–78, Paris.
   1945, 'Les ensembles cadastraux circulaires en Bretagne, Chronique géographique des pays celtes', *Annales de Bretagne*, 14–25.
   1945a, 'La commune rurale française', *Annales de Géographie*, 161–79.
   1966, 'La genèse du parcellaire breton', *Norois*, 595–610.
   1972, 'Parcellaires circulaires et elliptiques', *Norois*, 117–19.
Millet, E., 1935, 'Les pays meusiens', in *Géographie Lorraine*, pp. 262–310, Nancy.
Miquel, André, 1975, *La géographie humaine du monde musulman jusqu'au milieu du XIe siècle*, II, Paris.
Mirot, Léon and Albert, 1947, *Manuel de géographie historique de la France*, 2nd edn., 2 vols., Paris (1st edn. by L. Mirot, 1929).
Mongrédien, Georges, 1965, *Recueil des textes et des documents du XVIIe siècle relatifs à Molière*, 2 vols., Paris.
Moreau, J., 1972, *Dictionnaire de géographie historique de la Gaule et de la France*, Paris.
Morineau, M., 1970, 'La pomme de terre au XVIIIe siècle', *Annales, Economies, Sociétés, Civilisations*, 1767–85.
Mours, Samuel, 1957, 'Les églises réformées en France', *Bulletin de la Société d'Histoire du Protestantisme Français*, 103 (1–3) and separately, Strasbourg (1958).
Müller, Bodo, 1971, 'La bipartition linguistique de la France, mise au point de l'état des recherches', *Revue de Linguistique Romane*, 35: 17–30.
   1974, 'La structure linguistique de la France et la romanisation', *Travaux de linguistique et de littérature publiés par le Centre de philologie et de littérature romane de l'Université de Strasbourg*, 12(1).
Mulliez, J., 1982, 'La fixation de la race percheronne à la fin du XVIIIe siècle', in *Le cheval dans l'agriculture* (*Ethnozootechnie*, 30), pp. 3–14, Paris.
Musset, Lucien, 1965, *Les invasions: les vagues germaniques* (*Nouvelle Clio*, 12), Paris.
   1965a, *Les invasions: le second assaut contre l'Europe chrétienne* (VIIe–XIe *siècles*) (*Nouvelle Clio*, 12a), Paris.
Musset, René, 1919, 'Le Perche, nom de pays', *Annales de Géographie*, 342–59.
Navarro, J. M. de, 1925, 'Prehistoric routes between Northern Europe and Italy defined by the amber trade', *Geographical Journal*, 66: 481–507.
Nègre, E., 1963, *Les noms de lieux en France*, Paris.
Niel, Fernand, 1970, *La civilisation des mégalithes*, Paris.

Niemeier, Georg, 1961, 'Agrarlandschaftliche Reliktgebiete und die Morphogenese von Kulturlandschaften in atlantischen Europa', *Geografiska Annaler*, 229–35.
Nisard, Charles, 1864, *Histoire des livres populaires et de la littérature de colportage*, 2 vols., Paris.
Noin, Daniel, et al., 1984, *Atlas des Parisiens*, Paris.
Nora, S. and Minc, A. (eds.), 1978, *L'informatisation de la société*, Paris.
Olschki, Leonardo, 1913, *Paris nach den altfranzösischen nationalen Epen. Topographie, Stadtgeschichte und lokale Sagen*, Heidelberg.
Onde, Henri, 1942, 'L'émigration en Maurienne et en Tarentaise', *Bulletin de la Société Scientifique du Dauphiné*, 1–57.
Papy, Louis, 1948, 'L'ancienne vie pastorale dans la Grande Lande', in *France méridionale et Pays Ibériques, Mélanges offerts à Daniel Faucher*, II, pp. 629–40, Toulouse.
Parain, Charles, 1938, 'Les anciens procédés de battage et de dépiquage en France', in *Premier Congrès International de Folklore de Paris en 1937*, pp. 86–91, Tours.
 1957, 'Voies et formes de la différenciation dans les petits vignobles du centre de la France: techniques, coutumes, croyances', *Arts et Traditions Populaires*, 134–58.
La patrie gauloise d'Agrippa au VIe siècle. Actes du Colloque (Lyon, 1981), 1983, Lyons.
Pédelaborde, Pierre, 1957, *Le climat du Bassin parisien*, 2 vols., Paris.
Pélatan, Jean, 1985, *Le Perche. Un pays et ses hommes*, Paris.
Peltre, Jean, 1975, *Recherches métrologiques sur les finages lorrains*, 1 vol. + 1 atlas, Lille–Paris.
 1978, 'Les industries de la Lorraine ducale d'après les déclarations des communautés de 1708', *Comité des Travaux Historiques et Scientifiques, Bulletin de la Section de Géographie*, 82 (1975–77), 153–64.
 1980, 'Contribution à l'archéologie industrielle: recherches sur l'impact paysager d'industries anciennes en Lorraine', *Revue Géographique de l'Est*, 3–14.
 1985, 'Brandeviniers et bouilleurs de cru en Lorraine depuis 1700', in A. Huetz de Lemps and P. Roudié, *Eaux-de-vie et spiritueux* . . . , pp. 205–12.
Perpillou, Aimé, 1971, 'L'évolution du paysage rural de la Normandie depuis le début du XIXe siècle', in F. Dussart (ed.), *L'habitat* . . . , pp. 341–64.
 1977, *Utilisation agricole du sol – France – 1re moitié du XIXe siècle*, map, Paris,
 1977a, *Utilisation agricole du sol – France – 1re moitié du XXe siècle*, map, Paris,
Perrenot, Th., 1942, *La toponymie burgonde*, Paris.
Perrin, Charles-Edmond, 1935, *Recherches sur la seigneurie rurale en Lorraine*, Paris.
 1942, *Les classes paysannes et le régime seigneurial en France du début du IXe siècle à la fin du XIIIe siècle*, Paris.
Perrot, Jean-Claude, 1974, *Genèse d'une ville moderne. Caen au XVIIIe siècle*, Lille.
Petit-Dutaillis, Charles, 1950, *La monarchie féodale en France et en Angleterre, X^e–XIIIe siècles* (*L'Evolution de l'Humanité*, 41), Paris.
Petry, F., 1976, 'Observations sur les vici explorés en Alsace', in *VGR*, pp. 273–95.
 1981, '*Ager publicus* et *ager privatus*. Note à propos d'un abornement antique de la forêt de Saverne (Bas-Rhin)', in *FG*, pp. 21–7.
Le phénomène des grandes 'invasions'. Réalité ethnique ou échanges culturels. L'anthropologie au secours de l'histoire, 1983 (*Centres de Recherches Archéologiques. Notes et Monographies techniques*, 12), Valbonne.

Phlipponneau, Michel, 1956, *La vie rurale de la banlieue parisienne*, Paris.
Picard, Gilbert-Charles, 1970, 'Les théâtres ruraux de Gaule', *Revue archéologique*, 184–92.
Piganiol, André, 1962, *Les documents cadastraux de la colonie romaine d'Orange* (16th supplement to *Gallia*), Paris.
Pignon, J., 1960, *L'évolution phonétique des parlers du Poitou (Vienne et Deux-Sèvres)*, Paris.
Pijassou, René, 1980, *Le Médoc*, 2 vols., Paris.
Pinkney, D. H., 1950, 'Paris, capitale du coton sous le Premier Empire', *Annales, Economies, Sociétés, Civilisations*, 50–60.
Pirenne, Henri, 1909, 'Draps de Frise ou draps de Flandre? Un petit problème d'histoire économique à l'époque carolingienne', *Vierteljahrschrift für Social- und Wirtschaftsgeschichte*, 7: 308–15.
Pitte, Jean-Robert, 1978, 'Les origines et l'évolution de la châtaigneraie vivaraise à travers un document cadastral du XVIIIe siècle', *Comité des Travaux Historiques et Scientifiques, Bulletin de la Section de Géographie*, 82 (1975–77), 165–78.
 1983, *Histoire du paysage français*, 2 vols., Paris.
 1983a, 'Deux exemples de "renaissance" de l'urbanisme romain à l'orée du XVIIe siècle: Charleville et Henrichemont', in *Présence de l'architecture et de l'urbanisme romains (Caesarodunum*, 18a), pp. 249–59, Paris.
 1986, *Terres de Castanide. Hommes et paysages du châtaignier de l'Antiquité à nos jours*, Paris.
 1987, 'Une lecture ordonnée de la carte des fromages de France', in P. Brunet, *Histoire et géographie des fromages . . .*, pp. 201–7.
Plandé, R., 1938, 'La fonction politique de la frontière des Pyrénées', *Revue Géographique des Pyrénées et du Sud-Ouest*, 221–42.
Planhol, Xavier de, 1959, 'Essai sur la genèse du paysage rural de champs ouverts', in *Géographie et Histoire Agraires* (Mémoire no. 21 in *Annales de l'Est*), pp. 414–23, Nancy.
 1961, 'La genèse des paysages agraires européens au symposium de Vadstena', *Revue Géographique de l'Est*, 235–46.
 1965, 'L'openfield à noyers dans le Sud-Est du Bassin Parisien (Basse-Bourgogne, Nivernais septentrional, Sancerrois)', *Revue Géographique de l'Est*, 473–82.
 1968, 'Les limites septentrionales de l'habitat rural de type lorrain', in H. Jäger *et al.*, *Beiträge . . .*, pp. 145–63.
 1971, 'Aux origines de l'habitat rural de type lorrain', in F. Dussart (ed.), *L'Habitat . . .*, pp. 69–92.
 1972, 'Historical geography in France', in Alan R. H. Baker, (ed.), *Progress in Historical Geography*, pp. 29–44, Newton Abbot; the slightly abridged French text published as: 'Structures universitaires et problématique scientifique: la géographie historique française', in *La pensée géographique française contemporaine. Mélanges offerts au Professeur A. Meynier*, 1972, pp. 155–65, Saint-Brieuc.
 1978, 'Le paysage rural gallo-romain: état des connaissances et perspectives de recherches', *Geographia Polonica*, 38: 233–43.
 1982, 'Le paysage rural de la *villa* gallo-romaine', in *Actes du Colloque, 'La villa romaine dans les provinces du Nord-Ouest' (Caesarodunum*, 17), pp. 403–10, Tours.

Planhol, Xavier de and Lacroix, J., 1963, 'Géographie et toponymie en Lorraine', *Revue Géographique de l'Est*, 9–14.
Planhol, Xavier de and Pérardel, André, 1969, 'La répartition géographique des vestiges archéologiques gallo-romains en Lorraine', *Revue Géographique de l'Est*, 177–80.
Planhol, Xavier de and Popelard, Jean, 1976, 'Les limites alsaciennes de l'habitat rural de type lorrain', in R. H. Buchanan, R. A. Butlin and D. McCourt (eds.), *Fields, Farms and Settlement in Europe, Papers presented at a Symposium, Belfast, July 12–15, 1971*, pp. 152–61, Holyrood.
Plavinet, Pierre, 1974, 'La Brie. Essai de terminologie géographique', *Bulletin de la Société d'Etudes de la Région Parisienne* (April–June): 1–11.
Poète, Marcel, 1919–20, 'Les débuts d'un grand siècle de l'évolution urbaine: l'introduction de la grande industrie à Paris', *La Vie Urbaine* (1919), 413–56; (1920), 19–42.
Pomponi, Francis, 1979, *Histoire de la Corse*, Paris.
Porée, Ch., 1905, *La formation du département de l'Yonne en 1790*, Paris.
Pougnard, G., 1952, *Le parler franco-provençal d'Aiript*, La Rochelle.
Poupardin, René, 1901, *Le Royaume de Provence sous les Carolingiens (Bibliothèque de l'Ecole des Hautes Etudes, Sciences Historiques et Philologiques*, 1309), Paris.
  1907, *Le Royaume de Bourgogne (888–1038) (Bibliothèque de l'Ecole des Hautes Etudes, Sciences Historiques et Philologiques*, 163), Paris.
Pourcher, Guy, 1964, *Le peuplement de Paris*, Paris.
Pouthas, Charles-Henri, 1956, *La population française pendant la première moitié du XIXe siècle*, Paris.
Prêcheur, Claude, 1953, 'Grandeur et décadence du vignoble lorrain', in *Mélanges géographiques offerts à Philippe Arbos*, pp. 209–13, Clermont-Ferrand.
  1959, *La Lorraine sidérurgique*, Paris.
  1963, *La sidérurgie française*, Paris.
  1969, *1968: les industries françaises à l'heure du Marché Commun*, Paris.
Price, Roger, 1983, *The modernization of rural France*, London.
Prince, Hugh D., 1977, 'Regional Contrasts in Agrarian Structures', in H. D. Clout (ed.), *Themes . . .*, pp. 129–84.
Provost, Michel, 1981, 'Recherches sur les limites de la cité des Andes', in *FG*, pp. 180–3.
Pumain, Denise, 1982, *La dynamique des villes*, Paris.
  1982a, 'Chemin de fer – croissance urbaine en France au XIXe siècle', *Annales de Géographie*, 529–50.
Ramin, Jacques, 1965, *Le problème des Cassitérides*, Paris.
  1975, 'Le Problème de Corbilo et le rôle économique de l'embouchure de la Loire', in *Actes du Colloque 'Du Léman à l'Océan. Les eaux en Gaule' (Caesarodunum*, 10), pp. 119–23, Tours.
Réau, Louis, 1951, *L'Europe française au siècle des lumières (L'Evolution de l'Humanité*, 70), Paris.
Reclus, Elisée, 1879, *Nouvelle géographie universelle, II. La France*, Paris.
Reinaud, M., 1836, *Invasions des Sarrasins en France*, Paris.
Reitel, François, 1966, 'A propos de l'openfield lorrain', *Revue Géographique de l'Est*, 21–52.
Renard, J. and Chauvet, A., 1979, *La Vendée*, Les Sables-d'Olonne.

Renucci, Janine, 1974, *Corse traditionnelle et Corse nouvelle*, Lyons.
Reuss, Rod, 1920, *Histoire d'Alsace*, New edn., Paris.
Rey, G. de, 1971, *Les invasions des Sarrasins en Provence*, reprint, Marseilles.
Rey, M., 1929, 'Les limites géographiques de l'habitat perché dans les Alpes françaises', *Revue de Géographie Alpine*, 5–39.
Rézeau, P., 1976, *Un patois de Vendée. Le parler rural de Vouvant*, Paris.
Richard, J., 1954, *Les ducs de Bourgogne et la formation du duché du XIe au XIVe siècle*, Paris.
Ricolfis, Jean-Marie, 1980, *Essai de philologie, de toponymie et d'anthroponymie françaises; les noms de lieux, du pays niçois*, 2 vols., Lille–Paris.
Riedinger, A., 1950, 'Riquewihr, grande et petite exploitation', *Annales de la Société d'Ethnographie Française*, 111–20.
Riff, Adolphe, 1923, *L'art populaire en Alsace*, Strasbourg.
Rivals, Claude, 1984, 'Divisions géographiques de la France indiquées par une analyse de l'état des moulins en 1809', *Revue Géographique des Pyrénées et du Sud-Ouest*, 55: 367–84.
Robert-Muller, Charles and Allix, André, 1923, 'Un type d'émigration alpine: les colporteurs de l'Oisans', *Revue de Géographie Alpine*, 585–634.
Roblin, Michel, 1969, 'La frontière entre les Bellovaques et les Véliocasses à l'époque gallo-romaine et franque', *Bulletin de la Société Nationale des Antiquaires de France*, 295–320.
Rochefort, Michel, 1960, *L'organisation urbaine de l'Alsace*, Paris.
Rohlfs, G., 1970, *Le Gascon*, 2nd edn., Tübingen and Paris.
Roman, Y., 1977, 'Remarques à propos du commerce de l'étain en Gaule', in *Actes du Colloque: Géographie commerciale de la Gaule: ENS, juin 1976*, 2 vols., (*Caesarodunum*, 12), I, pp. 260–70, Tours.
Rondeau, A., 1964, *La Corse*, Paris.
Rose-Villequey, Germaine, 1970, *Verre et verriers en Lorraine au début des Temps Modernes (de la fin du XVe siècle au début du XVIIe siècle)*, Nancy.
Rossi, Marguerite, 1975, *Huon de Bordeaux*, Paris.
Rostaing, Charles, 1945, *Les noms de lieux*, Paris.
  1950, *Essai sur la toponymie de la Provence, depuis les origines jusqu'aux invasions barbares*, Paris.
Rotzler, Hermann, 1913, *Die Bezeichnungen der Milchstrasse im Französischen*, Erlangen (Diss. Phil. Basle, 1913).
Rouche, Michel, 1969, 'Le changement de nom des chefs-lieux de cité en Gaule au Bas-Empire', *Mémoires de la Société Nationale des Antiquaires de France*, 9th series, 4: 47–64.
  1977, *L'Aquitaine des Wisigoths aux Arabes*, Lille (and Paris, 1979).
Roudié, Philippe, 1985, 'Une vénérable entreprise bordelaise de liqueurs: Marie Brizard et Roger', in A. Huetz de Lemps and P. Roudié, *Eaux-de-vie et spiritueux ...*, pp. 215–300.
Rouillard, Dominique, n.d., *Le site balnéaire*, Brussels.
Roupnel, Gaston, 1932, *Histoire de la campagne française*, Paris.
Saint-Julien, Thérèse, 1982, *Croissance industrielle et système urbain*, Paris.
Salin, Edouard, 1950–59, *La civilisation mérovingienne d'après les sépultures, les textes et le laboratoire*, 4 vols., Paris.

Sansot, Pierre, 1985, *La France sensible*, Seyssel.
Schier, Bruno, 1939, *Der Bienenstand in Mitteleuropa*, Leipzig (separately and also in *Zeitschrift für Volkskunde*, 1938).
Schmitt, Paul, 1981, 'Les frontières de la Gaule d'après Ptolémée', in *FG*, pp. 5–16.
Schmittlein, Raymond, 1954, *La première campagne de César contre les Germains, 58 av. J.-C. (Travaux et Mémoires des Instituts Français en Allemagne, 6)*, Paris.
Schneider, Jean, 1939, 'La cité de Metz (962–1552)', in *Histoire de Lorraine*, pp. 162–83, Nancy.
Schulten, A., 1914, *Numantia, I. Die Keltiberer und ihre Kriege mit Rom*, Munich.
Schwab, Roland, 1980, *De la cellule rurale à la région. L'Alsace 1825–1960*, Paris–Strasbourg.
Sclafert, Th., 1961, *Cultures en Haute-Provence: déboisements et pâturages au Moyen Age*, Paris.
Secher, Reynald, 1986, *Le génocide franco-français: la Vendée-Vengé*, Paris.
Sée, Henri, 1891, *Louis XI et les villes*, Paris.
Sénac, Philippe, 1980, *Musulmans et Sarrasins dans le Sud de la Gaule du VIIIe au XIe siècle*, Paris.
Servais, J., 1889, *La Neustrie sous les Mérovingiens*, Turin.
Siegfried, André, 1913, *Tableau politique de la France de l'Ouest sous la IIIe République*, Paris (reprinted 1964).
Sinclair, Stéphane, 1985, *Atlas de géographie historique de la France et de la Gaule*, Paris.
Sion, Jules, 1909, *Les paysans de la Normandie orientale*, Paris.
 1934, *La France méditerranéenne*, Paris.
Smith, C. T., 1978, *An Historical Geography of Western Europe before 1800*, revised edition, London–New York.
Sorel, Albert, 1887–1911, *L'Europe et la Révolution Française*, 9 vols., Paris.
Soriano, Marc, 1968, *Les Contes de Perrault: culture savante et traditions populaires*, Paris.
Souillet, G., 1943, 'Chronologie et répartition des noms de lieux en -ière et -ais dans la Haute-Bretagne', *Annales de Bretagne*, 90–8.
Soutou, A., 1963, 'Typologie chronologique de quelques boutons du Midi de la France', *Bulletin de la Société Préhistorique de France*, 60: 372–84.
Soyer, J., 1933–52, *Recherches sur l'origine et la formation des noms de lieux dans le Loiret*, Orléans, separately and *Bulletin de la Société Archéologique d'Orléans*.
Soyer, Jacqueline, 1970, *La conservation de la forme circulaire dans le parcellaire français*, Paris.
 1973–74, 'Les centuriations de Provence', *Revue Archéologique de Narbonnaise*, 6: 197–232; 7: 179–99.
Specklin, Robert, 1979, 'Les contrastes Nord–Sud en France', *Regio Basiliensis*, 20: 16–63.
 1979a, 'Notes sur la genèse des paysages agraires en France', *Recherches Géographiques à Strasbourg*, 11: 45–60.
 1982, 'Etudes sur les origines de la France', *Acta Geographica* (Paris), 49: 21–34; 50: 37–43.
Stein, H. and Legrand, L., 1905, *La frontière d'Argonne (843–1659). Procès de Claude de la Vallée (1535–1561)*, Paris.

Subrenat, Jean, 1974, *Etude sur Gaydon, chanson de geste du XIIIe siècle*, Aix-en-Provence.
Sutcliffe, Anthony, 1970. *The Autumn of Central Paris. The Defect of Town Planning*, London.
Tanguy, Bernard, 1977, *Aux origines du nationalisme breton: le renouveau des études bretonnes au XIXe siècle*, 2 vols., Paris.
Tanière, M., 1954, 'La région industrielle de Lavelanet. Notes de géographie sociale', *Revue Géographique des Pyrénées et du Sud-Ouest*, 59–70.
Tardieu, Suzanne, 1950, *Les meubles régionaux datés*, Paris.
Tenèze, Marie-Louise, 1955, 'Etude de proverbe: "Niais de Sologne qui ne se trompe qu'à son profit"', *Arts et Traditions populaires*, 211–26.
Ternes, Ch. M., 1976, 'Le *vicus* d'époque gallo-romaine en pays trévire et rhénan', in *VGR*, pp. 18–31.
Theret, Marcel, 1981, 'Les concours de Poissy 1844–1867 (premiers concours d'animaux de boucherie)', in *Les concours de bétail* (*Ethnozootechnie*, 28), pp. 3–13, Paris.
  1983, 'L'introduction des races britanniques dans le cheptel bovin français et ses conséquences', in *Les bovins: origine; évolution de l'élevage* (*Ethnozootechnie*, 32), pp. 159–73, Paris.
Thévenot, E., 1949, *Histoire des Gaulois*, Paris.
  1952, 'Les origines du vignoble bourguignon. Les conditions de son établissement', *Annales de Bourgogne*, 245–57.
  1961, 'Quand fut créé le vignoble bourguignon?', *Revue Archéologique de l'Est*, 61–64.
  1969, *Les voies romaines de la cité des Eduens*, Brussels.
Thibaudet, Albert, 1938, *Réflexions sur la littérature*, Paris.
Thomas-Lacroix, P., 1954, 'Toponymes du type "Chez X"', *Revue Internationale d'Onomastique*, 261–70.
Thuillier, G., 1966, *Aspects de l'économie nivernaise au XIXe siècle*, Paris.
Timbal, Pierre-Clément, 1973, 'Civitas Parisius, communis patria', in *Economie et société au Moyen Age. Mélanges offerts à Edouard Perroy*, pp. 661–5, Paris.
Toujas-Pinède (Christine), 1960, 'Une émigration de survivance: les Vendéens en Aquitaine', *Revue Géographique des Pyrénées et du Sud-Ouest*, 399–423.
Tregaro, L., 1931, 'Les Maures et l'Esterel, étude de géographie humaine', *Bulletin de la Société Languedocienne de Géographie*, 1–14.
Tricart, J., Dirrig, Roger and Dugrand, R., 1949, 'Le vignoble alsacien', *L'Information Géographique*, 21–7.
Tristan, Frédérick, 1979, *Géants et gueux des Flandres. Dix siècles de mythes et d'histoire*, Paris.
Tulippe, O., 1934, *L'habitat rural en Seine-et-Oise*, Paris.
Uhlig, Harald, 1961, 'Old hamlets with infield and outfield systems in Western and Central Europe', *Geografiska Annaler*, 285–313.
Vadé, Yves, 1972–74, 'Le système des *Mediolanum* en Gaule', *Archéocivilisation*, 11–13: 87–109.
  1976, 'Le problème des *Mediolanum*', in *VGR*, pp. 50–8.
Van Tieghem, P., 1917, *Ossian en France*, 2 vols., Paris.
Vendryès, J., 1955, 'Notes sur la toponymie celtique', in *Recueil de travaux offerts à M. Clovis Brunel*, II, pp. 640–50.

Vercauteren, M., 1934, *Les civitates de la Belgique seconde*, Brussels.
Verhulst, Adriaan, 'La genèse du régime domanial classique en France au Haut Moyen Age', in *Settimane di studio del centro italiano di studi sull' alto medioevo di Spoleto, XIII: Agricultura e mondo rurale in Occidente nell'alto Medioevo* (= *Studia Historica Gandensia*, 47), pp. 135–60.
Veyret-Verner, Germaine, 1948, *L'industrie des Alpes françaises*, Grenoble.
VGR = Actes du Colloque 'Le vicus gallo-romain'. ENS 14–15 June 1975 (Caesarodunum, 11), Tours, 1976.
Vidal de La Blache, Paul, 1908, *La France. Tableau géographique*, Paris (also Book 1, Part 1 of E. Lavisse, *Histoire de France depuis* ...).
Viers, G., 1958, 'L'industrie et la population de Mauléon', *Revue Géographique des Pyrénées et du Sud-Ouest*, 97–119.
Villard, F., 1960, *La céramique grecque de Marseille (VIe–IVe siècle). Essai d'histoire économique*, Paris.
La ville d'hiver d'Arcachon, 1983, Paris.
Vincent, Auguste, 1937, *Toponymie de la France*, Brussels.
Viollet, Paul, 1893, *Histoire du droit civil français*, 2nd edn., Paris.
Vossler, Ch., 1953, *Langue et culture de la France*, French translation, Paris.
Wackermann, Gabriel, 1987, 'Le rôle d'un fromage dans la mutation socio-culturelle contemporaine: le Munster', in P. Brunet, *Histoire et géographie des fromages* ..., pp. 251–5.
Wandruszka von Wanstetten, M. W., 1939, *Nord und Süd im französischen Geistesleben* (*Berliner Beiträge zur Romanischen Philologie*, IX, 1–2), Iena and Leipzig.
Weber, Eugen, 1983, *La fin des terroirs. La modernisation de la France rurale 1870–1914*, Paris.
Werner, Karl Ferdinand, 1970, 'Les nations et le sentiment national dans l'Europe médiévale', *Revue Historique*, 244: 285–304.
 1984, *Les origines* (*Histoire de France*, under the direction of Jean Favier, I), Paris.
Wolff, Philippe, 1970, *Les origines linguistiques de l'Europe occidentale*, Paris.
Wolkowitsch, Maurice, 1946, 'L'élevage dans le Bazois', *Annales de Géographie*, 205–10.
 1980, 'Réseau ferré et structuration de l'espace français', in *Eventail de la spatiologie offert à J. E. Hermitte*, pp. 103–21.
Woronoff, Denis, 1970, 'Vers une géographie industrielle de la France d'Ancien Régime', *Annales, Economies, Sociétés, Civilisations*, 127–30.
Wright, John Kirtland, 1965, *The Geographical Lore of the Time of the Crusades*, 2nd edn., New York.
Zeldin, Theodore, 1980–81, *Histoire des passions françaises, 1848–1945*, 5 vols., Paris.
Zeller, Gaston, 1926, *La réunion de Metz à la France*, 2 vols., Strasbourg.
 1932, *La France et l'Allemagne depuis dix siècles*, Paris.
 1933, 'La monarchie et les frontières naturelles', *Revue d'Histoire Moderne*, 306–33.
Zumthor, Paul, 1954, *Histoire littéraire de la France médiévale, VIe–XIVe siècles*, Paris.

Index

In this index, geographical names are printed in roman lettering, subjects in italics, the names of individuals in small capitals and the names of groups in bold lettering. Both the notes and the figures are indexed. The titles of works and sources are included from the body of the text but not from the notes, except in the cases of anonymous works. The names of translators and editors of texts have not been indexed except where there were special reasons for doing so. 'France' has been indexed only very incompletely, as have the large-scale maps. The names of people are given in the form in which they appear, that is to say usually without forenames, except in cases of homonymy.

Abbeville, 270, 450
ABBON, 252
Abondance (breed of cattle), 358
Abrincatui, 174
-ac, 86
academies, 273, 274, 314, 316, 319, 408
Actes du Colloque de Rennes..., 494
-acum, 65, 86
ad -iacas, 63
ad Sanctos, 96
ADER (Guillaume), 314, 320
Adour, region, 230, 343, 369
aedificium (-a), 16–17, 18, 48, 53, 471
Aedui, 10, 11, 12, 20, 40, 43, 44, 56, 69, 176, 177, 180
Aeneid, 101
aeronautical (industries), 437, 438, 441, 450, 452
AESCHYLUS, 102, 204
AETIUS, 76
Africa, 255; North, 27; black, 20
AGACHE, 50, 51, 471, 474, 475, 476, 478, 487
Agde, 27, 30
Agen, 8, 43, 179, 225, 286, 318, 321; Agenais, 347; Aginnum, 36
Agriates (desert of the), 185
AGRIPPA, 37
Aigurande, 15

AIMOIN, 98, 481
Ain (department), 282, 283
air (lines), 450, 454, 455; *see aviation*
airports, 454, 455
-ais, 141, 166, 297
Aisne (department), 282, 377
Aix-en-Provence, 21, 42, 43, 143, 161, 183, 286, 314, 315, 402
Aix-la-Chapelle, 59, 83, 96; *Treaty of*, 115
Aix-les-Bains, 42, 287, 289, 290, 409; *see* Aquae
Ajaccio, 185, 186, 318
*al, 198
Alamans, 59, 66, 76, 79, 80, 95
Albi, 35, 68, 179, 225, 369; Albiga, 179; Albigeois, 179, 356; **Albigensians** (*crusade, heresy*), 117, 129, 144, 162, 184, 285, 290
ALCAN, 495
ALCUIN, 490
ALDUC - LE BAGOUSSE, 479
ALEMBERT (d'), 281
Alençon, 166, 175, 290
Alès, 271, 339, 450; neighbourhood, 476
Alesia, 12, 20, 23, 43
ALETHEUS, 72
ALEXANDER, 29
Algia, 490; *see* Auge (Pays d')
Alicante-Bouschet, 349

533

534   Index

Alionensis (Pagus); Alionis (castrum), 486
Alisacensis (Pagus), 486
Allemagne (De l'), 134
Allier, 177; plain of, 177; (department), 282, 395
ALLIX, 489, 491
Allobroges, 8, 10, 12, 34, 35, 42, 44, 55; *allobrogica*, 55
ALPHAND, 458
ALPHONSE DE POITIERS, 100, 146
Alpilles *(canal des)*, 346
Alps, 5, 7, 8, 16, 20, 31, 39, 45, 73, 85, 110, 111, 113, 114, 123, 130, 147, 152, 180, 183, 198, 223, 234, 242, 265, 331, 358, 359, 360, 366, 370, 382, 385, 387, 388, 390, 392, 393, 395, 396, 397, 400, 407, 428, 433, 435, 450, 455, 485, 489; Basses —(department), 282, 360; —Cottiennes, 39; —Grées, 39, 73; Hautes —(department), 282, 360, 399; —Maritimes, 27, 39, 74; department, 282, 360; —Pennines, 39, 73; *alp, alpe*, 198
Alsace, **Alsatians** Alsatian dialect, 36, 44, 56, 59, 66, 111, 114, 115, 136, 153, 161, 168, 173, 195, 207, 212, 215, 224, 227, 235, 272, 273, 280, 310, 311, 313, 321–2, 325, 328, 331, 335, 352, 376, 378, 385, 389, 400, 407, 433, 438, 445, 447, 483, 485; Alsace-Lorraine, 273, 442; Porte d', 66; Porte de Bourgogne et d'Alsace, 173
Alsegau, Alsegaugensis (Pagus), 165
altar-pieces, 196
Amavus (Pagus), 164
Ambazac (Monts d'), 177, 356
amber, 26
Ambert (region), 388
Ambiani, 12, 29, 30, 171–2
AMBIGAT, 23
Amboise, 270
America, 235, 349, 454; mid-west, 440; South, 322, 388; *see* United States
AMÉRO (Constant), 292
Amiens, 161, 235, 241, 251, 270, 402, 455, 478, 491
AMMIANUS MARCELLINUS, 251, 475
Ammophila arenaria (marram grass), 361
amphorae, 33, 56
Amsterdam, 455
ANATOLE, 496
ANCEL, 483
Ancien Régime, 153, 160, 162, 163, 188, 189, 240–1, 280, 285, 298, 313, 324, 326, 341, 374, 401, 419, 422, 435
Andecaves, Andecavi, 12, 14, 175
Andematunnum, Andemantunnum, 36, 43
ANDRÉ, 473
Andrézieux, 271
-ange, -anges, 61, 63

Angers, 42, 161, 225, 270, 376
ANGEVILLE (d'), 485
Angevins, 100, 189, 487
Anglo-Norman (monarchy), Anglo-Norman area, 192
Angoulême, 43, 176, 274; Angoumois, 126, 143, 176
-anicum, 62
Anjou, 87, 161, 168, 176, 214, 223, 225, 228, 256, 333, 340, 342; *duc d'*, 286; *see* **Angevins**
ANNE (duchess), 297
Annecy, 371, 455
-ans, 61, 63, 66
anthropological (data), 60–1, 155
anthroponymy, 66, 203, 214
Antibes, 27
ANTOR, 101
ANVILLE (d'), 114
Anzin *(Compagnie d')*, 403, 438
APOLLONIUS of RHODES, 8
apothecaries, 244
apples, 54; *apple trees*, 217
Aquae Augustae, 42; Sextiae, 36, 42, 43; Tarbellicae, 36: *see* Aix
Aquitaine, Aquitania, 7, 11, 39, 40, 43, 44, 61, 63, 66, 74, 75, 77, 79, 81, 93, 94, 106, 113, 124, 129, 143, 146, 147, 152, 153, 168, 171, 176, 177–8, 179, 180, 184, 212, 218, 228–9, 242, 314, 334, 343, 376, 378, 389, 390, 395, 397, 409; *see* south-west France; *Blonde d' (breed of cattle)*, 358; **Aquitani**, 8, 100, 105
Arab (influences), 132, 144, 147, 186; *(names)*, 82, 84, 85; *geography*, 9, 252; *see* **Muslims, Saracens**
Aragon, *Aragonese monarchy*, 110, 115
ARAMBOUROU, 498
Arbois, 228, 406
ARBOS, 489, 500
Arcachon, 7, 384, 439, 460
architecture, 132, 192, 193, 195, 243, 293, 310, 459, 460, 461
Ardèche (department), 221, 282
Ardenne, Ardennes, Arduenna, Arduinna, 20, 63, 88, 108, 140, 172, 198–9, 202, 234, 304; *counts of*, 88; department, 282; Ardennensis (Pagus), 198
ARDOUIN-DUMAZET, 412, 413–14, 467
Arecomici (Volcae), 12, 34
Arelate, 28, 36, 44; Arles, 28, 44, 47, 78, 143, 180, 251; kingdom of, 109; region, 309, 346
Arezzo, 38
ARGENSON (d'), 281
Argentomagus, 36
Argentorate, 36, 42, 69; *see* Strasbourg

Argonne, 108, 170
Ariège (department), 62, 282, 283, 328, 367, 387
ARIOSTO, 254
ARIOVISTE, 35
ARISTOTLE, 204
Arles, 47, 109, 143, 180, 203, 251, 309, 346
Armagnac, Armaniacus, 179, 230; *armagnac*, 230, 376
ARMAND, 212, 489, 501
ARMBRUSTER, 484
Armentières, 403, 405
Armorica, 5, 20, 24, 30, 37, 85, 89, 142, 146, 153, 174, 175, 258, 329, 340-2, 466; *see* Brittany, Maine, Anjou, Bocage Normand, Cotentin, Vendée; **Armoricans**, 12, 20, 24, 174; *Armoricain (L')*, 323
arms, armaments (industry), 60, 72, 93, 136, 166, 234, 369, 373, 437, 450
ARQUÉ, 499
Arras, 13, 15, 50, 100, 161, 282, 291
Arrée (Monts d', Montagne d'), 300, 305, 331
arrondissement, 281, 324
Arrouaise, 13, 15, 171, 204
art, 129, 192-6, 243, 293-4, 309, 310, 311; *see* architecture
ARTAGNAN (d'), 133
artichokes, 347
ARTMANNUS, 179
Artois, 6, 64, 66, 100, 161, 168, 171, 227, 249, 283, 390, 414
Arve (valley), 239
Arvernes, Arverni, 11, 12, 20, 23, 32, 33, 34, 40, 43, 177, 472, 474
Aryans, -ism, 94, 99
Asia, 297; Minor, 7, 27; *see* East, Levant
asparagus, 347
Assertio Normaniae, 107
Association Bretonne, 319, 324
Astérac, Asteriacus, ASTERIUS, 179
ASTRONOMER of LIMOGES, 95
ATEIUS, 38
Athens, 275; *charter*, 463-4
Atlantic (Ocean), 29, 44, 113; seaboard, 3, 5, 9, 16, 29, 37, 55, 56, 58, 72, 73, 76, 93, 98, 137, 152, 175, 179, 228-9, 237, 270, 340, 384, 435, 436, 456, 472
Atlas historique de la France contemporaine, 450, 501
Atrebates, 6, 12, 13, 171
Attuariorum (Pagus), Atuyer, 164
Aube, 368; department, 199, 282
Auberive (forest), 68
AUBERY (Antoine), 111
Aubrac, 199, 205; *breed of cattle*, 358; *Aubrac (L')*, 489, 499

Auch, 7, 12, 145, 470
Aude, 27; valley, 29, 368; lower plain of the, 349; department, 282, 350
Audierne, 305
AUDIN, 471, 472, 474, 475
Auge (Pays d'), 140, 175, 206, 215, 372, 377; *see* Algia; forest, 471
AUGÉ, 490
Augusta Praetoria, 36; —Rauricorum, 36, 44; —Suessionum, 36; —Taurinorum, 36; —Treverorum, 36, 42, 44; *see* Trèves; —Viromanduorum, 50
Augustobona, 36; —Tricassium, 42
Augustodunum, 36, 43, 69
Augustodurum, 36; —Baiocassium, 42
Augustomagus Silvanectum, 42
Augustonemetum, 36, 43
Augustoritum, 36, 69, 177; —Lemovicum, 42, 69
Augustum, 36
AUGUSTUS, 11, 36, 38, 39, 40, 41, 45, 55, 78
Aulerci, Aulerques, 10, 12, 175
Aulne, 298, 305, 307
Aunis, 161, 168, 176, 228, 288, 304, 342, 486
AUPHAN, 501
Auray, 306, 307
Aurelia (Via), 36
AURELIAN, 73
AURELIUS VICTOR, 56
Ausci, Auscii, 7, 12, 470
AUSONIUS, 53, 251, 476
Austrasia, **Austrasii**, Austria, 77, 81, 83, 94
Austria, 288; house of, 173; *see* **Habsburgs**
autonomism, 322, 325, 341
Autun, 43, 47, 56, 69, 181, 284, 478
Auvergne, 53, 71, 96, 161, 168, 177, 181, 189, 192, 193, 195, 199, 222, 256, 286, 303, 305, 314, 356, 360, 377, 378, 387, 388, 389, 390, 409, 410; **Auvergnats**, 133, 258; *Auvergne dialect*, 71, 124
aux-, 197
Auxerre, 70, 228, 270, 284, 304, 353
Avaricum, 12, 20, 21, 22, 43, 69, 436
Aven, 305, 342
Avesnes, Avesnois, 377
Aveyron (department), 282, 395
aviation, 373, 437, 458; *see* aeronautical (industries), air (lines)
Avignon, 228, 230, 270, 285, 286, 315, 316, 318, 347
Avranches, 174; Avranchin, 87
Ay (region), 232
AYMARD, 476

BABELON, 483
Baghdad *(Caliph of)*, 92
bailiff, bailiwick, 160, 167, 186, 188, 284

536 Index

Baïocasses, -i, 174, 471
-*bais*, 61
Baïse, basin, 230, 362
BAKER (Alan R. H.), xxii
Balagne, 185, 226
BALAGUER, 319
BALIBAR, 494
Balkans, 7
Baltic (countries), 3, 76, 235, 241, 288
BALZAC, 426
BANCAL, 486, 494
banks, 18, 407, 408, 409, 410, 416–17, 439
Banque de France, 408, 409, 417, 438
Bapaume, 14, 249
Bar-le-Duc, 91, 109, 286, 368, 375; Barrois, 169, 173, 367, 368
Bar-sur-Aube, 368; Bar-sur-Seine, 368
BARBIER, 501
Barcelona, 110, 319; *counts of*, 117
Barcelonnette (valley), 387, 388
BARDES, 189
BARDOS (Jean-Pierre), 495
barley, 54, 433, 449
BARRAL, 493
BARRÈS (Maurice), 275, 324, 427
BARRIÈRE, 492, 493
Barzaz-Breiz, 191, 319, 323
Basates, 8, 12, 179
basilica, 41
BASIN (count), Basiniacus (Pagus), 165
Basin Parisian, *see* Paris Basin
Basle, 44, 270; diocese of, 173
Basque (country), **Basques**, 7, 75, 152, 178, 179, 311–13, 316–17, 319, 322, 328, 329, 333, 335, 368, 470; *Basque language*, 63, 123, 127, 279, 376, 478
Bassigny, 165, 203
BASTARD OF VAURUS, 107
Bastelica, Bastelicaccia, 185
Bastia, 161, 185, 186, 318
bastides, 144–7, 167
BASTIÉ (Jean), 451, 499, 502, 503
Bastille, 249
Bataves, 164
BATTIAU, 503
Battle-axe (culture), 4, 6
Bauges, 183, 212, 239
BAUTIER, 492
Bavaria, **Bavarians**, 80, 93, 94
Bayart (magic horse), 243
BAYET, 472
Bayeux, 42, 88, 174, 401, 471; region, 84
Bayonne, 7, 12, 179, 218, 270, 286, 376
Bazas, 8, 362; Bazadais, 179, 213
beads (pottery), 26
beans, green, 347
Béarn, 110, 131, 153, 161, 168, 219, 304, 311,
312, 314, 316, 318, 328, 340; Béarn valleys, 148, 152, 155, 179, 329
bears (performing), 388
Beaucaire, 271, 369
Beauce, 64, 65, 140, 176, 200, 203, 206, 241, 249, 250, 267, 292, 357, 493
BEAUJARD, 471
Beaujolais, 234, 263; Monts du, 181, 214, 309; *Beaujolais* (wine), 232
BEAUNE (Colette), 480, 481, 482, 483
Beaune, 182, 228, 256; *côtes de*, 232
Beausset (basin), 346
BEAUSSIRE (Emile), 343
Beauvais, 43, 176, 227, 231, 235; Beauvaisis, 164
BEC, 484
-*bec*, 83, 87
-*becque*, 61
BÉDIER, 487
beds, 294
beeches, 360
beehives, 130, 484
beer, 217, 231, 267
beetroot, 389
beggars, 388
bekrr, 87
Belfort, 437; gap, 6; *see* Alsace (Porte d'); region, 437
Belgae, Belges, Belgica, Belgium, 6, 10, 11, 20, 22, 23, 24, 35, 40, 59, 73, 94, 114, 123, 124, 172, 194–5, 196, 199, 251, 389, 416, 434, 474
BELLAUD DE LA BELLAUDIÈRE, 315
Bell-Beaker (culture), 4, 6
Belleville, 250, 263
Bellovaci, Bellovaques, 11, 12; Bellovacensis (Pagus), 164
BENDA, 481, 482, 483
Benedictines, 166
BÉNÉVENT, 498
BENOÎT (F.), 472, 473, 474
BENOÎT XII, 285
BENOÎT-LÉVY, 461
BENONI DROCHON (A.), 489
BENSON, 318
BÉRARD, 472
beret, 312, 368
Bergerac, 145, 229
BERGERON, 257, 491, 492
BERLUC-PERUSSAS, 325
Berre (Etang de), 346, 450
Berri, Berry, **Berrichons**, 96, 97, 161, 168, 170–1, 176, 202, 217, 234, 235, 256, 284, 298, 299, 304, 305, 309, 329, 334, 363, 369, 382, 495; *dialect*, 126; *duke of*, 286, 287
Berte aux grands pieds, 481, 492
BERTILLON, 492

Besançon, 33, 43, 66, 69, 96, 161, 164, 182; see Vesontio
BESNIER, 36
Bessin, 174, 175, 215
BÉTEILLE, 502, 503
BÉTHEMONT, 503
-beuf, 87
Beuve (valley), 8
Beuvray (Mont), 12, 43
Béziers, 7, 8, 12, 21, 51, 209, 318
Biarritz, 384, 439, 452, 461
Bibliothèque Bleue, 279
Bibracte, 12, 20, 43, 181
bicycles, 369
BIDAULT, 499
Bièvre, 249, 251
Bigerriones, Bigorre, 12, 179, 311, 312
bigouden (pays), 297, 301, 302, 305
bishop, bishoprics, dioceses, 70, 83, 86, 111, 162, 173, 176
Bituriges, 10, 11, 12, 20, 21, 23, 43, 69, 170, 171; —**Cubi**, 12; **Vivisci**, 7, 12, 69; biturica, 55
Black Death, 16, 146, 222
black wheat, 218; see buckwheat
BLANCHARD (Marcel), 493
BLANCHARD (Raoul), 490, 500, 501
Blavet(ria), 300, 381
Blesensis (Pagus), 165
Blésois, 197
BLOCH (G.), 486
BLOCH (Marc), 202, 477, 484, 487, 489
Blois, 197, 224, 270, 271, 286; house of, comté, 171, 176
BLOTTIÈRE, 477, 388
Bluebeard, 278, 279
bocage, 53, 134, 141, 148, 152, 202, 447; Bocage, 201, 206; Bocage Normand, 87, 149, 152, 175, 202, 215, 218, 387; Bocage Vendéen, 202, 340
BODEL (Jean), 97, 105
Bohemia, 7, 29
BOHÉMOND, 105
Boiates, Boïens, Boii, 7, 8, 10
Boisbelle (principality), 382
Boischaut, 171, 202, 354
Boisgelin (canal), 346
Bolbec, 403; river, 381
BOLLÈME, 494
Bon Marché, 418
-bona, 21
bonapartism, 343
Bondy (forest), 268
Bonifacio, 185
Bonn, 42
BONNAUD (P.), 63, 124, 126, 192, 477, 484, 485, 488, 489

BONNEFONT (Gaston), 290, 292
bonnets, 312
Bonnets Rouges (revolt), 189, 331
BONNEVAL, (chevalier de), 483
Bontran, 80
BOQUET (Véronique), 469
Bordeaux, 7, 12, 20, 30, 31, 37, 44, 47, 56, 69, 123, 126, 161, 177, 180, 212, 213, 215, 225, 228–9, 231–2, 235, 236, 241, 251, 255, 259, 270, 271, 273, 274, 286, 314, 318, 328, 329, 345, 361, 362, 376, 384, 389, 402, 404, 410, 434, 440, 450, 452, 455, 459, 478; **Bordelais**, 191
Bornholm (island), 76
BORVONE, 42
bottles, 232, 267
Bouches-du-Rhône (department), 282
BOUCICAUT (Aristide), 418
BOUGLER, 499
Bouillon (counts of), 88
BOULARD, 497
Boulogne, 20, 37, 84, 161, 172, 251, 270, 272, 405, 414; see Gesoriacum, Portus Itius
Boulogne (Bois de), 268, 458
Boulonnais, 161, 167, 171, 283, 369, 479; Boulonnais (breed of horse), 358
BOUQUET, 480
Bourbon, 42
Bourbonnais, 129, 142, 161, 176, 177, 181, 305, 328, 334, 356, 364; see Sologne bourbonnaise
BOURDIER, 469
BOURDOT DE RICHEBOURG, 487
Bourg, 309
Bourges, 43, 69, 161, 170, 171, 177, 270; see Avaricum; kingdom of, 100, 107, 109, 283, 284, 285, 363; election, 160
BOURGIN (H. and G.), 370, 491
bourgs, bourgades, 146
BOURGUIGNON, 501, 502
BOURQUELOT, 479, 480
BOUTRUCHE, 484
Bouvines (battle of), 106
Bouzanville (abbey of), 478
BOVON DE COMMARCIS, 190
BOWEN, 82
braid, 372
BRAUDEL, 469, 471, 472, 483, 501, 502, 503
BRAVARD, 503
Bray (Pays de), 140, 330, 377
Brazil, 474
bread, 54, 218, 433
BRÉAL (Michel), 324
breeches, 297, 299
Breitstreifenfluren, 139
BRÉMONTIER, 361
Brenne, 140, 170, 176

Brenner (pass), 26
Bresle (valley), 87, 217
Bresse, 113, 129, 142, 180, 214, 283, 298, 304, 309, 364
Brest, 189, 211, 236, 302, 305, 382, 402, 452
Brétigny *(Peace of)*, 106
BRETTE, 484
Briançon, 388
Briare *(canal de)*, 230, 263
brick, 460
Brie, 140, 170, 201; *cheese*, 264, 377, 378
Brière, 207, 294, 298, 466
Briga, 114; *briga*, 201; *-briga*, 21
Brioude, 401; *pays*, 283
Brisach, 114; *see* Mons Brisiacus, Neuf-Brisach
British Isles, Great Britain, *influences*, 6, 9, 18, 24, 29, 30, 37, 44, 55, 84, 85, 89, 251, 297, 303, 416, 434, 464; *see* England
Brittany, **Bretons**, 5, 6, 29, 53, 71, 82, 85, 86, 87, 88, 89, 97, 100, 117, 139, 143, 146–7, 149, 152, 153, 161, 163, 165, 168, 189, 206, 208, 218, 228, 264, 283, 288, 294, 299–302, 303, 305, 309, 310, 311, 316, 319, 322–3, 325, 328, 331, 333, 334, 340, 341, 342, 355, 357, 376, 382, 383, 385, 389, 390, 395, 396, 400, 414, 428, 430, 433, 445, 448, 451, 452, 455, 479; *Breton (language)*, 82, 86, 88, 123, 279, 301; *see* Armorica
Brittonic language, 6, 89
briva, 21
broad beans, 224
BROC, 469
BROCHE, 473
BROËNS, 477
Bronze Age, 6, 17, 26, 29, 137
BROSSARD, 503
Brotonne (forest), 466
BRUN, 494
BRUNET (Pierre), 267, 447, 476, 493, 502
BRUNET (Roger), 490, 500, 502, 503
BRUNHES (Jean), 83, 193, 195, 272, 484, 486, 488, 494
BRUNO (G.), 290, 291, 292, 495
BRUNOT, 484
Brussels, 241, 270, 455, 459
BRUTAILS, 129, 484, 488
BRUYELLE, 503
buckwheat, 85, 118, 218, 433
BUFFON, 420, 436
Bugey, 113, 283
building (workers), 323, 389; *see masons*
BUR, 486
Burdigala, 7, 36, 44, 69; *see* Bordeaux
Burgondes, Burgundian Kingdom, Burgundia, 66, 71, 76, 77, 79, 80, 81, 112, 181

Burgundarholm, 76
Burgundy, **Burgundians**, 44, 55–6, 61, 68, 76, 77, 81, 94, 96, 97, 100, 103, 112, 132, 140, 142, 143, 161, 163, 164, 168, 172, 173, 180, 181, 192, 193, 203, 214, 229, 231, 263, 333, 352, 376, 389; *see* **Burgondes**; *canal*, 271; *duke of*, 181, 228, 281, 286, 287; Porte de, 173; Gate(way), 44, 165, 173; *see* Belfort (gap), Alsace (Porte d'); Montagne Bourguignonne, 364, 434; *wine*, 231, 232
BURIN, 471
BURKE (Edmund), 284, 494
BUSQUET, 473
BUSSET, 474
BUTLIN, 471
butter, 264, 344, 355

Cabillonum, 36, 43
cadastral survey (first), 209, 220, 236, 365
Cadurci, 12, 179
Caen, 61, 140, 170, 202, 215, 235, 270, 369, 450
CAESAR, 9, 11, 13, 16, 21, 22, 23, 38, 39, 41, 42, 55, 198, 469, 471, 472, 474, 488
Caesarodunum Turonum, 42, 44, 69; *see* Tours
Caesaromagus, 36
Cahors, 43, 145, 179, 229
Caisse des Dépôts et Consignations, 417, 462
Calais, 116, 117, 219, 270, 271, 405; Calaisis, 283, 490
Caleti, 10, 12, 174, 381; Caletus (Pagus), 163
California, 322, 347
CALLOT, 475
CALMETTE, 483
Calvados (department), 139, 206, 282, 374; *calvados (brandy)*, 374, 376, 378
Camaracum, 36, 50
Camargue, 84, 85, 203, 387; Petite, 349; Camarica insula, 203; CAMARS (Annius), 203
Camberlots, 389
Cambrai, 13, 14, 15, 79, 108, 136, 270; Cambrésis, 108, 249
camels, 73, 362, 478
Camembert *(village, cheese)*, 374, 377, 378
CAMERON (Rondo), 418
Camisards *(war)*, 338
campagne, Campagne, 140; — de Caen, 140, 175, 201–2, 354; *campania*, 140, 170, 202
CAMPS, 4, 469
campus, 52, 200, 476
canals, 116, 230, 263, 346, 347, 363, 366, 382, 416
candles, 371
Cannes, 404

Cantal, 177, 199, 205, 242, 387, 388, 390, 395, 396, 409; department, 282, 258; Châtaigneraie of, 220, 222, 397
Canti popolari corsi, 318
cantons, 188, 283, 285, 324, 328, 337, 407
Cap Corse, 185, 226
Capetiens, *early years*, 95–9, 108, 132, 171, 181, 252–3, 487; *see* HUGH CAPET
car making, 373, 437, 441, 442, 450, 452, 468
CARACALLA *(edict of)*, 55, 70
Carcasso, Carcassonne, 7, 12, 367, 478; comté, 179
CARCOPINO, 473
cardial (pottery), 4, 5
Carhaix, 301
CARIBERT I, 80, 252; II, 75, 80
CARLIER, 499
Carlist Wars, 319
Carmaux, 369, 425
CARMONA, 483
Carnutes, 11, 12, 15, 22, 24, 43, 292, 471
Carolingian, *texts, period*, 52, 60, 68, 70, 71–2, 75, 76, 77, 83, 84, 87, 90, 94–5, 96, 97–8, 105, 109, 110, 132, 142, 159, 163, 165, 179, 243, 487; *see* CHARLEMAGNE
CARON, 501, 502
Carpentras, 347
carpets, tapestries, 234, 372
CARRÉ, 502
carriages, 372, 478
CARRIÈRE, 501
carrots, 224
Carte géographique des postes ..., 271
Carthage, **Carthaginians**, 27, 29
cartwrights, 155, 157
Cassiterides, 30
Castagniccia, 202, 219, 222
CASTELA, 499
castellum (-a), 20, 251
Castelnau, 144, 145–6; *castelnaux*, 146
Castelnaudary, 166
CASTER, 490, 491
CASTEX, 503
Castille, 3; *Castillian (language)*, 123
castle(s), 135, 166, 168, 187, 232, 269, 486
Catalonia, 110; *Catalan (language)*, 123; *(movement)*, 319
CATHERINE DE MEDICI, 225, 286, 314
catholicism, **Catholics**, 94, 99, 152, 316, 323, 331, 334, 335, 342, 421, 429, 443
Cattes, 79
cattle, 217, 310, 353–4, 355–9, 364
CATULLUS, 470
Causse(s), 179, 200, 202, 239, 353, 355, 387, 395
Caux (Pays de), 87, 163, 174, 235, 330, 354, 397

CAVAILLÈS, 158, 270, 474, 483, 493, 499
Cavaillon, 7, 12, 347
cavale, 129, 131; *see mare*
CAVALLI (Marino), 244
Cavares, 8, 41
Cebenna, 198, 488
CÉLESTE, 503
Celtic (province), 39, 40, 44
Celts, *Celtic languages*, 6–7, 8–9, 13, 17, 21, 22, 24, 32, 35, 46, 53, 89, 125, 130, 134, 168, 239, 277–8, 474; **Proto-**, 6
Celtic fields, 18
CELTILL, 472
Celto-Ligurians, 21, 28, 147
celtomania, 319, 496
Cemenelum, 183
cemeteries, 20, 60, 67, 108, 137
Cenomanni, Cénomans, 10, 12, 175
central heating technicians, 393
centuriations, 47–8, 51, 70
Cerdagne, 110, 383
cereals, cereal cultivation, 49, 54, 55, 92, 129, 136–7, 140, 141, 147, 148, 169, 174, 175, 178, 179, 216, 217, 226, 241, 265, 346, 349, 353, 354, 355, 364, 387, 389, 392, 433, 434, 447, 448; *see barley, buckwheat, maize, millet, oats, rye, wheat*
Cette, 290; *see* Sète
Cévennes, 34, 55, 198, 219, 222, 239, 338, 383, 387, 395, 396; *national park*, 466
Cézallier, 177, 388
Chablais, 183, 393
Chablis, *wine*, 228, 232, 263, 353
CHABOT, 489
CHAHO (Augustin), 319
chaillot, 250
Chalcolithic, 5
chalk, chalky soil, 53, 171
Chalon-sur-Saône, 37, 43, 79, 81, 181, 270, 291, 409, 454
Châlons-sur-Marne, 227, 256, 270; *area*, 140, 146, 235
Chamavi, Chamavus (Pagus), 79, 94, 164
Chambéry, 183, 211, 212, 371, 409
CHAMBRELENT, 362
Chambres de Réunion, 173
CHAMLA, 485
Chamousey *(abbey of)*, 478
Champagne, Champenoise (Plain), 161, 170–1, 176, 192, 201, 202, 214, 235, 258, 265, 272, 348, 352, 364–5, 448; —Humide, 108, 140, 170, 200; 'pouilleuse', 364, 448; southern, 166; *dialect*, 124; *wine*, 231–2; *champagne*, 149, 232
Champagne Berrichonne, 140, 170, 189, 202
Champeigne Tourangelle, 176, 202

chansons de geste (period of), 96, 97, 101, 102, 105, 132, 190–1, 192, 242
Chanson d'Antioche, 481
Chanson de Roland, 96, 97, 101, 481, 482
Chanson des Narbonnais, 254
Chanson des Quatre Fils Aymon, 198, 202, 243, 488
Chanson des Saisnes, 97
Chantilly, forest, 268, 269
Chaouilley, 67
CHAPELLE and BACHAUMONT, 286, 287
CHAPELOT, 484
CHAPUIS, 503
CHARDONNET, 502, 503
Charente, regions, 84, 228, 233, 382, 434; department, 282, 343, 344, 349, 387;
—Inférieure, Maritime (department), 282, 343, 349, 387
Charenton, 249, 250
CHARLEMAGNE, kingdom of, *time of*, 92, 105, 109, 160, 202, 252; *see* **Carolingian**
CHARLES I OF GONZAGUE, 382
CHARLES III (OF LORRAINE), 382
CHARLES V, 228, 315
CHARLES VII, 109
CHARLES IX, 286, 287
CHARLES OF ORLÉANS, 107
CHARLES THE BALD, kingdom of, *reign of*, 90, 91, 93, 95, 96, 108, 190, 252
CHARLES THE BOLD, 105, 112
CHARLES THE SIMPLE, 96
CHARLES-BRUN (J.), 325
Charleville, 382
Charolais, 356; *(breed of cattle)*, 354, 358
CHARTIER (Alain), 107, 482
Chartres, 15, 22, 44, 252, 270, 271, 471
Chassean, **Chasseans**, 5
Château-Thierry, 166, 270
Châteaulin, region, 301, 305
Châteaumeillant, 283
Châteauneuf-du-Faou, 301
Châteauroux, 166, 171
CHÂTELAIN, 387, 500, 501
Châtillon, 166; -sur-Seine, 368; Châtillonnais, 32, 354
CHAUCHARD, 418
CHAUME, 479, 486
CHAUMIÉ (Jacques), 484
Chaumont, 368; *pays*, 166
CHAUNU (Pierre), 186, 477, 481, 482, 483, 487
CHAUSSÉE (de la), 484
CHAUVET, 498
Chavignol, *crottin*, 374, 377
Chaville, 250
cheese, 264–5, 353–6, 358, 374, 376–8, 449
chemical *(industries)*, 370, 371, 372, 373, 437, 441, 450

CHÉNON, 486
Cher, 305; (department), 282, 283, 284, 374
Cherbourg, 24, 383
cherries, 54, 347, 374, 376
Cherusques, 13
chestnut groves, trees, 219–22, 433
CHEVALIER (Louis), 257, 492, 493
CHEVALIER (Michel), 485, 490, 491, 499, 502, 503
CHEVALLIER (Raymond), 470, 474, 475
Chevreuse (forest), 268
chez-, 138, 139, 141
CHIFFRE, 139, 485
CHILDEBERT I, 80; II, 99
CHILPERIC, 80
chimney sweeps, 256, 388
chocolate, 232
Cholet, 225, 344, 405; Choletais, 344, 428
CHOLLEY, 492
chouannerie, 340, 341, 383, 428
CHOUQUER, 475
CHRÉTIEN DE TROYES, 102
christianity, **Christians**, 95, 98, 99–100, 103–4, 107, 152, 166, 191, 328, 330, 331, 421, 467; *see bishoprics, parishes*
CICERO, 473, 476
cider, 217, 374, 376
Cimiez, 183
cinema, 373, 441, 458; *cinematographical industry*, 437
Cirnea, 318
Cisalpine *see* Gaul
Cistercians, 166–7
city (Gallic and Gallo-Roman), 11, 16, 40, 42, 43, 46–7, 162, 164, 167–8, 169, 172, 175, 199
Claire, 381
Clamecy, 329, 331
clans, 297; *clan conflicts*, 317
claret, 232
CLARETIE, 321
CLARK, 472, 473
classicism, classical period, culture, 134, 274, 277, 279, 459
CLASSICUS, 45
CLAUDIUS, 37, 42
CLAUDE DE FRANCE, 286
CLAVAL (Paul), xxii, 487
CLAVEL, 470, 473, 475, 488
clay, clayey regions, soils, 53, 139, 140, 169, 170, 201, 223, 237, 269, 309, 354, 356
Clermont-Ferrand, 43, 99, 161, 177, 271, 371, 401, 441, 474
Clermont-l'Hérault, 383
clock making, 239, 367
CLODOMIR, 80
CLOTAIRE I, 80, 81, 252; —II, 81, 252

cloth, manufacture, merchants, 388
CLOUT (H.), 157, 213, 236, 404, 469, 489, 490, 491, 498, 499, 500, 501, 502
CLOVIS, 75, 78, 80, 97, 99, 100, 251, 252; —II, 83
Cluny *(abbot of)*, 98, 105; **Cluniacs**, 166
coal, coalfields, 116, 271, 273, 363, 365–6, 369, 370, 383, 404, 405, 416, 418, 432, 434–5, 440, 441, 442, 444, 450
cobblers, 289
cock, 9, 470
COCULA, 491
coffee, 232
Cognac, 230, 233; *cognac*, 230, 232–3, 376
COGNACQ (Mme), 418
COHEN, 491
coiffes, 297, 298, 299, 302
Cointreau, 376
COIRAULT (Patrice), 493, 494
COIRAULT (Yves), 494
COLBERT, 234, 241, 271, 372, 382, 383
COLBERT DE BEAULIEU (J.-B.), 487
COLIN, 499
collars, 297
Collège de Rhétorique, 314
Cologne, Colonia Agrippina, 36, 37, 42, 44, 80, 81, 172, 256, 455
Colomba, 318
Colomban, Saint, 201; see *Life of*
coloni, 70
Colonia Traiana, 96
COLUMELLA, 476
Comédie des Proverbes, 190
comitatus, 160
Commenicus (Pagus), 179
Commentry, 369
Comminges, 179, 280
Common Market, 415, 444, 447, 448
commune(s), 143–4, 188, 324, 328, 345, 397, 461; *communal land*, 349
Commune (of 1871), 341, 427
communism, communist, 334, 430, 431
Compagnie, des Indes Orientales (East India Company), 382
Compagnon du Tour de France (Le), 291
compagnonnage, 288, 424
competitions (animal), 357
Compiègne, 270
Comtat Venaissin, 8, 161, 182, 347, 387
COMTE (Auguste), 324
comté, 160, 162, 165, 176
Comtesse d'Escarbagnas (La), 274
Comtois (dialect), 126
Condate, 36, 69
Condé *(princes of)*, 269
Condom, Condomois, Condomagus, Condomum, 143, 179

CONDORCET, 281
confectionery, 409
conscripts, 155; *conscription*, 341
conservatives, 325, 328, 331, 333, 335, 344, 421, 427, 428, 429
Considérations sur le gouvernement ancien et présent de la France, 281
Consoranni, Consorani, 12, 179
CONSTANCE CHLORE, 164
Contes de Perrault, 277, 278
contredanse, 303–5
Convenae, 12; **Convenarum** (civ.), Convenicus (Pagus), 179
COORNAERT, 289, 494
copper, 234, 371
COQUILLE (Guy), 188
Corbeil, 513; *Treaty of*, 110
Corbières, 328, 360
Corbilo, 7, 20, 30, 470
Corded Ware (pottery), 6
Coriosolitae, Coriosolites, 12, 83, 85, 86
Cornouaille (Brittany), 298, 299, 301, 305
Cornwall, 29, 31, 85
Corrèze (department), 282
Corsica, 27, 131, 161, 184–6, 202, 219, 222, 226, 283, 317–18, 322, 356, 389, 395, 414, 466
Cortaillod *(civilization of)*, 5
Corte, 186
costume, 296, 302, 307, 309
Côte d'Azur, 347, 384, 389, 406, 414, 455, 461
Côte-d'Or, 53; (department), 282
Côte-Rôtie, 55, 230
Cotentin, 84, 87, 152, 174, 175
Côtes (plateau), 474
Côtes-du-Nord (department), 86, 282, 447, 449, 489
cotton (industry), 235, 366, 370, 372, 434, 436; *cultivation of*, 362
Counter-Reformation, 152
country dance, 303
country towns, 305
COURSELLES (Paul de), 292
COURSON (Aurélien de), 319
Court-, -court, 63, 64, 142
court (royal), 187, 225, 252, 254, 267, 269, 273, 297, 298, 303, 372
courtly romance, 102, 132, 243
Courtrai *(Treaty of)*, 115
Coutances, 174, 256
COVILLE, 478, 482, 494
CRAMPON, 487
CRAPONNE *(canal)*, 346
Crau, 204, 205, 346; *Cravus campus*, 204
Crédit Lyonnais, 409
CRESTON, 297, 301, 495

Creuse, 228; department, 282, 283, 389, 391
Crop-marks, 18–19
crop rotation, 136, 139, 364
Croquants (revolt), 189
Crozon, peninsula, 301, 302, 305
CRUBELLIER, 489
Crusades, 99, 105, 107, 132, 253; *see* **Albigensian** *(crusade)*
CUISENIER, 495
culture, cultural geography, 5, 102, 131–2, 153–4, 192–7, 243–4, 275, 276–81, 284, 291, 293–324, 326, 336, 467–8
CURTIUS, 481
customs posts, 116; *dues*, 241
CYRANO DE BERGERAC, 133

DAGOBERT, 80, 81, 94, 252
DAGUERRE, 458
DAINVILLE, 157
dairy farming, dairy produce, 261, 263, 344, 353–5, 356, 358, 434, 448, 449
dal, -dale, -dalle, 83, 87
Dam-, 142
dance, 303–7, 313
DANJON, 372
DANTON, 114
Danubian culture, 4, 5
Daoulas, 299, 301
DAREMBERG, SAGLIO and POTTIER, 36
DARMON, 494
D.A.T.A.R., 452
DAUBENTON, 357, 499
DAUDET (Alphonse), 133
DAUMAS, 502
Dauphiné, 100, 132, 143, 161, 167, 183, 211, 214, 242, 258, 328, 330, 340, 370, 399; *dauphinois (dialect)*, 124
DAUVIGNY, 190
DAUZAT, 65, 477, 479, 484, 488
DAVEAU, 483
Dax, 42, 362
de-, 197
Deauville, 384, 439
Débats (Les), 325
DEBESSE-ARVISET, 499
Decazeville, 369, 383, 404
decentralization, 336, 406, 451–2, 468, 497
DECHELETTE, 473
Deffence et Illustration de la Langue Françoyse, 104
DEFFONTAINES (P.), 193, 272, 474, 478, 488, 490
DEHN, 472
DEIMIER (Pierre), 316
DELAFOSSE, 489
DELAMARE, 166
DELARUE, 493

DELASPRE-PARQUET, 491
DELFOSSE, 500
Délices de la France (Les), 189
DELORME (Sixte), 292
Delphi, 24
DEMANGEON, 154, 195, 272, 487, 491, 493, 499
demography, 16, 47, 66, 68–9, 73, 135, 136, 137, 141, 144, 147–8, 149, 152, 195, 196, 222, 233, 259, 343, 364, 385, 390–2, 393, 396–405, 415, 457
DEMONET, 502
DENIS, 495
Denmark, **Danish**, 87, 88, 91, 92, 288, 449
department stores, 393, 418
departments, departmental (organization), 163, 281–5, 291, 324–5, 343, 395, 408, 494
DERRUAU, 485, 488
DESAUCHES, 158
DESBORDES (J. M.), 471, 472
DESCHAMPS (Eustache), 201
DESPREZ, 495
Deux-Sèvres (department), 282, 340, 343
DEVAILLY, 486, 487, 490
DEVÈZE, 499
Dévoluy, 397
DEYBER, 471
Diablintes, 10, 12, 13, 23, 175
dialects (disappearance of), 280; *see* patois
Dialogue des Nymphes, 314
Diaroritum, 36
Dictionnaire de l'Académie, 105
DIDEROT, 281
Dieppe, 166, 209, 211, 290
DIETRICH, 500
Dijon, 108, 161, 181–2, 256, 270, 288, 384
dioceses, see bishoprics
DIODORUS SICULUS 31, 472, 473
DION (Raymond), 500
DION (Roger), xxiii, 15, 31, 56, 91, 92, 93, 139, 248, 251, 469, 470, 472, 473, 474, 476, 480, 481, 483, 484, 485, 486, 487, 490, 491, 492, 493, 499, 503
DIO CASSIUS, 474
DIONNET, 374, 500
Dionomachia (La), 318
districts, 281, 283, 285
Djabal al-Fulal (Qilâl), 82, 84, 85, 481
DOISY, 163
Dol, 86
dolerite (axes), 5
dolmen, 19
Dom-, 142
DOM PÉRIGNON, 232
Dombes, 161, 167, 180, 283, 364, 409
domestic servants, 208, 231, 393, 405
DOMET, 499

Domfront, 82; region, 87; Domfrontais, 375
Dominus, 142
Domitia (via), 37
DOMITIAN, 55
Domrémy, 142
DONDAINE, 484
DONTENVILLE, 491
Donzac, 146
Dordogne (valley), 126, 229; department, 282, 335, 349
Douarnenez, 301, 302
Doubs (department), 115, 282; Haut-, 428
Dover, 107
dragons, 343, 491
Dreux, 270
Dreyfus (Affair), 421
Drôme, 8, 183; department, 282, 304, 360
DROUOT, 473
Druids, 23–4, 45, 292
DU BARTAS (Salluste), 314
DU BELLAY, 103, 104, 112, 191, 255, 483
DU BUS (Gervais), 482
DUBOIS (Maurice), 499
DUBOIS (P.), 502
DUBY, 478, 482, 484, 485, 502, 503
DUFOURNET, 495
DUGRAND, 489, 501, 503
DUMAS (Alexandre), 109, 133, 483
DUMONT (Louis), 491
DUMONT (Paul), 151, 157, 485, 502
DUMONT (René), 503
dunes, 359, 361
Dunes, 146
Dunkirk, 116, 236, 256, 450
DUNNING, 470
-dunos, -dunum, 21, 22
DUPÂQUIER, 486
DUPARC, 478
DUPEUX, 500, 501
DUPIN, 485
DUPONT-FERRIER, 486
DUPOUY, 484
DUPRONT (Alphonse), 494
DUPUY, 487, 491, 493
Durance, valley, 8, 184, 346, 347, 397, 399
DURAND, 459, 487
DURAND (Alfred), 501
DURAND (Yves), 486, 487, 488, 489, 491, 494, 495, 498
Durham (breed of sheep), 357
DURKHEIM, 428
Durocasses, 36
Durocortorum, 36, 69
DUVAL, 12, 470, 472, 473, 474, 492
dwarves, 278
dyes (vegetable), 347

-é, 86
EARDEN, 491
earthenware, 294
EAST, 490
East, 29, 54; *see* Levant
east (of France), 108, 213, 223, 358, 406, 428, 436, 440, 449
Ebro (basin), 7, 110
Eburones, Eburons, 12
Eburovices, 10, 12, 23, 174
Ecole des Beaux-Arts, 458, 459
Ecole Nationale d'Administration, 443
Ecole Polytechnique, 155, 157, 292
Ecolisma, 43, 176
Ecrins *(national park)*, 466
EIFFEL, 458
-ein, 63
Elbe (valley), 9, 26, 60, 95
Elbeuf, 209, 211, 403
elections (political), 160, 331, 334, 336, 337, 423, 430ff.
Elections, pays d', 188, 284, 327, 330, 486
electrical (industries), 450
electro-chemical, electro-metallurgical plants, 370
ELHAÏ, 211, 489
Elimberris, Elimberrum, 7, 36
ELISABETH (queen of Spain), 286
Elite, élites, 155, 157, 279, 322, 418ff., 443, 457, 464
Elliant region, 301
Elne, 7, 184
Elosedienses, 486
Elusates, 12
'embouche' (= rich pasture) 354
Embrun, 78; Embrunais, 387
EMMANUELLI, 133
Empire (Romano-Germanic), *emperor*, 76, 90, 100, 108, 111, 117; Empi, 109; *see* Germany
Empire (First), 274, 372, 385, 387; *style, mode*, 295, 298, 302, 435, 436, 438, 468; *see* Napoleon I
Empire (Second), 153, 259, 280, 291, 328, 349, 354,355, 359, 363, 373, 384, 409, 417–18, 432, 440, 441, 442; *see* Napoleon III
-en, 166, 197
en dro, 307
En Route avec l'oncle Mistral, 292
enclosures, 18, 52, 71–2, 134, 137, 141, 148, 149, 151, 202
Encyclopédie, 281, 364
-enge, 61
engineering, 441
England, **English**, 60, 87, 89, 98, 105, 106, 108, 109, 146, 227, 228–9, 231, 232, 256,

England (*cont.*)
288, 312, 348, 371, 373, 392, 415, 432, 434, 438, 449, 457
English Channel, 5, 6, 37, 113, 117, 172, 301, 305; *coast*, 461; tunnel, 456
ENJALBERT, 491, 499
ENLART (C.), 193, 195
Enlightenment, 116, 281, 329, 334, 421
-ens, 61, 63
Ensérune, 21
Entre-Deux-Mers, 126
Entremont, 21, 43
Epernay, 97, 227; pays d', 166
Epernon *(duc d')*, 314
Epinal, 290, 375
equa, 129, 131; see *mare*
equoranda, 15, 471
ERMOLD THE BLACK, 93, 480
Escaut, valley, delta, 5, 84, 90, 96, 108, 116, 223, 227
ESMEIN, 486, 487, 491
Essai sur la constitution et les fonctions des assemblées provinciales, 281
ESTIENNE (Charles), 167, 201, 267, 486, 488
ESTOURBEILLON (marquis d'), 325
Esuvii, 12, 174; *see* **Sagii**
Etampes, 250
Etangs, Pays des, 354
Etats, pays d', 163, 188, 327, 330, 486
ETCHEVERY (Michel), 496
Etruria, **Etruscans**, 27, 29
Etude sur les torrents des Alpes, 359
Eu (region), 217
Eumenius (pseudo-), 52, 55
Eure, 14; (department), 87, 282
Eure-et-Loir (department), 282
European Community, see *Common Market*
Evreux, 174
Ewig, 96, 481
exhibitions international, 373
EYCHART, 474

FABRE (Georges), 360
fairs, 166, 186, 206, 214, 312, 407
Falaise, 166, 401, 406
FALC'HUN, 480
family (structures) 331, 333, 336, 424–5
farandole, 304, 311
farmers, farming, 17, 52, 134, 135–7, 140–1, 148, 149, 153–4, 157, 265–7, 310, 343, 355–9, 363, 428, 444
FAUCHER, 485, 490
Faucigny, 388, 392
FAURE, 501, 503
FAURIEL, 132, 318
FAVIER, 492
FAVIÈRE, 495

FEBVRE (Lucien), 93, 117, 480
Fécamp, 209, 211
federalism, 325
Féderation Régionaliste Française, 325
FEL, 157, 185, 220, 485, 490, 491, 499, 500, 503
Félibre (movement), Félibrige, 124, 319, 321, 325
FELIX OF GERONA, SAINT, 77
FÉNELON, 281
FERRERO (Léon), 493
FERRY (Jules), 292, 336, 424; *see Laws*
fertilizer, 224, 447; *chemical*, 365, 468; *human*, 224; *urban*, 265; *vegetable*, 361, 365
FEUCHÈRE, 489
feudalism, feudal times, feudal estates, 135, 141, 143, 147, 167, 168, 169, 173, 335, 467
FÉVRIER, 478
fibula, ansated, 60, 62
Fidentiacus, FIDENTIUS, 179
FIERRO-DOMENECH, 469, 478, 483, 486, 494
FIÉTIER, 486
Fins, *fines*, 28, 41
Finistère (department), 10, 130, 131, 282, 307, 430, 449
FLACH, 475
FLAMENT, 499
Flanders, **Flemings**, 61, 91, 100, 104, 117, 161, 168, 171, 191, 204, 223, 227, 228, 256, 291; *Flemish (language, cultural area)*, 61, 103, 317; *flemish (breed of cattle)*, 358
FLATRÈS, 478, 479, 485, 487, 488, 489, 502, 503
flax, 224
FLÉCHIER, 189, 487
-fleur, 87, 479
FLEURIOT, 82, 479, 480
FLEURY (cardinal), 239
Fleury-sur-Loire *(abbey of)*, 98; *see* Saint-Benoît-sur-Loire
Florence, 259
FLORUS, 15
FLORY, 497
flour-mills, 240, 371
flowers, flower cultivation, 261, 263, 347, 389
FOHLEN, 499, 502
FOISIL, 487
Foix, 7, 12, 161, 312; comté, 161, 167, 179
FONCIN (P.), 325
Fontainebleau, forest, 268, 269
FONTENAY (de), 357
food, 215, 440, 442; *processing industries*, 371, 373, 374
FORESTIER, 462
forests, 13, 68, 71, 85, 236, 268–9, 359, 364
forêts de la France (Les), 359

Forez, 143, 230, 387; Monts du, 177, 214; plain, 109, 180, 181
Fort-Louis, 114, 382
forum(a), 20, 47
Forum, Julii, 36; — Segusiavorum, 36
FOSSIER, 475, 477, 484, 485
Foucon de Candie, 190
FOUILLÉE (Mme Alfred), 291
four rivers, the, 91, 100, 108–9
FOURASTIÉ (Jean), 415
FOURNIER, 475, 476, 477, 478, 484, 486
Fourvière, 44, 45
FOUTREIN, 503
Français du Nord et Français du Midi, 321
France (Pays de, plain), 136, 203, 249, 267, 268; — (Ile-de-), see Ile-de-France; *wines of* 227
France (La), 413
FRANCE (Anatole), 482, 493
Franche-Comté, 5, 10, 61, 113, 117, 142, 161, 164, 182, 280, 282, 288, 289, 328, 329, 333, 376, 387, 390
Franci, 94, 96, 98, 99
Francia, 81, 83, 94–5, 98, 99, 117
Franciscans, 100
Francisci, 96
Franco-Provençal (language), 123, 125
Franco-Prussian war, 153, 324, 427, 440
FRANÇOIS I, 273
Franconia, **Franconians**, 94, 95
francus, 94, 105
frankish (dialect), 71, 124, 131
Franks, 59, 61, 65, 66, 75, 76, 79, 80–1, 84, 90, 92, 93, 94–5, 96, 97, 98, 99, 100, 105, 134, 164, 177, 318, 335, 467, 479
FRÉDÉGAIRE (pseudo-), 98, 478
Fréjus, 41, 270
FRÉMONT (Armand), 457, 501
French (breeches, style), 298–9; *language* 103–4, 122ff., 131, 133–4, 168, 279ff., 298–9, 301, 310–13, 314, 315, 322, 421
FREYCINET *(plan)*, 273
Fribourg, 114
Friesland, 90; *Frisian cloth*, 92; *Friesian (breed of cattle)*, 358
FROISSART, 123
frontiers, 13–15, 35, 40–1, 52, 58–60, 86, 87, 100, 108–17, 133–4, 165, 168, 169, 173, 181, 182, 284, 382, 483
fruit, cultivation, 54, 92, 216, 225, 347, 434, 448
FURET, 151, 154, 485
furniture, 293–6, 309ff., 372
FUSTEL DE COULANGES, 48, 51, 322, 475, 478

Gabali, 12
GACHON, 500, 501

GADILLE, 499
Gai Savoir, 314
gaigneries, 139, 149
Gaillac, *vineyard*, 55, 229
GAIN, 487
Galates, Galatia, *Galatus*, 7, 9, 31
Galicia, 30
GALLERAN (capitaine), 341
GALLET DE SANTERRE, 472
Galli, 98
Gallia, 9, 34, 98
Gallicus, 106
GALLOIS, 480, 481, 488, 489
gallomania, 99
Gallo-Roman (linguistic group), 61–2, 64, 67, 76, 89, 123, 125, 162, 166, 167
GALTIER, 499
GAMBETTA, 329, 336
Ganges, 209, 339
GANSHOF, 93, 480
Gap, 388, 407
GARAT, 322
GARCIA (Carlos), 496
GARCIN (Eugène), 321
Gard (department), 282, 349, 440
gardens (vegetable), 216, 219, 261; *cities*, 461
Garden of France, 104, 293
Gargantua, 243, 491
GARNIER (Bernard), 500
GARNIER (Charles), 458
Garonne, valley, 3, 7, 8, 29, 31, 39, 73, 75, 124, 126, 127, 131, 271, 329, 335, 340, 343, 347, 362, 384; mid-Garonne region, 328, 347, 387, 389; (Haute-), department, 114, 283
GARROS (Jean de), 314
GARROS (Pey de), 314
gas, 444
GASCON, 489
Gascons, Gascony, 42, 105, 131, 133, 146, 161, 168, 178–9, 189, 312, 314, 316, 320, 389; *Gascon (dialect)*, 124, 127; *black-muzzled Gascon (breed of cattle)*, 358
-gate, 84
Gâtinais, 140, 142, 165, 170, 202, 249, 364
Gâtine, 148, 176, 202, 249, 304
GAUDRON, 473
Gaul, **Gauls**, *Gallic and Gallo-Roman periods*, 8ff., 22ff., 58ff., 69, 70, 71, 73, 75, 77, 80, 84, 85, 86, 93, 94, 98–9, 112, 113, 117, 122, 123, 130, 132, 133, 137, 159, 162, 180, 275, 426, 467, 470, 473; Cisalpine, 7, 9, 10, 34; Transalpine, 9, 34; Gauls (diocese of), 40, 73; Three Gauls, 39–40, 44, 45, 48, 163
GAUTIER (Léon), 105, 481, 482, 484, 487
GAUTIER (Marcel), 72, 478, 489

GAUZLIN, 98
Gaves (terraces, valley), 179, 304
GAY, 485
Gaydon, 190
Gélise, 362
Genabum, 12, 21, 43; *see* Orléans
Gendrey (region), 68
généralité, 161, 175, 177, 281, 284
Geneva, Genava, 8, 36, 157; city, 76, 79, 183, 239, 406, 409, 435, 454; Genevois, 183
Geneva, Lake, 34
Genoa, *influence of*, 185–6, 222, 226
gentian, 378
Gentilome Gascoun (Lou), 314
GEORGE, 501
Gergovia, 12, 20, 21, 43, 474
Germania, **Germans**, *germanic (invasions, influences, language)*, 12, 13, 15, 35, 36, 40, 42, 44, 45, 46, 56, 58, 60, 61, 63–4, 67, 68, 73, 79, 80, 83, 84, 93, 94, 95, 98, 99, 111, 122, 123, 125, 132, 134, 135, 143, 153, 164, 171, 177, 180, 181, 226, 251, 279, 313, 467, 468, 477
Germany, **Germans**, *German language*, 98, 130, 135, 232, 241, 256, 277–8, 288, 313, 317, 322, 352, 371, 415, 416, 432, 434, 435, 437, 440, 452, 457, 464, 473, 474
GERNAND, 484
Gers (basin), 230; department, 282
Gesoriacum, 36; *see* Portus Itius, Boulogne
Ghisoni, Ghisonaccia, 185
Giffre (Massif du), 389, 393
GILBANK, 374, 499, 500
GILCHRIST *(process)*, 440
GILDAS (GWELTAS), SAINT, 82
GILLE, 491, 502
GILLET, 501, 502
GILLIÉRON, 129, 484
gin, 231
GIORDAN, 133, 496
GIRALDUS CAMBRENSIS, 483
GIRARD D'ALBISSIN, 483
Girart d'Amiens, 481
Girart de Roussillon, 132, 190
Gironde, 7, 142, 143, 227, 340, 363, 369; (department), 282, 343, 349, 387, 389, 395
Girondins, 274, 324, 497
GISCARD D'ESTAING, 344
GIUSTINIANO (Marin), 487
glassware, 234, 436
glycerine, 371
goats, 85, 354, 377
GODFRIN, 484
GODOLIN, 315, 316
GOGUEL, 497
Goidels, 6
gold, goldwork, 372

Golo (department), 186, 282
GONDEBAUD, 79
GONDOVALD, 74, 75
GONTRAN, 81
Gothia, 75; *Gothic (art)*, 129, 132, 192, 195, 243; **Goths**, 79; *see* **Visigoths, Ostrogoths**
Gouarec, 301
GOUBERT (Pierre), 487
GOUDINEAU, 472, 475
Gouézec, 301
Gourin, 301
GOURVIL, 191
gouvernements, 160–1, 167–8, 172, 176, 178, 180
grain, *see* cereals
GRAND, 479, 480
Grande-Chartreuse (Massif de la), 390
Grandes Chroniques, 102, 106
grandes écoles, 418, 420, 421, 427, 432, 437, 458
Grands Jours, 189
grape harvests, 387, 389, 393
grapes (dessert), 347
GRAS (C.), 325, 497
GRAS (S. et C.), 189
Grasse (Préalpes de), 390
GRAVIER, 500
Greece, **Greeks**, *(influence)*, 7, 9, 16, 20, 21, 23, 31–3, 101–2, 147, 459
GRÉGOIRE (abbot), 279–80
GREGORY IX (pope), 99
GREGORY THE GREAT (pope), 99
GREGORY OF TOURS, 81, 199, 251
GRENIER, 475
Grenoble, 161, 183, 200, 211, 212, 286, 371, 405, 406, 409, 432, 437, 455, 457, 478
GRIGNION DE MONTFORT, 342
GROSCLAUDE, 337, 498
GROSHENS, 495
Gruyère, 353, 355
GUELLEC, 447, 494
GUENÉE (Bernard), 106, 243
Guérande (region), 294, 299, 301, 302
Guéret, 161
GUERMOND, 498
GUIBERT, 502
GUIBERT DE NOGENT, 99, 105
GUICHARD, 79, 478
GUICHONNET, 491, 500
Guide des chemins de France (Le), 167
GUILAINE, 469
GUILCHER (André), 478, 499
GUILCHER (Jean-Marie), 307, 495, 496
Guines *(comte de)*, 490
GUINET, 479
GULLEY, 469
GUY DE BAZOCHES, 255

Guyenne, 100, 161, 168, 178, 288
GUYONVARC'H, 24, 472

Habsburgs, 111
haga, hedge, 71; *see enclosures*
Hainaut, 161, 167, 249; French, 63
HALÉVY, 493, 495
Hallstatt, *Hallstatt,* 6, 31
-ham, 84
Hamburg, 288
hamlets, 134, 139, 141, 148, 149
Harfleur, 381
HARMAND, 471, 472
harvesters, 387, 389, 390
harvesting machine, horse-drawn, 17, 18, 51, 52
HASSELN (Robert de), 281, 283
hat industry, 297, 298, 299, 311, 368
Hattuarii, 164
HAUDRICOURT, 478, 479, 485
HAUSSMANN, 439, 458, 459, 460
HAWKES, 470
HECATAEUS, 8
Helena, 184
hellenism, see Greece, **Greeks**
Helvetii, 10, 12, 17, 20, 35, 473, 488
Helvii, Helvians, 12, 34
-hêm, 84
hemp, hemp fields, 224, 225, 387; *industry,* 234, 239, 240; *combers,* 387
Hénin-Liétard, 383, 403
Hennebont, 369
HENRI II, 111, 273
HENRI IV (HENRY OF NAVARRE), *reign of,* 113, 230, 234, 256, 314, 320, 382
Henrichemont, 382
HENRY V (of Germany), 105
Hérault, 27, 35; (department), 283, 349, 350
HERCULES, 23
HÉRIOT, 418
HEROLD, 92
HERRIOT (Edouard), 336
HERVIEU, 503
Hesse, 174
HIGOUNET, 478, 483, 485
HINCMAR, 99, 105, 252
Histoire de France, 426; *see Tableau de la France*
Histoire de la Gaule méridionale sous les conquérants germains, 318
Histoire de la poésie provençale, 318
ho, 84
HOFFMANN (STANLEY), 502
HOFMANN (B.), 38
Holland, **Dutch,** 115, 181, 230, 231, 234, 244, 288, 449; *cheese,* 377; *Dutch (breed of cattle),* 358

Honfleur, 381
HOPKINS, 481
hops, 224, 389
horse(s), 31, 85, 92, 129, 136, 187-8, 357, 358, 447
horticulture, 225, 261, 389; *see gardens*
-hou, 84
houses, rural, 15-16, 194-5, 310ff., 449; *urban,* 20-1, 460, 461, 462-3
HOVELACQUE, 325
HUBERT, 470, 471, 472
Huelgoat (wood), 298
HUETZ DE LEMPS (Alain), 374, 500
HUETZ DE LEMPS (Christian), 489, 490
HUGH CAPET, 96; *see* **Capetians**
HUGUES DE FLEURY, 96
humanists, 99, 105, 107, 274, 467
Hundred Years War, 103, 106, 117, 126, 142, 149, 217, 222, 255, 267, 285
Hungary, 244; Hungarian (invasions), 58, 135
Huningue, 114, 382
Huon de Bordeaux, 190
Hurepoix, 140, 203, 250
hydraulic (energy), hydro-electricity, 360, 370, 371, 416, 435, 450
Hyères, 8, 84, 286

-iacus, 63
Iberian peninsula, 3, 21, 23, 75, 242; *see* Spain
Iberians, *Iberian language,* 6, 7, 8, 127, 129, 179, 199, 312, 470
IBN HAUQAL, 481
-ic, 479
Ictis, 30, 31
-ien, 197
-ies, 63
Ifran*dja,* 97, 481
Ile-de-France, 97, 123, 132, 161, 170, 171, 172, 176, 192, 227, 234, 250, 261, 265, 267, 269, 350, 449
Iliberris, Illiberris, 7, 184
Ill, 173, 486; terraces, 173
Ille-et-Vilaine (department), 86, 282
Iluro, 7; Iloronensium *(civitas),* 179
imbrex, 130
Indo-Europeans, 6, 16
Indre (department), 282, 391
Indre-et-Loire (department), 282
industry, industrialization, 116, 233-4, 344, 365-6, 371-2, 392, 402, 420, 424, 428, 434, 435, 436, 437, 440, 443, 456
infield, 98
-ing, -ingen, -ingue, 61, 63
Ingrandes, Ingranne, 14, 15
innovation, inventions, 344, 436, 437, 441, 458, 460

548 Index

Intendances, 160–1, 284, 324
Ireland, **Irish**, 6, 17, 55, 85, 86, 87, 227, 325
Iron Age, 6, 18; *iron ore*, 234, 383, 440, 444, 450; *iron and steel industry*, 368, 369, 370, 438, 439, 440, 442; *see metallurgy*
irrigation, 204, 346, 347, 350
-is, 197
ISAMBERT (F.-A.), 327
Isère (basin), 81, 142, 183, 211; (department), 282
ISIDORE OF SEVILLE, 93, 476, 480
ISORE, 254
Italy, Italians, 5, 6, 7, 12, 21, 34, 37, 41, 45, 55, 90, 93, 95, 98, 104, 105, 110, 114, 133, 225, 230, 256, 317, 371, 436, 449; *Italian (language)*, 123, 135, 181
ITZIN, 488
-ius, 63

jacket, 297, 298
Jacobins, *Jacobinism*, 274, 324, 336
JACOUBET, 496
JACQUARD, 372, 457
JACQUES DE REVIGNY, 106
JALUZOT, 418
JANNORAY, 470, 472
Jansenism, 316
Japan, 416
JARRY, 478, 479, 486, 494
JASMIN, 318, 321
JEAN-BRUNHES-DELAMARRE, 485
JEAN COMNÈNE, 98
JEAN DE BUEIL, 107
JEAN DE JANDUN, 254
JEANNEAU, 500
JEANNENEY, 502, 503
JEANTON, 129, 484, 495
Jeunes Voyageurs (Les), 291
Jeux Floraux, Académie des, 314
jewelry, 60, 61, 372
JOAN OF ARC, 103, 107, 421
JOHN LACKLAND, 200, 217
JOHNSON (J.), 84, 477
JOHNSON (J. H.), 471
Joigny, 353
JOINVILLE, 106, 482
JONAS, 81
JOSEPHUS, 470
JOUAN (Abel), 494
JUILLARD, 138, 149, 485, 490, 499, 501, 502, 503
JULIAN (emperor), 46, 251, 476
Juliobona Caletorum, 36, 42, 381
Juliomagus Andecavorum, 42
JULLIAN (Camille), 12, 16, 33, 47, 164, 470, 471, 472, 473, 474, 475, 476, 478, 481, 486, 488, 489

July Monarchy, 280, 382, 420, 437, 439
Jura, 6, 66, 113, 115, 173, 180, 182, 198, 228, 239, 258, 265, 309, 353, 356, 366, 385, 387, 428, 488; (department), 282
jurists, 103; *see law*
justice, judiciary *(organization)*, 40–1, 162, 184, 187, 188–9, 207–8

*kal, 200
Kehl, 114
Kertzfeld, 138
KLÄUI, 484
KLEINCLAUSZ, 487
KLEINSCHMAGER (Richard), 497, 498
KLIMRATH, 484
KNAFOU, 503
KOLENDO, 471
KOLODNY, 185, 487
KRENZLIN, 484
KROEMER, 65, 477, 488
KRÜGER, 484
KRUTA, 471
KURTH, 480
Kurzkreuzgewannflur, 139

La-, 141
LA BORDERIE (Arthur de), 324
La Chapelle pass, 250
La Défense, 465, 466
La-Roche-sur-Yon, 383
La Rochelle, 107, 127, 161, 176, 189, 214, 228, 236, 270, 486
La Tène, 6, 31
LA VILLEMARQUÉ (Th. Hersart de), 191, 319, 322, 323, 496
LABASSE, 501
LABBE (Philippe), 113
LABORDE, 500
Labourd, 7, 179, 304
LABROUSSE, 501, 502
LACAM, 479
Lacaune (Monts de), 130, 356; *breed of cattle*, 358
LACOSTE, 497, 498
LACROIX, 476
LAFERRÈRE, 499, 502
Lafite, 232
LAFONT (R.), 133, 315, 337, 496, 497, 498
Lagny, 268
LAMBERT, 470, 471, 474
LAMBERT D'ARDES, 490
LAMENNAIS, 324
Lan-, 86
landes, 202, 363, 396
Landes, 202; (department), 282, 384, 385; of Gascony, 131, 178, 202, 218, 305, 359, 360ff., 465

LANDRIOT, 476
landscape, rural, 17–19, 47–52, 70, 134–8, 203–4, 265, 464; see *openfields, enclosures*
LANGLOIS (C.), 327
Langon, 8, 229, 362
Langres, 37, 43, 256; city, 164, 166, 172; region, 68; plateau 241; *pays,* 166; diocese, 284
Langstreifenflur, 139
Languedoc, 7, 8, 34, 48, 51, 68, 75, 96, 100, 110, 117, 143, 161, 163, 167, 184, 192, 193, 209, 229, 235, 242, 290, 313, 321, 325, 329, 333, 334, 335, 338, 340, 341–2, 348, 351, 367, 382, 389, 452, 455; coast, 384, 465; Haut-, 466; *canal,* 348; *languedoc (dialect), see occitan; langue d'oc see oc (langue d')*
LANSON, 134, 484
Laon, 96, 489; region 171, 227
Lapurdum 7, 179
Larchamp *(battle of),* 191
laridés, 307
Laroche-Migennes, 384
Larzac (Causse du), 355; *breed of cattle,* 358
LASSERRE, 497
LASTEYRIE (C. P. de), 499
LASTEYRIE (R. de), 480
LATINI (Brunetto), 482
Latin (authors), *Latin* (language), 9–10, 64, 89, 103–4, 105, 106, 122, 123, 125, 134, 192, 243, 314, 420
LATOUCHE, 479, 486
Lauraguais (pass, gap), 3, 8, 34, 184, 271
LAURENT, 499
Laval, 196
LAVALLÉE, 483
LAVISSE, 158, 479, 480, 482, 486, 487, 491, 493, 494, 498
LAVOISIER, 259
law: customary, 129, 131, 188; *written,* 129, 131, 188; *Roman law,* 188; *faculties of,* 420
Laws: CORNUDET, 461; GUIZOT, 280; LOUCHEUR, 442; MALRAUX, 464; *Salic,* 103; SIEGFRIED, 462; *of* Jules FERRY, 292; *of the* **Ripuares**, 478
Le-, 141
LE BOUVIER (Gilles) (the Herald Berry), 482
LE BRAS, 257, 327, 398, 497, 498
LE BRAZ, 496
LE CORBUSIER, 461, 464
Le Creusot, 366, 369, 383, 404, 435, 439
LE GONIDEC (J.-F.-M.), 319
Le Havre, 161, 209, 211, 236, 270, 330, 381, 402, 405, 449
LE LANNOU, 487
Le Mans, 161; (election), 160
LE PLAY, 333
Le Puy, 240, 409

LE ROUX (F.), 472
LE ROUX (Muriel), 497
LE ROUX DE LINCY, 492
LE ROY LADURIE, 151, 157, 485, 490, 498, 502
Le Touquet, 384
L'Hermitage *(wine),* 230
lead (ore), 234
League of Augsburg (war), 232
leather work, 238, 339
LEBEAU, 503
LEBEL, 471
LEBRUN (F.), 327
LECOUTEUX, 491
Lectora, Lectoure, 7, 143
LEDAY, 475
LEFEBVRE DES NOËTTES, 485
LÉGER, 503
legion, 162
legitimists, 343
LEGRAND, 486
LELONG (Father), 487
Léman (department), 282
Lemaniam (Pagum), Lemane (m) Arverniam, 199
LEMARIGNIER, 486
LEMONNIER, 491, 493
Lemovices, 10, 11, 12, 69, 175, 177
LENTACKER, 483
Léon, 301, 303, 305, 331
LÉON (Albert), 496
LÉON (Pierre), 491, 499, 501, 502
LEQUIN, 403
LERAT, 490
LEROUX, 448
LEROY-BEAULIEU, 422
Les-, 139, 141
LESTOCQUOY, 478, 482, 488
LETRONE, 281
Letters to a Provincial, 274
Lettres sur l'Histoire de France, 318
Leucate, 112–13
Leuci, 10, 12, 13, 23, 69, 172
LEVAINVILLE, 488
Levant, 235; *see* East
Lévezou, 199, 355, 360
LÉVY-LEBOYER, 501, 502
LEWIS (B.), 481
Lexovii, 12, 174, 471
Lézardes dans la maison (Les), 427
Liamone (department), 186, 282
liberals, 324, 334
Libourne, 146
LIÉBAUT, 201
Lieuvin, 174, 175
Life of Saint Colomban, 81
Life of Saint Didier, 81

Life of Saint Rémi, 99
Ligny, 368
Ligurian, Liguria, 6, 7, 8, 11, 20, 21, 27
Lille (built-up area), 116, 161, 235, 273, 289, 369, 404, 405, 406, 412, 432, 437, 439, 455, 457, 459, 460; region, 385
Lillebonne, 42, 174, 381
Lily *(fleur de lys)*, 103
lim, 200
Limagne (Grande), 33, 43, 52, 68, 148, 177, 199–200, 206, 395, 397; Limagnes, 177; — du Sud, 397, 399; Limania, 199–200
limestone, limestone soils, 51, 53, 97, 140, 169–77, 181, 182, 183, 203, 217, 235, 250, 267, 269, 329, 350, 353–4, 363, 377
Limoges, 42, 43, 69, 70, 161, 177, 270, 271, 402, 441
Limonum, 36, 43, 69
Limousin, 143, 153, 161, 168, 177, 213, 218, 222, 249, 256, 328, 329, 334, 356, 360, 387, 388, 389, 390, 391, 396; Limousine (Montagne), 356, 396; *Limousin (dialect)*, 124; *Limousin (breed of cattle)*, 358
line, Saint Malo–Geneva, 157, 406, 435
linen industry, 235, 366
Lingones, Lingoni, 11, 12, 23, 43, 182; city of, 172, 181
Lion (Golfe du), 3, 8
liqueurs, 374, 378
Lisieux, 20, 174, 251, 471
literacy, 151–3, 278–9, 280, 304, 322
Littré, 493
litus saxonicum, 82, 84
livestock (flocks, herds), 18, 52, 71, 136, 137, 148, 187, 206, 225, 265, 353, 356, 361, 447; *see cattle, sheep, stock raising*; *breeds of*, 205, 356–9, 447
Livet (Georges), 189, 325, 497
Livet (Roger), 139, 479, 485
living standards, 154–7, 440, 443
Livradois, 177, 310, 387, 396, 397
Livre des compagnonnages (Le), 291
Livre des Règles des Jolis Compagnons Tourneurs de Bordeaux (Le), 288
Livy, 471, 472, 473
Loc-, locus, 86
Lodève, 209, 239
loess, 92, 137, 140, 173
Loing, 247; *canal*, 230, 263
Loir-et-Cher (department), 282
Loire, Pays de la —, valley, 7–8, 14, 24, 30, 31, 37, 39, 43–4, 56, 60, 64, 67, 75, 76 (upper), 83, 84, 85, 86, 94, 96, 98, 100, 106, 124, 125, 127, 132, 153, 171, 175, 176, 177, 203, 213, 217, 224, 227, 228, 234, 255, 256, 263, 264, 271, 284, 285, 301, 340, 364, 369, 378, 383, 384, 390, 395, 406, 409, 414, 433, 440, 455, 479; department, 282, 373; — Inférieure (department), 282, 305, 340; (Haute-, department), 252, 283, 409
Loiret (department), 282
Lombardy, 193; *Lombard (chroniclers)*, 95
London, 232, 259, 318, 455
Longnon (Auguste), xxi, 477, 479, 480, 483, 486
Longwy, 382
Lorient, 307, 369, 381, 382
Lorraine, **Lorrainese**, 56, 65, 67, 95, 96, 105, 116, 117, 130, 131, 137, 161, 162, 173, 174, 191, 195, 196, 203, 217, 223, 237, 238, 273, 280, 289, 292, 309, 328, 329, 350, 354, 369, 374, 376, 377, 378, 383, 387, 395, 400, 434, 435, 440, 441, 450, 466, 478, 483, 487; *Lorrain (dialect)*, 61, 123
Lot (Ferdinand), 42, 64, 96, 470, 471, 475, 477, 479, 481
Lot (department), 282; 'vallons', 355
Lot-et-Garonne (department), 282, 335
Loth, 479, 480
Lothaire, 90, 93, 95, 108
Lotharingia, 88, 108, 172, 173
Louis (René), 96, 481, 487
Louis VII, 253
Louis IX, (Saint Louis) 100, 103, 106, 110, 146; *crown of*, 103, 106; *cult of*, 100
Louis XI, 110, 270
Louis XIII, 176; *style, fashion*, 297, 298, 301
Louis XIV, *reign of, style*, 113, 115, 207, 224, 232, 383
Louis XV *(style)*, 295, 309
Louis XVI *(style)*, 295–6, 388
Louis the Child, 95
Louis the Debonair, 90, 92, 95
Louis of Germania, 90, 92, 93, 95, 108
Louis VI the Fat, 105, 253
Louvain *(counts of)*, 88
Louviers, 209, 211
Louvre *(palace)*, 253; *(shop)*, 418
Lovie, 483
Lozère (department), 282, 395, 396, 409
Luchaire, 487
Lucius, 499
Lug, 44
Lugdunum, 36, 44; *see* Lyons; — Batavorum, 36; — Convenarum, 36; Lugdunensis 40, 180
Lugge, 480, 481
Lumière (brothers), 458
Lutetia, 12, 21, 22, 36, 37, 43, 47, 69, 248, 251–2; *see* Paris
Luxembourg, 196; *counts of*, 88
luxury goods, 182, 215, 234, 241, 372, 408
Lyons, **Lyonnais**, 37, 39, 44, 45, 47, 56, 73, 76, 83, 84, 109, 112, 117, 123, 125, 130,

Index   551

143, 161, 165, 172, 177, 180, 213–14, 234, 235, 236, 240, 242, 251, 259, 271, 273, 286, 357, 366, 367, 371, 388, 389, 391, 402, 405, 406, 409, 412, 432, 434, 436, 437, 438, 439, 441, 442, 450, 454, 455, 457, 459; Monts du—, 180, 214

Mâcon, 108, 182, 263, 270, 309; Mâconnais, 64, 229
MAÇOUDI, 252, 492
madder, 224, 387
Madeleine (Monts de la), 360
Madelaine au désert de la Sainte-Baume (La), 191
Madrid (château de), 372
MAGE, 494
MAILLART (J.), 200
Maine, 79, 87, 88, 161, 168, 176, 223, 333, 340, 389, 400
Maine-et-Loire (department), 282, 340
Mainz, 42, 44, 76; archbishopric, 173
maize, 218, 433, 447, 449
MALHERBE, 316
MALINVAUD, 502
MALOUET, 283
Man (Isle of), 6
Manche, (department), 207, 282, 399, 447
MANDROU, 338, 486, 489, 491, 494, 498
Manasque, 407
manse, mansionile, mansus, 64
Mantes, 97
MANTEYER, 479
Manuel d'Histoire de la Littérature Française, 134
Marais, 201, 204; —Breton, 201, 205, 305; of Poitou, 201, 205; (district of Paris), 259
marc, 376
MARCAGGI, 496
Marche, 129, 134, 143, 161, 167, 177, 181, 282, 283; —(Basse), 100, 356; —(Haute), 356; *marches*, 42, 109, 110, 169
mare (female horse), 129, 131
-mare, mare, 83, 87
MARÉCHAL, 474, 479
margarine, 371
Margeride, 199, 360, 396
MARGUERITE-MARIE ALACOQUE, SAINT, 421
MARGUERITE DE VALOIS, 314
MARIE, 503
Marie-Brizard, 376
market gardening, 216, 219, 224, 225, 259–61, 346, 353, 389
markets (rural), 166–7, 186, 206
Marne, 97, 250, 268, 366, 368; valley, 232, 247, 249; department, 282, 365; —(Haute-), department, 282, 364, 388, 395
MAROT, 483

Marres, 488
marriages, 152, 207, 333, 419, 420
Marseilles, 8, 9, 26–33, 83, 143, 183, 225, 271, 285, 386, 304, 315, 318, 329, 334, 336, 345, 371, 388, 402, 405, 406, 412, 449, 450, 457, 459, 472; basin, 346
MARTHELOT, 499
MARTIAL, 55
MARTIN (Jean-Clément), 498
MARTIN (M-M), 480, 481, 492
MARTIN (P. E.), 478
MARTIN (Pierre de), 495
MARTIN *(process)*, 369
MARTIN (R.), 471
Martyre de saint Baccus (Le), 32
masons, 222, 256, 329, 388, 390, 391
MASSEPORT, 497, 498
Massif Central, 3, 11, 29, 32, 34, 123, 126, 129, 130, 131, 141, 143, 148, 151, 152, 171, 175, 176, 177, 179, 182, 184, 198, 199, 200, 202, 214, 223, 229, 236, 237, 239, 242, 258, 265, 271, 273, 285, 286, 289, 328, 333, 338, 341, 353, 354, 356, 358, 360, 366, 369, 383, 385, 388, 389, 390, 391, 392, 395, 396, 397, 399, 400, 401, 404, 406, 409, 428, 430, 431, 433, 435, 440, 445, 449, 451, 452
Massilia, 36; *see* Marseilles
Maugis d'Aigremont, 243
Maures (massif des), 8, 84, 85, 479, 481
MAURIANGE, 495
Maurienne, 388, 392, 399
MAURIZIO, 479
MAURRAS, 324
Mayenne (valley), 13; department, 153, 282
Mazamet, 403, 405
MAZARIN, 111
measurement, agrarian, 24, 47ff., 149
meat, 433, 448
Meaux, 251, 378
mechanical (industries), 370; *see metallurgy*
mechanization of agriculture, 444–9
méchou, méjou, 139, 149
medicine (faculties of), 420
mediolanum, mediolanos, 15, 24–5
Mediolanum Santonum, 36
Mediomatrics, Mediomatrici, 12, 172
Mediterranean, basin, 3, 4, 5, 7, 54, 113; Mediterranean France, 26–33, 35, 56–7, 72, 73, 76, 90, 93, 108, 134, 135, 141, 147–8, 153, 180, 181, 183, 216, 229–30, 289, 346, 369, 376, 387, 389, 395, 397, 401, 406, 434, 436
Médoc, 8, 232; **Medulli**, 8
Meerssen, 108
megaliths, 4, 5, 14
Megontiacum, 36
Meillant, 15

Meldes, Meldi, 12, 16
Mélian, 15
MÉLINE, 441
melons, 347
Mélusine, 243
Menapii, 12, 20, 23, 471
MÉNARD, 488
Mende, 339, 409
MENDRAS, 502, 503
MENÉNDEZ PIDAL, 102, 481, 482
Ménilmontant, 250
Menton, 400
-*mer*, 87
Mercantour *(national park)*, 466
MERCIER (Louis-Sébastien), 492
MERIAN, 478
MÉRIMÉE, 318
MERLE, 485
MERLIN, 394, 398, 500, 501, 502
Merovingians, *Merovingian (period)*, 53, 58, 65, 68, 74, 75, 76, 77, 79, 81, 92, 94, 96, 99, 165, 168, 173, 178, 179, 180, 181, 251–2, 475, 487
MERTENS, 474
metallurgy, 72, 136, 239, 355, 367, 369, 372, 373, 402, 405; *see iron ore*
METHUEN *(Treaty of)*, 232
METTRIER, 494
Metz, 59, 69, 70, 79, 81, 92, 111, 137, 161, 167, 172, 173, 402; diocese, 164; *governement de*, 167
Meudon, 22, 250
Meurthe, 173; department, 282, 374
Meurthe-et-Moselle (department), 195, 282
Meuse, basin, region, 5, 36, 81, 90, 94, 96, 108, 116, 172, 173, 181, 198, 203, 483; department, 282, 378; Côtes de, 350, 352
Mexico, 322, 388
MEYER, 476
Meylan, 15
MEYNIER, 139, 144, 149, 471, 478, 485, 500, 501
Mézières, 108, 382
MICHEL (Francisque), 319
MICHELET, 103, 312, 413, 426, 496, 498
Middle Ages, xxii, 16, 48, 53, 70, 71, 96, 97, 98, 104, 105, 108, 109, 115, 123, 126, 127, 129, 130, 131, 134, 135, 140–6, 157, 166, 168, 178, 184, 187, 200, 204, 223, 226, 244, 256, 259, 287, 303, 318, 348, 363, 380, 426
migrations, 85, 152, 211–14, 215, 222, 242, 256, 310, 322, 329, 335, 343, 365, 385–400, 457
Milan, 24, 259
milk, see *dairy farming, dairy produce*
Milky Way, 73, 478
Millau, 143, 240, 304

millet, 54, 218, 361
mills, 129, 130–1, 136, 236, 237, 240, 363, 370
MINC, 503
miners, 342, 392
Minoan (civilization), 26
MIQUEL, 481
MIRABEAU, 283
Mireille, 319, 321
MIROT (L. and A.), 40, 79, 161, 469, 478, 480, 486, 487, 494
mirrors, 372
missions, missionaries, 328, 343–3, 421
MISTRAL (Frédéric), 124, 319, 320, 321, 323, 496
MITTERRAND (François), 334, 431
Moislains, 15
Moissac, 100, 347
Moissy-Cramayel, 267
Molain, 15
mole catchers, 387
MOLIÈRE, 274, 286, 287
MOMMSEN, 497
Monaco, Monoikos, 27
monarchists, 343, 421; *see legitimists, royalism*
monasteries, monks (in general), 114, 166–7, 226, 267, 364
MONET, 462
Mongols (invasions), 85
MONGRÉDIEN, 287, 494
MONNET *(plan)*, 443
monoculture, 216
Mons Brisiacus, 36
Mons Mauripensis, 199
Monsieur de Pourceaugnac, 274
mont, 64
Mont Blanc, 114, 384; department, 282
Mont-Cenis (plateau), 114
Mont Dauphin, 382
Mont-Dore, Monts-Dore, 177, 388, 396; *(cheese)*, 377
Mont Royal, 382
Mont Saint-Michel, 96
Mont-de-Marsan, 146, 362
Montagne Noire (south of Massif Central), 130
Montagne(s) Noire(s) (Brittany), 299, 305, 331
Montagne Sainte-Geneviève, 250, 251
MONTAIGNE, 255
Montauban, 143, 146; region, 328
Montbéliard, 437, 442; comté, 165; region, 383
Montceau-les-Mines, 369, 383
Montélimar, 48
Montereau, 30, 31, 454
MONTESQUIEU, 274

Index 553

MONTFORTAINS, 342–3
MONTLOSIER, 360
Montlouis, 382
Montluçon, 369, 404
Montmartre, 250
Montmélian, 15
Montmorency, 268; hill, 249; forest, 269
Montpellier, 209, 211, 325, 348, 452, 459; Généralité de —, 162, 184
morality, 428
Morbihan (department), 282, 449
MOREAU, 486, 488, 489
MORINEAU, 490
Morini, 10, 12, 20
Morlaix, 305; river, 298
Morvan, 43, 140, 142, 176, 181, 199, 329, 354, 356, 387, 391, 466; Morvensis, Morivensis (Pagus), 199; Morvois, 199
Moselle, valley, 13, 44, 56, 90, 94, 172, 173, 227, 353, 382, 454; department, 282; Côtes de, 173, 440
motorways, 451, 454
Moulins, 161, 177, 270
mountains (names of), 198
MOURS (S.), 338
MOUSKES (Philippe), 481
muleteers, 387
Mulhouse, 211, 235, 366, 370, 402, 432, 434, 439, 460; region, 407
MÜLLER, 484
MULLIEZ, 499
Murat (region), 388
Muret *(battle of)*, 110
mushroom cultivation, 261
Muslims *(invasions)*, 58, 75, 82, 84, 132, 135
MUSSET (L.), 59, 83, 477, 478, 479, 480, 481
MUSSET (R.), 488
Mycenean (civilization), 26
Mystery Plays, 316

-*naes*, 84
Namnetes, 12, 23, 86
Nancy, 173, 203, 211, 270, 324, 382, 402, 405, 406, 432, 437, 439, 455, 459; Forestry School, 359; *programme*, 324, 427
Nantes, 86, 87, 191, 207, 225, 235, 270, 291, 305, 341, 402, 434, 450, 457, 459; region, 294; *Edict of*, 338
Nantuates, 10
NAPOLEON, 285, 308, 319, 322, 408, 419; *see Empire (First)*
NAPOLEON III, 363, 377, 383, 458
Narbo, Narbonne, 7, 8, 12, 28, 30, 31, 34, 36, 47, 75, 84, 143, 270, 271, 347; Narbonnaise (Gallia Narbonensis), 34, 40, 43, 45, 69, 73, 163; *see* Province, Roman; first, 184; second, 73, 183

Narbonnais (Les), 481; *see Chanson des Narbonnais*
natio, nation, 11, 104–5, 106, 256, 320, 321–4, 326, 412–14, 421, 466–8; *national territory*, 380; *national-socialist ideology*, 325; *National Front*, 431; *nationalism*, 104, 195–6, 421; *nationalities*, 103
Navarre, 110; Lower, 179; marches, 90
NAVARRO, 472
navigation shipping, 5, 26–7, 29–31, 228–9, 418, 436, 439; *see ports*
Neapolitans, 388
Neckar, 174
NÈGRE, 477
Nemausus, 8, 36
Nemetacum 36, 50
Nemetes, 12
-*nemetis*, -*nemetos*, -*nemetum*, 21
Neolithic culture, 4, 5, 17
Nérac, 314
Nervii, 10, 12, 13, 15, 18–19
Neuf-Brisach, 114, 382
Neustria, Neuster, Neuestreich, 77, 81, 83, 94, 96, 490; **Neustrians**, 217
Nevers, Nevirnum, 161, 176, 228, 383
newspapers, 273, 278, 280
-*nez*, 84
Niais de Sologne (Le), 190
Nice, Nicaea, Nikaia, 27, 36, 318, 388, 390, 404, 454, 455, 472; comté de, 114, 183; *dialect of*, 127
NIEL, 4
NIEMEIER, 485
NIEPCE, 458
Nièvre (department), 258, 282, 387, 395
Nijmegen, 42; *Treaty of*, 113, 115
Nîmes, 8, 47, 75, 209, 251, 459; *see* Nemausus
Niort, 166, 270
NISARD, 494
Nitensis (Pagus), 164
NITHARD, 480
Nitiobroges, 8, 12, 179
Nivernais, 140, 161, 168, 176, 199, 217, 234, 235, 329, 335, 353, 389; *nivernais (dialect)*, 126; *Nivernais-Charolais (breed of cattle)*, 358
NOIN, 503
NOMÉNOÉ, 87
NORA, 503
Nord (department), 116, 235, 282, 330, 333, 334, 369, 373, 383, 384, 396, 403, 441
Normandy, **Normans**, 6, 64, 66, 79, 83, 84, 85, 87–9, 100, 105, 106, 152, 161, 165, 168, 174–5, 176, 189, 191, 192, 193, 209, 227, 228, 234, 235, 256, 264, 274, 283, 286, 288, 309, 328, 438, 439, 457, 461, 462, 479;

Normandy (cont.)
  Norman (dialect), 123; Normand (breed of cattle), 358
north (of France), 39, 48, 56–7, 66–7, 75, 79, 83, 100, 117, 122, 127, 132, 133–4, 137, 151, 152, 155, 165, 169, 172, 204–5, 216, 219, 241, 271, 285, 304, 328, 333, 335, 340, 341, 350, 352, 354, 355, 366, 369, 374, 376, 377, 378, 384, 389, 390, 395, 397, 400, 401, 406, 414, 433, 436, 438, 440, 445, 448, 450
North Sea, 90, 91, 223, 227, 248, 436
north-east (of France), 108, 111, 137, 151, 155, 164, 247, 264, 304, 395, 411, 450
Northmen, 87
north-west (of France), 20, 217, 235
Norwegians, 87
NOSTREDAME César de, 315
NOSTREDAME, Jean de, 315
notaires, 188, 215, 408, 409
Notes d'un voyage en Corse, 318
Notitia provinciarum..., 74; *dignitatum*, 84
Novempopulania, 40, 45, 74, 75, 178
Noviodunum, 21, 36; —Aeduorum, 176
Noviomagus, 36
Nuits (côtes de), 232
Nu-Pieds (revolt), 189
nurseries, 261, 263
nutrition, 155, 219, 223
nuts, 54

oath, non-swearers of (priests), 327, 328, 330, 341–2
oats, 354, 433
Oc (langue d'), 123–9, 131, 132, 133, 134, 153, 274, 279; *Occitan*, 124, 127, 129; Occitania, 133, 311, 313–14, 321; *poetry*, 314–16; *movement*, 318–19
Odet, 306
ogres, 243, 278
oil-producing plants, 92, 217
Oïl (langue d'), 123–9, 132, 133, 172, 192, 335
-ois, 166, 197
Oise, 14, 97, 247; department, 212
olive tree, olives, olive oil, olive groves, 55, 92, 147, 216, 346, 348, 371, 387
Oloron, 7, 179, 368; viscomté of, 179
OLSCHKI, 492
ONDE, 500
open fields, 18, 19, 52–3, 71, 135–40, 148–9, 169, 202, 217, 265, 271, 445, 469
Opera House (Paris), 458, 459
oppidum (a), 12, 16–17, 19–21, 42–3, 47, 147, 164
Orange, 48
orchard(s), 353, 376; *see fruit cultivation*
ORESME (Nicole), 482

Orge (valley), 249
oriflamme, 254
Orléans, 15, 21, 30, 31, 37, 43, 69, 79, 81, 96, 100, 106, 161, 176, 225, 252, 256, 270, 286, 292, 363, 378, 383; *canal*, 263; *forest*, 364; Orléanais, 161, 168; *vineyards*, 228, 263, 350
Ornain (valley), 367
Orne, 174; valley, 215; department, 282, 387, 399
ORRY, 155, 157, 271, 372
Osismii, 10, 12, 85, 86
OSSIAN, 319
Ostrogoths, 80
Otlinga Saxonia, 84
-ou, 84
Ouessant, 466
Ourcq, 247; *canal*, 249
over-shirt, 298, 299
OZOUF, 485

pagus (i), 11, 43, 75, 87, 159–60, 163–6, 167, 186, 197, 203, 488
painting, 196, 293
Palatinate, 36, 76, 90, 143
Paleolithic, 277
PALLADIUS, 471, 475
PALLUEL, 460
palm trees, 225
PANERAI, 503
Pantagruel 255
PAOLI, 322
Papier Timbré (revolt), 297
PAPY, 490, 499
PARAIN, 484, 495
Paris, 21, 22, 56, 69, 70, 79, 81, 88, 116, 161, 175, 176, 211, 213, 227, 230, 236, 241, 247–56, 285, 297, 303, 336, 345, 350, 364, 372, 384, 389, 391, 393, 397, 402, 406, 407, 414, 417, 418, 426, 428, 432, 437, 439, 441, 442, 451, 454, 458; *influence of*, 126, 152, 170, 256–75; *Parlement de*, 162, 316; *suburbs*, 259, 263, 452, 460, 462–3
PARIS (Gaston), 132
Paris, Basin, 5, 6, 13, 52, 79, 81, 97, 106, 126, 127, 131, 140, 142, 160, 170–1, 172, 178, 181, 199, 202, 258, 264, 269, 328, 329, 331, 333, 341, 354, 364, 378, 385, 387, 389, 391, 396, 397, 400, 401, 404, 434, 445, 448, 455, 456
Paris-Plage, 384
PARISET (G.), 498
parishes, 86, 135, 143, 147, 208, 281, 340, 341
Parisii, 6–7, 12, 22, 43, 69, 251
parks, 263, 458, 461, 462; *national and regional*, 466
parlements, parlement members, 162, 184, 209, 211, 267, 315, 316

PASCAL, 274
Pas-de-Calais (department), 96, 251, 282, 283, 330, 369, 395, 403, 441, 452; *coalfield*, 273, 383, 435, 441
PASTEUR, 458
pastoral (life), shepherds, pastures, 6, 70–2, 91, 183, 185, 186, 204, 222, 353–6, 360, 364
Pastourade Gascone, 314
patois, 127, 129, 131, 151, 279–80
patria, patriots, patriotism, 23, 46, 104–8, 190–1, 321, 413, 421, 470
patrie gauloise (La), 475
Pau, 161, 180, 312, 452
PAUL SAINT, 82
PAUL II, 99
PAUL (Pierre), 315
PAUL DE VENDÉE, 489
Pax Dei, 132
pays, 46, 159–61, 163–6, 167, 168, 178, 179, 186, 187–8, 197–206, 209, 283; *see pagus*
Pays basque, sa population, sa langue, ses mœurs (Le), 319
Pays Fort, 170, 202
peach, 54
pear (liqueur), 374, 376; *pear trees*, 217
peas, 347
peddling, 152, 278, 280, 317, 387, 392
PÉDELABORDE, 490
PÉLATAN, 488
pelota (Basque), 312
PELTRE, 238, 374, 469, 490, 491, 500
PÉRARDEL, 476
Perche, 15, 175, 200, 203, 249, 292, 471; *Percheron (breed of horse)*, 357, 358
PERDIGUIER (Agricol), 289, 291
Pére Goriot (Le), 426
PÉREIRE *(park, villa)*, 460
Périgord, 143, 179, 189, 192, 217, 234, 335, 343; — Blanc, 203; — Noir, 203
Périgueux, 179, 403, 478
Péronne, 14
Perpignan, 161, 184, 346, 452
PERPILLOU, 351, 354, 362, 485, 498
PERRAULT, 277, 278
PERRENOT, 477
PERRIN, 478
PERROT, 489
Persians, 27, 102
Pertensis, Perticus (Pagus), 166, 200; Perthes, 166; Perthois, 166, 200; pertica (silva), 201
PETER (Venerable), 98
PETIT-DUTAILLIS, 481, 483, 491
Petrocorii, 10, 12, 179
petroleum, 444; *refining*, 450
Petromantalum, 36
PETRY, 474, 495
PEUGEOT, 442

Peutinger's table, 50, 475
Pézenas, 209
PFISTER, 479, 480, 486, 487
Phalsbourg, 194, 292
pharmaceutical industry, 372, 373, 437, 441, 450
PHILIP I, 95, 253
PHILIP IV THE FAIR, 99; *reign of*, 109
PHILIP VI, 285
PHILIP AUGUSTUS *reign of*, 88, 96, 105, 109, 200, 253; *territory of*, 100
PHILIPPONNEAU, 260, 263, 492, 493, 497, 498
Phocis, **Phocians**, 27
Phoenicians, 27, 29
phosphates, 371
photography, 372, 373, 458; *aerial*, 18–19, 47, 49–51, 52, 65, 70, 139
phylloxera, 343, 347, 349–50, 352, 376, 440
PICARD (G.), 474
Picardy, 18, 29, 50, 51, 53, 63, 64, 66, 68, 70, 136, 168, 171–2, 227, 235, 256, 258, 265, 286, 304, 334, 384, 385, 389, 393, 455; **Picards**, 172; *Picard (dialect)*, 123, 172
Pictavi, Pictones, 10, 11, 12, 43, 69, 175
Piedmont, 173, 183, 242, 388; **Piedmontese**, 114, 123
PIERRE DE BELLEPERCHE, 253
PIERRE DE SAINT-LOUIS, 191
PIERRE LE PEINTRE, 191
PIGANIOL (A.), 73, 475
PIGNON, 484
pigs, 223, 261, 346, 356, 449
PIJASSOU, 491
PILET (Christian), 477
PINCHEMEL, 501
pine woods, 360–1, 363–5, 448
PIPPIN, 95
PINKNEY, 499
pirates, 79, 84
PIRENNE, 480
PITTE (J. R.), 264, 447, 472, 474, 475, 476, 477, 490, 492, 493, 500, 503
plains, 48, 51, 59, 65, 91, 137, 169, 199
Plaine, 202; Plaines du Neubourg, de Saint-André, 174, 202
PLANDÉ, 483
PLANHOL, 195, 469, 471, 475, 476, 478, 484, 485, 488, 490
Plantagenets, 106, 171, 178, 253; *see Anglo-Norman (monarchy)*
PLAVINET, 488
PLINY, 204, 471, 473, 476
Plogoff, 301
plot-patterns, 17–19, 47–8, 52–3, 135–40, 148, 149, 265–7, 364–5, 445–8, 464
Plou-, 86
Plougastel, 299, 301
plough, 136, 140

plums, prunes, 225, 347, 374, 376
PLUTARCH, 20, 470, 471, 473
Po, 7, 9, 24, 26, 111; *see* Gaul Cisalpine
Pocrinium, 36
Poème du Rhône, 319
POÈTE, 499
Poissy, 357; — Beauregard, 462
Poitiers, 37, 43, 56, 69, 84, 161, 176, 270, 273, 286, 304, 328
Poitou, **Poitevins,** 56, 129, 142, 161, 168, 176, 178, 192, 193, 228, 243, 249, 256, 304, 329, 340, 342, 343, 382, 390; *Poitevin (dialect),* 126
police, 389
polka, 305
POLYBIUS, 9, 31, 470, 471, 473
polyculture, 216–17, 224, 346, 347, 348, 352, 353, 433
pomegranate trees, 225
Pomerania, 76
POMPONI, 497
Pont-Audemer, 209, 211
Pont-Aven, 300, 306
Pont-l'Abbé, 297, 301
Pontivy, 302, 307, 383
POPELARD, 195, 488
poplar trees, 462
porcelain, 372
PORÉE, 494
pork butchers, 387
port *(wine),* 232
Port-Vendres, 199, 290
porters, 256, 388
Porto (Corsica), 185
Porto-Vecchio, 185
ports, 37, 69, 236, 270, 380–2, 402, 434, 435, 440, 441, 450
Portugal, 232, 235
Portus Itius, 20
Portus Namnetum, 36
POSIDONIUS, 472, 473
postal service, 270–1; *postal workers,* 389
potatoes, 222, 364, 396, 433
pottery industry, 371
POUGNARD, 484
Poujadism, 337
poultry, 219, 364, 449
POUPARDIN, 479, 488
POUTHAS, 489
Préalpes, 239, 353, 358, 366, 387, 389, 390, 393, 399
PRÊCHEUR, 368, 370, 502, 503
PRICE, 502
PRIMAT, 102, 106
PRINCE, 485
Printemps (shops), 418
printing, 294, 313, 314, 316

PROBUS, 55, 56
protectionism, 275, 441
Protestantism, Protestants, 152, 331, 338, 344, 420, 436; *see* Reformation
PROUD'HON, 324
Provence, 7, 8, 26–7, 32, 34, 48, 75, 85, 100, 117, 133, 143, 153, 161, 163, 168, 183, 191, 192, 193, 195, 235, 283, 288, 309, 315, 321, 328, 333, 334, 335, 346, 347, 385, 400, 428, 455, 456; *Provençal (dialect),* 84–5, 124, 127, 129, 132, 200, 315, 319, 479
Province, Roman, 9, 34–5, 41, 48, 55, 66, 69, 168, 184; *see* Narbo
provinces, provincialism, 162–3, 167–9, 188–9, 274, 285, 315, 326, 343, 427, 432
provincia, 162
PROVOST, 470, 471
PRUDENCE DE TROYES, 108
PTOLEMY, 7, 10, 472
public sector, 419, 422, 443
Puigcerda, 110
Puisaye, 140, 142, 176, 199, 364; Puseium, 199
PUISEULX (marquis de), 483
PUMAIN, 456, 501, 503
Puy-de-Dôme (department), 192, 282, 360, 388
Pyrenees, 8, 9, 16, 42, 45, 75, 100, 110, 113, 114, 123, 143, 148, 179, 184, 199, 213, 248, 265, 286, 304, 360, 367, 371, 382, 385, 395, 396, 400, 433; of Ariège, 62, 223, 239, 388; departments: Basses—, 153, 282, 356, 390; Hautes— 153, 282; Orientales, 282; *Treaty of,* 113
PYTHÉAS, 9, 29, 31

Quatre Fils Aymon, see Chanson des . . .
Quercy, 179, 219, 347
Queyras, 388, 466
Quiberon, 307
QUICHERAT, 482
Quimper, region, 86, 299, 301, 302, 305
Quimperlé, 305
quinine, 363

RABELAIS, 255, 267
RACINE, 274
radicals, 421, 425
Rahmatollâh, Ramatuelle, 84
railways, 116, 182, 250, 269, 271–3, 291, 326, 345, 346–7, 348, 349, 362, 373, 383–4, 399–400, 406, 409, 410, 411–12, 416, 418, 438, 439, 450, 454, 457, 459, 460, 467
rainbow, 73, 478
Rambouillet, forest, 268; *royal flock (of sheep),* 357
RAMIN, 470, 473

rape, 224
RAUCH (F.), 359
Raurici, 12
RAVENNA COSMOGRAPHER, 94
RAYNOUARD, 132
real estate, property, 417, 419, 422, 443, 445
REAU, 493
RÉBELLIAU (A.), 498
REBEYROL (Yvonne), 477
Récits des Temps Mérovingiens, 318
RECLUS (Elisée), 413, 497
records (of livestock breeds), 358
Red Piebald (breed of cattle), 358
Redon, 290; *cartulary of*, 71
Redones, 12, 13, 69, 86
Reformation, 183, 286, 314, 316, 339; *see* Protestantism
régénération de la nature végétale (La), 359
REGINON, 92
regionalism, 317, 320, 324–5; *regions*, 293–4, 309, 320, 324–6, 395, 413–14, 427–8, 429, 431, 432, 433, 466, 467
RÉGNIER (Mathurin), 112, 483
Reichenau *(glossary)*, 98
Reii, 36
REINACH (Joseph), 325
REINAUD, 479
REITEL, 485
Religion (Wars of), 152, 316, 363
Remi, 10, 11, 12, 22, 23, 69, 170
Renaissance, 273, 313, 416
RENAN, 428
RENARD, 498
Renaus de Montauban, 481
RENÉ D'ANJOU, 315
Rennes, 69, 70, 86, 87, 161, 301; Généralité de, 162; region, 294
RENUCCI, 490
Republicans, 324, 334, 341, 421, 424, 428; *Republic (Third)*, 329, 331, 413, 429
Restoration, 298, 334, 345, 383, 409, 427, 438
REUSS, 496
revolts, 189, 276, 297, 331, 338–44, 427
Revolution, French, 111, 116, 160, 162, 167, 178, 183, 215, 235, 274, 279, 280, 298, 312, 322, 323, 326, 330, 334, 335, 341, 342, 376, 408, 412, 418, 421, 424, 429, 435, 436, 438, 468, 470
Revolution, Agricultural, 219, 224, 265; Second, 153, 365, 468
Revolution, Industrial, 385, 405, 406, 412, 416, 424, 438, 467; *see industry*
Revue de l'Armorique, 319
Revue de Paris, 325
REY, 479, 485
RÉZEAU, 489

Rheims, 22, 37, 45, 47, 69, 70, 79, 81, 140, 170, 172, 227, 235, 241, 251, 252, 289, 402, 405, 452, 455, 459, 478
Rhine, Rhineland (régions), 6, 16, 35–6, 39, 44, 45, 56, 58, 59, 60, 69, 76, 79, 80, 81, 83, 90, 93, 94, 95, 96, 98, 99, 111, 112, 113, 114, 115, 172, 173–4, 182, 192, 193, 195, 202, 223, 224, 227, 232, 242, 270, 288, 382, 454, 472; Bas-Rhin (department), 39, 153, 282, 407; Haut-Rhin (department), 153, 282, 407; Rhineland Massif, 172
Rhône Rhône gap, Rhône route, 3, 7, 8, 11, 26, 31, 34, 35, 44, 45, 48, 55, 62, 67, 69, 76, 84, 90, 108, 109, 110, 127, 129, 133, 152, 180, 184, 203, 204, 215, 223, 228, 230, 243, 251, 265, 270, 271, 272, 304, 315, 336, 339, 340, 369, 383, 384, 387, 395, 414, 450, 455, 466, 483; (department), 282, 395; plains of lower, 309, 406, 445, 447
RICCI ZANONI, 163
rice processing, 371
RICHARD, 486
RICHARD I, 88
RICHELIEU, (Cardinal), 111, 114, 382
RICHELIEU (town), 382–3
RICHER, 78, 182
RICOLFIS, 472
RIEDINGER, 499
RIFF, 495
Riom, 177, 401, 406
RIOUFFREYT, 470, 471, 474
riparia, 350
Ripuares, 94, 478; *see laws*
-ritos, -ritum, 21
RIVALS, 129, 484
roads, 23, 36, 37–9, 48, 68, 69, 76, 155, 158, 172, 176, 177, 180, 181, 183, 184, 223, 228, 230, 248–50, 263, 264, 269–71, 276, 291, 295, 304, 313, 329, 342, 345, 371, 397, 416, 451, 453–4, 467
Roanne, 263, 270, 271, 391, 405, 409
ROBERT-MULLER, 491
Rochefort, 382
ROCHEFORT (Michel), 501
ROCHET, 476
Rocroi, 382
Rodez, 35, 179, 271
ROHLFS, 484
ROLAND, 101; *see Chanson de*
ROLLO, 88
ROMAN, 29
Roman de Rou, 88, 202
Roman du comte d'Anjou, 200
Romance (languages), 125
Romanesque (art), 129, 132, 192, 193, 195
Romani, 94
Romanticism, 109, 134, 277, 317, 318, 319, 459

558   Index

Rome, **Romans,** *Roman (influence, period gallo-)*, 6, 8, 11, 16, 17ff., 31–3, 34ff., 45ff., 53ff., 60, 79, 85–6, 94, 97, 117, 122, 130, 132, 137, 159, 164, 226, 270, 474
Rome, Treaty of, 444
ronde, 303–7
RONDEAU, 490
RONSARD, 102, 225, 255, 267
roofs, 20, 129–30, 195, 196, 459, 460
Roquefort *(cellars, cheese)*, 354, 355, 358, 362, 377
Roquefort (department), 282
Rosalinsis (Pagus), 165
ROSE-VILLEQUEY, 491
Rosporden, region, 299, 301, 302
ROSSI, 487
ROSTAING, 63, 472, 477, 479, 488
ROSTAND (Edmond), 133
Rotomagus, 36, 69, 174
ROTZLER, 478
Roubaix, 369, 402, 403, 405
ROUCHE, 74, 77, 478
ROUDIÉ, 500
Rouen, 69, 88, 161, 174, 175, 209, 211, 235, 241, 253, 259, 270, 271, 289, 330, 369, 370, 402, 434, 435, 436, 452; region, 87
Rouergue, 35, 199, 222, 225, 304
ROUGIER DE BERGERIE, 359
ROUILLARD, 502, 503
ROUPNEL, 484
ROUSSEAU (Jean-Jacques), 274
Roussillon, 7, 110, 161, 167, 183, 184, 312, 334, 346, 351, 455
routes (overland), see *roads*
royalism, royalist, 329, 341; see *monarchists*
Rue Sain-Jacques, 249, 250
Rue Saint-Martin, 249, 250
Ruette, 195
RUFFI (Robert), 315
Ruscino, 36, 184
Russia, 388, 440
Ruteni, 10, 11, 12, 35, 179
rye, 202, 217–18, 361, 433
Ryswick (*Treaty of*), 114

Sagii, 15, 174, 471; see **Esuvii**
SAGNAC, 494
Saint-, 143
Saint-Amand et Raismes (forest), 466
Saint-Benoît-sur-Loire, 254; see Fleury-sur-Loire
Saint-Brieuc, region, 86, 301
Saint-Clair-sur-Epte *(Treaty of)*, 83, 97
Saint-Claude, 239, 387
Saint-Denis, *basilica, monks of, monastery*, 103, 106, 252, 253, 254

Saint-Etienne, 234, 271, 369, 391, 402, 405, 406, 437, 450; basin, 383
Saint-Florent, 185, 290
Saint-Flour, 177, 390, 409; '*planèze*' *(inclined plateau)*, 390, 409
Saint-Fulgent (canton), 344
Saint-Gall, 98
Saint-Georges (gardens), 347
Saint-Germain, forest, 269, 359
Saint-Germain-des-Prés (abbey, abbot of), 70, 252
SAINT-JULIEN, 503
Saint-Malo, 86, 157, 406, 435
Saint-Martial (abbey), 70
Saint-Médard (abbey), 70
Saint Michael's Mount (Cornwall), 30, 31
Saint-Nazaire, 86, 340
Saint-Omer, 190
Saint-Pierre d'Almanarre, 27, 84
Saint-Pierre-du-Dorat (sanctuary), 100
Saint-Pol, 86
Saint-Pourçain, 228
Saint-Quentin, 13, 14, 43, 403, 405; *canal*, 116
Saint-Rémy (abbey), 70
Saint-Sauveur, 114, 304
Saint-Sernin (cathedral), 314
SAINT-SIMON, 281
Saint-Valérien (abbey), 70
Sainte-Baume (la), 286, 289
Sainte-Marthe-de-Tarascon (sanctuary), 100
Saintes, 37, 56, 177, 270
Saintonge, 53, 56, 126, 142, 168, 176, 178, 228, 304, 328, 343, 486
saints (names of), 131, 142; *cults of*, 82, 100, 421; *statues of*, 294; *lives of*, 316; *patron saints*, 308
Saint-Maurice and Saint Lazarus (order of), 114
Salers (breed of cattle), 358; (*liqueur*), 378
SALIN, 62, 477, 478, 479
SALLUST, 35
Salluvii, 8, 12, 21, 43, 94
Salodurum, 36
salt, salt-flats, 32, 92, 301
salt-tax, 189
Samaritaine (shop), 418
Samarobriva, 36, 50
SAMSON D'ABBEVILLE, 271
Sancerre, 228, 353, 374, 377; Sancerrois, 142, 171
Sanctuaries, 20, 70
SAND (George), 291
sands, 26, 53, 169, 170, 177, 178, 199, 203, 218, 223, 234, 237, 269, 361, 364, 382
SANSOT, 494
Santiago de Compostela, 361
Santones, 11, 12, 176

Saône, 108; valley, 31, 34, 44, 45, 53, 69, 73, 90, 108, 109, 130, 180, 228, 263, 270, 272, 454; plain, plateaux of upper, 68, 172, 434, 445, 447
Saône, Haute- (department), 282
Saône-et-Loire (department), 282, 395
Saracens, 85, 102, 147, 254, 479; *see* **Muslims Arabs**
sardines (tinned), 302
Sarre (valley), 44, 165; *coal field*, 366
Sarrelouis, 382
Sartène, 186
Sarthe (department), 282, 341, 390
Sauldre *(canal)*, 363
Saulx (valley), 367
Saumur, 161, 166, 270; Saumurois, 161, 329
sauternes, 232
Sauvetés, 144, 145
Saverne (pass), 172
SAVINIEN D'ALQUIÉ, 189, 191
Savoy, 42, 62, 76, 103, 112, 114, 123, 143, 183, 214, 256, 258, 282, 289, 302, 328, 330, 333, 370, 388, 390, 392, 399, 409; Haute-Savoie, 367, 395, 455; **Savoyards**, 242, 256; *Savoyard monarchy*, 126
sawmills, 237, 240, 363
Saxons, 82, 83, 84–5, 95; *Saxon (language)*, 87
Scandinavians, *invasions*, 58, 64, 79, 83, 84, 87, 88, 135
SCHIER, 484
SCHMITT, 470
SCHMITTLEIN, 473, 474
SCHNEIDER (brothers), 439
SCHNEIDER (J.), 487
schools, 280, 292, 313, 320, 322, 326, 420, 421, 423–4, 428, 430
SCHULTEN, 470
SCHWAB, 489, 503
Scilly Isles, 30
SCLAFERT, 499
Scotland, 6
SCRIBE, 493
SECHER, 498
Sedan, 108, 161, 235
SÉE, 483
seedsmen, 388, 393
Sées, 15, 174, 471
Ségala(s), 199, 200, 202, 355, 396, 397; *segala*, 218
Segodunum, 36
SEGUIN (Marc), 457
Segusiaves, Segusiavi, 12, 40, 44, 177
Sein (island), 299, 302
Seine, valley, region, 6, 11, 30, 31, 37, 39, 43, 56, 63, 67, 76, 79, 81, 83, 85, 86, 87, 88, 89, 94, 96, 97, 174, 181, 203, 214, 227, 236, 247, 253, 256, 263, 272, 354, 365, 380, 384, 406, 437, 442, 450, 454, 470; department, 258, 282, 372, 373
Seine Inférieure (department), 282
Seine-et-Marne (department), 267, 282
Seine-et-Oise (department), 282, 395
SÉNAC, 82, 479
Senart (forest), 268
seneschal, seneschalsies, 160, 167, 188
Senlis, 22, 42, 250, 251, 252
Senones, 11, 15, 16, 181, 471
Sens, 12, 15, 70, 96, 97, 100, 284, 471, 478
separatism, 321, 323
Septimania, 75, 84, 110, 184
Séquanaise, 40, 73; **Sequani**, 10, 12, 43, 69, 182, 470, 488
SERVAIS, 479
servi casati, 70
Sète, 348, 382, 405; *see* Cette
settlement: hilltop, 22, 43, 85, 139, 141, 147; rural, 16–19, 49, 60, 61, 73, 134–51, 178, 261, 342
Sèvres, 22, 372
sharecropping, -croppers, 149, 153–4, 301, 363, 419, 443
sheep, 91, 235, 354, 355, 357, 365
shepherds, 52, 137, 361
shoe making industry, 368
shops (large), 393
Sicambres, 79, 94
Sicily, 267
SIDONIUS APOLLINARIS 46, 52, 199, 476
SIEBURG (Friedrich), 292
SIEGFRIED (André), 430, 497, 498
SIEYÈS, 283
SIGEBERT I, 80, 81
SIGEBERT III, 83
siliceous (soils, crust), 13, 15, 53, 140, 170, 174, 218, 219; *see sands*
silk industry, 234, 235, 242, 366, 367, 371–2, 434; *silkworm producers*, 387
silt, silty (soils), 51, 137, 169, 171, 174
silva, 140
Silvanectes, 22
SINCLAIR, 469, 483
Sion, 67
SION (Jules), 83, 487, 490, 499
Sizun (Cap), 299, 302
Sketches of Corsica, 318
slaughterhouses, 372
slaves, 17, 33, 48, 70, 85, 135
SMITH, 4
soap making, 371, 372
SOBOUL, 497
socialism, socialists, 336, 421, 425, 430
Société Générale, 409
Society of Jesus, 113
Soissons, 43, 70, 79, 81, 227, 251, 270, 478

Sologne, 140, 170, 176, 189, 202, 218, 363, 364, 488; —bourbonnaise, 139
Somme, 14, 50; department 282
Songe du Verger (Le), 100
songs, 276–7, 315, 317
SORIANO, 494
SOUILLET, 485
Soule, 179, 316
south (of France), 7, 11, 20, 48, 58, 66–7, 75, 83, 100, 106, 117, 122, 130, 131, 143, 151, 178, 180–1, 216, 286–7, 304, 334, 335, 336, 346, 353, 355, 360, 431, 456
south-east (of France), 7, 11, 152, 153, 180, 213, 390, 416, 438, 455
south-west (of France), 7, 8, 11, 39, 100, 106, 140, 151, 153, 192, 333, 334, 376, 395, 406, 428, 432, 438, 445, 449, 451, 452
SOUTOU, 473
SOYER (J.), 477
SOYER (Jacqueline), 19, 471, 475
Spain, **Spaniards**, 6, 27, 34, 37, 41, 55, 93, 95, 97, 113, 133, 325, 322, 388, 390; *see* Iberian peninsula; Spanish marches, 90, 110; *Spanish (language)*, 113; *see* Castillian
sparkling wines, see *champagne*
spas, 42, 304, 439
SPECKLIN, 129, 134, 477, 478, 484, 487, 489
spirits, 230, 232, 348, 374, 442
STAËL (Madame de), 134, 484
stationers, 409
steam engine, 233, 365, 457
steel, 369, 373; *see steelworks, metallurgy*
steelworks, 272, 273, 418, 437; *see metallurgy*
STEIN, 486
STEPHEN (Pope), 92
stirrup, 72
Stock Exchange, 407, 417, 439
stock-raising, 149, 353–9, 364, 433, 445–9; *see livestock, cattle, goats, sheep, pastoral (life)*
stone masons, 389
STRABO, 3–4, 44, 45, 55, 204, 251, 469, 470, 472, 473, 474, 476, 492, 503
Strasbourg, 42, 69, 161, 173, 211, 212, 216, 273, 289, 402, 405, 407, 451; Stratae Burgus, 69
strikes, 425
stud farms, 357, 362
styles, regional 308ff.
SUBRENAT, 487
suburbs, 259, 261, 404, 442, 449, 456, 460; *see* Paris (*suburbs*)
SUCHIER (Hermann), 96, 481
Suentensis (Pagus), 203
Suessiones, 10, 12, 16, 20, 23
SUETONIUS, 476
Suevi, 13, 35

sugar beet, 373, 449
sugar refineries, 371, 372, 373, 433
SUGER, 105, 200, 253, 254, 482
Suggentensis (Pagus), 203
SULLY, 271, 382
Sûre (valley), 44
SURELL, 359
SUTCLIFFE, 503
Suze (liqueur), 378
Sweden, 312
Switzerland, 5, 114, 123, 142, 228, 235, 288, 388, 434, 483
Sybillates, 179
symbols (of France), 103–4
Système de politique positive, 324

Tableau de la France, 413, 426
Tableau de la géographie de la France, 414
Tableau politique de la France de l'Ouest, 430
tacit communities, 142
TACITUS, 71
TACKETT (T.), 327
TAILLARD (Constant), 291
TAINE, 427
TANGUY, 496, 497
TANIÈRE, 499
Tarascon, 7, 12, 143, 406
Tarasque, 243, 491
Tarbes, **Tarbelli**, 7, 12, 42, 312, 362; *tarbellica arva*, 53
TARDIEU, 495
TARGET, 325
Tarine (breed of cattle), 358
Tarn (valley), 55, 329, 347; (department), 282
Tarn-et-Garonne (department), 233, 282
Tartarin de Tarascon, 133
TASTU (Amable), 290, 291
TAVERNIER (Melchior), 270
tea, 232
teachers, 388
Tech, 184, 347
Tectosages (Volcae), 7, 12, 34
TEDBALD DE BOURGES, 191
tegula, 130
telecommunications, 455, 468
Temps (Le), 413
TENÈZE, 487
TERNES, 475
Terrefort, 202; *terrefort*, 178
TERRENOIRE (J.-P.), 327
Terret noir, 349
Terror, White, 329, 334
TESSIER, 499
textile(s), industry, products, 234–5, 239, 242, 344, 365, 366, 367, 370, 372, 402, 405, 434, 436, 437, 438, 441, 452; *cultivation*, 216, 225, 347; *see hemp, linen*

thatch, 20, 129, 130
THEODORIC I, 80
THEODULF, 226
THÉRET, 499
THÉVENOT, 56, 471, 474, 476
THIBAUDET, 482
Thiérache, 140, 170, 171, 377
THIERRY (AUGUSTIN), 318
THIERS, 234, 238
Thirty Years War, 137, 196, 217, 232
THOMAS *(process)*, 369, 440
THOMAS-LACROIX, 485
THOURET, 283
threshing, 129, 484
THUILLIER, 502
-thun, 84
Thuringians, 80
tiles, 129, 130; *tileworks*, 237
TIMBAL, 492
tin, 7, 25–6, 29–32
TISSERAND, 492
TIXIER-VIGNANCOUR (J.-L.), 337
tobacco, 224, 347
TOCQUEVILLE, 274, 279, 494
TODD, 498
Tolosa, 7, 36
Tolvedunensis (Pagus), 164
tomatoes, 347
tombstones, discoid, 312
TOMMASEO (Niccolo), 318
Tongres (city), Tungrorum (civitas), 198
Tonnerre, 228, 353
toponymy, 7–10, 14–15, 20, 21, 41, 48–9, 56, 60, 61–5, 83, 84, 85–7, 97, 141–4, 149, 159–60, 163–6, 197–204, 474, 479
-tot, 83, 87
TOUJAS-PINÈDE, 498
Toul, 69, 111, 161, 167, 172, 173; Gouvernement de, 167, 197, 270, 454; Toulois, 197, 352
Toulon, 7, 8, 12, 34, 163–4, 286, 289, 402; Toulonnais, 164
Touloubre (region), 346
Toulouse, 7, 8, 12, 43, 75, 129, 131, 143, 161, 179, 180, 184, 251, 259, 271, 283, 286, 314, 315, 316, 318, 348, 402, 410, 452, 455; marches of, 110; (region), 34, 225, 312
Tour de France (cycle race), 292
Tour de la France par deux enfants (Le), 290, 291, 293
Tour de France d'un Petit Parsien (Le), 292
'*Tour of France*', 285–93, 424
Touraine, 161, 168, 176, 217, 225, 235, 255
Tourcoing, 369, 402, 403, 405
tourism, 384, 404, 442, 465
Tournai, 256
Tournus, 70, 309

Tours, 42, 44, 69, 161, 256, 270, 383
towns, urbanization, town-planning, 20–2, 46–7, 49, 70, 73, 104, 143–4, 164–7, 168, 172–3, 186, 187, 205, 208–16, 223–6, 234, 251, 252–3, 285, 370, 380–414, 420, 425, 431, 432, 437, 441, 443, 454, 455, 456–7, 458–9, 460, 461–6
trade, 25, 26, 32–3, 37, 49, 56, 76, 214–15, 225, 226–33, 240–2, 371, 374–9, 407, 420, 439, 441; *wholesale*, 409, 438; *itinerant tradesmen*, 152, 387–8; *see markets, fairs*
trade unions, 344, 425, 445
Traité de la Police, 166
TRAJAN, 42
Tramontana (A), 318
Transalpine *see* Gaul
transhumance, 183, 185, 204, 353
Tré-, treba, tref, 86
Trégor, 301, 303, 305
Tréguier, 86, 316
TRESSAN, 318
Treuga Dei, 132
Treveri, 11, 12, 20, 23, 45
Trèves, 37, 42, 44, 47, 56, 59, 130, 172, 180, 251
Triboci, 12
TRICART, 499
Tricasses, *12*
Tricastini, 8
TRISTAN, 491
Troia, 96; Troy, *Trojan (legend)*, 96, 98, 467
troubadour(s), 124, 127, 132, 133; *genre*, 318
Troyes, 42, 161, 170, 235, 241, 279, 402, 405, 406; 'city of', 199
TRUDAINE, 271, 357
TUILLERIE (Augustine), 291
TULIPPE, 477
Tullum, 36, 69
-tûn, -tun, 84
TURGOT, 241
Turin, 270
Turnacum, 36
turnips, 224
Turones, Turoni, 10, 12, 69, 176
Tyrrhenian Sea, 27

Ubii, 12
UHLIG, 485
Unelli, 12, 174
Union Régionaliste Bretonne, 325
United States, 281, 416, 435, 457
Universities, xxi, xxii, 107, 292
URBAN II, 99
Urnfield (culture), 4, 6
Utrecht *(Treaty of)*, 115
UXELLES (maréchal d'), 321
Uxellodunum, 12

Uzès, 73, 209

VADÉ, 25, 472
Vaison, 47
-val, 64
Val de Germigny, 171, 354
Valence, 182, 270, 409; Valentia, 36; Valentinois (comté de), 182
Valence d'Agen, 146
Valenciennes, 270; Généralité, 162; region, 63
VALENTINIAN, 251
VALÉRY, 318
Vallis Subolae, 179
VAN DE WALLE (E.), 398
VAN TIEGHEM, 496
VANDREUIL, 283
Vangiones, 12
Vannes, Vannetais, 86, 89, 298, 299, 302, 305, 307, 316, 319, 325; *(dialect)*, 89
Vanoise (national park), 468
Var, 27, 113; department, 139, 282, 346, 360
Vascones, Vascons, 7, 127, 312; *see* **Basques**
VAUBAN, 382
VAUCANSON, 457
Vaucluse (department), 41, 282, 360, 389, 395
Vaud (region), 198, 243, 288, 393
Vectis, 31
vegetables, 347, 448; *see gardens (vegetable)*
Velay, 32, 40, 240, 283, 390
Veliocasses, Veliocassi, 12, 69, 174
Vellaunioi, 30, 32; **Vellavi**, 12, 40
Vendée *(revolt)*, 176, 340–4, 383, 428; department, 282, 304, 305, 328, 340, 383, 385, 395, 396, 400
vendettas, 317
Vendôme, Vendômois, 163
VENDRYÈS, 472
Veneti, 10, 12, 85, 86
Venice, 259; *glassware*, 234
VERCAUTEREN, 478
VERCINGETORIX, 43, 472
Vercors, 212, 390, 466
Verdun, 109, 111, 161, 167, 172, 173; Gouvernement de, 167; *Treaty of*, 76, 90–1, 93, 95, 97, 108, 117, 173, 181, 182, 244, 253
VERHAEGUE, 500
VERHULST, 478
Vermand, Vermandois, **Viromandi**, 13, 14, 15, 171, 172
Versailles, 256, 258, 383, 462
Vesontio, 12, 21, 36, 43, 69, 182
VEYRET-VERNER, 491
VIALE (Salvatore), 318
VIARD, 503

Vichy, 439
VICTOR-EMMANUEL, 114
vicus(i), 16–17, 18, 42, 48, 49, 51, 425
VIDAL DE LA BLACHE (Paul), xxi, xxii, 80, 158, 325, 326, 414, 469, 474
Viducassï, 174
Vie de Louis VI, 200
Vienna, Vienne, 12, 36, 47, 55, 182, 183, 230, 251, 270; (diocese of), 40, 73–4
Vienne (river, 177; department, 282, 343
VIERS, 499
Vierzon, 403, 441
VIGIER (Philippe), 325
VIGUERIE (Jean de), 498
villa, 40, 48, 49, 51, 53, 70, 137, 267, 460, 461, 474, 475
village: clustered, grouped, 16, 49, 64, 127, 134, 135, 139, 141, 142–8, 153, 167, 169, 196, 271, 304, 330; *central*, 207, 215, 407, 408
Villar, Villard, Villars, 64
VILLARD, 473
Villard-de-Lans, 358
-*ville*, *Ville-*, 63, 64, 65, 142
Ville d'Avray, 249, 250
ville d'hiver d'Arcachon (La), 460, 502, 503
Villefranche-de-Rouergue, 146
Villefranche-sur-Meuse, 382
Villeneuve-sur-Lot, 146, 347
Villeron, 267
-*villers*, 63, 64
Villers-Cotterets *(Edict of)*, 133, 243
Villers-sur-Mer, 139
-*villiers*, 63, 64, 65
VILLON (François), 107, 255, 256, 492
Vincennes, Bois de, 268, 458
VINCENT, 477
Vindocinus (Pagus), 163
vineyards, see *wine*
VIOLLET, 491
VIRGIL, 101–2
VIRGILE MARON, 200
Virodunum 36
Visigoths, *Visigothic (state)*, 71, 74, 75, 76, 80, 84, 134, 178
Vitry-en-Perthois, 382; -le-François, 382; *pays*, 166
Vivarais, 34, 67, 109, 182, 219, 222, 235, 387
VIVIEN DE SAINT-MARTIN, 487
Viviers, 91, 109
Vix, 30, 32, 33
Voconti, 10, 12
VOILLIARD, 325
Volcae, 8, 10, 12, 34, 41
Volcans d'Auvergne (parc des), 466
Vollerand (farm), 267
VOPISCUS, 480

Index   563

Vosges, 40, 59, 66, 140, 172, 173, 211, 234, 237, 304, 310, 352, 353, 366, 370, 377, 378, 399, 414, 452, 466; department, 282; VOSEGUS, 199
VOSSLER, 492
Voyage en Corse, 318
Voyage en France (d'Ardouin-Dumazet), 413–14
Voyage en France (Le), (d'Amable Tastu), 290, 291
Voyage en Navarre pendant l'insurrection des Basques, 319
Voyage en zigzags de deux jeunes Français en France, 290, 292

WACE (Robert), 88, 202
WACKERMANN, 500
waggoners, 387
waistcoat, 297, 299
Waldensis (Pagus), 198
Wales, 85, 86; *Welsh (dialect)*, 86
WALLON, 502
Walloon (dialect), 124; *Walloon (linguistic area)*, 124, 243
walnut trees, 217
waltz, 304
WANDRUSZKA, 133, 484
warcry, 190, 254
Wastinensis (Pagus), 165, 198
water buffalo, 362
water carriers, 388, 389
WATTENBACH, 492
weavers, 235, 341, 366, 383
WEBER (Eugen), 423, 501, 502
weights and measures, 24, 241
-weiler, 64
WENDEL (de), 440
WERNER, 105, 479, 481, 482
Weser (valley), 26
west (of France), 100, 134, 140, 141, 143, 148, 151, 196, 202, 322, 328, 340, 341, 383, 387, 406, 428, 431, 432, 435, 438
Westphalia *(Treaty of)*, 111, 173
wheat, 51, 52, 70, 97, 200, 217–18, 354, 433, 434, 438, 440, 448
Wight (Isle of), 30, 473
WILLIAM OF ORANGE, 132

WILLIAM X OF POITIERS, 228
WILLIAM THE BRETON, 175, 200, 217, 253, 490
WILLIAM THE CONQUEROR, 117, 174
WINWALOE (GWENOLË), SAINT, 82
wine, wine growing, 32–3, 53–4, 55–6, 71, 92, 97, 147, 217, 225, 226–30, 261–2, 263, 308, 347, 348ff., 367, 387, 389, 393, 433, 440, 442
woad, 225, 241, 491
WOLFF, 484
women, 151, 152, 297–8, 299–302, 399, 403, 405, 457
wood (for building, heating, charcoal), 22, 130, 234, 237–9, 241, 269, 294, 361, 389; *woodwork*, 239; *floating of logs*, 329, 331, 387, 391
woods, 51, 71, 151, 169, 173, 269, 361–3, 445
wool: industry, 234, 239, 366; *carders of*, 387
workers, 231, 389, 424, 425, 429–31, 437, 440, 443, 453
World War: First, 206, 273, 329, 352, 355, 364, 373, 378, 384, 392, 413, 435, 437, 441, 459, 461; Second, 114, 153, 157, 289, 331, 355, 358, 359, 365, 368, 373–4, 377, 384, 406, 415, 420, 429, 438, 442–3, 462, 468
WORONOFF, 491
WRIGHT, 492

Xaintois, 195, 203
Xanten, 42, 96

-y, 63
YOLE (Jean), 498
Yonne, valley, basin, 228, 247, 258, 263, 329, 391; (department), 258, 282, 283, 387
Yorkshire, 7
YOUNG (Arthur), 219, 224, 274, 312, 485, 493, 496
YVES DE SAINT-DENIS, 100
Yvetot, 209, 211
Yurande, 15

ZELDIN (Theodore), 423, 493, 501
ZELLER (Gaston), 111, 483
zinc (roofs), 459
ZUMTHOR, 481

1. Period and place: research methods in historical geography. *Edited by* A. R. H. BAKER and M. BILLINGE
2. The historical geography of Scotland since 1707: geographical aspects of modernisation. DAVID TURNOCK
3. Historical understanding in geography: an idealist approach. LEONARD GUELKE
4. English industrial cities of the nineteenth century: a social geography. R. J. DENNIS*
5. Explorations in historical geography: interpretative essays. *Edited by* A. R. H. BAKER and DEREK GREGORY
6. The tithe surveys of England and Wales. R. J. P. KAIN and H. C. PRINCE
7. Human territoriality: its theory and history. ROBERT DAVID SACK
8. The West Indies: patterns of development, culture and environmental change since 1492. DAVID WATTS*
9. The iconography of landscape: essays in the symbolic representation, design and use of past environments. *Edited by* DENIS COSGROVE and STEPHEN DANIELS*
10. Urban historical geography: recent progress in Britain and Germany. *Edited by* DIETRICH DENECKE and GARETH SHAW
11. An historical geography of modern Australia: the restive fringe. J. M. POWELL*
12. The sugar-cane industry: an historical geography from its origins to 1914. J. H. GALLOWAY
13. Poverty, ethnicity and the American city, 1840–1925: changing conceptions of the slum and ghetto. DAVID WARD*
14. Peasants, politicians and producers: the organisation of agriculture in France since 1918. M. C. CLEARY
15. The underdraining of farmland in England during the nineteenth century. A. D. M. PHILLIPS
16. Migration in Colonial Spanish America. *Edited by* DAVID ROBINSON
17. Urbanising Britain: essays on class and community in the nineteenth century. *Edited by* GERRY KEARNS and CHARLES W. J. WITHERS
18. Ideology and landscape in historical perspective: essays on the meanings of some places in the past. *Edited by* ALAN R. H. BAKER and GIDEON BIGER
19. Power and Pauperism: the workhouse system, 1834–1884. FELIX DRIVER
20. Trade and urban development in Poland: an economic geography of Cracow from its origins to 1795. F. W. CARTER
21. An historical geography of France. XAVIER DE PLANHOL

Titles marked with an asterisk * are available in paperback

www.ingramcontent.com/pod-product-compliance
Ingram Content Group UK Ltd.
Pitfield, Milton Keynes, MK11 3LW, UK
UKHW032324190125
453752UK00011B/178